MAGILL'S ENCYCLOPEDIA OF SCIENCE

ANIMAL LIFE

MAGILL'S ENCYCLOPEDIA OF SCIENCE

ANIMAL LIFE

Volume 2
Estivation–Learning

Editor
Carl W. Hoagstrom, Ph.D.
Ohio Northern University

Project Editor
Tracy Irons-Georges

Salem Press, Inc.
Pasadena, California
Hackensack, New Jersey

Some of the updated and revised essays in this work originally appeared in *Magill's Survey of Science: Life Science*, edited by Frank N. Magill (Pasadena, Calif.: Salem Press, Inc., 1991), and in *Magill's Survey of Science: Life Science Supplement*, edited by Laura L. Mays Hoopes (Pasadena, Calif.: Salem Press, Inc., 1998).

∞ The paper used in these volumes conforms to the American National Standard for Permanence of Paper for Printed Library Materials, Z39.48-1992 (R1997).

Library of Congress Cataloging-in-Publication Data

Magill's encyclopedia of science : animal life / editor, Carl W. Hoagstrom.
p. cm.
Includes bibliographical references (p.).
ISBN 1-58765-019-3 (set : alk. paper) — ISBN 1-58765-020-7 (v. 1 : alk. paper) — ISBN 1-58765-021-5 (v. 2 : alk. paper) — ISBN 1-58765-022-3 (v. 3 : alk. paper) — ISBN 1-58765-023-1 (v. 4 : alk. paper)
1. Zoology—Encyclopedias. I. Hoagstrom, Carl W.

QL7 .M34 2002
590′.3—dc21

2001049799

Second Printing

PRINTED IN THE UNITED STATES OF AMERICA

TABLE OF CONTENTS

MAGILL'S ENCYCLOPEDIA OF SCIENCE

ANIMAL LIFE

ESTIVATION

Types of animal science: Behavior, ecology, physiology
Fields of study: Ecology, herpetology, invertebrate biology, physiology, wildlife ecology

Estivation, like hibernation, is a dormant state that enhances an animal's ability to survive when environmental conditions are harsh. In hot, arid climates, estivating animals reduce water losses and energy requirements during long periods of drought when little water or food is available.

Principal Terms

EPIPHRAGM: covering or sealing membrane
HYGROSCOPIC: able to retain moisture
NEMATODE: a long, cylindrical worm; some are parasitic
POIKILOTHERM: cold-blooded or ectothermic; any organism having a body temperature that varies with its surroundings; in general, reptiles, amphibians, fish, and invertebrates
TORPID: dormant, numb, sluggish in action

During true estivation, metabolic processes including oxygen consumption, respiration, heart rate, and neurological activity decrease substantially, but the term is often used loosely to describe the activity of any animal that spends part of the warmer, drier season in a torpid state. Except for the environmental stimuli that trigger them, estivation and hibernation appear similar, although they differ physiologically, and occur most often in the poikilotherms. Among the animals most likely to estivate are mollusks, arthropods, fish, amphibians, and reptiles, although some small desert mammals estivate as well.

Invertebrate Estivation

A seemingly lifeless desert may be teeming with estivating life underground, waiting for seasonal rains that will awaken them to resume their life cycles. Snails, slugs, earthworms, insects, spiders, and nematodes, along with cocoons, eggs, grubs, larvae, and pupae, may all lie dormant in the soil, in building foundations and rock crevices, or under rotting logs or other vegetation. The animals are not just dormant; some of them will also be in an arrested state of sexual development called diapause. Some tropical snails can estivate for years at a time. To prepare, a snail digs a deep burrow in moist ground or under rocks. Next, it forms an epiphragm (a sealing membrane for the shell) to prevent evaporation and desiccation. Finally, its metabolism slows dramatically until it detects favorable environmental cues such as increased moisture.

Lungfish Estivation

Among the fishes, the process of estivation is best known in the air-breathing lungfishes of Africa (*Protopterus*) and South America (*Lepidosiren*). Adult fish have paired lungs and vestigial gills; in fact, they will drown if held underwater. Both abilities—to estivate and to breathe air—have contributed to their survival in areas that experience severe seasonal droughts. When the rivers, lakes, and marshes they inhabit dry up, lungfish dig burrows deep in the mud, leaving air passages to the surface. They curl up inside and wait out the arid conditions until rain fills up their burrows with water. *Protopterus* secretes a mucous coating that hardens, forming a tough, cocoonlike hygroscopic chamber with only one opening, connected to its mouth. When the rains come and flood the passage, the dormant fish awakens with a cough. *Lepidosiren* burrows more deeply than *Protopterus* and plugs the entrance to its air tube with perfo-

Many turtles and tortoises, especially those that live in the desert such as this Sonoran wood turtle, estivate through the hottest summer months, emerging only to browse and drink after infrequent thunderstorms. (Rob and Ann Simpson/Photo Agora)

rated clay. Its burrow is somewhat larger and usually contains some water. It coats the walls with a jellylike substance to maintain moisture. During estivation, lungfish are so torpid that they make easy prey for local fisherman, who spear them in their burrows.

During estivation, energy required for reduced metabolism is provided by the breakdown of tissue protein. The waste product is urea, which is excreted in large amounts once the fish is again submerged in water. Lack of oxygen can become a problem for lungfish sealed in their burrows, but the aerobic metabolism of an estivating *Protopterus* is only 20 percent of its normal resting metabolism. Similarly, *Lepidosiren* and the swamp eel (*Synbranchus*) can survive a lack of oxygen for long periods during estivation.

Estivation in Reptiles and Amphibians

During the summer, many reptiles and amphibians estivate. Some frogs and toads insulate themselves in cocoons composed of many layers of unshed skin. The eastern spadefoot toad digs backward into sandy soil with its spade-shaped hind feet to estivate for periods of weeks; experiments indicate that the sound of rain falling rather than moisture itself may trigger the toad's arousal. The mud turtle *Kinosternon* abandons its drying pond for a burrow where it estivates up to three months. Desert tortoises of the North American Southwest spend much of their summers estivating in burrows but emerge to drink and browse after infrequent thunderstorms. In the same region, the lizard called the chuckwalla stops eating and estivates in rock crevices, emerging every third day for about an hour at sunset.

Estivation in Mammals and Birds

Among the few estivating mammals are small desert rodents, lemurs, and hedgehogs. Some ground squirrels remain dormant in their burrows from late summer, merging estivation with hibernation. They reappear in early spring to take advantage of new growth after winter rains. The round-tailed ground squirrel of southern Arizona (*Citellus tereticaudus*) avoids the hot, dry autumn by disappearing in August and September. Generally, the period of estivation coincides with the period most prone to scarcity of vegetation.

While a few birds may hibernate, it is unlikely that they estivate. They can migrate to more attractive regions, or, like many reptiles and most mammals, they cope with high temperatures by confining their activities to cooler parts of the day.

—Sue Tarjan

See also: Deserts; Hibernation; Insects; Lungfish; Metabolic rates; Respiration and low oxygen; Respiratory system.

Bibliography

Badger, David. *Frogs*. Stillwater, Minn.: Voyageur Press, 1995. Easy-to-understand and well-illustrated introduction to frogs and toads. Contains a short section on hibernation and estivation.

Cloudsley-Thompson, J. L. *Ecophysiology of Desert Arthropods and Reptiles*. New York: Springer-Verlag, 1991. Challenging content; well illustrated. Particularly good discussion of survival strategies of inhabitants of temporary pools which dry up in the summer.

Heatwole, Harold. *Energetics of Desert Invertebrates*. New York: Springer-Verlag, 1996. Estivating to conserve energy and outlast scarcity is one of many adaptive mechanisms described in detail. Technical detail but clear and understandable.

Moyle, Peter B., and Joseph J. Cech, Jr. *Fishes: An Introduction to Ichthyology*. 4th ed. Upper Saddle River, N.J.: Prentice Hall, 2000. Excellent introductory textbook that includes a complete account of lungfish estivation.

Schmidt-Nielsen, Knut. *Desert Animals: Physiological Problems of Heat and Water*. Reprint. New York: Dover, 1979. Clearly written conversational style makes this scientific study easy to understand. Excellent chapter on estivation in desert mammals.

Warburg, Michael R. *Ecophysiology of Amphibians Inhabiting Xeric Environments*. New York: Springer-Verlag, 1997. In-depth study of adaptations of amphibians to a desert environment. Overview of cocoon formation to prevent desiccation. Technically challenging.

ESTRUS

Types of animal science: Behavior, physiology, reproduction
Fields of study: Anthropology, ethology, reproduction science

Estrus is the recurring period of sexual excitement and receptivity in female mammals. A central feature of most mammalian species' reproductive lives, it helps structure social and communicative behavior as well.

Principal Terms

COPULATION: mating; the insertion of the male's penis into the female's vagina to fertilize her ova

ESTROGENS: a group of female sex hormones which regulate the estrous cycle

ESTRUS CYCLE: the cycle of females' bodily changes related to reproductive potential

OVA: eggs released from the females' ovaries at the height of estrus

PROGESTERONE: a female sex hormone produced chiefly in the latter half of the cycle

The physiological and behavioral events which make up estrus are complex and intertwined. For successful mating to occur, it is usually necessary that male and female cooperate in the act. Estrus ensures that both sexes are impelled to do so at the same time, when the female's body is ready for a pregnancy.

During the days prior to estrus, ova in follicles ripen rapidly within the ovary. Estrogens, produced by cells in the follicle walls, increase rapidly in this phase, causing not only follicular growth but thickening of the vagina walls and the various signs of "coming into heat," and then of estrus itself. In most species, the ova spontaneously burst from their follicles at the peak of estrus and enter the oviduct, which leads to the uterus, ready to join with sperm. Females of several species, however, including the rabbit and the cat, require the added stimulus of mating for release of their ova. Once the ova are released, the ruptured follicles begin secreting progesterone in place of estrogen; this hormone prepares the uterus to support a pregnancy. This internal process is regulated through feedback connections between the ovaries and the hypothalamus in the brain. The hormonal effects which make females eager for mating also originate largely in the hypothalamus, sensitizing the nervous system to respond strongly to the presence of and stimulation by their male counterparts.

Variations Between and Within Species

External estrus signals vary greatly by species. In every case, they are largely produced by hormonal influences, and are keyed to evoke specific sexual response from the male of the species. Such signals may be visual, behavioral, auditory, or scent-based, or may possibly draw on senses imperceptible to human observers.

Visual signals are often found among social mammals. These have been most observed in primates. Chimpanzee and bonobo females show a spectacular pink swelling of their genitals during estrus, which recurs every five weeks if pregnancy does not occur. Certain monkeys develop an estrus flush on their faces as well as their buttocks.

Scent and sound signals can alert males who are not initially within sight range of the receptive female. As such, they are effective for species who do not live in groups, although these signals are not confined to such species. Urine markings are among the most common scent signals. A female black rhino in estrus leaves a long trail of scent posts, which the male follows. Other scent signals are partly airborne, like those of domestic cats and dogs. Mating calls are given by many females, such as the female gibbon's ascending call, which is then answered by the male.

Behavioral changes during estrus are almost universal, as both sexes concentrate on the quest for one or several partners. The initial stages, which may suggest female coyness to an observer, are part of the courtship process. For example, the female cheetah leads several males on a headlong run across the plain, finally selecting one with whom to mate. Pet owners notice a restlessness in their dogs and cats; female cats in heat are especially likely to roam. The penultimate female signal in many species is lordosis, an arched-back posture which allows the male to mount. Mammalian species are normally either monestrous—having a single estrus period a year—or polyestrous—with several estrus periods recurring annually. The latter situation is more common.

Environmental changes can cause variations in these patterns. Such factors include climate shifts, changes in available light, nutrition, and the presence of a new male. In monestrous mammals, the estrus period usually falls at a time when resulting births will take place in the optimal season for young animals to thrive and grow. Domestication may disrupt this pattern. For instance, a female dog typically comes into heat twice a year with no regard to seasons. Her ancestor the wolf has one estrus period a year, sometime between January and April, depending upon latitude.

The length of estrus also varies by species: It lasts four to nine days in the mare, and only fourteen hours in the rat. In a few species, such as the ferret, estrus lasts for several weeks if copulation has not occurred. There are also infrequent phenomena such as "silent heats," in which the animal ovulates but the usual external signs are absent.

Estrus, Evolution, and Ethology

Sexual reproduction provides genetic diversity and adaptability to a species. However, in order to proceed successfully, it requires a fairly complex series of events. Estrus is central in this process. It communicates to males a female's readiness at the same time that it impels that female to mate and primes her body for pregnancy. Nature plays many

The Northern Fur Seal

Estrus behavior has seldom been studied closely by biologists. More often it has been a practical concern of livestock breeders and pet owners, or subsumed in general studies of reproductive behavior. However, as part of a study of the northern fur seal, Roger L. Gentry and colleagues made detailed observations on estrous females over a six-year span.

Like many animals, the seals have a definite seasonal breeding time, typically July 7 through 14 of each year. Females come ashore and huddle in large groups on breeding beaches, where mating occurs. The seals' known sex ratio is sharply skewed, with forty or more females to each male. The male in the territory works his way through the group, mating with many females as each one reaches the peak of her receptivity. This arrangement is efficient for both sexes.

Several unusual species-specific features were discovered by Gentry's study. Fur seals go into estrus only five days after giving birth. Any resulting pregnancy enters an embryonic diapause, or delayed implantation, until the newborn seal is weaned the following fall. The actual period when female seals are receptive is brief, usually only twenty-four hours.

Unlike many other female animals, who choose mates based on status, appearance, or other traits, fur seal females seem indifferent to which male they mate with. Some are almost comatose by the time copulation occurs. Gentry concludes that female choice cannot be an evolutionary factor in this species; males are selected for reproduction solely by direct competition among themselves. The pattern seems to be an optimum one, though, for a species which spends most of its time at sea without opportunity to mate.

The Human Puzzle

Humans are unusual, but not unique, among mammals in lacking a perceptible estrus period. Did humans, or their hominid ancestors, once have estrus and lose it? Is this absence just another oddity in the amazingly wide spectrum of mammalian reproductive traits? The fossil record gives no clues, nor are they likely to be found in the future.

This silence has not kept anthropologists, ethologists, and others from speculating. The prevailing theory until the 1980's asserted that loss of estrus was essential to becoming human, because only a female's continuous sexual availability secures a monogamous pair bond and keeps the male from wandering away.

This theory has many problems. Monogamous pair bonding now appears to be a "hard-wired" trait in some species, unrelated to continual sexual contact. The bonobo, and to some extent other primate species, copulate even when the female is not in heat. Humans (and presumably their primate ancestors) are innately social animals. Without group backup, a lone prehistoric or hominid hunter would not fare well. Loss of estrus would also seem to have disruptive social possibilities. Rape, fetish obsessions because males have no instinctual guide to what is sexy, and a double standard denying that females have sexual desires—all are made possible by the lack of clear female signals.

Anthropologist Sarah Hrdy suggests that the human pattern may simply resemble that of species such as orangutans, who show no visible estrus signals. Study subjects ranging from Connecticut college women to Kalahari hunter-gatherers have reported peaks of female sexual interest around their time of ovulation.

If hominid or prehistoric human females did have full estrus, what was the evolutionary advantage or mechanism of its loss? The subject is one of those puzzles unlikely to be solved, but fascinating to speculate about.

variations on this basic theme, each connecting the species' estrous cycle to its environment and its whole life cycle.

Because the survival of young mammals requires maternal care, and often that of the father and/or other adult animals as well, estrus patterns are significant for social structure. In some species, estrus may help secure a pair bond, a polygynous herd, or bonds within a troupe or pack. It can also cause disruption, as when males battle for access to estrous females, or a new male takes over a harem and kills the young, hastening their mothers' next heat.

Humans have known of estrus ever since they first domesticated animals. Charles Darwin's *On the Origin of Species* (1859) credits estrous females' mate choices with being an evolutionary mechanism, giving many examples. Ethologists' field studies have recently described many details about species-specific estrus behavior, as well as its role in life cycles and social behavior.

However, direct estrus observations are hard to make and sometimes hard to interpret, especially in wildlife.

—*Emily Alward*

See also: Asexual reproduction; Breeding programs; Cleavage, gastrulation, and neurulation; Cloning of extinct or endangered species; Copulation; Courtship; Determination and differentiation; Development: Evolutionary perspective; Fertilization; Gametogenesis; Hermaphrodites; Hydrostatic skeletons; Mating; Parthenogenesis; Pregnancy and prenatal development; Reproduction; Reproductive strategies; Reproductive system of female mammals; Reproductive system of male mammals; Sexual development.

Bibliography

Gentry, Roger L. *Behavior and Ecology of the Northern Fur Seal*. Princeton, N.J.: Princeton University Press, 1998. The chapter on "Estrus and Estrous Behavior" in this large ecological study describes northern fur seals' estrus events in detail. Field observations were supplemented by nondestructive experiments on site, and data are given for both.

Hrdy, Sarah Blaffer. "Why Women Lost Estrus." *Science '83* 4 (October, 1983): 72-77. Discussion of several theories that have been advanced on this topic. Hrdy offers her own recasting of the question, along with some data that tend to support it.

Kevles, Bettyann. *Females of the Species*. Cambridge, Mass.: Harvard University Press, 1986. A survey of female reproductive behavior throughout the animal kingdom. Full of interesting examples. It focuses on females' active roles in the reproductive process.

Small, M. F., and F. B. M. de Wall. "What's Love Got to Do with It?" *Discover* 13, no. 6 (June, 1992): 46-51. An article exploring bonobo sexual behavior. It notes that this primate, so close to humans on the evolutionary tree, differs drastically in having prolonged estrus swellings. Includes photographs of estrus phenomena.

Young, J. Z. *The Life of Mammals: Their Anatomy and Physiology*. 2d ed. Oxford, England: Clarendon Press, 1975. The chapter "The Oestrous Cycle of Mammals" contains standard and precise descriptions of the internal process of the mammalian estrous cycle. Includes many drawings and diagrams for clarity.

ETHOLOGY

Type of animal science: Ethology
Fields of study: Genetics, neurobiology

Ethology is the study of animal behavior from the perspective of zoology. The information acquired through ethology has helped scientists better understand animals in all their variety.

Principal Terms

ADAPTATION: a structure, physiological process, or behavioral pattern that gives an organism a better chance of surviving and of reproducing

BEHAVIORAL ECOLOGY: the systematic study of the strategies animals use to overcome environmental problems and the adaptive value of those strategies

BEHAVIORISM: the school of psychology that focuses on the investigation of overt behaviors and rejects allusion to inner processes as a means of explaining behavior

COMPARATIVE PSYCHOLOGY: the branch of psychology that uses comparative studies of animals as a means of investigating phenomena such as learning and development

EVOLUTIONARILY STABLE STRATEGY: a behavioral strategy that will persist in a population because alternative strategies, in the context of that population, will be less successful

IMPRINTING: a specialized form of learning characterized by a sensitive period in which an association with an object is formed

SOCIOBIOLOGY: the scientific discipline that examines social behavior in the context of evolutionary theory

Ethology is the branch of zoology that investigates the behavior of animals. Behavior may be defined as all the observable responses an animal makes to internal or external stimuli. Responses may be either movements or secretions; however, the study of behavior is much more than a descriptive account of what an animal does in response to particular stimuli. The ethologist is interested in the how and why questions about the behaviors observed. Answering such questions requires an understanding of the physiology and ecology of the species studied. Those who study animal behavior are also interested in the ultimate or evolutionary factors affecting behavior.

The Roots of Ethology

Ethology is a young science, yet it is also a science with a long history. Prior to the late nineteenth century, naturalists had accumulated an abundance of information about the behavior of animals. This knowledge, although interesting, lacked a theoretical framework. In 1859, Charles Darwin published *On the Origin of Species*, and with it provided a perspective for the scientific study of behavior. Behavior was more central to two of Darwin's later books, *The Descent of Man* (1871) and *Expression of the Emotions of Man and Animals* (1873). By 1973, the science of ethology was sufficiently well developed to be acknowledged by the presentation of the Nobel Prize for Physiology or Medicine to Nikolaas Tinbergen, Konrad Lorenz, and Karl von Frisch for their contributions to the study of behavior. The work of these men was central to the development of modern ethology.

The experimental studies of Frisch revealed the dance language of the honeybee and ways in which the sensory perception of the bees differed from our own sensory world. An awareness of species-specific sensory abilities has provided an important research area and has emphasized a factor that must be considered in the experimental design and interpretation of many types of behavioral research.

Tinbergen studied behavior in a variety of vertebrate and invertebrate organisms. He was good both at observation of animals in their natural habitat and in the design of simple but elegant experiments. His 1951 book, *The Study of Instinct*, is a classic synthesis of the knowledge that had been gained through the scientific study of animal behavior at that time.

Konrad Lorenz is considered by many to be the founder of ethology, because he discovered and effectively publicized many of the classic phenomena of ethology. Pictures of Lorenz being followed by goslings are almost a standard feature of texts that discuss the specialized form of learning known as imprinting. In natural settings, imprinting allows young animals to identify their parents appropriately. Another contribution of Lorenz was his book *King Solomon's Ring: New Light on Animal Ways*, published in 1952. This extremely readable book raised public awareness of the scientific study of animal behavior and kindled the interest of many who eventually joined the ranks of ethologists.

Ethology and Neurobiology

Many of the features of ethological research characteristic of the work of Lorenz, Tinbergen, and von Frisch have continued to be characteristic of the field. They were concerned that the behavior of animals be understood in the context of the species' natural habitat and that both proximate and ultimate levels of explanation would be examined. Their research strategies have been supplemented by an increase in laboratory-based research and by the introduction of new types of experimental design. These developments have softened the distinctions between ethology and another field of behavioral study, comparative psychology. The focus of comparative psychology is comparative studies of the behavior of nonhuman animals. Initially, questions about learning and development were the major problems investigated in comparative psychology. Although the animals most frequently studied were primates and rodents, those doing the research were interested in gaining insight into the behavior of humans. Comparative psychology was long dominated by behaviorism, a school of thought that assumes that the ultimate basis of behavior is learning. The behaviorists employed rigorous experimental methods. Because such methods require carefully controlled conditions, behavioral research is typically laboratory based, and animals are therefore tested in surroundings remote from their natural environments. Over time, comparative psychology has broadened both the questions it asks and the organisms it studies. The boundaries between comparative psychology and ethology have been further blurred by the rising number of scientists crossing disciplinary lines in their research. Each discipline has learned from the other, and both have also profited from knowledge introduced through neurobiology and behavioral genetics.

Neurobiology investigates the structure and function of the nervous system. One area of ethology that has been directly enriched through neurobiology is the study of sensory perception in animals. The techniques developed in neurobiology allow the investigator to record the response of many individual neurons simultaneously. The neurobiologist examines phenomena such as stimulus filtering at the level of the cell. Stimulus filtering refers to the ability of nerve cells to be selective in their response to stimuli. For example, moths are highly sensitive to sounds in the pitch range of sounds made by the bats that are their chief predators. Neurobiology provides a powerful tool for understanding behavior at the proximate level.

Behavioral Genetics

Another source of information for the ethologist is behavioral genetics. Because of the evolutionary context of ethology, it is important to have an understanding of the genetic basis of behavior. If there were no genetic component in behavior, behavior would not be subject to natural selection. (Natural selection refers to the process by which some genes increase in frequency in a population while alternates decrease because the favored genes have contributed to the reproductive success of those organisms that have them.) While the ethological approach to behavior assumes that behavior patterns are the result of interactions between genes and environment, investigators often ask questions about the genetic programming of behavior.

Early ethologists performed isolation and cross-fostering experiments to discover whether behaviors are learned or instinctive. If a behavior appears in an individual that has been reared in isolation without the opportunity to learn, the behavior is considered instinctive. Observing the behavior of an individual reared by parents of a different species is similarly revealing. When behavior patterns of conspecifics appear in such cross-fostered individuals, such behaviors are regarded as instinctive. Instinctive behaviors typically are innate behaviors that are important for survival. For example, one very common instinctive behavior is the begging call of a newly hatched bird. Isolation and cross-fostering experiments are still a part of the experimental repertoire of ethologists, but behavioral genetics permits the asking of more complex questions. For example, a behavior may accurately be labeled instinctive, but it is more revealing to determine the developmental and physiological processes linking a gene or genes to the instinctive behavior.

The ethologist is also interested in determining whether behavior is adaptive. It is not sufficient to identify what seems to be a commonsense advantage of the behavior. It is important to show that the behavior does in fact contribute to reproductive success in those that practice the behavior and that the reasons the behavior is adaptive are those that are hypothesized. When behaviors are tested, they frequently do turn out to be adaptive in the ways hypothesized. This type of research, however, has provided many surprises. Research on the adaptive value of behaviors in coping with environmental problems that affect reproductive success is known as behavioral ecology; this is a major area of modern ethology.

Behavioral ecology addresses a variety of questions, in part because the process of evolution is opportunistic. For any environmental problem there are alternate solutions, and the solution a particular species adopts is dependent upon the possibilities inherent in its genes. Questions addressed include such things as whether a species is using the optimum strategy or how a species benefits from living in a group. Because alternate strategies are possible even within a species, behavioral ecologists are interested in evolutionarily stable strategies. An evolutionarily stable strategy is a set of behavioral rules that, when used by a particular proportion of a population, cannot be replaced by any alternative strategy. For example, the sex ratio present in a particular population will determine the optimum sex ratio for the offspring of any individual.

Sociobiology

Sociobiology is another major area of modern ethology. Sociobiology examines animal social behavior within the framework of evolution. Animal species vary in the degree of social behavior they exhibit; other variables include group size and the amount of coordination of activities occurring within the group. The sociobiologist is interested in a number of questions, but prominent among them are the reasons for grouping. Hypotheses such as defense against predators or facilitation of reproduction can be tested. The particular advantage or advantages gained by grouping varies among species. Two important concepts in sociobiology are kin selection and inclusive fitness.

Kin selection refers to the differential reproduction of genes that affect the survival of offspring or closely related kin. Behavior such as the broken-wing display of the killdeer is an example. The behavior carries risk but would be promoted by selection if the offspring of individuals using the display were protected from predators often enough to compensate for the risk. Inclusive fit-

As a result of imprinting in the moments immediately after hatching, baby geese automatically follow their mother in a line wherever she goes. (Corbis)

ness is the term used to recognize the concept that fitness includes the total genotype, including those genes that may lower the individual's survival as the price of leaving more genes in surviving kin. The concepts of kin selection and inclusive fitness help to address one problem raised by Darwin, the question of altruistic behavior. Ethology is a young science but a very exciting one, because there are so many questions that can be asked about animal behavior within the context of evolution.

Studying Ethology

The methods and tools of ethology cover the entire spectrum of complexity. One simple, but demanding, method is to collect normative data about a species. In its simplest form, the scientist observes what an animal does and writes it down in a field notebook. Finding and following the animal, coping with field conditions such as bad weather and rugged terrain, and keeping field equipment in operating condition add challenge and variety to this approach. The ethologist uses various techniques to get data as unbiased as possible. One of these is to choose a focal animal at random (or on a rotation) and observe the focal animal for a specific amount of time before switching observation to another member of the population. This prevents bias in which individuals and which behaviors are observed. The sampling of an individual's behavior at timed intervals is an even more effective way of avoiding bias.

When all or most of an animal's behavioral repertoire is known, a list known as an ethogram can be constructed. This catalog can be organized into appropriate categories based on function. Ethograms provide useful baseline information about the behavior of a species. For animals that are difficult or impossible to follow, radio-tracking techniques have been developed. Collars that emit radio signals have been designed for many animals. Miniaturization has made it possible for radio tracking to be used even on relatively small animals.

In field studies, animals are often marked in some way so that observers are able to follow individual animals. A number of techniques have been developed, including banding birds with colored acrylic bands. Color combinations can be varied so that each member of the population has a unique combination. Marking allows the observer to get information such as individual territory boundaries and to determine which animals interact.

Models are frequently used in experiments. For example, a model can be used to determine whether individuals in a species need to learn to identify certain classes of predators. Models were used in many of the classic experiments in ethology. Modern technology has allowed the development of much more sophisticated models. One of the most interesting is a "bee" that can perform a waggle dance (used by bees to indicate location) so effectively that its hivemates can find the food source. Whether a model is simple or sophisticated, it can provide a tool to determine the cues that trigger an animal's response.

Neurobiologists use electrodes and appropriate equipment to stimulate and record the responses of neurons. They can also stimulate specific regions of the nervous system by using tiny tubes to deliver hormones or neurotransmitters. Genetic technology has made it possible to examine the deoxyribonucleic acid (DNA) of individuals in a species. This tool can be used, for example, to determine whether females in monogamous species are completely monogamous or whether some of their offspring are fathered by males other than their mates.

Tape recorders have become very important in studies of animal vocalizations. Recorders are used in two ways. The animal's vocalizations may be recorded and the recording used to make sound spectrographs for analysis. The recordings may also be used to determine whether individuals can discriminate between the vocalizations of neighboring and nonneighboring individuals in their species. Playbacks can also be used to simulate intruders in the territory of an individual and can be applied to many other experimental situations in both field and laboratory.

The methods used by ethologists are as varied as the problems they investigate. Because the skill of the observer is still a vital link in the investigation of animal behavior, ethology remains one of the more approachable areas of scientific investigation.

Uses of Ethology

The investigation of animal behavior has a number of benefits, both practical and abstract. Some animals are pests, and knowledge of their behavior can be used to manage them. For example, synthetic pheromones have been used to attract members of some insect species. The insects may then be sampled or killed, depending upon the application. To the extent that researchers develop behaviorally based pest management strategies and reduce pesticide use, they will be promoting our own safety as well as that of other species.

It is sometimes important for humans to be able to understand the communication signals of other species and the characteristics of their sensory perception. The knowledgeable individual can recognize the cues that indicate risk that a dog might bite, for example, and can also avoid behavior that the dog will regard as threatening. Understanding the behavior of the wild animals most likely to be encountered in one's neighborhood is an important factor in peaceful coexistence.

The study of animal behavior is providing one of the more fascinating areas of evolutionary biology. Ethology has demonstrated more effectively than most fields of study how diverse the solutions to a given problem can be and has provided insight into human behavior from a biological perspective. Knowledge of animal behavior also enriches human lives simply by satisfying some of our natural curiosity about animals.

—Donna Janet Schroeder

See also: Altruism; Coevolution; Communication; Communities; Competition; Domestication; Emotions; Evolution: Animal life; Evolution: Historical perspective; Grooming; Groups; Habituation and sensitization; Hormones and behavior; Imprinting; Insect societies; Isolating mechanisms in evolution; Language; Learning; Mammalian social systems; Migration; Offspring care; Pheromones; Predation; Reproductive strategies; Rhythms and behavior; Sex differences: Evolutionary origin; Territoriality and aggression; Zoology.

Bibliography

Allen, Colin, and Mark Bekoff. *Species of Mind: The Philosophy and Biology of Cognitive Ethology*. Cambridge, Mass.: MIT Press, 1997. An interdisciplinary introduction to the comparative, evolutionary, and ecological study of animal thought processes, beliefs, rationality, information processing, and consciousness, by a philosopher and an ethologist. Illustrated.

Barrows, Edward M. *Animal Behavior Desk Reference*. 2d ed. Boca Raton, Fla.: CRC Press, 2000. A dictionary of terms in animal behaior, evolution, ecology, genetics, biogeography, systematics, psychology, and statistics. Includes pronunciations, chronological evolution of terms, identification of obsolete and nontechnical definitions.

Clutton-Brock, T. H., and Paul Harvey, eds. *Readings in Sociobiology*. San Francisco: W. H. Freeman, 1978. This anthology provides short selections from a number of the scientists who have made contributions to the theoretical base of sociobiology. The text is sparsely illustrated, with some graphs, tables, drawings, and diagrams. References are provided at the end of each chapter. Index.

Daly, Martin, and Margo Wilson. *Sex, Evolution and Behavior*. North Scituate, Mass.: Duxbury Press, 1978. A well-written synthesis that examines the phenomenon of sex and sexual strategies in an evolutionary context. Illustrations include graphs, tables, and a small number of black-and-white photographs. Glossary, bibliography, and index.

Goldsmith, Timothy H. *The Biological Roots of Human Nature: Forging Links Between Evolution and Behavior*. New York: Oxford University Press, 1991. An attempt to clarify common misunderstandings and misconceptions about the evolutionary approach to explaining human and animal behavior, written by an academic for the general public.

Lehner, Philip N. *Handbook of Ethological Methods*. 2d ed. New York: Cambridge University Press, 1996. Addresses the practical, methodological, and analytical questions that arise in the study of behavior both in the field and in the lab.

Lorenz, Konrad. *King Solomon's Ring: New Light on Animal Ways*. New York: Thomas Y. Crowell, 1952. A delightful book touching a number of different animals. Lorenz writes out of his personal experience as a keen observer of animals. He communicates good information in an informal and anecdotal style. The vocabulary is nontechnical. Illustrations consist of line drawings on page margins. Index.

Wilson, Edward O. *Consilience: The Unity of Knowledge*. New York: Alfred A. Knopf, 1998. Wilson proposes that it is possible to construct a unified scientific understanding of the most fundamental questions of science through sociobiology.

_______. *Sociobiology*. Cambridge, Mass.: The Belknap Press of Harvard University Press, 1975. A classic, synthesizing a massive amount of information to illustrate the key concepts of sociobiology. The style is succinct. The book is challenging to read, but more because of the format and type size than the style. Illustrations include black-and-white photographs, drawings, tables, and graphs. Glossary, bibliography, and index.

EVOLUTION: ANIMAL LIFE

Types of animal science: Evolution, fields of study
Fields of study: Developmental biology, evolutionary science, genetics, human origins, zoology

Evolution is the change in the gene pool of a population over time by such processes as natural selection, genetic drift, mutation, and migration.

Principal Terms

GENE: the basic unit of heredity

GENE FLOW: the movement of genes from one population to another

GENETIC DRIFT: change in gene frequencies in a population owing to chance

INTERBREEDING: the mating of closely related individuals, which tends to increase the appearance of recessive genes

MIGRATION: the movement of individuals, resulting in gene flow, changing the proportions of genotypes in a population

MUTATION: alteration in the physical structure of the DNA, resulting in a genetic change that can be inherited

NATURAL SELECTION: the process of differential reproduction in which some phenotypes are better suited to life in the existing environment and thus are more likely to survive

SPECIATION: the formation of new species as a result of geographic, physiological, anatomical, or behavioral factors

SPECIES: the basic category of biological classification representing a group of potentially or actually interbreeding natural populations which are reproductively isolated from other such groups

TAXON (pl. TAXA): group of related organisms at one of several levels such as the family Canidae, the genus *Canis*, or the species *Canis lupus*

Current consensus holds that the so-called big bang—the high-temperature, high-density event that marked the beginning of the universe—occurred some fifteen billion years ago, with the sun and Earth formed about four and a half billion years ago. Four billion years ago, the relatively newly created sun shone with only 70 percent of its current strength. The atmosphere had no free oxygen. No bacteria, no viruses, no plants, and no animals were in existence. Subsequently, as chemical processes are assumed to have created oxygen and an organic "soup," microbial life in the form of the simplest cells without a nucleus, prokaryotes, developed out of this primordial ooze. These bacteria were the only living organisms for about two billion years. After that time, about 1.5 billion years ago, more complex cells with nuclei, eukaryotes, appeared. Thus, in all, for some 5.5 billion years, bacteria were the only existing animal organisms.

The Beginnings of Animal Life

Eventually, the two kingdoms, the botanical and the zoological, started to diverge and millions of animals came into being, multicelled and with specialized body parts, distinguished from plants primarily but not exclusively by their methods of feeding, locomotion, and reproduction. Most of the phyla seem to have appeared during the Precambrian period, an immense span of geological time that ended about 590 million years ago. Few fossils have remained from these prehistoric times, but the most explosive period for the development of life was the Cambrian, some 590 to 505 million years ago. In a relatively short span of ten million years, all the animal phyla currently known came into being, perhaps encouraged by an increase in oxygen in the seas, where animal life began.

Eventually, animals developed a nervous system enabling them to control their movements more appropriately, as well as sense organs to help them find suitable food. At the margins, however, the dichotomy between the botanical and the animal world has remained ambiguous, since there are many microorganisms which defy clearcut classification At times, these difficult cases are known as Protista, or Protists.

In all forms of life, including the animal kingdom, no phylum has been produced by a single evolutionary event. Nor have different animal orders appeared as a result of sudden evolutionary changes. Rather, all have come about, whether in gradual or punctuated manner, by the cumulative effect of small steps in different directions. This, at least, is the theory posited by Charles Darwin's explanation of evolution.

The time line of the animal kingdom is closely connected with this evolutionary chronology. Darwin's theory of the origin of species through natural selection, nearly universally accepted in the scientific community but at times opposed on Biblical creationist grounds and in some circles of the lay community, has it that animals, like plants, have changed since the beginning of life on earth and are still evolving today. In this view, there was no sudden creation of all species. Instead, over long periods of time, new species have evolved from isolated populations of existing species. These came to occupy new niches separate from the niches of the original species. Thus, all current species are changed descendants of others that existed previously. If there are fewer apparent links between phyla than between families further down the classification ladder, the reason is that phyla have had a longer history and so have experienced more opportunities for the elimination of intermediate forms. From an evolutionary viewpoint, the difference between species down an evolutionary line are even more recent than those between families, and so on. On average, it takes about 500,000 generations for one species to evolve into another. For species to have survived in their environments—with simultaneous changes in ecology, climate, and flora—many animal forms are now more complex and efficient than their ancestors used to be.

Genes and Evolution

Life began in the seas, so for animals to live in freshwater, let alone on dry land or in the air, many obstacles had to be overcome. The impetus to conquer these inhospitable realms came from competition, according to Darwin's theory of the survival of the fittest. Those fauna that managed to surmount the problems were the ones that underwent waves of adaptive measures and evolved into new kinds of animals.

It was only in the twentieth century that it was discovered that the characteristics of a species are passed on from parents to offspring by genes. Genes provide cells for particular features, such as webbed feet. With such characteristics inherited by offspring from parents, there is a resemblance among generations. However, at times a parent may produce quite a different offspring because of genetic change. The young, in turn, may replicate this difference in their own descendants in a process known as mutation, which may occur spontaneously for unknown reasons, but also may occur due to known causes, such as exposure to radiation. At times, mutations may be useful in allowing the species to adapt better to its environment; for instance, darker moths have a better chance of survival in a forest than lighter ones because the latter are more visible to their predators than the former. Other mutations may be harmful, such as larger size that slows down a species, making its flight from danger more difficult.

Whatever the case, within a time frame of 1 to 10 million years, animals may remain the same, may evolve, or may become completely extinct, as did the dinosaurs about 60 million years ago, after being dominant for some 350 million years.

Adaptation to Environment

Natural selection leading to a new species may be accelerated when members of the original species move to a new environment, whether voluntarily or driven by the elements. Separated populations may develop different traits as they adapt to their new condition. Eventually, they will become sufficiently different to be unable to produce offspring with members of their original population. This process has repeated itself many times, over millions of years, and accounts for the large diversity in the animal world, not to mention the additional diversity consequent on artificial breeding by humans, widely observed among domesticated animals such as horses, cattle, and dogs.

Animals occupied new environments as species living in water moved to the land or later to the air. Thus, the step from fish to amphibian was essentially one from living in water for the whole life to living on land for the adult stage of the life cycle. The step from amphibian to reptile was one of increasingly proficient adaptation to land life at all stages of the life cycle.

Birds and mammals evolved in different directions from the reptiles, the first in adaptation to an arboreal and finally a flying life and the second as a further advance in the maintenance of an even and high body temperature—homeothermy—by combining an insulating external layer such as hair with a variety of physiological thermostats.

Events in geology, climate, and flora also determined the geographic distribution of species. Thus, marsupials are currently found almost entirely in Australasia and South America. The tiger exists only in India and Southeast Asia. The lion is restricted mainly to Africa. This pattern reflects the way in which these groups have evolved in relation to the physical world.

New animal groups evolve into many different forms, especially when they become dominant. For instance, when mammals came to occupy the dominant position, some became meat-eaters while others became vegetarians; some became smaller while others became larger; some became runners while others ended up as burrowers or flyers; still others returned to the water. This trend allowed the descendants of the original type to

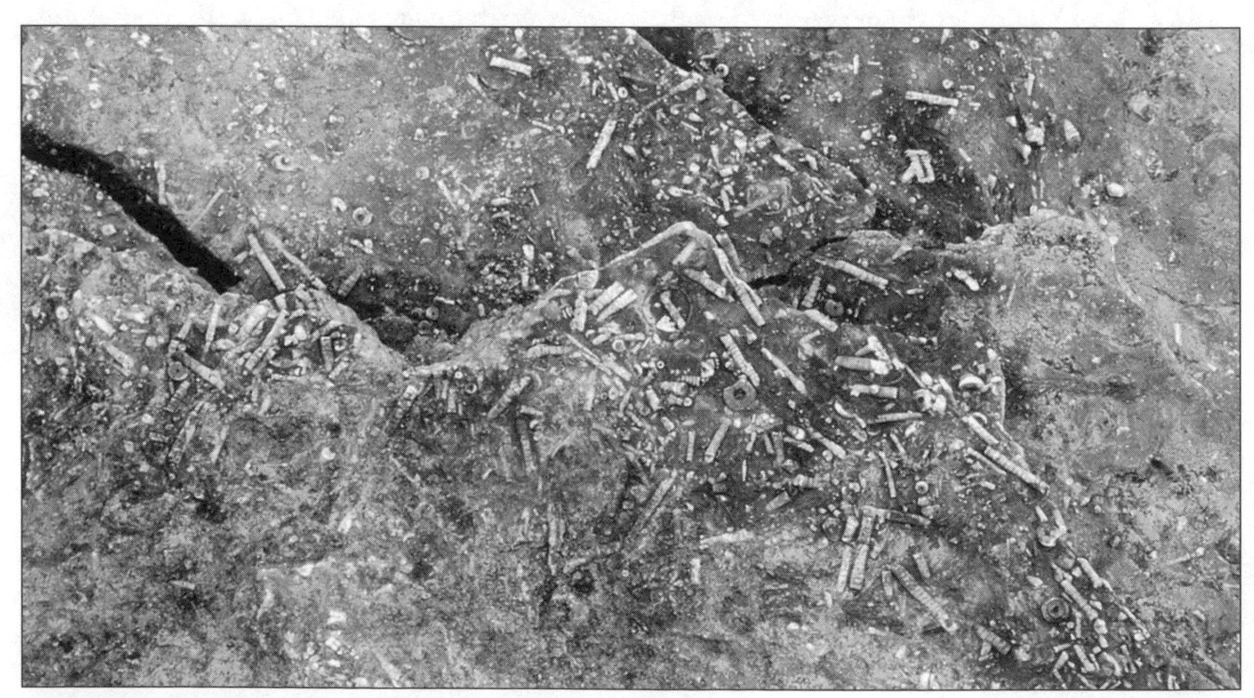

Fossils provide our only source of knowledge for the morphology of extinct animals. Through studying fossils, paleontologists can begin to trace the evolution of today's animals from extinct forms, as well as identifying forms that have no contemporary descendants. (Kenneth Layman/Photo Agora)

exploit a much greater range of environments and resources. Essentially, those species whose sense organs or brain morphology and functions improved the most ended up being dominant—primates in general and humans in particular.

Time Frame

The exact time of the origins of animals during earth's evolutionary history is not known because the early species were soft-bodied, at first single-celled and later multicellular life forms, that did not fossilize well. Fossils are the best material evidence of archaic times. Fossils do not appear earlier than 650 to 500 million years ago, not only because the animal life of the time was inappropriate for fossilization, but also because continued crustal shifts in the ensuing eons disturbed the very early rock formations. Accordingly, fossil evidence is unavailable for the entire early history of animals, which must consequently remain speculative. Current taxonomic interrelationships suggest the early history, and taxonomic diagrams may be regarded as presumptive evolutionary diagrams as well. However, a ball of carbon discovered in a cavity etched in a rock some 3.86 billion years old suggests that some life on earth was already possible at that time.

Knowledge is also limited by the fact that, even though over a million different species of animals have been identified, it is suspected that a similar number remain to be discovered or became extinct before such identification could be made. In the United States alone, some forty species of birds, about thirty-five species of mammals, and twenty-five other species have become extinct in the last two hundred years alone—less than a blip on earth's time scale—as a result of human activities such as the destruction of animal habitats through urbanization, the clearing of land for agricultural purposes, pollution, the introduction of new species from other parts of the world which turned out to be predatory to domestic specimens, hunting, and especially human population growth. It is widely predicted that climatic change triggered by greenhouse gases will continue, even enhance, this process, thereby endangering more animal species. Whatever the future, however, evolutionary biologists estimate that some 99 percent of all species that have ever lived on earth are now extinct.

Despite these and other caveats, here is a very approximate timetable of the evolution of animal life:

Life Form	Date of Emergence
Simplest single-celled Protozoa	3.5 billion years ago
Invertebrates evolving from Protozoa	670 to 640 million years ago
First vertebrates evolving from invertebrates	500 million years ago
First mammals	200 million years ago
Hominids (modern man) from apelike hominoids	200,000 to 25,000 years ago

Translated in terms of a single year:

January 1: Big Bang
March 22: Bacteria, the first living animals
November 9: Invertebrates
November 22: Vertebrates
December 16: Mammals
December 28: Primates, the highest order of mammals
December 31, a few minutes before midnight: Modern man, the dominant primate

Evolution of Existing and Extinct Human Species and Australopithecines

The root of the hominid evolutionary tree is still imperfectly known. The earliest australopithecine species, *Australopithecus anamensis*, is believed to be over four million years old, by which time that branch had diverged from African apelike ancestors. This species was followed by the *Australopithecus afarensis* nearly 3.5 million years ago. Much later came *Homo habilis*, called "skillful man" since they could presumably produce primitive tools, some two million years ago. They were followed by *Homo erectus*, "upright man," about one million years ago. Finally, *Homo sapiens*, "knowing man," emerged about 200,000 years ago. In the meantime, the australopithecine branch, after evolving through a number of intermediate species such as *A. africanus*, *A. aethiopicus*, and *A. robustus*, died out about one million years ago. To date, the earliest unearthed fossil, that of Lucy, a three-foot-tall female discovered in Ethiopia, is about four million years old. Modern humans are believed to have radiated out of Africa into Asia and Europe. Subsequently, cultural evolution became more prominent than biological evolution, but as modern humans evolved over the last four million years to their current condition, they developed manipulative skills, bipedalism, a change from specialized to omnivorous feeding habits, and especially, a threefold increase in cranial capacity from *H. afarensis* to *H. neanderthalensis*, together with behavior appropriate to the control of the environment.

Although humans are not the only animals capable of conceptual thought, they have refined and extended that ability until it has become their hallmark. Thus, thanks to the symbolic language of *Homo sapiens*, modern humans make possible the accumulation of experience from one generation to the next. Such cultural evolution is possessed by few, if any other animal species. It is for this reason that humans, more than other animals, have found ways to mold and change their environment according to need rather than in response to environmental demands. Because of this ability and humans' control of technology, the species has more say about their biological future than any other.

—Peter B. Heller

See also: Adaptive radiation; Apes to hominids; Clines, hybrid zones, and introgression; Coevolution; Convergent and divergent evolution; Development: Evolutionary perspective; Evolution: Historical perspective; Extinction; Extinctions and evolutionary explosions; Gene flow; Genetics; Hardy-Weinberg law of genetic equilibrium; Heterochrony; *Homo sapiens* and human diversification; Human evolution analysis; Isolating mechanisms in evolution; Natural selection; Neutral mutations and evolutionary clocks; Nonrandom mating, genetic drift, and mutation; Paleoecology; Paleontology; Phylogeny; Punctuated equilibrium and continuous evolution; Sex differences: Evolutionary origins; Systematics.

Bibliography

Carroll, Sean B., Jennifer K. Grenier, and Scott D. Weatherbee. *From DNA to Diversity: Molecular Genetics and the Evolution of Animal Design*. Malden, Mass.: Blackwell Science, 2001. Covers general principles of the genetic basis of morphological change, including the history of animal evolution, model system developmental genetics, genetic regulatory mechanisms, and case studies of evolutionary change. Color diagrams and images, glossary.

Conway-Morris, Simon. *The Crucible of Creation: The Burgess Shale and the Rise of Animals*. Los Angeles: The Getty Center for Education in the Arts, 1999. A fascinating study of one of the richest fossil deposit sites on earth, focusing on what can be learned of evolution from fossil remains.

Lavers, Chris. *Why Elephants Have Big Ears: Understanding Patterns of Life on Earth*. New York: St. Martin's Press, 2001. Explores the evolving interaction of form and function in animals.

Long, John A. *The Rise of Fishes: Five Hundred Million Years of Evolution*. Baltimore: The Johns Hopkins University Press, 1996. A detailed history of fish evolution. Color photographs, glossary.

EVOLUTION: HISTORICAL PERSPECTIVE

Type of animal science: Evolution
Fields of study: Ecology, embryology, evolutionary science, genetics, paleontology, population biology, zoology

Evolution is the process of change in biological populations. Historically, it is also the theory that biological species undergo sufficient change with time to give rise to new species.

Principal Terms

ADAPTATION: the possession by organisms of characteristics that suit them to their environment or their way of life

CATASTROPHISM: a geological theory explaining the earth's history as resulting from great cataclysms (floods, earthquakes, and the like) on a scale not now observed

DARWINISM: branching evolution brought about by natural selection

ESSENTIALISM (TYPOLOGY): the Platonic-Aristotelian belief that each species is characterized by an unchanging "essence" incapable of evolutionary change

GENOTYPE: the hereditary characteristics of an organism

GEOFFROYISM: an early theory of evolution in which heritable change was thought to be directly induced by the environment

LAMARCKISM: an early evolutionary theory in which voluntary use or disuse of organs was thought to be capable of producing heritable changes

SCALE OF BEING (CHAIN OF BEING): an arrangement of life forms in a single linear sequence from "lower" to "higher"

UNIFORMITARIANISM: a geological theory explaining the earth's history using processes that can be seen at work today

Evolution is the theory that biological species undergo sufficient change with time to give rise to new species. The concept of evolution has ancient roots. Anaximander suggested in the sixth century B.C.E. that life had originated in the seas and that humans had evolved from fish. Empedocles (fifth century B.C.E.) and Lucretius (first century B.C.E.), in a sense, grasped the concepts of adaptation and natural selection. They taught that bodies had originally formed from the random combination of parts, but that only harmoniously functioning combinations could survive and reproduce. Lucretius even said that the mythical centaur, half horse and half human, could never have existed because the human teeth and stomach would be incapable of chewing and digesting the kind of grassy food needed to nourish the horse's body.

For two thousand years, however, evolution was considered an impossibility. Plato's theory of forms (also called his "theory of ideas") gave rise to the notion that each species had an unchanging "essence" incapable of evolutionary change. As a result, most scientists from Aristotle to Carolus Linnaeus in the eighteenth century insisted upon the immutability of species. Many of these scientists tried to arrange all species in a single linear sequence known as the scale of being (also called the chain of being and the *scala naturae*), a concept supported well into the nineteenth century by many philosophers and theologians as well. The sequence in this scale of being was usually interpreted as a static "ladder of perfection" in God's creation, arranged from higher to lower forms.

The scale had to be continuous, for any gap would detract from the perfection of God's creation. Much exploration was devoted to searching for "missing links" in the chain, but it was generally agreed that the entire system was static and incapable of evolutionary change. Pierre-Louis Moreau de Maupertuis, in the eighteenth century, and Jean-Baptiste Lamarck were among the scientists who tried to reinterpret the scale of being as an evolutionary sequence, but this single-sequence idea was later replaced by Charles Darwin's concept of branching evolution. Georges Cuvier finally showed that the major groups of animals had such strikingly different anatomical structures that no possible scale of being could connect them all; the idea of a scale of being lost most of its scientific support as a result.

The Struggle to Conceptualize Evolution

The theory that new biological species could arise from changes in existing species was not readily accepted at first. Linnaeus and other classical biologists emphasized the immutability of species under the Platonic-Aristotelian concept of essentialism. Those who believed in the concept of evolution realized that no such idea could gain acceptance until a suitable mechanism of evolution could be found. Many possible mechanisms were therefore proposed. Étienne Geoffroy Saint-Hilaire proposed that the environment directly induced physiological changes, which he thought would be inherited, a theory now known as Geoffroyism. Lamarck proposed that there was an overall linear ascent of the scale of being but that organisms could also adapt to local environments by voluntary exercise, which would strengthen the organs used; unused organs would deteriorate. He thought that the characteristics acquired by use and disuse would be passed on to later generations, but the inheritance of acquired characteristics was later disproved. Central to both these explanations was the concept of adaptation, or the possession by organisms of characteristics that suit them to their environments or to their ways of life. In eighteenth century England, the Reverend William Paley and his numerous scientific supporters believed that such adaptations could be explained only by the action of an omnipotent, benevolent God. In criticizing Lamarck, the supporters of Paley pointed out that birds migrated toward warmer climates before winter set in and that the heart of the human fetus had features that anticipated the changes of function that take place at birth. No amount of use and disuse could explain these cases of anticipation, they claimed; only an omniscient God who could foretell future events could have designed things with their future utility in mind.

The nineteenth century witnessed a number of books asserting that living species had evolved from earlier ones. Before 1859, these works were often more geological than biological in content. Most successful among them was the anonymously published *Vestiges of the Natural History of Creation* (1844), written by Robert Chambers. Books of this genre sold well but contained many flaws. They proposed no mechanism to account for evolutionary change. They supported the outmoded concept of a scale of being, often as a single sequence of evolutionary "progress." In geology, they supported the outmoded theory of catastrophism, an idea that the history of the earth had been characterized by great cataclysmic upheavals. From 1830 on, however, that theory was being replaced by the modern theory of uniformitarianism, championed by Charles Lyell. Charles Darwin read these books and knew their faults, especially their lack of a mechanism that was compatible with Lyell's geology. In his own work, Darwin carefully tried to avoid the shortcomings of these books.

Darwin's Revolution in Biological Thought

Darwin brought about the greatest revolution in biological thought by proposing not only a theory of branching evolution but also a mechanism of natural selection to explain how it occurred. Much of Darwin's evidence was gathered during his voyage around the world aboard HMS *Beagle*. Darwin's stop in the Galápagos Islands and his study of tortoises and finchlike birds on these islands are usually credited with convincing him

Comte de Buffon (Georges-Louis Leclerc)

Born: September 7, 1707; Montbard, France
Died: April 16, 1788; Paris, France
Fields of study: Anthropology, biology, botany, chemistry, geology, mathematics, paleontology, zoology
Contribution: The greatest naturalist of the eighteenth century, Buffon popularized zoology and botany through his publications. He tried to separate science from religious and metaphysical ideas and rejected teleological reasoning and the idea of God's direct intervention in nature.

Born to a noble family in Dijon, Buffon was admitted to the French Academy of Sciences in 1734 and the Académie Française in 1753. In 1739 he was appointed director (*Intendant*) of the Jardin du Roi, the royal botanical garden. Between 1749 and his death, he published (with associates) thirty-six volumes of the *Histoire Naturelle* (44 vols., 1739-1804; *A Natural History, General and Particular*, 10 vols., 1807) which took in the formation of the earth, geology, paleontology, zoology, and botany. The last eight volumes were posthumous. A master of style, his books were among the most popular in the eighteenth and early nineteenth centuries.

Buffon believed that the study of Earth was a necessary prerequisite to botany and zoology, and wrote two important texts on geology and paleontology. The first, *Théorie de la terre*, was published in 1749, the other, *Époques de la nature*, in 1778. From experiments on the cooling of globes, he estimated the age of the earth to be 85,000 years, significantly at variance with his contemporaries' estimation of an origin around 4000-6000 B.C.E. Buffon's cosmogony replaced the intervention of God by a cause whose effects are in accord with the laws of mathematics.

Buffon postulated that new varieties of plants and animals (including humans) were produced in nature by external geographical influences. Such influences could also cause degeneration. He was against classifying nature, as "... everything that can be, is." He gave "species" a purely biological definition: animals that by means of copulation perpetuate themselves and preserve their similarity. He thought families were artificial creations made by man. He therefore thought Linnaeus' classification of plants based on sexual characters was too rigid. He arranged animals in order of their utility to humans (later he rearranged them according to distinctive characteristics) and believed that some forms might have degenerated from others over time—thus, the ass might be a degenerate form of the horse. Buffon thus alluded to a form of "evolution" where physical characteristics produced by external influences could be passed down the generations.

The Comte de Buffon was perhaps the greatest naturalist of the eighteenth century, rejecting religious explanations of the state of the world in favor of scientific research. (Library of Congress)

Buffon tried to separate science from religious and metaphysical ideas, and rejected teleological reasoning and the idea of God's direct intervention in nature. His theories went against the accepted theological belief of immutability of species and he was reprimanded by the Faculty of Theology at the University of Paris for this. Buffon apologized but did not change his views.

Buffon was made a count in 1773 by Louis XV. He was greatly respected by his contemporaries and was made a member of almost every learned society in Europe.

—*Ranès C. Chakravorty*

that evolution was a branching process and that adaptation to local environments was an essential part of the evolutionary process. Adaptation, he later concluded, came about through natural selection, a process that killed the maladapted variations and allowed only the well-adapted ones to survive and pass on their hereditary traits. After returning to England from his voyage, Darwin raised pigeons, consulted with various animal breeders about changes in domestic breeds, and investigated other phenomena that later enabled him to demonstrate natural selection and its power to produce evolutionary change.

Darwin's greatest contribution was that he proposed a suitable mechanism by which permanent organic change could take place. All living species, he said, were quite variable, and much of this variation was heritable. Also, most organisms produce far more eggs, sperm, seeds, or offspring than can possibly survive, and the vast majority of them die. In this process, some variations face certain death while others survive in greater or lesser proportion. Darwin called the result of this process "natural selection," the capacity of some hereditary variations (now called genotypes) to leave more viable offspring than others, with many leaving none at all. Darwin used this theory of natural selection to explain the form of branching evolution that has become generally accepted among scientists.

Darwin delayed the publication of his book for seventeen years after he wrote his first manuscript version. He might have waited even longer, except that his hand was forced. From the East Indies, another British scientist, Alfred Russel Wallace, had written a description of the very same theory and submitted it to Darwin for his comments. Darwin showed Wallace's letter to Lyell, who urged that both Darwin's and Wallace's contributions be published, along with documented evidence showing that both had arrived at the same ideas independently. Darwin's great book, *On the Origin of Species by Means of Natural Selection*, was published in 1859, and it quickly won most of the scientific community to a support of the concept of branching evolution. In his later years, Darwin also published *The Descent of Man and Selection in Relation to Sex* (1871), in which he outlined his theory of sexual selection. According to this theory, the agent that determines the composition of the next generation may often be the opposite sex. An organism may be well adapted to live, but unless it can mate and leave offspring, it will not contribute to the next or to future generations.

Acceptance of Darwinism in the Twentieth Century

In the early 1900's, the rise of Mendelian genetics (named for botanist Gregor Mendel) initially resulted in challenges to Darwinism. Hugo de Vries proposed that evolution occurred by random mutations, which were not necessarily adaptive. This idea was subsequently rejected, and Mendelian genetics was reconciled with Darwinism during the period from 1930 to 1942. According to this modern synthetic theory of evolution, mutations initially occur at random, but natural selection eliminates most of them and alters the proportions among those that survive. Over many generations, the accumulation of heritable traits produces the kind of adaptive change that Darwin and others had described. The process of branching evolution through speciation is also an important part of the modern synthesis.

The branching of the evolutionary tree has resulted in the proliferation of species from the common ancestor of each group, a process called adaptive radiation. Ultimately, all species are believed to have descended from a single common ancestor. Because of the branching nature of the evolutionary process, no one evolutionary sequence can be singled out as representing any overall trend; rather, there have been different trends in different groups. Evolution is also an opportunistic process, in the sense that it follows the path of least resistance in each case. Instead of moving in straight lines toward a predetermined goal, evolving lineages often trace meandering or circuitous paths in which each change represents a momentary increase in adaptation. Species that cannot adapt to changing conditions die out and become extinct.

Studying Evolution

Evolution is studied by a variety of methods. The ongoing process of evolution is studied in the field by ecologists, who examine various adaptations, including behavior and physiology as well as anatomy. These adaptations are also studied by botanists, who examine plants; zoologists, who examine animals; and various specialists, who work on particular kinds of animals or plants (for example, entomologists, who study insects). Some investigators capture specimens in the field, then bring back samples to the laboratory in order to examine chromosomes or analyze proteins using electrophoresis. Through these methods, scientists learn how the ongoing process of evolutionary change is working today within species or at the species level on time scales of only one or a few generations.

The long-term results of evolutionary processes are studied among living species by comparative anatomists and embryologists. Extinct organisms are studied by paleontologists, scientists who examine fossils. Biogeographers study past and present geographic distributions. All these types of scientists make comparisons among species in order to determine the sequence of events that took place in the evolutionary past. One method of reconstructing the branching sequences of evolution is to find homologies, deep-seated resemblances that reflect common ancestry. Once the sequences are established, functional analysis can be used to suggest possible adaptive reasons for any changes that took place. The sequences of evolutionary events reconstructed by these scientists represent the history of life on the earth. This history spans many species, families, and whole orders and classes, and it covers great intervals of past geologic time, measured in many millions of years.

The Historical Context of Evolutionary Theory

The historical development of evolutionary theory should be viewed in two contexts: that of biological science and that of cultural history. The concept of evolution had been talked about for many years before 1859 and was usually rejected because no suitable mechanism had gained widespread acceptance. The fact that the phenomenon of natural selection was independently discovered by two Englishmen shows both that the time was ripe for the discovery and that the circumstances were right in late nineteenth century England.

Evolutionary biology is itself the context into which all the other biological sciences fit. Other biologists, including physiologists and molecular biologists, study how certain processes work, but it is evolutionists who study the reasons why these processes came to work in one way and not another. Organisms and their cells are built one way and not another because their structures have evolved in a particular direction and can only be explained as the result of an evolutionary process. Not only does each biological system need to function properly, but it also must have been able to achieve its present method of functioning as the result of a long, historical, evolutionary process in which a previous method of functioning changed into the present one. If there were two or more ways of accomplishing the same result, a particular species used one of them because found it easier to evolve one method rather than another.

Everything in biology is thus a detail in the ongoing history of life on the earth, because every living system evolves. Living organisms and the processes that make them function are all products of the evolutionary process and can be understood only in that context. As biologist Theodosius Dobzhansky once said, "Nothing in biology makes sense, except in the light of evolution."

—*Eli C. Minkoff*

See also: Adaptive radiation; Apes to hominids; Clines, hybrid zones, and introgression; Coevolution; Convergent and divergent evolution; Development: Evolutionary perspective; Evolution: Animal life; Extinction; Extinctions and evolutionary explosions; Gene flow; Genetics; Hardy-Weinberg law of genetic equilibrium; Heterochrony; *Homo sapiens* and human diversification; Human evolution analysis; Isolating mechanisms in evolution; Natural selection; Neutral mutations and evolutionary clocks; Nonrandom mating, ge-

Jean-Baptiste de Lamarck

Born: August 1, 1744; Bazentin-le-Petit, Picardy, France
Died: December 18, 1829; Paris, France
Fields of study: Evolutionary science, invertebrate biology, paleontology, zoology
Contribution: Lamarck established the division of animal life into the vertebrate and invertebrate categories, and he formulated an evolutionary theory based on the premise that acquired traits are inheritable.

After studying briefly for the priesthood at the Jesuit seminary in Amiens, Jean-Baptiste de Monet, chevalier de Lamarck, served as an army officer in the Seven Years' War. Following an accident in 1768, he began to study botany and medicine. In 1778, he published his three-volume *Flore français* (French plants), which was widely used as a manual of identification.

Jean-Baptiste de Lamarck is best known for his theory that acquired physical traits may be inheritable. (Library of Congress)

Lamarck was then employed as assistant botanist at the royal botanical gardens of Paris, and he was also appointed to the prestigious Academy of Sciences. Count Georges-Louis de Buffon engaged him as tutor to his son, which allowed him to tour European botanical gardens for two years. When the Jardin des Plantes (the National Museum of Natural History) was founded in 1793, he was placed in charge of the collection of invertebrates (a term that he coined).

During the early nineteenth century, Lamarck published numerous books about invertebrates, paleontology, and biological evolution. His *Système des animaux sans vertèbres* (1801; system of invertebrate animals) presented a systematic basis for the classification of the lower animals. His *Hydrogéologie* (1802; *Hydrogeology,* 1964) interpreted the history of the earth as a series of inundations, each resulting in organic deposits that built up the continents. The book was especially noteworthy for its recognition of the vastness of geologic time.

Lamarck was not the first to propose a theory of biological evolution, but his theory was more systematic and coherent than previous versions. He gave the clearest explanation for his theory in *Philosophie zoologique* (1809; *Zoological Philosophy,* 1873), presenting a two-part process. First, a change in the environment forced organisms to change their behavior. If particular organs were used, they would increase in size and strength; in contrast, disuse or disease would weaken and shrink organs. Second, Larmark argued that such changes would be inherited, so that the characteristics of a species would change gradually over many generations.

Lamarck's scientific work culminated in an exhaustive study, *Histoire naturelle des animaux sans vertèbres* (1815-1822; natural history of invertebrate animals).

Charles Darwin acknowledged the great contribution of Lamarck's work, and Darwin's own theory of natural selection never entirely rejected the possibility that some acquired traits might be inherited. With advancements in the science of heredity during the twentieth century, the concept of Lamarckian inheritance has been largely abandoned. Late in the twentieth century, nevertheless, Edward J. Steele and other biologists found evidence that the acquired immunities of organisms might be passed on to their offspring.

—Thomas Tandy Lewis

netic drift, and mutation; Paleoecology; Paleontology; Phylogeny; Punctuated equilibrium and continuous evolution; Sex differences: Evolutionary origins; Systematics.

Bibliography

Bowler, Peter J. *Evolution: The History of an Idea*. Rev. ed. Berkeley, Calif.: University of California Press, 1989. A comprehensive history of the evolutionary theory for both specialist and nonspecialist.

_______. *Life's Splendid Drama: Evolutionary Biology and the Reconstruction of Life's Ancestry, 1860-1940*. Chicago: University of Chicago Press, 1996. A history of evolutionary morphology and its relationship with paleontology and biogeography. Covers scientific debates over the emergence of vertebrates, the origins and extinctions of animal species, and the role and influence of Darwin. Biographical appendix, bibliography.

Brandon, Robert N. *Concepts and Methods in Evolutionary Biology*. New York: Cambridge University Press, 1996. A collection of essays spanning two decades, addressing problems in the philosophy of biology, particularly the conception of relative adaptedness and the principle of natural selection.

Darwin, Charles R. *On the Origin of Species by Means of Natural Selection: Or, the Preservation of the Favoured Races in the Struggle for Life*. London: John Murray, 1859. This is the original edition, still worth reading. It is better than the more widely reprinted sixth edition, in which Darwin's more forceful statements were toned down as a response to criticism that is no longer greatly valued by biologists. Some knowledge of zoology, geology, and geography would definitely increase any reader's understanding and appreciation of this book. Darwin provided no bibliography, but some modern editors have supplied one.

Dobzhansky, Theodosius G. *Genetics of the Evolutionary Process*. New York: Columbia University Press, 1970. Although somewhat technical in places, this book is extremely well written, and a careful reader should be able to understand it all without any formal background. It is an excellent (and very detailed) outline of the evolutionary process in terms of genetic changes. It contains much information about the genus *Drosophila* (fruit flies), on which Dobzhansky was an expert. Very comprehensive bibliography.

Gould, Stephen J. *Ever Since Darwin*. Reprint. New York: W. W. Norton, 1992.

_______. *The Flamingo's Smile*. New York: W. W. Norton, 1985.

_______. *The Panda's Thumb*. Reprint. New York: W. W. Norton, 1992. These three books all consist of essays reprinted (and occasionally updated) from *Natural History* magazine. All are well written and directed to a general audience; no previous background is assumed. Although Gould has occasionally supported unorthodox viewpoints, most of the views represented here have become accepted into the mainstream with the passage of time. Gould's easy, familiar style makes for lively reading and he uses esoteric cases and seemingly inconsequential details to make important points about evolution in general. The bibliographies are wide-ranging but are confined to the topics of the individual essays.

Grant, Verne. *The Evolutionary Process: A Critical Study of Evolutionary Theory*. 2d ed. New York: Columbia University Press, 1991. A comprehesive and critical review of modern evolutionary theory. Focuses on whole organisms and general principles rather than molecular changes and mathematical models.

Rose, Michael A. *Darwin's Spectre: Evolutionary Biology in the Modern World*. Princeton, N.J.: Princeton University Press, 1998. Written for the general reader. Outlines the fundamental ideas of evolutionary biology and its influence in other fields such as agriculture, medicine, and eugenics.

Wills, Christopher, and Jeffrey Bada. *The Spark of Life: Darwin and the Primeval Soup*. Cambridge, Mass.: Perseus, 2000. Describes theories of the origins of terrestrial life and its evolution. Written for the general reader.

EXOSKELETONS

Type of animal science: Anatomy
Field of study: Invertebrate biology

The exoskeleton is one of the distinctive primary features of members of the phylum Arthropoda. The evolution of the exoskeleton cuticle is thought to have been important in this group's successful adaptation to many diverse environments. The exoskeleton's intricate structure has provided advantages enabling the arthropods to thrive for hundreds of millions of years.

Principal Terms

CHITIN: a cellulose-like, crystalline material that makes up 25 percent to 60 percent of the dry weight of the cuticle

CUTICLE: the outer arthropod exoskeleton consisting of several layers of secreted organic matter, primarily nonliving chitin

ENDOCUTICLE: usually the thickest layer of the cuticle, found just outside the living epidermal cell layer and made of untanned proteins and chitin

EPICUTICLE: the outermost and thinnest layer of the arthropod cuticle, composed mainly of the hardened protein cuticulin

EPIDERMIS: a living cellular layer that secretes the greater part of the cuticle and is responsible for dissolving and absorbing the cuticle during molting (also termed the hypodermis)

EXOCUTICLE: a thick middle layer in the cuticle made up of both chitin and rigid, tanned proteins termed sclerotin

INTEGUMENTARY PROCESSES: surface outgrowths from the cuticle, primarily rigid nonarticulated processes or movable articulated processes

SCLEROTIN: a hard, horny protein constituent of the exocuticle found in arthropods such as insects; it is superficially similar to vertebrate horn or keratin

The evolution of invertebrate animals possessing rigid, hard exoskeletons represented a great advance for members of the phylum Arthropoda. The development of such exoskeletons—in comparison with less solid structures, such as the hydrostatic skeleton of coelenterates—gave arthropods several distinct evolutionary advantages over other invertebrate phyla. Hydrostatic skeletons, such as those possessed by sea anemones, operate by the animal's musculature being arranged in a pattern that surrounds an enclosed volume of fluid. Contraction of any one section of the muscular system creates a fluid pressure in the central cavity that is consequently transmitted in an omnidirectional manner to the rest of the body. Arthropodic exoskeletons, on the other hand, are consistently rigid and much harder because they are composed to some extent of crystalline substances. Flexibility of movement is attained by multiple jointings in the limb system and in other appendages in the body such as feeding and sensory apparatuses.

Exoskeletons and Arthropod Success

As a consequence of two distinctive features—a hard, rigid exoskeleton and jointed appendages and other body parts—arthropods have become one of the most successful of all animal groups; indeed, it is by these two features that they are taxonomically defined. Biologists sometimes term the enhancement of the annelid (worm) body plan of segmentation by arthropod improvement "arthropodization." Because of it, the ancestors of the present immense spectrum of arthropod species successfully adapted to myriad ecological niches in the sea, on land, and in the air. Arthropod species account for more than three quarters of all known animal species. In fact, the class Insecta, one of a number of classes within the arthropod phylum, numbers at least 700,000 known species, with new species being discovered yearly, mostly in the tropics.

The immense success of arthropods is, to a great extent, the result of the advantages provided by the composition and structure of the seemingly simple surface architecture that is the arthropod exoskeleton. This exoskeleton not only provides a substantial chemical and physical barrier between the animal and the external environment, protecting the internal organs and fluids, but also allows a degree of temperature and osmotic regulation. In addition, the exoskeleton helps deter predation, provides a solid base of attachment for an internal muscular system, and offers a good site for the location of various sense organs. One of the most noteworthy evolutionary advantages of the exoskeleton is its service as a solid base of muscle attachment. The arthropod limbs act as a system of mechanical levers that is a much more efficient locomotive system than that of evolutionarily older and less sophisticated invertebrate locomotive systems such as that of the annelids. Because of exoskeletons and the structurally strong, jointed appendages that exoskeletons permit, arthropods possess an internal, muscular body wall broken down into separate muscles having an arrangement allowing contractions that are more localized in time and space than annelid or coelenterate muscle behavior. This more modular approach to the musculature allows arthropods to react to their environments to use energy more efficiently and with greater precision of movement and response. In fact, the inner surface of the exoskeleton acts as a limited type of endoskeleton, or inner skeleton, in that it provides good anchoring sites for muscle attachment, thus further increasing the leverage power of arthropod limbs and appendages.

The Parts of the Exoskeleton

The exoskeleton itself can be divided into several distinct units based on function and composition. These are composed of consecutive layers that surround the animal in an arrangement similar to a medieval knight's suit of armor. Like the armor suit, the outermost exoskeleton is, in a typical arthropod, very rigid and hard; movement is possible only because both protective systems are composed of plates or body-contoured segments that incorporate narrow, flexible jointings allowing motion. The motion is usually narrowly defined in extent and direction, and it is this quality that gives both armored humans and many arthropods their often distinctively awkward and ungainly mode of movement. The larger terrestrial beetles and marine forms such as crabs and lobsters are ready examples of this. Some arthropods are nevertheless adroit and delicate in their movements, as shown by various arthropod aerialists such as the dragonflies and butterflies.

Insects can, in many ways, be considered typical arthropods and are therefore useful as models for a discussion of the exoskeleton as found among all arthropods. All the various layers of the exoskeleton, both living and nonliving, are as a whole variously termed the integument or cuticle. These layers are the skin or surface of the animal. The innermost layer of the exoskeleton is termed the epidermis, or hypodermis, and is made of living cells. The epidermis is immediately external to a basement membrane that separates the epidermis from the inner body, with its organs and fluids. The epidermis is responsible for secreting the layers external to it, which are organic but actually nonliving material. From the epidermis outward, these layers in insects consist of the endocuticle, the exocuticle, and the epicuticle. (Some biologists use the term procuticle to describe the endocuticle and use epicuticle to mean both the epicuticle and the exocuticle.) Whatever precise terms may be employed, the general concept is that layers of material closer to the epidermis are more flexible and less chemically hardened than layers that are found closer to the actual exterior of the animal. The endocuticle is usually the thickest cuticle layer and is constituted of protein mixed with a material called chitin. Chitin is a cellulose-like, crystalline material that makes up anywhere from 25 percent to 60 percent of the dry weight of the cuticle. It has many useful properties, such as resistance to concentrated alkalies and acids. In chemical composition, chitin is a nitrogenous polysaccharide. Chitin itself can be a relatively soft and flexible material that gains hardness in the outermost arthropod exoskeleton in several ways. One way is by using the presence of a material termed sclerotin. The process of hardening through the agency of sclerotin is called sclerotinization and involves a molecular change in the organization of the protein part of the cuticle. The outermost chitin found in the exocuticle of insects, for example, is thoroughly sclerotinized, which characteristically results in a darkening of the chitin. The other method by which chitin hardens is the deposition of calcium carbonate, primarily in the form of calcite. This is the process found among marine arthropods such as the Crustacea: crabs, lobsters, and shrimp, for example. This process, called calcification, occurs among Crustacea, starting in their epicuticle, or outermost exoskeletal layer, and works inward to the exocuticle and finally the endocuticle.

Besides the darkening caused by sclerotinization, coloration of the cuticle is effected in two basic ways. One is simple pigmentation caused by the presence of colored compounds found within the cuticle itself. The other is through the presence of extremely fine parallel ridges found on the epicuticle. These ridges break normal white light into its constituent wavelengths by prismatic diffraction in the same way that raindrops create rainbows. It is by this means that the effect of spectacularly iridescent rainbow hues found on many insects' wings and bodies is achieved.

Adding to the complexity of the cuticle are great numbers of sensory organs that project from or extend through the various exoskeletal layers. Prominent among these sensory structures are tactile hairs, bristles, and spines found all over the general body surface and on limb surfaces. These sensory structures, or setae, are movable and are set into thin, flexible disks on the cuticle surface itself. When one of these projections is moved, its base mechanically stimulates one or more sensory cells, setting off stimuli to which the arthropod can respond.

The Drawbacks of Exoskeletons

While the exoskeleton has evolved wonderfully to protect the arthropod and to enhance its locomotive and sensory abilities, its overall rigid structure presents some inherent drawbacks. Perhaps chief among these is the fact that its formidable rigidity and solidity are limitations to an individual's physical growth throughout its lifetime. Growth, in fact, is probably an arthropod's single most difficult physiological problem. This is true because once formed and hardened, an exoskeleton cannot be enlarged as the animal within enlarges with time. The physiologic solution among arthropods is the process termed ecdysis, or molting.

This process is intrinsically dangerous to the arthropod, as it leaves each individual extremely vulnerable to predation during, and immediately following, molting. It has been estimated that as much as 80 to 90 percent of arthropod mortality occurs during ecdysis. The process takes place in stages. Prior to the shedding of the old exoskeleton, a new, soft cuticle is formed beneath the old one. The new cuticle has not started along either the sclerotinization or calcification process and therefore is still soft and pliable. As the new cuticle is forming, the lower section of the old cuticle is partially dissolved by corrosive fluids secreted by cutaneous glands situated below the new cuticle. Immediately prior to the shedding, also termed casting, of the old exoskeletal cuticle, the arthropod stops feeding and absorbs more than the usual amount of water and oxygen. Its body begins to swell, and the animal makes spasmodic movements to shake off the old cuticle, most of the base of which has been removed by the corrosive process. Eventually, the old exoskeleton is effectively disconnected from the arthropod's body, and the animal extricates itself from the remains. At this point, its new exoskeleton is soft and very pliable, and its movements are limited. Conse-

Exoskeletons offer protection, but not much room to grow. Most animals that have exoskeletons, such as this stone fly, must periodically shed their exoskeletons to make room for increases in size. (Larry and Rebecca Javorsky/Photo Agora)

quently, the individual is extremely vulnerable to both predation and serious damage from tearing through abrasive or sharp-edged materials in its environment. It takes some time before the animal's cuticle has hardened and thickened enough for it to resume its normal activities. In the meantime, the exoskeleton—which normally provides a great degree of protection and mobility—acts as a hindrance and danger to the arthropod.

Studying Exoskeletons

The two main approaches used to study the arthropod exoskeleton are the same types of study used by researchers in nearly all branches of the life sciences: field studies and laboratory studies. In the field approach, living arthropods are observed in nature. The specific techniques employed include both still and motion photography in various light—normal, infrared, and ultraviolet. It is important to combine the observational database with later structural analyses of specific arthropod body parts, such as exoskeletons, to ascertain how the anatomical components actually function in the natural setting. Actual physical collections are necessary for study by laboratory workers, who subject the specimens to a range of tests to determine their qualities and features in comparison with similar species and with the normal parameters that are known for previously collected members of the same species.

Specimens are often dissected—or in some cases, vivisected (disassembled while alive)—in order to record useful data such as chemical composition of exoskeletons, metabolic rates, and the estimated age of the sample. In the case of specimens raised in captivity, more precise data can be gained, as the precise age, food type, and daily or hourly intake are known with great precision.

A wide range of techniques are employed in the laboratory to analyze the structural components of exoskeletons. Among them, optical microscopy has traditionally been the primary approach. Working with dissected parts, frequently cut and chemically stained to facilitate viewing or bring out certain features selectively, researchers have used powerful microscopes capable of magnifying by a factor of many hundreds to see tiny subcomponent structures found within exoskeletal tissue. Optical microscopes have also been used to take a close look at associated cellular and noncellular organic matter, such as chitin and sclerotin. An exponential increase in magnification for study, however, has arrived with the advent of scanning electron microscopes (SEMs). These are instruments that use a beam of focused electrons to scan an object and form a three-dimensional image on a cathode-ray tube. The SEM reads both the pattern of electrons scattered by the object and the secondary electrons produced by it. This greatly enhanced ability to see smaller objects with great clarity allows scientists to see very small target sections of exoskeletal tissue, measured in microns (millionths of a meter) in circumference. Researchers can examine in minute detail the structures and interrelationships of the various layers of the arthropod cuticle.

Exoskeletons in the Fossil Record

Because the hard arthropod exoskeleton fossilizes more readily than the remains of many other animals, the evolutionary history of the phylum Arthropoda is abundantly represented in the fossil record. Much more is known about this phylum than other invertebrate phyla because of this phenomenon. Entire classes of arthropods that have left no modern descendants are known today because their substantial body armor appears in various marine strata. Examples of this are well documented in the remains of the extinct marine groups of the trilobites (similar to modern horseshoe crabs) and the eurypterids (giant "water scorpions"). In the case of the trilobites, many fossils are actually the result of cast-off exoskeletal moltings rather than the carcasses of the dead animals. This illustrates the fact that, for hundreds of millions of years, arthropods have maintained a lifestyle and evolutionary approach to physiological problems that are similar to those of modern forms. X-ray photography has been employed successfully to penetrate the hard, mineralized fossils of extinct and ancestral forms of modern arthropods, showing in good detail the internal structures of exoskeletons and other tissues. Radiographic images produced by this technique demonstrate the continuity of structure and life shared by members of this extremely successful phylum for a period extending beyond the early Cambrian period (500 million years before the present).

The exoskeleton is an evolutionary advantage shared by all arthropods. This advantage, along with their body and limb segmentation, has allowed them to move into myriad ecological niches, first in the sea and later on land, in freshwater, and finally in the air. As a phylum, arthropods are arguably the most successful of all metazoan animal phyla; they exceed all others combined in terms of the number of species, diversity, and the number of individual organisms. This ubiquity in all biomes and climates and in virtually every conceivable niche in every ecosystem makes them a force that has a constant influence on human life.

—Frederick M. Surowiec

See also: Arthropods; Bone and cartilage; Cell Types; Circulatory systems of invertebrates; Endocrine systems in invertebrates; Endoskeletons; Evolution: Historical perspective; Hydrostatic skeletons; Invertebrates; Insects; Molting and shedding; Metamorphosis; Muscles in invertebrates; Scales; Shells.

Bibliography

Anderson, D. T. *Atlas of Invertebrate Anatomy*. Sydney, Australia: Univerity of New South Wales Press, 1996. Biological drawings illustrate the internal and external anatomy of eighty species frequently studied in zoology and marine biology class labs. Includes notes on classification, life cycle, and habitat. Indexed by taxonomy and anatomical feature.

Boardman, Richard S., Alan H. Cheetham, and Albert J. Rowell, eds. *Fossil Invertebrates.* Palo Alto, Calif.: Blackwell Scientific Publications, 1987. An extremely thorough and exhaustively detailed treatment of all known types of fossil invertebrates, with an emphasis on metazoans. The chapter on the phylum Arthropoda is of prime interest because it includes a comprehensive description of exoskeletons and what is known of their evolution from the fossil record in extinct and extant taxonomic classes. Features numerous expert drawings, diagrams, and black-and-white photographs. It is suitable for readers who have a solid foundation in natural history.

Clarkson, E. N. K. *Invertebrate Palaeontology and Evolution.* 4th ed. Boston: Allen & Unwin, 1998. Intended as a college-level textbook for introductory courses in invertebrate paleontology, this book nevertheless is a very useful source of information on exoskeletons for all readers having a serious interest in the subject. Chapter 11, "Arthropods," is the main focus of useful information, with subsection 11.2, "Classification and General Morphology," treating the subject in the greatest detail.

Ruppert, Edward E., and Robert D. Barnes. *Invertebrate Zoology.* 6th ed. Fort Worth, Tex.: Saunders College Publications, 1994. A college-level textbook, emphasizing the correlations of structure and function in living invertebrates. Illustrations, glossary.

Savazzi, Enrico, ed. *Functional Morphology of the Invertebrate Skeleton.* New York: John Wiley & Sons, 1999. A wide-ranging collection of essays by an international roster of experts, covering the current understanding of the functional morphology of the invertebrate skeleton.

Stachowitsch, Michael. *The Invertebrates: An Illustrated Glossary.* Illustrated by Sylvie Proidl. New York: Wiley-Liss, 1992. Covers seventy-seven living invertebrate taxa, with full descriptions of major anatomical features. Intended as a guide to identification and description. Illustrated.

EXTINCTION

Type of animal science: Evolutionary science
Fields of study: Conservation biology, evolutionary science, genetics, population biology

Extinction is the total disappearance (by dying out) of all members of a species. Two types of extinction are recognized by scientists: background extinction, in which a single species disappears for one or more reasons, and mass extinction, which occurs when many different species disappear in a relatively short period of time.

Principal Terms

CATASTROPHISM: a scientific theory which postulates that the geological features of the earth and life thereon have been drastically affected by natural disasters of huge proportions in past ages

FOSSIL: a remnant, impression, or trace of an animal or plant of a past geological age that has been preserved in the earth's crust

GENE: an element of the germ plasm that controls transmission of a hereditary characteristic by specifying the structure of a particular protein or by controlling the function of other genetic material

GENE POOL: the whole body of genes in an interbreeding population that includes each gene at a certain frequency in relation to other genes

SPECIES: a category of biological classification ranking immediately below the genus or subgenus, comprising related organisms or populations potentially capable of interbreeding

UNIFORMITARIANISM: a scientific theory that all processes which have affected the earth and living creatures thereon in the past are presently at work and observable by scientists

In 1796, a French naturalist, Georges Cuvier, demonstrated incontrovertibly that many species of once-living plants and animals had completely disappeared from the earth. Cuvier was also the first to recognize that many of the extinct species he identified had disappeared at approximately the same time—a mass extinction. He and his successors attributed extinctions to catastrophic events, such as the biblical Flood. The catastrophists held the field of scientific opinion concerning extinction until the publication of Charles Lyell's *Principles of Geology* (1830-1833), which proclaimed the doctrine of uniformitarianism. Lyell maintained that no processes have affected the earth (including its flora and fauna) that are not presently observable. He denied that any spectacular cataclysms had occurred.

Climate, Evolution, and Extinctions

Lyell's arguments convinced geologists, but the problem of extinctions remained: What could cause a species to die out? In 1859, Charles Darwin offered a biological explanation for extinctions that seemed to answer all questions. In *On the Origin of Species*, Darwin suggested that all life-forms engage in a perpetual struggle for survival. The best-adapted species, therefore, survive and perpetuate themselves. The less-adapted species are outcompeted and disappear. Darwin's ideas seemed to fit well with the doctrine of uniformitarianism, but the problem of mass extinctions remained. He suggested no reasons why great numbers of very different species should disappear at approximately the same time.

Although his ideas were not accepted for many years after Darwin, another scientist proposed a

Coelacanths

Bony fishes are divided into two lineages, the Actinopterygii, or ray-finned fishes, which are by far the dominant group today, and the Sarcopterygii, or lobe-finned fishes, which are now very rare although they were a much more significant group in the past. The sarcopterygians include the modern lungfish and coelacanths, as well as the extinct rhipidistians, which gave rise to the tetrapods or land animals in the Devonian period (350 million years ago). All sarcopterygians are characterized by the presence of a muscular fin base that contains a stout bony skeleton, and it is this structure that became the limbs in land animals.

Coelacanths are known from the Early Devonian era (410 million years ago) but have not been found in sediments younger than the end of the Cretaceous era (70 million years ago). They are fairly large fish, up to five feet long, with deep bodies covered by large, rough scales, paired fins that are conspicuously lobed, and symmetrical tails with small central lobes. As they were thought to be extinct, it was a great surprise when a specimen was caught off the southeast coast of South Africa in 1938. It was recognized as a coelacanth by Marjorie Courtenay-Latimer, who brought it to the attention of the ichthyologist J. L. B. Smith. He named it the following year as *Latimeria chalumnae*, after its discoverer and the Chalumna River, near the estuary of which the specimen had been caught. The next specimen was not discovered until 1952, this time almost one thousand miles farther north, in the Comoro Islands, which lie between East Africa and the northern tip of Madagascar. Since then, numerous specimens have been caught around the two main Comoro islands, and it is clear that a breeding population of these fish lives around the islands at a depth of between one hundred and five hundred meters. Because *Latimeria* lives at depth, it does not survive being caught on a line and pulled rapidly to the surface. However, in 1987, it was filmed swimming free in its natural environment by Dr. Hans Fricke, using a submersible. His film shows them moving slowly and hovering above the bottom, using the paired lobe fins in a regularly alternating pattern to scull themselves along. Although no film has been taken of *Latimeria* feeding, it is clear from stomach contents that it feeds on small fish, and its body shape indicates that it is a fast swimmer over a short distance, suggesting that it is probably an ambush predator.

No additional populations were discovered until 1997, when a specimen was recognized in a fish market in North Sulawesi, Indonesia, by Dr. Mark Erdmann. North Sulawesi is about eight thousand miles from the Comoros, with no apparent water current interactions. That this was a separate population was subsequently confirmed when a second specimen was caught, and the fish was then named *Latimeria menadoensis*. This discovery raises the hope that other populations may exist, and that although still a "living fossil," the coelacanth may be more widely distributed than previously thought.

—David K. Elliott

possible solution to the puzzle of mass extinctions as early as 1837. Louis Agassiz produced evidence that the earth has periodically undergone periods of extreme cold, with much of its surface covered by glaciers. Agassiz's glaciers are presently in existence, move very slowly, and thus fit well with the uniformitarian view. His glaciers might also explain mass extinction in evolutionary terms, since plants and animals that could not adapt to changing climate would be outcompeted by other species that could adapt, and, thus, the less-flexible species would disappear.

Uniformitarian views about extinction received a further boost in the twentieth century with the proposal of the theory of continental drift, which has evolved into the modern theory of plate tectonics. Scientists have demonstrated that the landmasses of the earth are in constant motion relative to one another and to the poles of the planet. Over millions of years, the present landmasses have occupied very different positions on the earth's surface, often drifting quite near the polar regions, causing massive climatic changes. In addition, scientists have demonstrated that landmasses have subsided to be covered by the seas and have risen to create dry land from ocean floors. These massive changes would also have a profound effect on flora and fauna, creating constant competition and struggle for survival.

Genetic Factors in Extinctions

Many contemporary biologists believe that background extinctions may be the result of genetic rather than strictly climatic factors. In simple terms, extinction results from an excess of deaths over births in a given species. This excess represents, in biological terms, a reduction in the characteristics of a species that allow it to adapt to its environment. Those characteristics, biologists argue, are brought about through natural selection, a process that favors the genes that give the organism an advantage in the struggle for survival. Since natural selection has produced not only the species itself but its ancestors as well, any change in environment must work against the existing organisms. In the new process of changed environment, new gene combinations will result that will produce an organism better adapted to the changed environment. If, however, these genetic changes do not occur rapidly enough, adaptation will not take place and the species will become extinct.

Another possibility results if an organism is too well adapted to its environment. The gene pool of a species usually combines those genes that work well with one another to produce a well-adapted individual. Some biologists have observed a tendency within the gene pools of some species to resist genetic adaptation when environmental changes occur. Thus, genetic changes that could impart greater chances for survival of the individual in times of environmental turbulence do not occur. This effect also occurs when groups of genes are linked together (usually by a chromosomal inversion) in such a way as to prevent new gene combinations that might impart the ability to adjust to changed environmental conditions. This phenomenon actually retards the ability of a species to survive and may lead to extinctions.

Genetics may also explain mass extinctions. One biologist has theorized that species which live in environments that change very little over long periods of time probably have low genetic variability (a gene pool in which nonutilitarian genes have disappeared). Conversely, species in

One theory for the extinction of dinosaurs is that a meteor or asteroid struck the earth, probably creating the Chicxulub Crater in the Yucatán Peninsula of Mexico; the amount of debris and dust blasted into the atmosphere triggered a violent climatic change to which the dinosaurs could not adapt. (©John Sibbick)

changeable environments should have much more genetic diversity, allowing them to cope with rapidly altering living conditions. If that is true, then mass extinctions might result from a rapid environmental change after a very long period of stability. When the change came, the theory goes, widespread extinctions of many species resulted. Most geneticists, however, reject this theory of mass extinctions. The species in unstable environments may actually have less genetic variability than those in stable environments. Biologists therefore have not been able to advance a plausible genetic explanation for mass extinctions.

Mass Extinctions

Geologists have identified a number of what they call "mass extinction events." The first such event known to paleontologists occurred 440 million years ago at the end of the Ordovician period, during which more than 22 percent of all families and 57 percent of all genera disappeared. Another mass extinction occurred during the Devonian period, 370 million years ago, during which more than 20 percent of all marine families disappeared. The greatest of all mass extinctions occurred 248 million years ago, at the end of the Permian period. During that event, 52 percent of all marine families, 83 percent of all genera, and a frightening 95 percent of all species became extinct. During this event, land animals and plants vanished, along with marine flora and fauna. Yet another "great dying" took place during the Triassic period, approximately 215 million years ago, when 20 percent of all marine families and 48 percent of all genera disappeared.

The mass extinction known most widely outside the scientific community occurred at the end of the Cretaceous period, about 65 million years ago. During this event, the dinosaurs vanished, along with 50 percent of all marine genera. A number of less-spectacular mass extinctions have taken place in the earth's long history, including a relatively recent one at the end of the Pleistocene epoch, which included among its victims such well-known extinct animals as the woolly mammoth and the so-called saber-toothed tiger.

Some paleobiologists and archaeologists are convinced that many of the Pleistocene extinctions were caused by the activities of human beings. They point out that the extinction of most of the large North American land mammals coincided with one theoretical date for the appearance of humans in the Western Hemisphere, approximately 11,500 years ago. According to these scientists, especially efficient human hunters were responsible for those extinctions; however, this theory seems unlikely as an explanation for all the Pleistocene extinctions. Many species in areas other than North America disappeared at the same time, most of which would have had little or no value as game. In addition, there are huge "boneyards" containing fossils of the extinct species in areas as far separated as Alaska and Florida that apparently died at the same time from causes that seem to be related to some great natural cataclysm.

Asteroids as Agents of Mass Extinctions

The species that disappeared in each of these mass extinctions apparently died at approximately the same time, and scientists were at a loss to explain them until 1980. In that year, a scientific team led by Nobel laureate Luis Alvarez presented what seems to be irrefutable evidence that the extinction of the dinosaurs coincided with the collision of a huge asteroid or comet with the earth. The evidence is based on a layer of clay that separates the rock formations associated with the dinosaurs from the overlying formations, which contain fossils from the era of mammals. The clay contains large amounts of iridium and other elements that are scarce in the crust of the earth but common in asteroids and comets.

Many scientists now contend that the collision between the earth and a large extraterrestrial body would have thrown enormous quantities of dust into the atmosphere, sufficient to block out the sun's radiation for an extended period of time. The resulting subzero weather would have had devastating effects on flora and fauna, including the seemingly invincible dinosaurs. Shortly after the evidence for the collision appeared, other scientists presented evidence for an even more frightening phenomenon—periodicity of extinction events.

A scientist charting the occurrence of mass extinction events showed that they seem to occur at regularly spaced intervals, approximately every 26 to 30 million years. Almost immediately after the presentation of the evidence for periodicity, new scientific studies demonstrated more evidence that several mass extinctions other than the one during which the dinosaurs disappeared are also associated with unusually high concentrations of iridium. Taken together, these data seem to indicate that most, perhaps all, mass extinctions are caused by extraterrestrial agents and recur on a regular basis. If that is indeed the case, the implications for all the sciences, including evolutionary biology, are profound.

An Interdisciplinary Study

Scientists from several different disciplines are currently studying extinctions, including mass extinctions, in a variety of ways. Many biologists believe that the most effective way to understand background extinctions is to examine those that have taken place in historic times. Geologists and paleontologists are subjecting the fossil record to a new and rigorous examination, armed with new, supersensitive techniques for ascertaining the ages of the rocks in which fossils occur in an attempt to understand mass extinctions better. These new techniques are the products of research in nuclear physics. Ecologists are particularly examining currently endangered flora and fauna, which may soon disappear. Even some astronomers are actively engaged in research into mass extinctions, scanning the heavens with powerful telescopes in search of an extraterrestrial agent that might explain the apparent periodicity of mass extinction events.

Biologists studying recent background extinctions conclude that virtually all of them are the results of the activities of humankind. The great auk, the last known specimens of which were killed by Icelandic fishermen in 1844; the Tasmanian tiger cat, the last known specimen of which died in captivity in 1934; and many other species are examples of human-caused extinctions. These studies lend validity to the theory that at least some of the many extinctions during the Pleistocene resulted from the hunting activities of prehistoric peoples. More disturbing are studies which show that such modern phenomena as acid rain and ozone depletion, both results of industrialization, may be doing irreparable damage to the environment, which could result in another mass extinction event in the very near future. Indeed, some biologists and ecologists believe that such an event has already begun.

A number of geologists and paleontologists, using new dating techniques based on the rate of radioactive decay in rocks, are reassessing the ages traditionally assigned to fossils by less sophisticated techniques in the past. These studies should eventually reveal whether the mass extinction events of the remote past occurred in a very short or over a relatively longer period of time. Other geologists are searching for impact craters, to lend further credence to the theory that at least some and perhaps all mass extinctions resulted from periodic collisions between the earth and large celestial bodies.

A team of physicists (including astrophysicists), engaged in an ongoing search for a hypothetical dark companion to our sun, has postulated that the orbit of this presently undiscovered body may periodically disrupt the comet cluster on the outer fringes of the solar system, resulting in many comets being diverted into an intersection with the earth's orbit. If their search is successful, it will provide powerful substantiating evidence for the collision theory, with sobering implications. Some physicists have even proposed ways to prevent future collisions between the earth and large heavenly bodies.

The implications of background and mass extinctions are profound. If, as an overwhelming mass of evidence seems to indicate, the activities of humankind are a major cause of recent background extinctions, then those activities may soon lead to an ecological disaster of gigantic proportions. At present rates of disappearance, as many as two million species currently in existence will disappear by the middle of the twenty-first century. Unless immediate steps are taken, plant extinctions on the scale envisioned by many botanists may also cause massive climatic changes. Some ecologists and geophysicists warn that if the tropical forests disappear, the result will be the greenhouse effect. Of less immediate, but nevertheless great, concern is the theory that mass extinction events in the fossil record have resulted from collisions between the earth and asteroids or comets. The implications of the collision theory for evolutionary biology, however, are far-reaching. If the theory is correct, then the struggle for survival is not the most important feature of evolution. No matter how well adapted to its environment a species may be, survival of a collision would be largely a matter of chance. If the theory proves to be correct, biologists will need to rewrite the textbooks on evolution.

—Paul Madden

See also: Adaptive radiation; Cloning of extinct or endangered species; Ecological niches; Ecology; Ecosystems; Endangered species; Evolution: Historical perspective; Extinctions and evolutionary explosions; Food chains and food webs; Fossils; Habitats and biomes; Hardy-Weinberg law of genetic equilibrium; Natural selection; Neanderthals; Paleoecology; Paleontology; Pollution effects; Population fluctuations; Predation; Punctuated equilibrium and continuous evolution; Reproductive strategies; Veterinary medicine; Zoology; Zoos.

Bibliography

Donovan, Stephen K., ed. *Mass Extinctions: Processes and Evidence*. London: Belhaven Press, 1989. A series of chapters authored by specialists covers all the major mass extinctions and speculates on their causes. The first two chapters, on the historical aspects of studies of extinctions and their recognition, are particularly useful. Extensive bibliography.

Leakey, Richard E., and Roger Lewin. *The Sixth Extinction: Patterns of Life and the Future of Humankind*. New York: Doubleday, 1995. Argues that human activity is creating the sixth great extinction of life on earth, which is now under way.

McGhee, George R. *The Late Devonian Mass Extinction*. New York: Columbia University Press, 1996. An excellent synthesis of one of the less-studied mass extinctions in the history of life on earth. Argues that the disappearance of 22 percent of all families of marine animals, which occurred 365 million years ago, was caused by an asteroid impact that set off a chain of events that lasted close to three million years.

Martin, Paul S., and Richard G. Klein, eds. *Quaternary Extinctions: A Prehistoric Revolution*. Tucson: University of Arizona Press, 1984. Thirty-eight articles analyzing the most recent mass extinction event from many perspectives. Various articles examine the influence of humankind on extinctions in many parts of the world, the influence of climatic changes, and the influence of seafloor rising/land subsidence. The articles demonstrate the complexity of the problem of extinctions and the lack of a scientific consensus on their causes.

Muller, Richard. *Nemesis: The Death Star*. New York: Weidenfeld & Nicolson, 1988. Muller provides a fascinating account of the origin and evolution of the currently controversial theory that the dinosaurs became extinct because of the collision of the earth with an asteroid or comet. He then details the development of the theory that our sun has a presently undiscovered dark companion (he has named it Nemesis) that causes the disruption of either the asteroids or (more likely) the comet cluster every twenty-six million to thirty million years. This disruption results in periodic collisions between the celestial objects and the earth, causing mass extinctions of terrestrial life.

Raup, David M. *Extinction: Bad Genes or Bad Luck?* New York: W. W. Norton, 1991. A text that deals with the reasons for extinction events in an informative and readable style. Suitable for high school and college students as well as the general reader.

Stanley, Steven M. *Extinction*. New York: Scientific American Library, 1987. An excellent account of extinctions in general and major mass extinctions in particular. Stanley makes a strong case for gradual climatic changes being the primary agent in extinctions but acknowledges that catastrophic events and the activities of humankind may have played significant roles in some extinction events.

Stearns, Beverly Petersen, and Stephen C. Stearns. *Watching, from the Edge of Extinction*. New Haven, Conn.: Yale University Press, 1999. Offers accounts of eleven endangered species from the perspectives of scientists and others who work with them. Written for ecologists, conservation biologists, and wildlife biologists, but useful for everyone concerned with the threat of species extinction.

EXTINCTIONS AND EVOLUTIONARY EXPLOSIONS

Type of animal science: Evolution
Fields of study: Environmental science, evolutionary science, paleontology, zoology

The history of life has been punctuated by episodes of great change, some marked by the loss of large numbers of organisms, others by explosive development. Explanations proposed for these fluctuations have a bearing on current extinction levels and the extent to which they can be controlled.

Principal Terms

ADAPTIVE RADIATION: the rapid production of a new species following invasion of a new geographic region or exploitation of a new ecological opportunity

BOLIDE: an extraterrestrial object (for example, a meteorite) that hits the earth

DIVERSITY: the number of fossil taxa (classification groups) associated with a particular place and time

LAZARUS TAXA: groups that apparently disappear during a mass extinction only to appear again later

MASS EXTINCTION: an event in which a large number of organisms in many different taxa are eliminated

PERIODICITY HYPOTHESIS: the proposal that mass extinctions have occurred approximately every 26 million years over the past 250 million years

PHANEROZOIC: an era of geologic time beginning approximately 544 million years ago at the start of the Cambrian period, when animals with mineralized skeletons became common

REGRESSION: the migration of the shoreline and associated environments toward the sea

Extinction of species is a continuous process, and evidence of its occurrence abounds in the fossil record. It has been estimated that marine species persist for about four million years, which translates into an overall loss of about two or three species each year. This is the "background" extinction rate, and it is balanced by speciation events that result in the development of new species. Mass extinctions are events during which the rate of extinction rises dramatically above this background rate, and a number of these have been recognized in the Phanerozoic era. In each of these events, at least 40 percent of the genera of shallow marine organisms were eliminated. Using statistical methods, it has been estimated that at least 65 percent of species became extinct at each of these events, with 77 percent being eliminated at the event at the end of the Cretaceous period and 95 percent at the event at the end of the Permian period. These mass extinctions were balanced by periods of explosive development that often followed, as organisms moved into vacant adaptive zones during periods of adaptive radiation. The most important of these was at the base of the Cambrian period, 544 million years ago, when all the major groups in existence originated, but other radiations occurred in the Early Triassic period and at the start of the Tertiary period.

Causes of Mass Extinctions

Attempts to explain the causes of mass extinctions have centered on terrestrial phenomena such as sea level changes, climatic changes, or volcanic activity. The sea level has shown regular fluctuations on a global level during the Phanerozoic era, and these appear to be related to the melting or formation of polar ice caps or to major tectonic events such as continental splits or the collision and uplift or subsidence of ocean ridges. Extinction events appear to be correlated mostly with periods of marine regression. During such a regression, the withdrawal of the ocean leaves a much smaller habitat for shallow marine organisms. This leads to increased crowding and competition and ultimately to an increased extinction rate. Reduction of large terrestrial vertebrates during these regressions, as happened during the events at the end of the Permian and Cretaceous periods, may be related to increased seasonality caused by the loss of the ameliorating influence of the shallow epicontinental seas.

It has also been shown that some extinctions are related to transgressive events (the spread of the sea over land areas), possibly resulting from the spread of anoxic (oxygen-poor) waters across epicontinental areas. Climatic changes seem to be correlated with eustatic events (worldwide changes in sea level), and the evidence implicating temperature as the main cause of extinctions seems weak. For example, the most important extinction event at the end of the Permian period occurred at a time of climatic amelioration marked by the disappearance of the Gondwanaland ice sheet. Volcanic activity has been presented as a possible cause of the extinctions that occurred at the end of the Cretaceous period. The Deccan Traps of northern India were erupting at that time and would have produced large quantities of volatile emissions that could have resulted in global cooling, ozone-layer depletion, and changes in ocean chemistry. However, no evidence exists as yet for the involvement of volcanic activity in other extinction events.

Although various extraterrestrial causes for mass extinction events have been suggested in the past, these ideas have gained greater credence since the publication in the early 1980's of work by Luis and Walter Alvarez, who ascribe the end-Cretaceous extinction event to the effects of the impact of a large bolide, or extraterrestrial object, perhaps ten kilometers in diameter. The impact of such a large object would have resulted in some months of darkness because of the global dust clouds generated, and this would have halted photosynthesis and resulted in the collapse of both terrestrial and marine food chains. Although cold would initially have accompanied the darkness, greenhouse effects and global warming would have followed as atmospheric gases and water vapor trapped infrared energy radiating from the earth. Physical evidence for an impact rests on the presence in the period boundary layers of high concentrations of iridium and other elements generally rare at the earth's surface but abundant in asteroids. In addition, these layers often contain shocked quartz grains, otherwise found only in impact craters and at nuclear test sites, and microtectites, glassy droplets formed by impact. Although the evidence for extraterrestrial impacts having caused the other major extinction events is slight, this causal factor has been linked with the apparently regular 26-million-year periodicity exhibited by extinctions. Scientists suggest that the regular passage of an unidentified planetary body by the Oort Cloud of comets and the subsequent perturbation could result in increased asteroid impacts and extinction events on earth.

Historical Mass Extinctions

The first mass extinction event that can be recognized in the fossil record occurred in the Middle Vendian period, about 650 million years ago, when microorganisms underwent a severe decline. This event has been linked to climatic cooling related to glaciation. The extinction in the Late Ordovician period was a major event in which 22 percent of marine families became extinct. As there were two main pulses of extinction and no iridium anomaly was found, an extraterrestrial cause seems unlikely. However, sea level and temperature changes have been cited as likely causes. In addition, biologically toxic bottom waters might have been brought to the surface during periods of climatic change. The event at the end of the

Devonian period had a devastating effect on brachiopods, which lost about 86 percent of genera, and on reef-building organisms such as corals. Shallow-water faunas were most severely affected; only 4 percent of species survived, although 40 percent of deeper-water species did, and cool-water faunas also survived better. This event has been linked to a significant drop in global temperatures of unknown cause.

The mass extinction event at the end of the Permian period was the most severe of the Phanerozoic era and resulted in the extinction of up to 95 percent of all marine invertebrate species. On land, amphibians and mammal-like reptiles were both badly affected, and plant diversity fell by 50 percent. No iridium anomaly was found, and the most likely explanation is climatic instability caused by continental amalgamation and the simultaneous occurrence of marine regressions. These occurrences would have disrupted food webs on a major scale. The event that occurred at the end of the Triassic period was much less severe but still involved extensive reductions in marine invertebrates and reptiles. On land, a major faunal turnover took place. Primitive amphibians, early reptile groups, and mammal-like reptiles died out and were replaced by advanced reptiles and mammals. No evidence of an impact event has been found, and the extinctions are generally correlated with widespread marine regressions.

The extinction that took place at the end of the Cretaceous period has become the most hotly debated, in large part because of the bolide impact hypothesis. Although the broad pattern of extinction among marine organisms is known, the detailed picture only encompasses microorganisms such as planktonic foraminifera and calcareous nannoplankton. Study of the ranges of these microorganisms shows that the extinctions occurred over an extended period, starting well before and finishing well after the boundary. Although much has been made of the extinction of ammonites at the end of the Cretaceous period, there are too few ammonite-bearing sections to show whether it was gradual or abrupt. On land, evidence of an increase in the population of ferns just above the boundary suggests the presence of wildfires, as ferns are usually the first plants to recolonize an area devastated by fire. However, in many sections, a return of the Cretaceous vegetation is seen above the fern increase, indicating little extinction.

Post-Extinction Recoveries

Among the vertebrates, a picture of gradual change is seen for mammals, with drastic reductions occurring only in the marsupials. The boundary also does not seem to have been a barrier for turtles, crocodiles, lizards, and snakes, all of which came through virtually unscathed. The dinosaurs did become extinct, and much argument has centered on whether this was abrupt or occurred after a slow decline. In this context, it must be noted that there is only one area where a dinosaur-bearing sedimentary transition across the boundary can be seen, and that is in Alberta, Canada, and the northwestern United States. Records of dinosaurs in this area during the upper part of the Cretaceous period show a gradual decline in diversity, with a drop from thirty to seven genera over the last eight million years. Although explanations of the extinction of dinosaurs have ranged from mammals eating their eggs to terminal allergies caused by the rise of flowering plants to the current ideas about bolide impacts, the answer may be climate related. A major regression of the oceans occurred at this point, resulting in a drop in mean annual temperatures and an increase in seasonality. The bolide impact may have served as the death blow to taxa (animals in classification groups) that were already declining.

The main period of evolutionary expansion in the Phanerozoic era is at the base of the Cambrian period, 544 million years ago. Termed the "Cambrian explosion," it marks the development of all the modern phyla of organisms, and as many as one hundred phyla may have existed during the Cambrian period. This period seems to have lasted only about 5 million years, and the subsequent history of animal life consists mainly of variations on the anatomical themes developed during this short period of intense creativity. This period is represented in the fossil record by the remarkably well-preserved Burgess Shale fauna of British Columbia, which has been extensively described, and faunas of similar age from China and Greenland. Why the Cambrian explosion could establish all major anatomical designs so quickly is not clear. Some scientists believe that the lack of complex organisms before the explosion had left large areas of ecological space open, and when experimentation took place, particularly with the advent of hard skeletons, any novelty could find a niche. Also, the earliest multicellular organisms may have maintained a genetic flexibility that became greatly reduced as organisms became locked into stable and successful designs. Why some of the innovations were successful in the long term and others were not is unknown, as no recognized traits unite the successful taxa. It has even been suggested that success may be due to no more than the luck of the draw.

In contrast, the recoveries after the major extinctions at the end of the Permian and Cretaceous periods did not result in the development of new phyla. The earliest Triassic ecosystems were more vacant than at any time since the Cambrian period, yet no new phyla or classes appear in the Triassic period. This suggests that despite the overwhelming nature of the extinctions, the pattern was insufficient to permit major morphological innovations, in part probably because no adaptive zone was entirely vacant. Hence, despite the fact that the mass extinction at the end of the Permian period triggered an explosion in marine diversity described as the Mesozoic marine revolution, persisting species may have limited the success of broad evolutionary jumps.

Reading the Fossil Records

All understanding of extinction events or of evolutionary explosions depends on the fossil record. The study of the diversity of organisms through time—the number of different types of organisms that occur at a particular time and place—is therefore very important. The basic data consist of compilations of extinctions of taxa plotted against similar compilations of originations of taxa. Periods when either extinction or origination was unusually high show as peaks or troughs on a graph. Unfortunately, biases in the preservation, collection, and study of fossils have conspired to obscure patterns of change in diversity.

Geological history of patterns of diversity is obscured by a variety of filters, many of which are sampling biases that cause the observed fossil record to differ from the actual history of the biosphere. The most severe bias is the loss of sedimentary rock volume and area as the age of the record increases because the volume and area correlate strongly with the diversity of organisms described from a stratigraphic interval. The quality of the record also tends to fall with increasing age because the rocks are exposed to changes that may destroy the fossils they contain. The differences in levels of representation among the paleoenvironments in the stratigraphic record also influence the composition of the fossil record; for example, shallow-marine faunas are much better represented than are terrestrial faunas.

Diversity patterns are studied at a variety of levels, from the species upward, that vary in their quality and inclusiveness. A basic problem is that many of the processes that are of interest occur at the species level or even below it, but the biases of the fossil record mean that data are best at higher levels. Diversity of shallow-marine organisms for the Phanerozoic era cannot be read directly at the species level because the record is too fragmentary. The record at the family level is much more complete because the preservation of one species in a family allows the family to be recorded. For this reason, paleodiversity studies are often conducted at the family level. However, higher taxon diversity is a poor predictor of species diversity. For example, an analysis of the mass extinction at the end of the Permian period indicates that the 17 percent reduction in marine orders and 52 percent reduction in marine families probably represent a 95 percent reduction in the number of species. Another problem with the study of fossils is that soft-bodied and poorly skeletonized groups may leave little or no record. It has generally been assumed that the ratio of heavily skeletonized to

Georges Cuvier

Born: August 23, 1769; Montbéliard, France
Died: May 13, 1832; Paris, France
Fields of study: Anatomy, paleontology, zoology
Contribution: Cuvier, the founder of functional anatomy and the father of paleontology, was the first to devise a systematic natural classification of the animal kingdom and to document extinctions of ancient animals.

George Cuvier studied at the Académie Caroline in Stuttgart from 1784 to 1788. In 1795, he was invited to Paris by Étienne Geoffroy Saint-Hilaire, where he was appointed as a professor of animal anatomy at the French National Museum of Natural History.

After dissecting and studying many large animals, Cuvier concluded that organisms are functional wholes, in which the form and function of each part is integrated into the entire body. Each part of an organism bears signs of the entire organism. Thus, any change in an organism's anatomy would render it unable to survive. Consequently, Cuvier did not believe in organic evolution.

Based on rational principles and his knowledge of the comparative anatomy of living organisms, Cuvier reconstructed a number of organisms from fragmentary fossils. His studies led him to believe that any similarities between organisms were due to common functions, not to common ancestry. Paradoxically, some of his findings were later used to help support the theory of evolution.

During Cuvier's lifetime, some scientists interpreted fossils as remains of living species. Others thought that the unusual organisms known as fossils must still survive in unexplored parts of the world. Many could not believe that God would allow any species to become extinct. Cuvier published detailed studies about elephant anatomy showing that the African and Indian elephants were distinct species and that the fossil mammoths of Europe and Siberia were different from any living elephant species. He also published documentation of the past existence of large mammals, including the ground sloth, the Irish elk, and the American mastodon, that resembled no living species. Through these studies, Cuvier established the fact that some animal species had become extinct. The results of his studies, such as his *Discours sur les révolutions de la surface du globe* (1825; discourse on the revolutions of the globe), launched the field of modern vertebrate paleontology.

Georges Cuvier was the first scientist to document the existence of extinctions of animal life. (National Archives)

Cuvier believed that the Earth was very old and that periodic upheavals, rather than the inability of a species to adapt, had annihilated a number of species. He viewed the upheavals as catastrophic events produced by natural causes that were linked to geological conditions, including the periodic rise and fall of sea level. As a result, Cuvier inferred that extinctions of species documented by the fossil record were the result of abrupt changes that caused the strata of the continents to be dislocated and folded.

Eventually, the theory of catastrophism was supplanted in favor of uniformitarianism. However, in addition to the slower natural processes suggested by uniformitarianism, the Earth has also been shaped by occasional catastrophic events, which appear to be the source of major extinctions of animal species.

—*Alvin K. Benson*

non-skeletonized species has remained approximately constant through the Phanerozoic era; however, there are no data to support this and there is some evidence that skeletons have become more robust through time in response to newly evolving predators. The net result of these biases is severe. Only 10 percent of the skeletonized marine species of the geologic past and far fewer of the soft-bodied species are known.

Despite these problems, it has been possible to show that diversity of organisms has varied in a number of ways during the Phanerozoic era. Tabulations of classes, orders, and families have been used to show that there were significant periods of increased extinction or increased evolutionary rates. One of the most important uses of these data has been the tabulation at the family level that appears to show a regular periodicity of about 26 million years for extinction events and that has been used to support ideas about periodic extraterrestrial events. However, although fluctuations occurred, it has also been possible to show that the number of marine orders increased rapidly to the Late Ordovician period and has remained approximately constant since then.

The Ebb and Flow of Life on Earth

Mass extinctions and evolutionary explosions are the opposite faces of the pattern of diversity of organisms through time. During periods of mass extinction, the diversity of organisms on earth has dropped drastically, and in some cases entire lineages have been wiped out. Evolutionary explosions, on the other hand, resulted in enormous innovation, particularly at the beginning of the Cambrian period, and the development of new variations on established morphotypes (animal and plant forms and structures) later in the geologic record. Understanding the processes that caused these events is of major importance because people have reached the point where they are capable of influencing their environment in drastic ways.

Studies of extinction events have shown that they have a variety of causes, some of which appear to be environmental changes brought about by natural processes while others may be the result of extraterrestrial forces. The most severe of these extinction events occurred at the end of the Permian period, 245 million years ago, and resulted in the loss of up to 95 percent of marine invertebrate species. The cause of this extinction appears to be primarily that continents were amalgamating and oceans were retreating, which resulted in a major reduction in the habitat of shallow-marine organisms. Terrestrial habitats were also affected as the increase in continental area and loss of the ameliorating effect of extensive areas of shallow ocean brought about climatic changes. Although climatic changes are thought to be the main culprit in the majority of extinction events, some scientists believe that large bolides, or extraterrestrial bodies, struck the earth with such force as to create major changes in the environment that significantly reduced diversity. This theory has enjoyed the most popularity as the explanation for the event at the end of the Cretaceous period, during which the dinosaurs became extinct, but evidence for an extraterrestrial body's involvement in any of the other events is slight.

Whatever the cause, environmental change that results in habitat reduction is the main reason for species decline. As humans have risen to dominance over other species, the extinction rate has accelerated, and in the last half-century, this rate has climbed considerably above natural attrition as populations have increased and habitats have been altered. Although the levels of extinction have not yet reached those recorded during major extinction events of the past, some scientists believe people may be facing an ecological disaster. A better understanding of the processes surrounding past extinction events and the rebounds that followed them will help people prepare for and deal with the future.

—David K. Elliott

See also: Adaptations and their mechanisms; Adaptive radiation; Apes to hominids; Biodiversity; Clines, hybrid zones, and introgression; Demographics; Evolution: Animal life; Evolution: Historical perspective; Extinction; Human evolution analysis; Isolating mechanisms in evolution;

Natural selection; Nonrandom mating, genetic drift, and mutation; Paleontology; Population genetics; Punctuated equilibrium and continuous evolution.

Bibliography

Allen, Keith C., and Derek E. Briggs, eds. *Evolution and the Fossil Record*. Washington, D.C.: Smithsonian Press, 1989. Chapter 2, "Catastrophes in the History of Life," by George McGhee, Jr., provides a useful overview of extinction through the Phanerozoic era and possible causes of such events.

Briggs, Derek E., and Peter R. Crowther, eds. *Palaeobiology: A Synthesis*. Oxford, England: Blackwell Scientific Publications, 1990. Contains a series of short chapters dealing with the main extinction events and their causes plus sections covering adaptive radiations following the events. Chapters written by experts in the field provide excellent coverage; aimed at the advanced college student. Extensive bibliography.

Donovan, Steven K., ed. *Mass Extinctions: Processes and Evidence*. New York: Columbia University Press, 1989. A series of chapters authored by specialists covers all the major mass extinctions and speculates on their causes. The first two chapters on the historical aspects of studies of extinctions and their recognition are particularly useful. Extensive bibliography.

Drury, Stephen. *Stepping Stones: Evolving the Earth and Its Life*. New York: Oxford University Press, 1999. A synopsis of life on earth, incorporating geology, chemistry, physics, anthropology, and biology. Complex and controversial theories are clearly explained for the nonspecialist.

Frankel, Charles. *The End of the Dinosaurs: Chicxulub Crater and Mass Extinctions*. New York: Cambridge University Press, 1999. Presents the scientific detective story of how Chicxulub Crater, in the Yucatán Peninsula, was discovered, and the possible role of the asteroid that created the crater in the extinction of the dinosaurs.

Gould, Stephen J. *Wonderful Life: The Burgess Shale and the Nature of History*. New York: W. W. Norton, 1989. An informative and entertaining account of the collection and interpretation of data concerning the Cambrian period explosion. Particularly well illustrated and with a concluding discussion of the author's ideas on contingency.

McMenamin, Mark A., and Dianna L. McMenamin. *The Emergence of Animals: The Cambrian Breakthrough*. New York: Columbia University Press, 1989. A comprehensive account of the initial metazoan diversification in the Cambrian period with interesting speculations on the major controls affecting it. Glossary and bibliography.

Officer, Charles, and Jake Page. *The Great Dinosaur Extinction Controversy*. New York: Addison-Wesley, 1996. A text for the general reader that explains the background to the impact hypothesis for dinosaur extinction and explores alternative explanations. Extensive bibliography.

Raup, David M. *Extinction: Bad Genes or Bad Luck?* New York: W. W. Norton, 1991. A text that deals with the reasons for extinction events in an informative and readable style. Suitable for high school and college students as well as the general reader.

Runnegar, Bruce, and James W. Schopf, eds. *Major Events in the History of Life*. Boston: Jones and Bartlett, 1992. Offers a very clear synopsis of the events surrounding the development of the Metazoa with a useful background section and extensive references. Suitable for the general reader as well as students up to and including college level.

Ward, Peter D., and Don Brownlee. *Rare Earth: Why Complex Life Is Uncommon in the Universe*. New York: Copernicus, 2000. Written by a paleontologist and an astronomer, who argue that simple life, such as bacteria, may be common throughout the universe, but that complex life, such as has evolved on earth, is rare. Outlines the special features of earth that make the evolution of life possible, and the consequent tragedy of species extinction.

EYES

Type of animal science: Anatomy
Fields of study: Anatomy, biochemistry, cell biology, neurobiology, physiology

The eyes are the sensory organs that allow animals to visualize the world around them. Many different types of eyes have evolved, from very simple light-gathering eye spots to the complex compound eyes of insects. Each type of eye is used to convert a light signal into a useful piece of information for the animal.

Principal Terms

BINOCULAR VISION: the ability to utilize image information from both eyes to form a single image with depth information

CHROMOPHORE: the molecule which interacts with opsin; absorption of light changes the interaction and starts the phototransduction cascade

OMMATIDIUM: individual unit of the multifaceted compound eye

OPSIN: a membrane-bound protein, or pigment, which absorbs light

OPTIC NERVE: the main nerve taking information from the eyes to higher processing areas

PHOTORECEPTOR: cell containing membranes which house light-sensitive pigments

RETINA: the light-sensitive membrane at the back of the eye

There are many different kinds of eyes in the animal kingdom, each type with its own set of benefits for a particular species. The most basic function of the eye is to act as a light-sensitive organ, to detect the presence or absence of light. In addition to this, elaborations are often made, such as the ability to form an actual image on a retina, or the ability to control the amount of light entering the eye; each elaboration is made to provide the visual needs of individual species. The animal kingdom comprises eyes from the very rudimentary to exceptionally complex. The two basic types of eye are the simple eye and the compound eye. The simple eye is made up of a single light-sensitive region, whereas the compound eye comprises several such elements.

Simple Eyes

The most elementary simple eye consists of a photosensitive membrane, or eye spot. Light should enter the membrane from only one direction, and so eyes have a pigmented backing to stop entry of light from the wrong side. In order to give the light entering a sense of directionality, the photosensitive membrane is often shaped like a cup; this basic eye type is called a pigmented cup eye. An example of an animal which possesses a simple eye of this nature is the cephalopod, *Nautilus pompilius*, which is known as a living fossil as it is thought to exemplify the behavior and physiology of ancient organisms. *Nautilus* has a pinhole eye with a pigmented backing; the pinhole aperture is a primitive method to restrict the amount of light entering the eye. This eye has no other formal optics. An elaboration of this type of eye would be to add a spherical lens; this would allow the light entering to be focused and to form an image on the photosensitive membrane. In many animals, the lens can change shape (become thinner or fatter) using surrounding musculature, in order to properly focus the image on the retina; this ability is called accommodation. The lens changes the angle at which light bends, and hence the focal length of the eye. The focal length of the eye is the distance from the eye to the part of the retina on

Optimization of Eyes

In order to adapt to the particular needs of individual species over evolutionary time, simple eyes have had to change and become optimized to deal with unique circumstances. Two major aspects of superior vision are resolution and sensitivity. Resolution, or spatial acuity, refers the ability to tell that two points are separate, and not just one point. Thus, better resolution is beneficial to an animal that needs to resolve images at a distance. To increase resolution, one can build more receptors per retina. However, there is a limit to this; to fit in more photoreceptors, they would have to become smaller (thinner) and at some point they would be too small to trap photons efficiently. This can be counteracted by making the receptors longer to catch more photons of light. Also, with too many receptors the nervous system becomes overloaded; for this reason, throughout most of the retina, many photoreceptors converge onto a single ganglion. Nevertheless, there are regions of high resolution at which each photoreceptor has its own ganglion; this region of the retina is called the fovea and is best described in vertebrate eyes. Increasing the eye size also increases the resolution, since it increases the amount of light getting to each receptor. Some species need to maximize sensitivity of their eyes, for example, animals who are active at night. To do this, the aperture needs to be large. This allows for maximum light entry into the eye. To the same end, the receptors could be made bigger, as could the entire eye. Thus, nocturnal animals, such as the opossum, have much larger eyes and apertures than diurnal animals, such as primates, which use a smaller eye and a smaller lens to give better optics.

which the image is focused. Other animals actually move the entire lens forward or backward in order to accommodate. One drawback that comes along with spherical lenses is that the entire image is rarely totally focused on the retina because, as a result of the spherical shape of the lens, the light rays become focused over several focal lengths (this is called spherical aberration). To correct this, many lenses have a refractive index gradient across their length; basically, the density of the lens is different across its length, which causes the light rays to bend differently and corrects for spherical aberration, allowing proper focusing of the entire image on the retina. In spiders, a different kind of optics is found; instead of a lens, these creatures focus light rays with their corneas. The structure, unlike the lens, is fixed and does not change shape to accommodate.

Compound Eyes

Compound eyes are the most abundant type of eye in the animal kingdom, and can be thought of as being derived from the simple pigmented-cup type of eyes. The simplest compound eye consists of several pigmented-cup-type units, each of which samples a different angle of visual space. Found only in invertebrates, such as certain types of worm, this kind of eye gives very poor quality images, but does give the animal a sense of the direction from which light is coming. A more complicated version of this eye is the apposition compound eye, found in many insects, where each cup has its own optics. Each individual unit is called an ommatidium, comprising the rhabdom (containing the light-sensitive cells), directly contacting a light-focusing apparatus (either a lens or a cornea). In order to fit as many ommatidia into the eye as possible, the facets are hexagonal. Another elaboration is the superposition compound eye; this eye has a space between the cornea (or lens) and the rhabdom of each ommatidia. As such, this allows light from many corneal facets to converge onto each rhabdom. This increases the sensitivity as compared to the apposition eye. Several mechanisms are used to bend the light from each ommatidium; the simplest is the reflecting superposition eye, found in shrimp, which uses a series of mirrors along the edge of each facet to reflect the light onto the rhabdoms.

Photoreceptors

The photoreceptive element common to all eyes differs from species to species, from the cup-shaped retinas of simple eyes to the complex rhabdom structures of compound eyes. In simple eyes, the light-sensitive element is platelike, with projections containing flat layers or discs of membranes. In vertebrates, the two photoreceptor types are the cones (cone-shaped projections), which have layers of photosensitive membrane, and the rods (rod-shaped projections), which contain free-floating discs. Compound eyes have rhabdomeric microvillar photoreceptors; the microvilli are fingerlike structures that project from the rhabdomeres and are light-sensitive. Within each of these structures lies the actual light-sensitive pigment, a membrane-bound protein known as opsin. In the rhabdoms, the orientation of the opsin molecule is parallel to the axis of the microvilli; this fact aids in the perception of polarized light. There are many types of opsin protein, and they can be categorized according to which wavelength of the light spectrum they preferentially absorb. The opsin protein interacts with a Vitamin A-derived molecule called the chromophore. Its chemical name is 11-cis retinal. The absorption of a photon of light changes the chemical interaction between the opsin and the chromophore, and it is this chemical change which initiates the cascade of events that leads to a nerve signal, known as phototransduction.

The actual biochemistry of phototransduction is very complex, but involves the amplification of the signal of reception of light, and its transformation into a nerve impulse. The passage of the information through the nervous system is also very complicated and details are different in different species, but in general, the receptors converge onto axons and the nerve impulse travels through the optic nerve for higher processing. Of the two types of photoreceptor existing in vertebrates, rods are very light-sensitive and are used in situations where light is limited (scotopic conditions). Cones are less sensitive and are used in situations where light is abundant (photopic conditions). The vertebrate retina almost always possesses both rods and cones, although some nocturnal animals or animals living in an environment where light is scarce, for example deep-dwelling fishes, have all-rod retinas. Mammals have rod-dominated retinas, which are probably remnants of the time millions of years ago when mammals were nocturnal. There are several kinds of cones, which are characterized according to which opsins they carry within the membrane layers; this dictates which wavelengths of light the cone preferentially absorbs. Absorption of light by combinations of at least

Other Features of Photoreceptors: Pigments and Oil Droplets

Some vertebrate and invertebrate photoreceptors have been found to possess filtering mechanisms. The classic example of this is oil droplets in the cones of bird eyes. The cone possesses a droplet of oil positioned just behind the outer nuclear layer of the retina, where the photoreceptor layer containing the opsin molecules is located. The droplet is carotenoid in nature and as such can be colored blue, yellow, orange, or red. As light travels through this droplet, it is filtered in intensity and in wavelength. Thus, the droplet acts as a selective spectral filter, only allowing certain wavelengths of light to get through the pigment to be absorbed at the photoreceptor level. It therefore increases the spectral sensitivity of individual receptors, and could enhance the color vision of such animals. Since most of the carotenoid oil droplets block out the shorter wavelengths of light, it is thought that they could also be a mechanism for protecting the receptors from the damage of ultraviolet light. In invertebrates, a similar filtering system has been found, although it is not in the form of an oil droplet but in the form of pigmented vesicle bundles. In the same way, these vesicles filter the incoming light, narrowing the absorbance spectrum of the photoreceptor layer above them.

The placement of the eyes in the head, so that each eye captures a slightly different angle on the same visual field, creates binocular vision and depth perception. (Digital Stock)

two different cone types will allow a species color vision, given the cognitive ability to process such information.

Another important feature of eyes is their position on the head relative to each other. Most vertebrates and some invertebrates have two eyes, allowing them a certain amount of binocular vision depending on the angle between them. In fact, both eyes project a slightly different image onto each retina, but the brain is able to compute this as a single image. This results in a larger field of view and also a sense of depth; these are definite advantages, for example for a hunting animal, which needs to know how far it is from the prey target.

—*Lucy A. Newman*

See also: Anatomy; Beaks and bills; Bone and cartilage; Brain; Claws, nails, and hooves; Digestive tract; Ears; Endoskeletons; Exoskeletons; Eyes; Fins and flippers; Immune system; Kidneys and other excretory structures; Lungs, gills, and tracheas; Muscles in invertebrates; Muscles in vertebrates; Nervous systems of vertebrates; Noses; Physiology; Reproductive system of female mammals; Reproductive system of male mammals; Respiratory system; Sense organs; Skin; Tails; Teeth, fangs, and tusks; Tentacles; Vision; Wings.

Bibliography

Baylor, D. "How Photons Start Vision." *Proceedings of the National Academy of Science* 93 (January, 1996): 560-565. This review explores the biochemistry of phototransduction, and is fairly detailed.

Dowling, J. E. *The Retina: An Approachable Part of the Brain.* Cambridge, Mass.: Belknap Press of the Harvard University Press, 1987. This work approaches the retina as a part of the brain, and discusses much of the neurobiological research available on this topic.

Goldsmith, T. "Optimization, Constraint, and History in the Evolution of Eyes." *Quarterly Review of Biology* 65, no. 3 (September, 1990): 281-320. Discusses the evolution and adaptation of eyes and vision.

Hubel D. H. *Eye, Brain, and Vision.* 2d ed. New York: Scientific American Library, 1995. Approaches vision from a neurobiological standpoint; discusses in detail processing of visual information by the brain.

Solomon, E., L. Berg, D. Martin, and C. Villee. *Biology.* 3d ed. Fort Worth, Tex.: Saunders College Publishing, 1993. Chapter 41 covers image formation and, briefly, phototransduction.

FAUNA: AFRICA

Types of animal science: Ecology, geography
Fields of study: Conservation biology, ecology, environmental science, wildlife ecology, zoology

Africa is renowned for the richness of its wildlife, having a greater variety of large ungulates (some ninety species) and freshwater fish (two thousand species) than any other continent. However, many of these animals are being hunted to near extinction or endangered due to habitat loss.

Principal Terms

ARTIODACTYLS: mammals with an even number of toes

PREDATOR: a carnivorous animal that obtains food by hunting

UNGULATE: a hoofed mammal

Known for the enormous diversity and richness of its wildlife, Africa has a greater variety of large ungulates, or hoofed mammals (some ninety species), and freshwater fish (two thousand species) than any other continent. However, probably no group of animals is more identified with Africa than its flesh-eating carnivore mammals, of which there are more than sixty species. In addition to the better-known big cats, such as lions, leopards, and cheetahs, there are wild dogs, hyenas, servals (long-limbed cats), wildcats, jackals, foxes, weasels, civets, and mongoose.

The African Enigma

There are many theories as to why Africa has such an abundance of wildlife, and large wildlife at that. While early North American human societies drove mammoths, giant beaver, and sabertooth tigers to extinction, early Europeans wiped out lions and rhinos, and Asians domesticated their landscapes, Africans lived in relative accord with creatures that were no less grand or ferocious. Some African folklore places animals on the same footing with people.

Another theory is that the tsetse fly, by spreading the sleeping sickness, made much of tropical Africa uninhabitable by humans and protected the wilderness and wildlife from human depredation. Still another possibility may have been the constancy and small size of the African population in comparison with European and Asian numbers—too few people to either overhunt large mammals or exhaust their habitat. One more possibility is that the savannas of Africa—grassy plains with scattered tree cover—provide good habitat for so many ungulates. Humans cannot hunt ungulates easily under these conditions because the humans can be seen and outrun easily. At the same time, a large population of ungulates supports an appreciable population of predators such as lions and scavengers such as hyenas.

At one time, most African fauna was believed to have originated in the Palearctic regions, that is, Europe, northwest Africa, and much of Asia. There is no doubt that as recently as fifteen thousand years ago, a milder Saharan climate allowed typically Ethiopian forms, such as clariid catfish, to reach the river systems of North Africa. Similarly, northern animal life and vegetation seem to have extended far south into the Sahara. The white rhinoceros evidently coexisted with elklike deer.

The spread of forests during the wetter epochs created separate northern and southern wooded grasslands. This led to the evolution of such closely related northern and southern species of antelope as the kob and puku, the Nile and common lechwe, and the northern and southern forms of white rhinoceros. In earlier periods, the animal life was even more remarkable than in modern times. Fossil deposits have revealed sheep as big as present-day buffalo, huge hippopotamuses, giant baboons, and other types similar to existing species. These "megafauna" probably lived in wetter periods and died out as the climate became drier.

Effects of Human Population Growth

The fine conditions for Africa's fauna in the mid- to late nineteenth century started to come to an end when European settlers arrived in many parts of Africa. Technologies, in the form of Western

Habitats and Selected Vertebrates of Africa and Madagascar

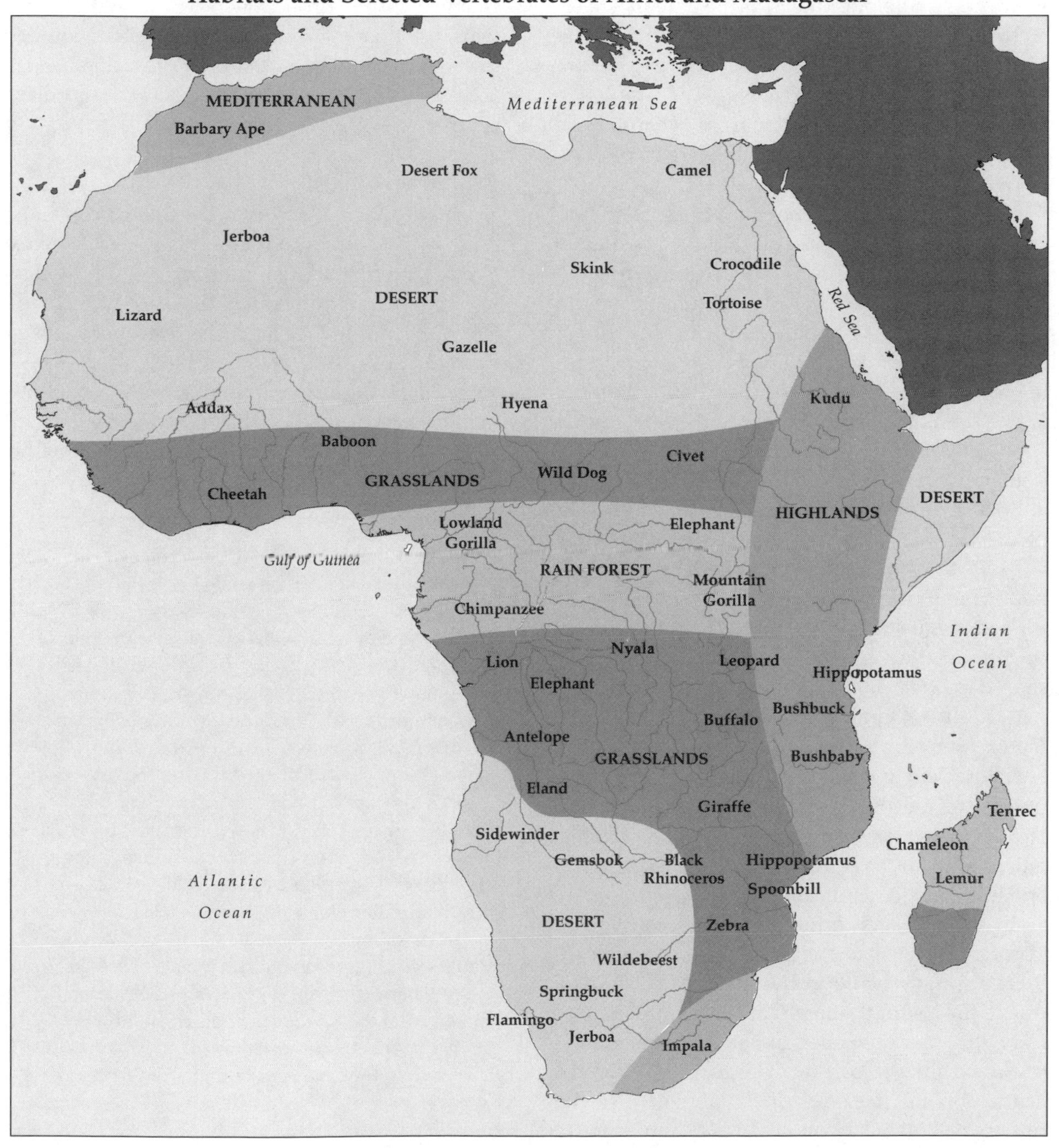

medicine and sanitation, sparked a demographic revolution. In places such as Rhodesia (now Zimbabwe), the human population exploded twenty-fold during the ninety-year reign of white settlers.

Since 1940, the combined pressures of hunting and habitat destruction have cut wildlife numbers greatly. The antelope known as the Zambian black lechwe, believed to have numbered one million in 1900, had been reduced to less than eight thousand by the late twentieth century. The population of African elephants declined from two million in the early 1970's to 600,000 by 1990, largely because of poaching for the ivory trade.

Since the 1960's, poaching has caused a 95 percent decline in the world's black rhino population to fewer than 2,600 in 1997. The African white rhinoceros reached the verge of extinction in 1980. In West Africa, the continual southward advance of the Sahara desert has amplified the twin pressures of habitat destruction and human population. The larger fauna that lived there, caught between the desert and the burgeoning population, are largely gone now.

In Kenya, farmers have long since cleared most of the central part of the country that was once a densely forested region inhabited by wild animals. Some people have invaded national parks for commercial purposes such as logging and cattle grazing, thus forcing wildlife out of their preserved habitats. Some animals have had no alternative but to fight with humans for food and water. Along the Kenyan coast, many people have been attacked and killed by charging hippopotamuses and crocodiles in search of food and space. Similar cases have been common in the highlands, where Kenyans have lost their lives to charging elephants or to leopards and buffalo.

In recent years, human-elephant conflicts in Cameroon have become a major issue. Such conflicts are more acute in the savanna ecosystem, due to the loss of the elephant's range and habitat following the conversion of natural vegetation to farmland and the logging of large tracts of forests. Cameroon still has a relatively large herd of elephants, estimated at about twenty thousand in 1997. Approximately 75 percent of these elephants live in the dense equatorial forest.

Another issue in Cameroon in 1999 was a 652-mile (1,050-kilometer) pipeline proposed to traverse tropical rain forests and link oil fields in landlocked Chad to an export facility in Kribi, Cameroon. The original route of the pipeline was changed to go through two less-fragile ecosystems, but the new route was still designed to cut through tropical areas and provide easier human access to endangered species such as gorillas, chimpanzees, and elephants.

African Mammals

The main group of herbivores is the African antelope, which belong to four subfamilies of the ox family. The first subfamily is further subdivided into the African buffalo and the twist-horned antelope, including the eland (the largest of all antelope), kudu, nyala, and bushbuck. The second subfamily is the duiker, a small primitive antelope that lives in the thickets, bush, and forests. Other well-known large African herbivores include the

Endangered Chimpanzees Source of HIV

A type of chimpanzee that is feared to be headed for extinction by the year 2010 carries a unique genetic blend of human immunodeficiency virus (HIV) and simian immunodeficiency virus (SIV). Scientists have stated that, in the genetic family tree of all known HIVs and SIVs, this virus appears to be the strain that could be the source of all HIVs. Once the strain known as SIVcpz made its way into the human species, probably in the 1950's, it mutated into the existing human strain, adapting naturally to its new host.

The chimpanzees are being driven from their homes by loggers in the rain forests of the Congo River basin. They also are being slaughtered for their meat. However, scientists consider it is crucial that this species of chimpanzee remain in existence so they may study them for critical HIV research.

zebra, giraffe, hippopotamus, rhinoceros, and African elephant.

Africa's large number of endemic or native mammal species is second only to that of South America. These include several families of the ungulate order Artiodactyla (mammals with an even number of toes), such as the giraffe and hippopotamus. Some carnivores—such as civets, their smaller relations, the genets, and hyenas—are chiefly African. The rodent family of jumping hares is endemic, and one order, the aardvark, is exclusively African. The Malagasy Republic (Madagascar) has a remarkable insect-eating family. These are the tenrecs, animals with long, pointed snouts. Some tenrecs are spiny and tailless.

The primates include about forty-five species of Old World monkeys and two of the world's great apes: the chimpanzee and the world's largest ape, the gorilla. The gorilla is present in two subspecies: the lowland gorilla of Central and West Africa and the mountain gorilla of East Africa. The rare mountain gorillas live only in the upland forest on the borders of Uganda, Rwanda, and Congo-Kinshasa (Zaire). There are two populations of roughly equal size. One is in Uganda's Bwindi Impenetrable Forest National Park, where a 1998 census counted 292 gorillas. The second is in the Virunga Mountains, on the borders of the three countries. The last census there, completed in 1989, estimated 324 gorillas in the Virungas, but war has prevented a recount.

Presimian primates include pottos or African lemurs and galagos, bush babies, or small arboreal lemurs. These and other African lemurs tend to be small and nocturnal. In the Malagasy Republic, where there are no true monkeys, the lemurs have occupied all ecological niches, both diurnal and nocturnal, that the monkeys would have taken. Accordingly, the world's most diverse collection of presimian lemurs survives in Madagascar.

African Reptiles, Birds, and Insects

Most African reptiles have their origins elsewhere—mainly in Asia. These include lizards of the agamid family, skinks, crocodiles, and tortoises. Endemic reptiles include girdle-tailed and plated lizards. Large vipers are common and diverse. Certain species have extremely toxic venom, but they are rarely encountered. One of the most noted is the black mamba. Amphibians also belong mainly to Old World groups. Salamanders and toothed tree frogs are confined to the Palearctic northwest Africa. Abundant and more common frogs and toads include such oddities as the so-called hairy frog of Cameroon, whose hairs are auxiliary respiratory organs.

The birdlife south of the Sahara includes almost fifteen hundred resident species. An additional 275 species either reside in northwestern Africa or are winter migrants from Europe. Once there may have been as many as two billion individual migrants, but their numbers have been reduced considerably by severe droughts and by human land use and predation. The few endemic bird species include the ostrich, shoebill, hammerkop, and secretary bird. The many predators of land mammals include eagles, hawks, and owls. Many more, such as storks, waders, and a few species of kingfishers, prey on fish. Even more feed on insects.

Insects include large butterflies, stick insects, mantises, grasshoppers, safari ants, termites, and dung beetles. Spiders abound throughout the continent, and scorpions and locusts can be plentiful locally. Huge swarms of locusts periodically spread over wide areas, causing enormous destruction to vegetation. Mosquitoes that carry malaria are present wherever there is a body of water. Female blackflies transmit the nematode *Onchocera volvulus*, a parasitic filarial or threadlike worm. This organism eventually collects in many parts of the body, including the head near the eye. Nematode clusters around the eyes cause a blindness known as "river blindness." This disease has prevented any significant human habitation in many of Africa's river valleys.

Tsetse flies carry the parasite that causes African sleeping sickness in humans and nagana in livestock. These flies are found in all tropical portions of sub-Saharan Africa. The controversial chemical pesticide, dichloro-diphenyl-trichloro-

Tsetse Fly: Winged Guardian of Wilderness

The tsetse fly acts as a carrier for trypanosomiasis, a parasite that causes sleeping sickness, which is a wasting disease. When sleeping sickness is acute, the symptoms are high temperature, anemia, fitful appetite, and swollen limbs. The victim eventually sinks into an irritable haze, then slips into a coma and dies.

Sleeping sickness has had a long history in Africa. One subspecies of the trypanosomiasis parasite lives amicably in the bloodstream of wild animals such as antelope. However, when it gets into a cow, it makes the animal waste away and die. As a consequence, raising livestock has been difficult or impossible in much of sub-Saharan Africa. More than three million square miles (eight million square kilometers) of that region are off-limits to cattle, and in places where they are raised, three million cows die of the disease each year. The tsetse fly is cheered by conservationists because it makes so much of Africa largely uninhabitable for humans and cattle, thus preserving wildlife. However, it is estimated that the sleeping sickness spread by the tsetse fly threatens fifty million Africans.

ethane (DDT), which is banned in the West, is being used in Zimbabwe and other countries to eliminate the tsetse fly. DDT has adversely affected birds and fish there and has even been found in nursing mothers' breast milk.

Changing Human Attitudes Toward Conservation

If people can benefit from wildlife, their attitudes and actions toward wildlife will improve. Starting in Namibia in 1967 and extending to Zimbabwe in 1975, lawmakers put the idea into action. Large landowners were allotted ownership rights to wildlife, an idea totally alien to the European and the United States' tradition of exclusive state ownership. Landowners, for the first time since the imposition of colonial rule, were free to make economically informed—and, as it turned out, ecologically desirable—market decisions on how best to use their land. By 1990, 75 percent of Zimbabwean ranchers in areas too dry to support crop production had shifted partly or entirely to wildlife ranching. That change was due to the nearly quadruple net profit per acre advantage held by wildlife over cattle.

Rinderpest, a highly contagious bovine plague that has killed millions of Africa's cattle, buffalo, and wildlife in the past century, is finally being brought under control. The disease has been eradicated in West and Central Africa, and is contained in most of East Africa. Rinderpest, caused by morbillivirus, attacks ungulates such as cows, sheep, buffalo, and giraffe and is almost always fatal. It is easily transmitted through direct contact and by drinking water contaminated with the dung of sick animals. It is believed to persist in the Sudan and possibly Somalia, and about four other places in the world.

In Swaziland, pastoralists known as Shewula are breaking with tradition by giving more than 7,400 acres (3,000 hectares) of land used for grazing cattle to a large new game reserve. It will be part of a new transnational reserve with neighboring South Africa and Mozambique. In return, donors have provided funds to build tourism facilities on the land and to train the community in conservation, management, and marketing skills. The goal is to develop the reserve's tourism potential to benefit the rural villagers who have chosen wildlife over cattle.

Similar arrangements have been made between an international hotel chain and a village community outside Tanzania's Serengeti National Park. The agreement involved more than 25,000 acres (10,000 hectares) of land in the Loliondo buffer zone between the Serengeti and the Masai Mara of neighboring Kenya. The Loliondo corridor is used by thousands of stampeding wildebeest during their famed migrations between the Serengeti and the Mara.

Progress has been made in efforts to increase the elephant population. In East Africa, the numbers of elephants are slowly increasing after the poaching rampages of the 1980's. The areas around Tsavo National Park in Kenya and in neighboring Tanzania have reported a count of eighty-one hundred elephants, as opposed to about six thousand elephants in the late 1980's. In 1972, however, there were about twenty-five thousand elephants in the greater Tsavo area. The elephant conservation record of Namibia, Botswana, and Zimbabwe has been impressive. As of the summer of 1999, they had an estimated 200,000 elephants.

—Dana P. McDermott

See also: Aardvarks; Antelope; Baboons; Birds; Camels; Cats; Cattle, buffalo, and bison; Chaparral; Cheetahs; Chimpanzees; Cranes; Crocodiles; Deserts; Dogs, wolves, and coyotes; Ecosystems; Elephants; Endangered species; Fauna: Madagascar; Fish; Flamingos; Flies; Foxes; Giraffes; Gorillas; Grasslands and prairies; Habitats and biomes; Hippopotamuses; Horses and zebras; Hyenas; Hyraxes; Insects; Lakes and rivers; Lemurs; Leopards; Lions; Lizards; Mammals; Manatees; Meerkats; Monkeys; Mosquitoes; Mountains; Ostriches and related birds; Parrots; Primates; Rain forests; Rhinoceroses; Savannas; Ungulates; Vultures; Weasels and related mammals; Wildlife management.

Bibliography

Disilvestro, Roger L., ed. *The African Elephant: Twilight in Eden*. New York: John Wiley & Sons, 1991. Covers all aspects of elephant life, including in-depth coverage of threats to conservation. Numerous color photographs.

Estes, Richard Despard. *The Behavior Guide to African Mammals: Including Hoofed Mammals, Carnivores, Primates*. Berkeley: University of California Press, 1991. Rather than simply providing an identification checklist, this book offers insight into the behavior of the main large mammals of Africa and what it means.

Halliburton, Warren J. *African Wildlife*. Columbus, Ohio: Silver Burdett Press, 1992. A guide to African fauna written for a juvenile audience.

Jones, Robert F. *African Twilight: The Story of a Hunter*. Gallatin Gateway, Mont.: Wilderness Adventure Press, 1995. Former hunting and outdoors writer for *Sports Illustrated* reflects on the changing face of African wildlife between 1964 and 1990, moving from sport hunting to conservation.

Kingdon, Jonathan. *The Kingdon Guide to African Wildlife*. San Diego, Calif.: Academic Press, 1997. An incredibly complete guide to African animal life, illustrated with remarkable paintings by an inspired wildlife artist.

FAUNA: ANTARCTICA

Types of animal science: Ecology, geography
Fields of study: Conservation biology, ecology, environmental science, evolutionary science, wildlife ecology, zoology

Antarctica, the name meaning "opposite to the Arctic," is the coldest and windiest continent on earth. It is also one of the driest, despite being covered in ice sheets up to four kilometers thick. The faunas of the Antarctic display indistinguishable diversity. Many species of fauna are rare and/or endemic, particularly the invertebrates.

Principal Terms

ANTARCTICA: the continent surrounding the South Pole, almost entirely covered by an ice sheet

COLLEMBOLA: primitive, wingless insects

HIGHER INSECTS: generally larger insects with increasing levels of morphological complexity

PACK ICE: floating islands of ice that form flows measuring from a few feet wide to ice islands miles across

SUB-ANTARCTIC: Antarctic landmasses of lesser southern latitude exhibiting more moderate climatic conditions

The history of life in the Antarctic is a fascinating one, largely because, in geologic terms, Antarctica has been very different from the way it is today. For much of Antarctic history, the continent was fully vegetated and carried all the animal life that depends on plants for a living. This is very different from today. Antarctica's gradual glaciations and long isolation from other continental landmasses has impeded the establishment and development of land-based flora and fauna. Other significant factors that have hampered terrestrial biotic evolution on Antarctica include the harsh climate, the ice cover, and the limited number of ice-free land areas. Consequently, the terrestrial flora and fauna of Antarctica are few.

The Antarctic is an extreme environment for any organism to survive in, yet both marine and terrestrial habitats of the Antarctic contain fauna that have adapted to the extreme conditions in order to effectively utilize available resources. Generally speaking, Antarctic regions are devoid of a wide array of faunal diversity due to lack of habitat. The sub-Antarctic and cool temperate islands have much greater diversity, including shrubs and trees. With this higher plant life comes a correspondingly complex array of animal life. The sub-Antarctic and cool temperate islands also possess high levels of endemism as a consequence of their long-standing geographical and ecological isolation from each other and the surrounding continental landmasses.

Antarctic Fauna

The terrestrial fauna consists of a few invertebrate species of protozoans, rotifers, nematodes, tardigrades, collembola, and a species of mite. These life forms are restricted mainly to moist beds of moss. The diversity of marine mammals and birds in the coastal areas and associated pack ice is dependent on marine food chains in the adjacent seas. Because of the vast ice, the continent supports only a primitive indigenous population of cold-adapted land plants and animals. The surrounding sea is as rich in life as the land is barren.

The resource of Antarctic and sub-Antarctic waters supports vast numbers of a variety of seabirds, which play an important role in the marine ecosystem. Nesting grounds are limited, being confined to the scattered sub-Antarctic islands and ice-free localities during the summer on the Antarctic continent and Antarctic Peninsula.

Few terrestrial vertebrates are resident in Antarctica, and those that do occur are limited to sub-Antarctic islands. There are no naturally occurring mammals, reptiles, or amphibians, although humans have deliberately or accidentally introduced a range of animals such as rats, mice, fish, chickens, rabbits, cats, pigs, sheep, cattle, and reindeer to the sub-Antarctic, many impacting native species. Terrestrial animals of the sub- and maritime Antarctic include arthropods, earthworms, and mollusks. Higher insects include spiders, beetles, and flies, most of which are confined to the less severe areas. Microinvertebrate groups such as nematodes, tardigrades, and rotifers are also numerically well represented. The terrestrial fauna of the severe areas of the Antarctic continent are even more simplified. No higher insects are present; the smallest arthropods are restricted to limited areas of vegetation. Instead, groups such as nematodes become dominant. In the most extreme continental cold deserts, simple food webs consist of as few as one to three nematode species, only one of which may be predatory.

Threats to the Antarctic

Much of the Antarctic bears the distinctive imprint of human modification, particularly through the deliberate introduction of animal pests and predators. Of particular concern are the introduced mammals, notably rodents, cats, rabbits, sheep, cattle, and reindeer. In recent years, there have been several cases of successful eradication of alien mammals from islands, and such efforts are continuing. This is important, since islands are vital breeding and resting grounds for seabirds.

—*Jason A. Hubbart*

See also: Birds; Ecosystems; Fauna: South America; Fish; Habitats and biomes; Marine animals; Penguins; Seals and walruses; Tundra.

Bibliography

Di Prisco, G., E. Pisano, and A. Clarke Milano. *Fishes of Antarctica: A Biological Overview.* New York: Springer, 1998. An enlightening source of information which focuses on evolutionary adaptation of Antarctic ichthyofauna, how it has been achieved, and mechanisms involved.

Mohammad, Arif. "Animal Life in the Antarctic: The Seventh Continent." *Journal of Applied Zoological Research* 4, no. 1 (1993): 13-20. An interesting and insightful glimpse of the fauna of the Antarctic.

Murphy, E., and J. King. "Icy Message from the Antarctic." *Nature* 389, no. 6646 (1997): 20-21. This article contains many recent studies on climate trends in the global warming process and emphasizes the importance of conserving this ecosystem, in a well-researched and easily understood manner.

Sayre, A. P. *Antarctica.* Brookfield, Conn.: Twenty-first Century Books, 1998. A great book for young readers describing unique characteristics of Antarctic continent, including geology, ice, climate, plants, animals, scientific research and environmental conservation.

Stone, L. M. *Birds of Antarctica.* Vero Beach, Fla.: Rourke, 1995. A highly recommended book for younger children, giving a user-friendly overview of biology and behavior of Antarctic birds.

FAUNA: ARCTIC

Types of animal science: Ecology, geography
Fields of study: Conservation biology, ecology, environmental science, evolutionary science, wildlife ecology, zoology

The Arctic, the name derived from the Greek arktos, *meaning, "bear," may appear to be desolate and lifeless, but it is far from these things. Plants exhibit surprising richness even though generally limited to a few inches in height and sparse distribution. The shallow, often sterile soil, abrasively high winds, low soil temperatures, and frequent freezing and thawing fluctuations have forced the development of a wide variety of fauna that have adapted to these harsh conditions.*

Principal Terms

ADAPTATION: the processes of change in a living organism that enable it to adjust to the conditions of the environment in which it lives

ARCTIC: the surface of earth lying above 66.5 degrees North latitude, including the Arctic Ocean between North America and Russia

PERMAFROST: any soil in the arctic regions in which temperatures below freezing have existed continuously for a long time

TUNDRA: a treeless, level or gently rolling plain of the arctic region; characterized by a marshy surface from which mosses, lichens, and low shrubs grow

The Arctic is the northernmost region of the earth, centered on the North Pole. The arctic landscape is mainly tundra. This frozen ground lies above tree line and is covered with hearty plants such as moss and lichen, from which caribou and musk ox feed. Arctic temperatures average 10 degrees to 20 degrees Fahrenheit on an annual basis. Due to its extreme northern latitude, the Arctic gets varying amounts of sunlight at different times of year. Precipitation is very low, with average annual rainfall of only seven inches.

Besides extreme fluctuations between summer and winter temperatures, the arctic landscape contains permanent snow and ice in the high country; grasses, sedges, and low shrubs in the lowlands; and permafrost in almost all locations. The permafrost limits drainage and retains moisture for plant growth within the active layer. This can result in wetland formation, as occurs in level areas of the coastal plains of the Arctic. Here, extensive wetlands are home to aquatic vegetation, invertebrate fauna, waterfowl, and shorebirds. The plant growth season is usually less than one hundred days. However, plants do take advantage of constant sunlight in summer months, which enables them to optimize photosynthesis without the usual nighttime cost of respiration.

Arctic Fauna

Arctic ecosystems lack the diversity and richness of species that characterize temperate or tropical ecosystems. Animal as well as plant species decline in number with increasing latitude. Vertebrate species of the Arctic are limited to approximately twenty species of mammals and more than one hundred species of birds. Most are closely related species, such as the caribou of North America and the reindeer of Eurasia. This similarity in arctic mammalian fauna is a result of the lower sea levels of the Pleistocene glaciations, when a broad land connection, known as the Bering Land Bridge, connected present-day Alaska and Siberia. Some arctic mammalian fauna, primarily herbivores and carnivores, rarely occur outside the Arctic and have adapted to life in this environment. Other fauna, such as species of ground squirrels, voles, shrews, red fox, ermine, wolverine, wolf, and brown bear, are common to other ecosystems but are distributed widely throughout the Arctic.

Wet sedge meadows often associated with lake margins, estuaries, and seacoasts are favored nesting habitats for millions of migrating waterfowl and shore birds during the summer months in arctic regions. In addition to shorebirds and waterfowl, finches, buntings, and sparrows are the most abundant species nesting in the Arctic. Terrestrial avian fauna of the Arctic include only a few resident species, among them the ptarmigan, snowy owl, gyrfalcon, and raven. Nonresident species include raptorial birds that commonly nest in the Arctic, such as the peregrine falcon, rough-legged hawk, short-eared owl, the golden eagle, and the white-tailed eagle. These remaining species are present in the Arctic only in summer to breed and rear young, migrating to southern latitudes during winter.

A wide range of marine mammals and fish live in the severe conditions of the Arctic Ocean. Whales, dolphins, walruses, and seven species of seals all make their home in this northern polar region, which is covered by ice most of the year.

Threats to the Arctic

Human activity has exercised a strong influence on the wildlife of arctic regions. Polar bear, walrus, musk ox, and caribou all have been greatly reduced in numbers through hunting alone. This danger was recognized in 1973 when protective legislation was approved for the polar bear. Still, species diversity continues to decrease at an alarming rate, and measures to protect the wildlife of the Arctic must be taken.

—Jason A. Hubbart

See also: Bears; Birds; Dogs, wolves, and coyotes; Dolphins, porpoises, and toothed whales; Ecosystems; Elk; Fauna: Asia; Fauna: Europe; Fauna: North America; Fish; Foxes; Geese; Habitats and biomes; Mammals; Marine animals; Moose; Owls; Polar bears; Reindeer; Seals and walruses; Shrews; Tundra; Weasels and related mammals; Wildlife management.

Bibliography

Arms, M. *Riddle of the Ice: A Scientific Adventure into the Arctic*. New York: Anchor Books, 1998. An investigative report containing nautical, scientific, and environmental perspectives educating one to the complexity of the Arctic system and the fragility of the environment.

Berg, G., ed. *Circumpolar Problems: Habitat, Economy, and Social Relations in the Arctic*. New York: Pergamon Press, 1973. This is a captivating source of information covering many aspects of the Arctic, including human interactions within this ecosystem.

Binney, G. *With Seaplane and Sledge in the Arctic*. London: Hutchinson, 1925. This book contains at least forty pages of illustrations including maps. It is very informative and invaluable for the historic content.

Gruening, Ernest. *The Alaska Book: Story of Our Northern Treasureland*. Chicago: J. O. Ferguson, 1960. This book is a pleasure to read and loaded with numerous illustrations and maps as well as information about the gold rush, natural history, hunting, and people.

Soper, Tony. *The Arctic: A Guide to Coastal Wildlife*. Guilford, Conn.: Globe Pequot, 2001. A field guide to arctic wildlife.

FAUNA: ASIA

Types of animal science: Ecology, geography
Fields of study: Conservation biology, ecology, environmental science, wildlife ecology, zoology

Asia is a huge continent with widely varying climate and geography. As might be expected, its animal life is equally diverse. Since Asia has areas of some of the most dense human population on earth, pressure on animal species from habitat loss is particularly dangerous.

Principal Terms

GRAZER: an animal that eats grass; some are wild, but many grazers have also been domesticated

HERBIVORE: an animal that only eats plants

NOCTURNAL: active at night and dormant or asleep during the day

PREDATOR: an animal that obtains food by hunting other animals

TAIGA: a moist, subarctic forest, primarily coniferous

TUNDRA: a rolling arctic or subarctic plain, located too far north for trees to grow

Asia is a vast and diverse continent with many different types of climate, temperature extremes, and large and small human populations. Many portions of Asia, such as Siberia, are almost empty of people. Some areas are quite densely populated, such as Java in Indonesia, Japan, India, and much of China. Unfortunately for the fauna of Asia, in those areas where the human population is thin, the extremes of climate do not allow for an abundance of wildlife. Many species are near extinction, endangered, or at risk because those areas of Asia that have a climate good for fauna are also densely populated. Some species, however, may have become too specialized for survival. The giant panda eats only bamboo and has an inefficient digestive system. This requires a giant panda to eat large quantities of bamboo. The panda also reproduces slowly. This characteristic may have evolved because the panda has few or no natural enemies—except humans. This type of adaptation can prove to be fatal for a species when factors such as climate, habitat, and predators change.

Much of Asian fauna is under pressure as a result of loss of habitat and overhunting. Most primates worldwide live only in tropical rain forests. The tropical rain forest is being rapidly cleared for timber, firewood, and land for agriculture. Species such as the rhinoceros are disappearing because they are hunted down just for their horns, which have been used for dagger handles in Yemen and for medicinal purposes. Many animals of all species are slaughtered because they are considered to be pests or just in the way of the ever-growing human population.

The Tundra, Taiga, and Steppes

Northern Siberia can be characterized as tundra. Because the tundra is partly free from snow only during the short summer, conditions for life are poor. The principal animals of the tundra are the reindeer, arctic hare, arctic fox, wolf, and lemming. With the exception of the lemming, they live in the tundra in the summer only and migrate in autumn. Birds can be found in the tundra, but with the exceptions of the willow grouse and ptarmigan, they also desert the tundra in winter. Many species of waders, the gray plover, and several kinds of sandpipers migrate to the tundra and breed there in the summer. The snow bunting and the Lapland bunting are also found there. Gyrfalcons (a type of large arctic falcon), buzzards, and skuas feed on these smaller birds and lemmings.

The taiga takes in much of the rest of Siberia and is forested mostly by pine trees. Fauna are richer and more diverse in the taiga than in the tundra, because the greater degree of vegetation provides more food and cover. Mammals found there include the brown bear, wolf, glutton (a kind of wolverine), otter, ermine, sable, lynx, elk, and forest reindeer. The rivers of northern Asia have many species of freshwater fish and several types of sturgeon.

The steppes are the southern edges of Siberia, portions of Kazakhstan, western China, and northern Tibet. Those areas are relatively treeless and similar to the northern Plains states of the United States. The steppes were the place of origin of the northern cattle, the horse, and the Bactrian (two-humped) camel. The animal life of the steppes includes burrowing rodents, jerboas, marmots, and piping hares, and larger animals such as diverse types of antelope. Wild sheep and goats live in the mountains and the plateau areas north of the Himalayas. Tibet is the home of the wild yak, which is an endangered species.

East and Southwest Asia

Northeastern and eastern China, Korea, and Japan have several native species of deer. The giant

Habitats and Selected Vertebrates of Asia

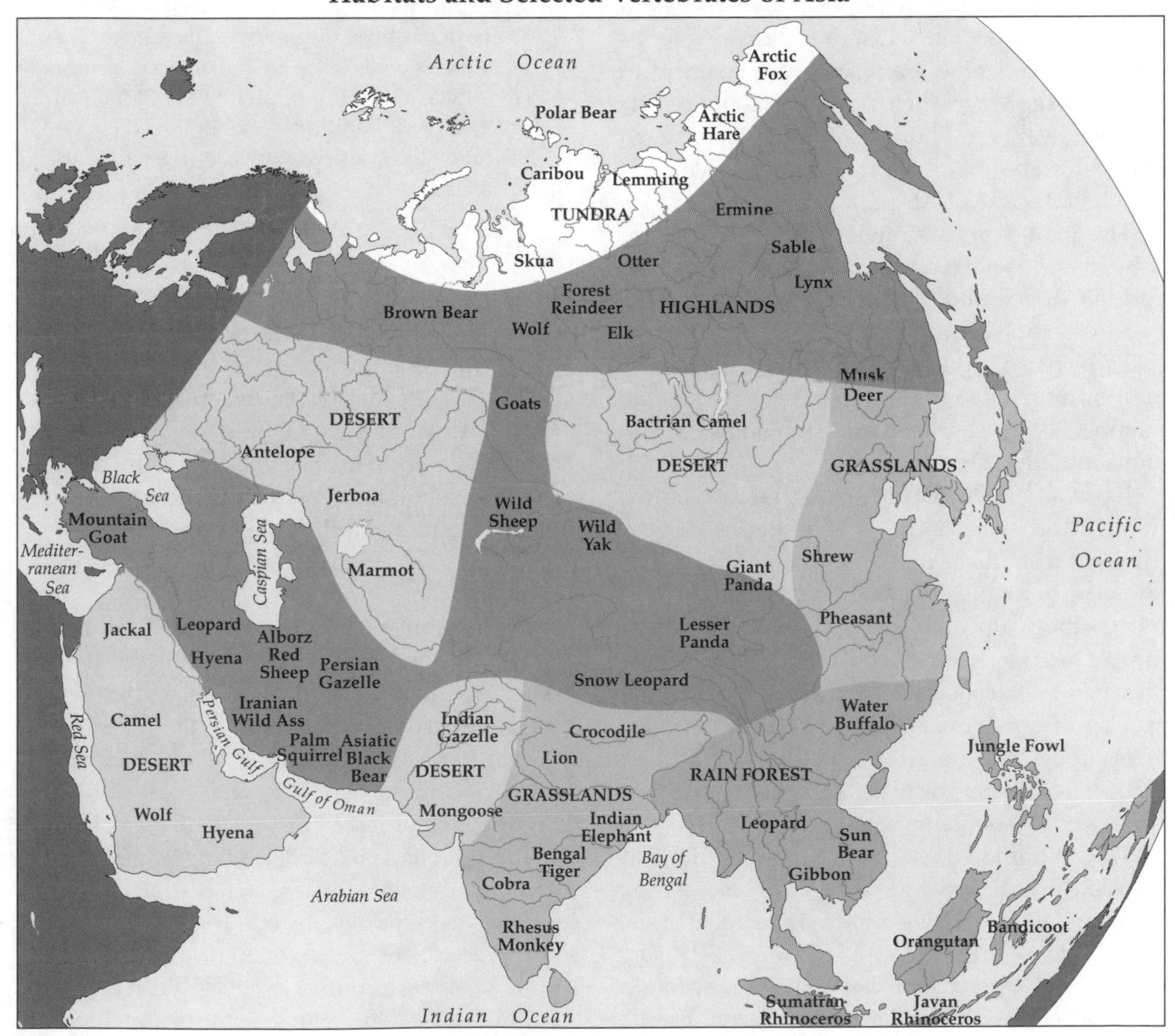

panda lives in the lower mountain area of China near Tibet. The lesser panda, a member of the raccoon family or perhaps its own distinct family, is native to the Himalayas. A rare animal in East Asia is the Siberian tiger, which feeds on elk and inhabits a corner of the Russian Far East and possibly small portions of China and North Korea.

As of 1997, there were estimated to be 360 to 406 wild Siberian tigers. Their numbers are declining, evidently due to overhunting of their main food source, the elk. Another endangered animal is China's true dragon, the Chinese alligator. The remaining populations of these timid reptiles, fewer than 150 in the wild, are limited to just a few village ponds in heavily populated southeastern Anhui Province. One of just two alligator species in the world, Chinese alligators are believed to have diverged from their American counterparts at least twenty million years ago. They reach lengths of about six feet (two meters), which is only half the size of American alligators.

The great rivers of China have a rich variety of fish. In the Yangtze and Huang Ho Rivers, the paddlefish is found. Its only close relative is the paddlefish of North America. The giant salamander, which can grow to a length of 4 feet (1.2 meters) or more, is found in Japanese waters. Most members of the carp family are in Southeast Asia and southern China.

Japan's fauna include the bear, wild boar, fox, deer, and antelope. Some of these species are very different from those on the Asian mainland. The Japanese macaque inhabits many areas. Those at the northern tip of Honshu form the northern limit of monkey habitat in the world. There are eagles, hawks, falcons, pheasant, and more than 150 species of songbirds. Waterbirds include gulls, auks, grebes, and albatrosses. Reptiles include sea turtles, freshwater tortoises, sea snakes, and two species of poisonous snakes.

The Philippines has about 220 species of mammals, including as many as fifty-six species of bats. There are more than five hundred species of birds, including jungle fowl (related to the chicken). The rare and endangered monkey-eating eagle is found in a few locales on Mindanao and Luzon. Fossils show that elephants once existed in the Philippines. It is possible that the climate of the Philippines was much drier in the somewhat distant past. If so, there may have been savanna-like grassland that would have favored elephants. As the climate subsequently became wetter, tropical rain forest grew, which would have reduced the elephants' habitat and thus their population.

Fashion Victim: The Tibetan Antelope

Known by its elegant, lyrate horns and the striking black markings on its face and legs, the male Tibetan antelope, or chiru, is considered one of the world's most beautiful mammals. It inhabits the windswept Tibetan steppe. At the end of the twentieth century, the population of the chiru was down to perhaps seventy-five thousand animals, from an estimated several million earlier in the century. The chiru's wool, called shatoosh, is considered to be the finest in the world and a growing status symbol in Western fashion. Tibetan antelope must be slaughtered to harvest the wool, and it takes an average of three to four antelope to make a single scarf. Ironically, sale of shatoosh has been illegal under the Convention on International Trade in Endangered Species since 1975. Nonetheless, despite the CITES ban, as late as 1998 fashion magazines touted the virtues of shatoosh scarves, priced in the thousands of dollars.

West Asia

Fauna in Iran include the leopard, bear, hyena, wild boar, ibex, and gazelle that inhabit the forested mountains. Seagulls, ducks, and geese line the shores of the Caspian Sea and the Persian Gulf. Buzzards nest in the desert. Deer, hedgehogs, foxes, and more than twenty species of rodents live in the semidesert high-altitude regions. Palm squirrels, Asiatic black bears, and perhaps a few lions inhabit Baluchistan in the southeast. Amphibians and reptiles include frogs, salamanders, boas, racers, rat snakes, cat snakes, and vipers. More than two hundred varieties of fish are found in the Persian Gulf, along with shrimp, lobsters, and turtles. Sturgeon is one of thirty species found in the Caspian Sea.

Arabia has camels, both wild and domesticated, sheep, goats, and Arabian horses (now rare there). Gazelles, oryx, and ibex are becoming rare. Other wild animals are the hyena, wolf, and jackal. The baboon, fox, ratel, rabbit, hedgehog, and jerboa are among the smaller animals. Reptiles include the horned viper, a species of cobra; striped sea snakes; and the large desert monitor. Common birds include eagles, vultures, owls, and the lesser bustard. Flamingos, pelicans, and egrets live on the coasts.

Common insects include locusts, which can descend on fields like a biblical plague. Turkey has the wolf, fox, boar, wildcat, beaver, marten, jackal, hyena, bear, deer, gazelle, and mountain goat. Game birds are the partridge, wild goose, quail, and bustard. The rest of Southwest Asia—Israel, Syria, Lebanon, Jordan, and Iraq—have fauna that are a mix of those found in Arabia and Turkey.

The Oriental Region

The oriental region includes India and extends eastward from India over the mainland and much of insular Southeast Asia. A major portion of the oriental region is tropical. That climate supports malaria-bearing mosquitoes and water-borne flukes carrying the schistosomiasis-causing parasite. These two problems are present in much of tropical Asia as well as in Africa.

In tropical rain forest areas, monkeys are common. Larger primates are found only in tropical rain forests because the kind of cover and food supply that they seem to need exists only there. Gibbons are found in Assam in northeast India; Myanmar (formerly Burma); the Indochinese peninsula—Thailand, Laos, Cambodia, and Vietnam; and the Greater Sunda Islands—Java, Sumatra, and Borneo. The orangutan is found only in Sumatra and Borneo. Indonesia is home to the world's two most endangered rhinoceros species, the Javan and Sumatran. The largest group of Javan rhinoceros is in Ujung Kulon National Park in West Java, but the number of rhinoceros there was fewer than seventy in the late 1990's. The only known wild population of Javan rhinos outside Indonesia, a small group of fewer than fifteen animals, is found in Cat Loc Nature Reserve in Vietnam.

India is an important part of the Oriental zoogeographic region. Almost all orders of mammals are found in India. The primates there include diverse types of monkeys, including the rhesus monkey and the Hanuman langur. Wild herds of Indian elephants can be found in several areas such as the Periyar Lake National Park in Kerala and Bandipur National Park in Karnataka. The Indian rhinoceros is protected at Kaziranga National Park and Manas Wildlife Sanctuary in Assam. There are also four species of large cats: the leopard, snow leopard, Bengal tiger, and lion. The Asian lion, once ranging into West Asia including the Levant, is now found only in the Gir Forest Lion Sanctuary in the Kathiawar Peninsula of Gujarat. Tigers are found in the forests of the Tarai region of Uttar Pradesh, Bihar, and Assam; the Ganges delta in West Bengal; the Eastern Ghats; Madhya Pradesh; and, eastern Rajasthan. The snow leopard is found only in the Himalayan regions.

More than twelve hundred species and perhaps two thousand subspecies of birds are found in India. Herons, storks, ibises, and flamingos are well represented, and many of these are found in the Keoladeo Ghana National Park in Rajasthan. The Rann of Kach forms the nesting ground for one of the world's largest breeding colonies of flamingos.

Crocodiles are found in India's rivers, swamps, and lakes. The estuarine crocodile, which can grow as large as thirty feet (nine meters), feeds on the fish, birds, and crabs of muddy delta areas. The long-snouted gavial or gharial, which is similar to the crocodile, is found in several large rivers, including the Ganges and Brahmaputra. There are almost four hundred species of snakes. One-fifth of these are poisonous, including kraits and cobras. The Indian python inhabits marshy areas and grasslands. More than two thousand species of fish are found in India, 20 percent of which are freshwater species. Commercially valuable insects include the silkworm, bees, and the lac insect. The lac insect secretes a sticky, resinous material called "lac," from which shellac and a red dye are made.

Southeast Asia is located where two important divisions of the world's fauna come together. It constitutes the eastern half of the Oriental zoogeographic region. Bordering on the south and east is the Australian zoogeographic region. The eastern part of the Southeast Asian islands—Sulawesi (Celebes), the Moluccas, and the Lesser Sunda Islands (Bali, Sumba, Flores, and Timor)—forms an area of transition between these two faunal regions. Southeast Asia thus has a considerable diversity of wildlife throughout the region. The region has placental mammals as opposed to the marsupials of Australia, but has hybrid species such as the bandicoot of eastern Indonesia. Small mammals such as monkeys and shrews are the most common. Larger mammals have been pushed into remote areas and national preserves.

Indonesia is located in the transitional zone between the Oriental and Australian faunal regions. The so-called boundary between these two zones is known as Wallace's Line. The line runs between Borneo and Sulawesi in the north and Bali and Lombok in the south. A unique species of proboscis monkey lives only in Kalimantan (southern Borneo). The babirusa (a hoglike animal with curved tusks) and anoa (a small, wild ox with straight horns) are found only in Sulawesi. A giant lizard, the Komodo dragon, occurs only on two small islands, Rinca and Komodo. Insect life in Indonesia includes giant walkingsticks that attain 8 inches (3.2 centimeters) in length, walkingsticks, large atlas beetles, luna moths, and bird-wing swallowtails.

Mammals in Vietnam include elephants, tapirs, tigers, leopards, rhinoceros, wild oxen such as gaurs and koupreys, black bears, sun bears, and several species of deer such as the small musk deer and barking deer. In Cambodia, small populations of elephants, wild oxen, rhinoceros, and several deer species can still be found, along with tigers, leopards, and bears. Snakes abound, with the four most dangerous species being the Indian cobra, the king cobra, the banded krait, and Russell's viper. The fauna of Myanmar and Thailand is similar to that found in Cambodia.

—Dana P. McDermott

See also: Antelope; Baboons; Bears; Birds; Camels; Cattle, buffalo, and bison; Chameleons; Cranes; Crocodiles; Deer; Deserts; Dogs, wolves, and coyotes; Ecosystems; Elephants; Elk; Endangered species; Fauna: Arctic; Fauna: Europe; Fauna: Pacific Islands; Fish; Flamingos; Forests, coniferous; Forests, deciduous; Foxes; Goats; Grasslands and prairies; Habitats and biomes; Hyenas; Insects; Lakes and rivers; Leopards; Lions; Lizards; Mammals; Marine animals; Monkeys; Mosquitoes; Mountains; Orangutans; Pandas; Parrots; Rain forests; Reindeer; Rhinoceroses; Salmon and trout; Snakes; Tidepools and beaches; Tigers; Tundra; Vultures; Weasels and related mammals; Wildlife management.

Bibliography

Laidler, Liz, and Keith Laidler. *China's Threatened Wildlife*. Poole, Dorset, England: Blandford Press, 1999. Profiles twenty endangered species of China, with photographs.

Mishra, Hermanta, and Margaret Jeffries. *Royal Chitwan National Park: Wildlife Heritage of Nepal*. Seattle, Wash.: Mountaineers Books, 1991. A guide to Nepalese wildlife.

Schaller, George B. *Wildlife of the Tibetan Steppe*. Chicago: University of Chicago Press, 1998. Detailed study of the flora and fauna of the Chang Tang steppe. Focuses on Tibetan antelope. Appendices of common names and scientific taxonomy.

Whitten, Jane. *Tropical Wildlife of Southeast Asia*. Boston: Periplus Editions, 1998. A guidebook to the animals of Southeast Asia.

Willis, Terri. *Sichuan Panda Forests*. Orlando, Fla.: Raintree Steck-Vaughn, 1995. Written for a juvenile audience, an introduction to giant pandas and their environment.

FAUNA: AUSTRALIA

Types of animal science: Anatomy, classification, ecology, reproduction
Fields of study: Anatomy, ecology, evolutionary science, wildlife ecology, zoology

The animals of the island continent of Australia are diverse and in a number of instances bizarre and strange in appearance and behavior. A large number of mammals are marsupials; native placental mammals are rare.

Principal Terms

ABORIGINE: European name for the Native Australians

MARSUPIAL: a type of mammal where the females possess an external abdominal pouch in which the young are suckled and protected

MONOTREME: a primitive mammal that lays eggs, such as the duck-bill platypus

PLACENTAL: a mammal in which the developing young are nurtured via a blood-rich tissue (the placenta) within the mother's body.

The European exploration of Australia and discovery of its rich but strange animal life was late in coming. The British naval officer, Captain James Cook, sailed around Australia and charted the east coast of Australia, named it New South Wales, and claimed it for England between 1768 and 1771. In 1788, the first Europeans were transported to what is now Sydney. They were prisoners who had been sentenced by the British courts and incarcerated in prisons in Australia.

Mammals

The groups of large hopping animals that the aborigines called kangaroo became a favorite source of meat to the prisoners. Like all the other native mammals, the kangaroos are marsupials. Several other marsupials are also hopping animals. They range from the slightly smaller wallaby (which resembles the great gray kangaroo) to the tiny, flat-headed marsupial mouse, which weighs a mere sixth of an ounce. The koala is a well-recognized marsupial, although it is commonly called a bear because of its resemblance to a child's teddy bear. Mainly a tree dweller, its favorite food is the highly fragrant leaves of eucalyptus trees.

In addition to the diverse marsupial mammals, the fauna of Australia includes several species of placental mammals, the majority brought to the island continent by humans. The aborigines, who came to Australia about forty thousand years before the present, were foragers or hunter-gatherers. They brought no animals with them. About eight to ten thousand years ago, fishermen from the islands of Indonesia regularly plied the waters off the north coast of Australia in search of sea cucumbers, a large, marine, wormlike animal. These were dried on board the ship and returned to the home islands as food. These ships apparently also carried the companion of humans through the centuries, the rat. The rats, inadvertently, were among the first placental mammals other than humans to find a foothold in Australia, and quickly established themselves wherever there was food and water. These fishermen also were responsible for taking dogs with them, probably as food. Some of the dogs escaped, becoming the dingoes or wild dogs of Australia.

Other placental mammals were introduced by Europeans for hunting, such as the rabbit and the red fox. Both soon escaped into the wild and became serious pests, especially the rabbit. With no predators to attack them, the rabbits reproduced prodigiously, overwhelming control efforts. The fox had been introduced to serve as quarry for hunters riding to hounds, but it never reached the pest status of the rabbit. Visitors to Australia also

Habitats and Selected Vertebrates of Australia

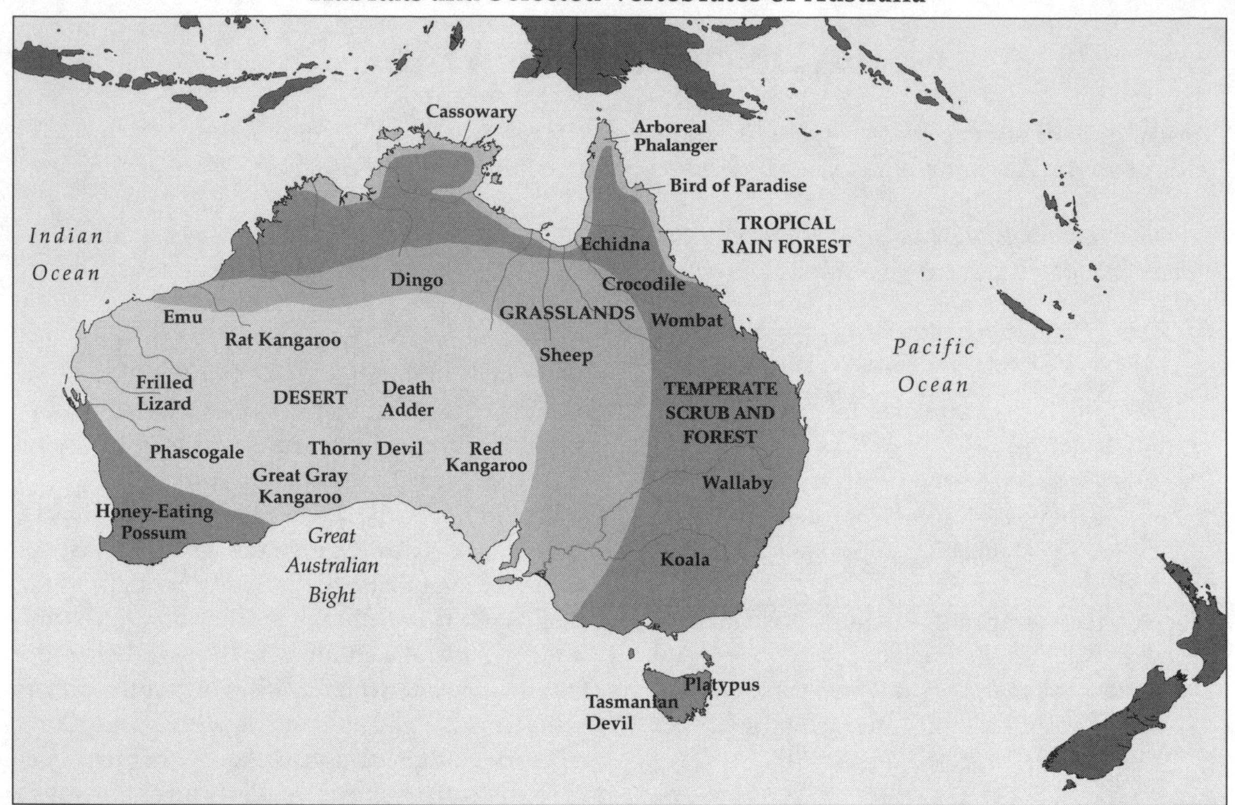

are often startled to see camels, apparently wild, striding over the sandy wastes of the Outback and other desert areas. These animals, now wild, are the descendants of domestic dromedary camels that had been used to carry freight across the deserts. A few domestic animals, principally dogs and cats, have escaped and established a feral existence. No doubt some cats were easily and swiftly caught and eaten by dingoes, crocodiles, and feral dogs. The dogs formed packs and preyed on a variety of small marsupials. They are a particular predator of the koala. Some feral dogs were preyed on by dingoes but also joined dingo packs, interbred with them, and quickly became part of the predatory fauna of Australia.

The Life Cycle of Marsupials

Marsupials, as mammals, have milk-producing glands and suckle their young. They reproduce similarly to nearly all other mammals: One or more eggs, produced internally by the female, are fertilized internally by the male and develop within the female's body. After a period of developmental stages lasting weeks or months, the fertilized egg becomes an embryo, then a fetus, and finally is ready to be born. Marsupials differ because after only a short development period, the young fetus leaves the female's uterus and painstakingly crawls over her abdomen to enter the brood pouch (marsupium). Here the fetus continues its development until it is a fully formed and viable organism ready to take up an external existence.

The female has prepared the way for the fetal animal to move over her abdomen by vigorously licking the surface hairs from birth canal to pouch. Observers had long believed the licking laid down a scent trail of saliva for the fetus to follow to the entrance of the marsupium. Recent studies suggest the licking is performed primarily to cleanse the path to be followed by the fetus.

The marsupium is endowed with nipples or teats—the number varies according to the species from one or two to a dozen or more—which the fetus takes into its mouth. This action causes the teat to swell, thus effectively sealing it into the fetus's mouth. The quality of the mother's milk varies during the development of the young; the milk fat increases with time to enhance the nutrition available to the developing young.

Kangaroos, Wallabies, Wombats, and Koalas

The most familiar and readily recognized Australian marsupial are the kangaroos. They range in size from the seven-foot-tall red kangaroo, which weighs up to 200 pounds, to the slightly smaller gray kangaroo, which weighs up to 165 pounds and may be as much as six feet tall. In both species, the heavy, muscular tail is nearly as long as the body. It serves as a prop when the animal is sitting and helps to balance it when it is fleeing. The kangaroo moves forward by a series of powerful hops at speeds of up to thirty miles per hour. The small front legs are used primarily in holding objects.

Wallabies are medium-sized members of the kangaroo family. For the most part, they feed in open grasslands at night. The body size, from nose to root of the tail, is eighteen to forty-one inches. The tail ranges from thirteen to thirty inches long. Because of their smaller size, wallabies are preyed on to a greater extent by dingoes, feral dogs, and, in years past, the Tasmanian tiger.

Before Europeans colonized Australia and introduced sheep and beef cattle, wombats and kangaroos were the major grazers in the grasslands of the island continent. Although it is a marsupial, the wombat looks nothing like any of the other marsupials. It has been described as resembling a fur-covered barrel with four legs. An adult may weigh up to one hundred pounds and measure four feet in length. It is a burrower, tunneling passageways one hundred feet long, six feet under the earth's surface.

The tunnels help the wombat conserve energy and water, and thus reduce its need for food. Its chief foods include grasses, generally of poorer quality than required by sheep. Thus, wombats are able to survive in areas where sheep do not thrive, but if there is better quality forage available, the wombat becomes a competitor with the sheep. As a result, sheep ranchers do not like wombats. They also do not like the cylindrical, muscular animals because they burrow under fences, damaging the fence lines and opening routes for rabbits.

The rotund wombat is not a very swift animal and many fall easy prey to dingoes and large, feral dogs. The most serious threat to the animals, however, is the automobile. Many wombats are killed on the highways as they amble across the pavement at night in search of water or new grazing.

Along with the kangaroo, the most readily recognized Australian animal is the koala, another marsupial. The koala produces one young at a time. When it is able to spend time out of the pouch, it clings to the female's fur and is carried along with the mother on her foraging expeditions. Much of the adult's time is spent high in the treetops, especially eucalyptus, whose leaves are a favorite food of the animal.

The average koala male weighs from fifteen to thirty pounds and is about two feet long. The fur is thick and soft

Comparison of Ecological Functions Between Placental Mammals of North America and Marsupial Mammals of Australia

Ecological Function	*North America*	*Australia*
Large grazer	American bison	gray kangaroo
Small grazer	cottontail	wallaby
Tree dweller	porcupine	koala
Burrowing grazer	woodchuck	wombat
Glider	flying squirrel	sugar glider
Large carnivore	gray wolf	Tasmanian wolf
Small carnivore	wolverine	Tasmanian devil

and the animals were hunted in the fur trade. The last fur season occurred in 1927, when six thousand koalas were slaughtered. They were part of the aborigine diet and undoubtedly had a strong eucalyptus flavor as a result of their diet of eucalyptus leaves.

Although there is no longer an open season on koalas, the mortality rate is high. Loss of habitat in land clearing is a major cause of its decline. In addition, many are killed by automobiles while crossing roadways. Dingoes and domestic dogs kill many of them and a significant number drown in backyard swimming pools. Pool owners often suspend a rope in the water so that koalas that fall in can climb out.

Tasmanian Tigers and Devils

Most of the marsupials in Australia have maintained a somewhat shaky coexistence with the European settlers. Two, however, have not. The Tasmanian tiger (more commonly called the thylacine) was the largest living carnivorous marsupial known. It resembled a large, long dog with a long, stiff tail, and brown fur marked with thirteen to twenty dark brown to black stripes on the rear portion of the body. It was active mostly at night and preyed on small animals and birds. The thylacine also occasionally killed and ate sheep and chickens kept by European settlers. It was a fearsome-looking animal, but it was shy and secretive, avoiding contact with humans.

Thylacine females had a back-opening pouch with three to four young in a litter. Pouch life is presumed to have been about four months. The Europeans feared the thylacine, probably because of its bizarre appearance, its nocturnal habits, and its presumed attacks on domestic animals. A bounty was placed on the animal and it was trapped, poisoned, and shot. The last known thylacine died in an Australian zoo in September, 1936. Occasional sighting of thylacines have been reported, but in 1986 the species was declared extinct.

The Tasmanian devil, although it is a relative of the thylacine, differs in a number of ways. Devils are smaller, about two feet long with a tail about one foot long. They weigh up to twenty-six pounds and are heavily built, with a broad head and a short, thick tail. The devil's powerful jaws and strong teeth help it to completely devour its food, bones, fur, and all. It is mainly a scavenger and will eat whatever is available. Wallabies, small mammals, and birds are included in its diet, either captured alive or scavenged as a carcass. It has been suggested that the Tasmanian devil helps maintain countryside sanitation by cleaning up carcasses.

The female devil has a back-opening pouch with four nipples. Since more than four young are born in a litter, the extra young die. On average, about two or three young survive in the pouch.

It is the coloration and behavior of the devil that have given it a fearsome reputation. The fur is black, sometimes marked with white patches, and it makes eerie screeches as it ambles about at night in search of food. At one time the Tasmanian devil was widely distributed on mainland Australia, but in modern times it is restricted to the island of Tasmania, off the southeast coast of the mainland. It is believed that the dingo ousted the devil from the mainland. The Tasmanian devil's population, under government management, appears to be stable.

Reptiles

Australian reptiles consist of a number of species of snakes—some poisonous—several monitor lizards (related to the Komodo dragon), and the most feared reptile, the crocodile. The largest monitor lizard, the giant goanna, reaches a length of 8.5 feet. It lives in rock crevices in desert regions, feeding on any animals that come within reach.

The estuarine or saltwater crocodile is the largest and probably most dangerous Australian reptile. Specimens thirty feet long have been captured, and there may be a few larger than that in some isolated coastal marshes. It can live in freshwater billabongs, brackish coastal estuaries, and has been observed swimming 150 miles out at sea. Its diet consists of any animals that come to waterways to bathe or drink, and it will attack and eat humans. Indeed, many Australian rivers, especially those in the Northern Territory, are marked with signs that warn visitors to beware of crocodiles in the water. The animals once were actively hunted for their hides, to be made into expensive shoes and bags. Prehunting populations were estimated at more than 150,000 crocodiles from Western Australia to Queensland. Hunting pressure reduced their numbers to about seven thousand individuals, but it rebounded to more than seventy thousand after hunting was banned. Today, the crocodile population is closely monitored and a limited number can be taken for their hides and meat.

Birds

The best known of Australian birds are the parrots. Many species of parrots are collected for the pet trade, the most popular being the small budgerigar, or parakeet. They are aptly called "love birds" because of the apparent kissing that goes on between pairs of the "budgies." Some of them can be taught to mimic human speech. Some Australian birds are not as well known as the diminutive (6.5 inch) budgerigar. Bowerbirds build elaborate structures of branches, grasses, and bits of bone or shell to attract females for mating. The kookaburra, though not often seen, is aptly named the "laughing jackass bird" because of its loud and very distinctive call.

Australia's birds also include two large, flightless species—the cassowary and the emu. Cassowaries are about fifty-two to sixty-five inches tall and weigh up to 140 pounds. The skin of the head and neck, in both males and females, is bare of feathers, with bright red, yellow, blue, and purple skin. The top of the head bears a bony structure called a casque. Cassowaries inhabit the tropical rain forests of the Australian northeast coastal region. They are well-muscled and dangerous; they can kill a human by leaping feet first and raking the intruder with sharp claws.

Like the cassowary, the emu is a large, flightless bird. At five to six feet in height, it is topped only by the African ostrich. Emus are found throughout Australia feeding on fruits, seeds, roots, and occasionally insects. They are pests in farming country because of their liking for cereal grains. They are eaten by aborigines and Europeans in the Outback. Attempts to farm them for the American market as a novelty food have not proved to be very successful.

Placental Mammals

Prior to the arrival of humans, especially Europeans, the Australian mammalian fauna consisted almost entirely of marsupials. One major exception was the flying fox, the large fruit bat. Undoubtedly the earliest bats flew, or were blown by storm winds, from the Asian mainland. With few natural enemies, the bats thrived in the tropical and subtropical forests. They continue to be abundant today. Almost all other placental mammals, such as the dingo, were introduced to Australia with the arrival of humans.

Fossil remains indicate the dingo has been a feature of the outback for about four thousand years. Contrary to earlier beliefs, it probably did not arrive with the aborigines forty thousand years ago but is a relative newcomer. Research by Dr. Laurie Corbett, a scientist with the Australian Commonwealth Scientific and Industrial Research Organization (CSIRO) in Darwin, suggests the dingo ancestors were transported, perhaps as food, aboard Malaysian fishing boats. Once in the wild, they soon found a good supply of game in the populations of ground-dwelling marsupials. When Europeans began to raise sheep and beef cattle, the dingo quickly found them a ready source of food. Research on dingo behavior revealed the animals often simply scatter the flocks or, at worse, ravage the animals without eating the carcasses.

To control the dingo, the Australian government has erected a wire mesh fence six feet high, extending nearly four thousand miles across the country. Many ranchers attempt to control the dingo by any means, including poison, traps, and guns. Dr. Corbett reports that the animal, as a separate species, may be naturally sliding toward extinction by being absorbed into the domestic dog population through interbreeding.

The interior sections of Australia, well away from the moist climate of the coastal regions, resemble what most people think of as desert. Vast wastelands of sand and gravel are interspersed with billabongs and other water courses. To complete the desert picture, groups of wild camels troop across as they seek favorable grazing and water. The camels are not native, but were introduced in the nineteenth century to serve as beasts of burden. The trails they followed served as lifelines for most of inland Australia. The animals brought food and other supplies to the settlements, mines, and sheep stations.

The settlers chose the one-humped dromedary, or Arabian camel, rather than the two-humped Bactrian camel. The Bactrian is an animal of the cold Eurasian steppes, whereas the dromedary is common in the hot deserts of Africa and thus better able to survive in the conditions of Australia. Their cloven hooves have cushioned pads underneath that spread with each step and provide good traction in the loose sand and gravel. The hump stores energy-rich fat and the animals conserve water by minimizing perspiration and recycling moisture exhaled through the nostrils.

At their peak there were between ten thousand and twenty thousand dromedaries on the desert trails of Australia. However, when road improvements and motor trucks came along, the camels were no longer needed. The camel owners slaughtered some for food and their hides, the rest were driven into the Outback and set free. Here they thrived and multiplied. Modern estimates of their numbers place the population of feral camels as high as four hundred thousand animals.

The now-wild camels are a nuisance to sheep and cattle station owners. The camels, ambling along, knock down fences, releasing the livestock and also providing entry ways for predatory dingoes. The resourceful Australians are turning a nuisance to a profit. Many of the feral camels are being harvested for food, wool, and leather. In visitor centers, such as Alice Springs, Uluru (Ayer's Rock) National Park, and Ratatjuta, camels are available for rides, either as "picture opportunities" or for guided tours through the Outback.

While the camel and dingo may be considered nuisances, the European rabbit turned out to be an obnoxious pest. In 1859, a homesick Australian stock grazer imported two dozen European rabbits from England and released them on his land. The rabbits bred prodigiously, and six years later he killed twenty thousand on the ranch. In 1895, twenty million rabbits were killed, and still they increased. Five rabbits eat as much grass as one sheep. The rabbits spread over the countryside like a plague. Rabbit-proof fences were erected, but usually the rabbits had already gotten through. By 1953, over a billion rabbits were inhabiting 1.2 million square miles of Australian grassland.

After years of fruitless efforts to control the rabbits, scientists introduced a disease organism, myxoma, a virus, and within one year, 95 percent of the rabbits were dead. However, in the long run, the control agent was not completely successful. A few rabbits were naturally immune to the disease, and others developed immunity and continued to reproduce. Unfortunately, the virus has evolved to an avirulent form. Thus, the biological control of the rabbit is no longer as effective as it once was.

—Albert C. Jensen

See also: Bats; Birds; Camels; Chaparral; Coral; Cranes; Crocodiles; Deserts; Dogs, wolves, and coyotes; Ecosystems; Endangered species; Fauna: Antarctica; Fauna: Pacific Islands; Fish; Forests, coniferous; Forests, deciduous; Grasslands and prairies; Habitats and biomes; Insects; Jellyfish; Kangaroos; Koalas; Lizards; Mammals; Marine animals; Marsupials; Mollusks; Ostriches and related birds; Parrots; Penguins; Platypuses; Reefs; Reptiles; Sheep; Snakes; Spiders; Tasmanian devils; Tidepools and beaches; Wildlife management.

Bibliography

Barboza, Perry S. "The Wombat Digs In." *Natural History* 104, no. 12 (December, 1995): 26-29. Offers good information about the life history of this unusual-looking Australian marsupial. It is a burrower and often upsets farm fences and provides entryways for unwanted farm pests.

Doubilet, David. "Australia's Saltwater Crocodiles." *National Geographic* 189, no. 6 (June, 1996): 35-47. Excellent photographs and clear text describe Australia's most feared predator. Some brief mention of commercial hunting of the crocodiles for meat and hides.

Drollette, Dan. "The Next Hop." *Scientific American* 277, no. 4 (October, 1997): 4. This brief article suggests that pouch wallabies (not yet free of the mother) might serve as laboratory animals, especially for neurological studies.

Frith, H. J., and J. H. Calaby. *Kangaroos*. New York: Humanities Press, 1969. These two researchers provide extensive material on a variety of kangaroos, especially on the life history aspects of feeding, digestion, reproduction, and social relationships. Detailed but clearly written.

Grzelewski, Derek. "It's Camelot in the Desert." *Smithsonian* 31, no. 10 (January, 2001): 84-94. Well-written, detailed, and sometimes amusing description of the author's excursion, via camel, across the Outback. Interesting sidelights of camel behavior and how they came to be in Australia.

Stix, Gary. "Broken Dreamtime: Will the Koala Go the Way of the Dodo?" *Scientific American* 272, no. 2 (February, 1995): 14-15. The author discusses environmental events, especially wildfire, that are impacting populations of the koala. Forest fires as well as loss of habitat in building developments appear to imperil the numbers of koalas.

Thwaites, Tim, et al. "Everything You Always Wanted to Know About Kangaroos." *International Wildlife* 27, no. 5 (September/October, 1997): 34-43. Four writers, two Australian and two American, bring together a readable, informative article.

FAUNA: CARIBBEAN

Types of animal science: Ecology, evolution, geography
Fields of study: Conservation biology, ecology, evolutionary science

Biologists have exploited the tremendous diversity and phenomenally high population densities of some West Indian animals to better understand evolutionary, ecological, and biogeographic principles.

Principal Terms

ECOLOGICAL NICHE: the sum of environmental conditions necessary for the survival of a population of any species, including food, shelter, habitat, and all other essential resources

ECOMORPH: species of different phyletic origins (at most distantly related) with similar structural and behavioral adaptations to similar niches

ENDEMISM: the condition of a species that has evolved in a given area, adapted to local conditions, and is found nowhere else in the world

SPECIATION: an increase in the number of species, usually resulting from descendants of a common ancestor adapting to different environmental conditions in different geographic areas

WEST INDIES: a body of islands including the Greater Antilles (Cuba, Jamaica, Hispaniola, Puerto Rico), the Lesser Antilles (an island arc extending from east of the Virgin Islands to the north of Trinidad and Tobago), associated smaller island banks, the Bahamas, and sometimes the islands of the western Caribbean east of Nicaragua, but not the continental islands of Central and South America

The Caribbean Sea is bordered to the west and south by Central and South America, respectively, to the east by the Lesser Antillean island arc, and to the north by the Greater Antilles. When used in a biogeographic context, however, the Caribbean Region is usually defined as the islands that have been called the West Indies since the days of Christopher Columbus. These islands vary greatly in size and age. The oldest islands sit on the Caribbean Plate, which first emerged in what is now the eastern Pacific Ocean, drifted between ancient North and South America before the Isthmus of Panama had formed, and eventually collided with the Bahama Platform and the Atlantic Plate. Throughout this passage, rising and falling sea levels repeatedly inundated entire low-lying islands and at least the lower elevations of the higher ones. In addition, to the east, a subduction zone creates a band of intense and often explosive volcanic activity, which has been responsible for the creation of many Lesser Antillean islands that constitute the newest additions to the archipelago. Islands vary in size from barely emergent rocks to Cuba (over 110,000 square kilometers) and Hispaniola (over 75,000 square kilometers), but sizes have varied as rising sea levels during interglacial periods have fragmented island banks and inundated smaller landmasses.

The island fauna has, like that of any region, been defined by a combination of historical events and current realities. Although transient contact with Central America may have occurred at one time, the vast majority of these islands have never been connected to a continent. Because seas are generally effective barriers to dispersal, relatively few terrestrial animals successfully reached the islands, and even fewer became established. Another factor limiting the number of species is the small number of ecological niches available on relatively small islands. Generally speaking, larger islands with greater elevations are more ecologically complex and contain more niches than smaller islands with less topographical relief, but even the largest islands pale in comparison with the ecological diversity of continents. However, when the region is considered as a whole, the fragmentation and isolation of the islands results in a phenomenally high species richness (of at least some animals) and a very high degree of endemism.

A Fauna Dominated by Reptiles and Amphibians

Studies of West Indian invertebrates are largely limited to those of specific sites or groups, and an overview would be cursory at best. As for terrestrial vertebrates, the quantity of mammalian fauna is relatively impoverished and most birds are migratory; consequently, amphibians and reptiles dominate the fauna in terms of both diversity and numbers. On islands with limited resources, the greater energetic needs of birds and mammals limit population sizes. In contrast, amphibians and reptiles do not require large amounts of energy to sustain stable body temperatures and, as a result, phenomenally high population densities are possible.

Over fifty species of bats and over one hundred species of other nonmarine mammals are known from the region. However, the vast majority of the ninety-four species of nonflying species, all endemic, are extinct, victims of habitat destruction, competition with introduced exotics, or exploitation by humans. Consequently, the mammalian fauna is currently dominated by bats and introduced species. Most of the over 550 West Indian species of birds are migrants, and another large number are sea birds with broad distributions throughout the world. Of the endemic groups, as many as a dozen appear to be relicts of genera that were formerly widespread on adjacent continents. Consequently, the avifauna merely mirrors the greater mainland diversity.

In contrast, almost seven hundred frogs (but no salamanders or caecilians), turtles, crocodilians, and squamate reptiles (lizards and snakes) are known from the region. The vast majority is endemic. Population densities include the highest reported for any vertebrate anywhere in the world (approaching 70,000 per hectare for at least two species of dwarf geckos).

Evolutionary and Ecological Studies

The diversity and population densities of amphibians and reptiles have provided biologists with many opportunities to investigate evolutionary, ecological, and biogeographic questions. In particular, the most diverse groups have demonstrated repeatedly the remarkable abilities of animals to speciate by adapting to varying local conditions on isolated islands or disjunct habitats created by topographic relief on larger islands. West Indian lizards in the genus *Anolis* have provided the best insights into the principle of ecomorphology. Structurally and behaviorally similar species on different islands are more closely related to very different species on the same islands than to their ecological counterparts elsewhere. This phenomenon provided evidence for evolutionary adaptations to local conditions and divergence from relatives that adapted to other available habitats. Many fundamental biogeographic principles have also been developed or supported by studies on West Indian animals. Over-water dispersal of large terrestrial vertebrates, long a staple of biogeographic hypotheses, was first demonstrated in 1995 when green iguanas were transported over hundreds of kilometers of open ocean from Guadeloupe to Anguilla on a floating mass of debris. Other important biogeographic concepts emerging from West Indian studies include positive relationships between biodiversity and area, habitat heterogeneity, and proximity to the nearest large body of land—and between habitat fragmentation and extirpation or extinction.

Conservation Concerns

Studies of habitat fragmentation, unfortunately, also provide many examples of the negative impact humans have had on the survival of the other species on the planet. Although Amerindians began the process, and may have been responsible for the extinction of the few large land mammals known to have lived in the islands, the most harm has resulted from humans of European and African descent, to whom the islands have been subjected longer than any other area in the Western Hemisphere. In addition, human population growth is rapid and island areas are limited. Finally, economic development, often related to tourism, is accelerating to accommodate the growing needs of humans living in a region with few natural resources.

Many West Indian animals have small ranges and are extremely vulnerable to habitat alteration and the impact of introduced exotics. A large percentage of frogs and some lizards are restricted to forests that are disappearing rapidly. Other species, notably the critically endangered iguanas of the genus *Cyclura*, the most endangered reptiles in the world, and hutias, large endemic rodents, have suffered from exploitation by humans for food. Many species in both groups are extinct, and all populations are threatened or endangered. Introduced predators, notably feral cats and the mongoose, which was introduced to control rats in sugar cane fields, have eliminated populations of ground-dwelling snakes and lizards on many islands. Goats denude vegetation on which native species depend. The outlook is grim for many species, but hard lessons learned in the West Indies might lead to more enlightened policies elsewhere.

—Robert Powell

See also: Amphibians; Bats; Birds; Coral; Crocodiles; Ecosystems; Endangered species; Fauna: Central America; Fauna: North America; Fauna: South America; Fish; Flamingos; Frogs and toads; Habitats and biomes; Lizards; Mammals; Manatees; Marine animals; Mollusks; Rain forests; Reefs; Snakes; Sponges; Tidepools and beaches; Turtles and tortoises; Wildlife management.

Bibliography

Censky, Ellen J., Karim Hodge, and Judy Dudley. "Over-water Dispersal of Lizards Due to Hurricanes." *Nature* 395 (1998): 556. This brief article first documented the over-water dispersal of large vertebrates across a vast expanse of open ocean.

Crother, Brian I., ed. *Caribbean Amphibians and Reptiles*. San Diego, Calif.: Academic Press, 1999. This compilation of original works features a history of herpetological work in the region, extensive coverage of the major islands and island groups, and chapters on ecology, evolutionary relationships, biogeography, and a comparison with the Central American herpetofauna.

Powell, Robert, and Robert W. Henderson, eds. *Contributions to West Indian Herpetology: A Tribute to Albert Schwartz*. Ithaca, N.Y.: Society for the Study of Amphibians and Reptiles, 1996. This symposium volume honors Albert Schwartz, whose career was dedicated to elucidating the herpetofauna of the region. It includes a history of West Indian herpetology, and sections covering evolution, systematics, and biogeography, ecology and behavior, and conservation.

Raffaele, Herbert, James Wiley, Orlando Garrido, Allen Keith, and Janis Raffaele. *A Guide to the Birds of the West Indies*. Princeton, N.J.: Princeton University Press, 1998. This comprehensive guide to the West Indian avifauna covers 564 species known to occur in the region with particular emphasis on endangered and threatened birds and the often island-specific conservation efforts to conserve birds and their habitats.

Woods, Charles A., ed. *Biogeography of the West Indies: Past, Present, and Future*. Gainesville, Fla.: Sandhill Crane Press, 1989. This volume synthesizes the geology of the islands with the effects of humans, historical perspectives of vegetation and faunal communities, and conservation trends and strategies.

FAUNA: CENTRAL AMERICA

Type of animal science: Geography
Fields of study: Ecology, environmental science, zoology

Central America is host to a tremendous diversity of animal life, a fauna quite different from other parts of the world. Central America is a land bridge that connects North America and South America, and it has fauna with similarities to both of these continents, as well as endemic species that have evolved within its own borders.

Principal Terms

ENDEMIC: confined and indigenous to a particular region or locality

HABITAT: the physical and chemical factors of the environment

MARSUPIAL: a pouched mammal that gives birth to embryonic young that complete development in a pouch, attached to the mother's nipples

NEOTROPICAL FAUNA: the geographic faunal region that includes Central and South America

OMNIVOROUS: a diet that consists of both plants and animals

PLACENTAL: a mammal whose young develop in the uterus and are nourished through the mother's blood vessels in the placenta

Central America, according to geographers, is the region that commences at the narrow Isthmus of Tehuantepec, in southern Mexico, and extends all the way down to the Atrato River valley, just east of the Panama border in Colombia, South America. Situated in the Western Hemisphere, it is a tapering isthmus, often called the land bridge that joins the two continents of North and South America. By definition, Central America is part of North America, and it includes the countries of Costa Rica, Panama, Belize, Honduras, Guatemala, Nicaragua, and El Salvador.

The world has been divided into six biogeographic regions based on the characteristics of the different types of animals that reside there. The Neotropical area begins at the southern border of the Mexican desert, and continues through Central America to the sub-Antarctic zone of South America to the south, and the West Indies to the east. There is significant mixing and influence from the northern continent, as well as endemic species that evolved in western Panama, eastern Costa Rica, and the highlands of Guatemala, Honduras, and northern Nicaragua.

The marsupials are the older indigenous mammals. After the land bridge was created, there was an influx of North American mammals into the once isolated ecosystem of Central and South America. This explains why many creatures of the Central American regions seem to have close ties to North American animals. The more recent arrivals (about three million years ago) are the rodents, including guinea pigs, and porcupines, tapirs, bats, deer, and guanaco, as well as carnivores such as the margay and jaguarundi. These are the more efficient placental mammals that have adapted and survived in the evolutionary and predatory struggle.

Warm-Blooded Vertebrates

The forests of Central America have a relatively meager population of mammals compared to reptiles, and it hosts widely varying species of birds and insects. The isthmus is the natural habitat of the jaguarundi, margay, jaguar, and ocelot. These species are excellent examples of the adaptation process of North American mammals that migrated south, eventually evolving into unique and well-situated species. The jaguarundi is a small, unspotted cat found only in the New World. It is also called the otter-cat because of its otterlike countenance and swimming ability. It lives in the forested and brushy regions near water and is very rare north of Mexico. Similarly, the coati or coatimundi is an animal like a raccoon (family Procyonidae), with a body that is slimmer, has a longer tail, and an elongated flexible snout. Like an anteater, this animal uses its snout to find insects in the ground, yet it can also climb trees to acquire birds and eggs, with a preference for fruits and berries. The Central American species is classified as *Nasua narica*. Other animals that have migrated to the central region but have not changed much from their North American counterparts are the coyote, gray fox, and puma.

Habitats and Selected Vertebrates of Central America

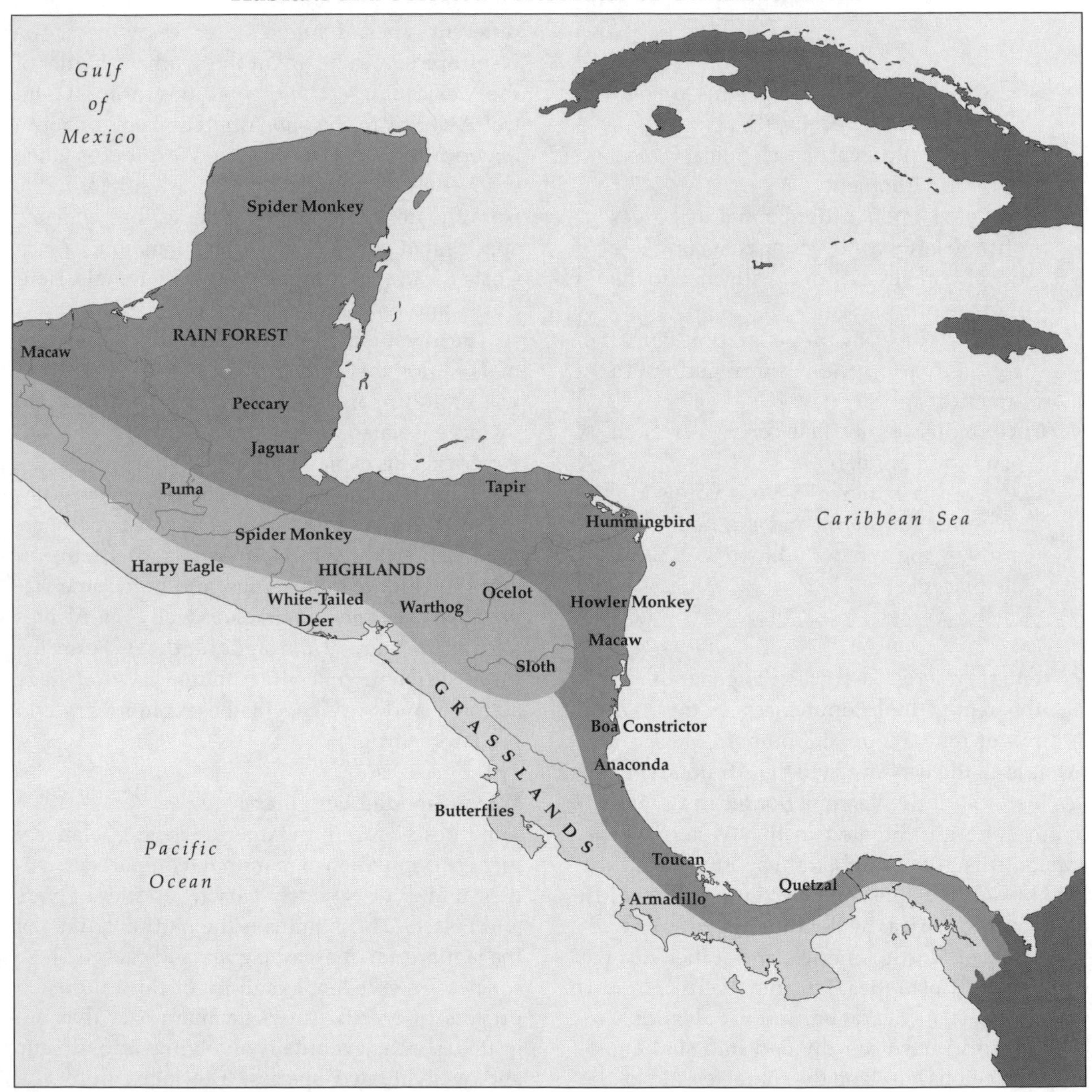

Howler Monkeys: Voices to Be Heard

Howler monkeys (primates of the family Cebidae) are among the most fascinating animals of Central America. These heavy-bodied monkeys are the largest primates of the New World tropical rain forests, with a combined head and body length of over 550 millimeters and a prehensile tail that may be even longer. They spend virtually their entire lives in the treetops, where they find food and water and may come down to drink only during the dry season. The fur is long, soft, and luxurious, and the animal almost looks as though it is wearing a cape. Both sexes have a beard, but it is much more developed in males. Color of the fur may range from a dark, glossy black to a dark reddish brown to a light ochre.

As impressive as howler monkeys look, it is not their appearance that makes them so distinctive. Males produce vocalizations which are so strong and loud that the reverberations can be heard up to several miles. If the source of the sound were not known, one would think that a much larger animal, such as a jaguar, was producing it. When making this extraordinary sound, the howler monkey's tongue is dilated and forms a gobletlike chamber. The chorus may begin with an old male, perhaps the leader of the band. The sound builds up, and there does not appear to be any pause. Other males may join in to create a loud cacophony.

Unfortunately, as is true of many animals in the tropical rain forests of Central America, loss of habitat is threatening the species.

On the other hand, some South American animals used the isthmus as a conduit to the northern continent. Some examples of these northbound animals are the armadillo, marley, opossum, anteater, and kinkajou. This north-south migration is why the Central American region has a mixture of Neartic (northern) and Neotropical (southern) animals. Armadillos are mammals that are related to the sloth and anteater, and inhabit open areas as well as forests. These animals have an armor plating on their bodies as a shield against predators. Although most anteaters are arboreal, with a prehensile tail, the giant anteater is not a climber. The opossum is an arboreal marsupial with an ability to swim. The kinkajou is a relative of the raccoon but behaves like a primate. In fact, for a long time scientists mistakenly grouped them with lemurs. Kinkajous are carnivores with a prehensile tail and are both arboreal and good swimmers. The giant tree sloth, a slow-moving arboreal mammal, makes Panama its home.

Primates of Central America are another group of interesting mammals. The New World monkeys are typified by their prehensile tails. They reside in forested areas. Honduras is home to the howler, spider, and capuchin monkeys. The red-backed squirrel money is found in Panama and Costa Rica. This monkey is valued by the biomedical industry.

Other unusual mammals found in this region are the manatee and the river otter. The spectacled bear is found in Panama. Bats are diverse and plentiful. The bloodsucking vampire bat is an endemic species of this area.

South America is called the bird continent. It is no surprise that the avian population of Central America harbors an enormous number of bird species. They are a unique mixture of birds found in both the northern and southern continents, as the isthmus serves as a migratory path between the two. Hummingbirds and parrots are perfect examples of this mixing.

The resplendent quetzal is a type of trogon with iridescent feathers. It is the national emblem of Guatemala, where the monetary unit is also called a quetzal. Brilliantly colored toucans and macaws are also characteristic of this region. Among the predatory birds, the giant harpy is the most powerful of all eagles. It hunts macaws, sloths, and monkeys. The crested eagle of El Salvador is becoming more rare due to extensive logging and the loss of its natural habitat.

Cold-Blooded Vertebrates

The fish life is prolific in Central America. The majority of fish belong to the following groups: catfish, characins, the pacus, the air-breathing eel, and the cichlids. Catfish make up about half of the total fish fauna, the characins are noted for containing the flesh-eating piranha, eel stun their prey with electric shocks, and the cichlids are similar to North American sunfish. One of the more interesting freshwater fish is the landlocked shark, found in Lake Nicaragua. The lowlands of Belize, with their mangroves and swamps, are the prime environment for the lungfish, which breathe through their lungs instead of gills.

Central America flourishes with a multitude of reptiles and amphibians. The effects of solitary evolution and divergent speciation are very noticeable among these animals. In Costa Rica alone, there are about 130 species of frogs. The golden tree frog is native to Panama. From voracious mammal-eating frogs to beautiful but deadly venomous ones, Central America is host to all. Some of the largest snakes in the world, such as the boas and the anacondas, make the tropical forest their home. Some of these constrictors can devour caimans and mammals. There are also colubrids or coral snakes, green snakes that have adapted to living in trees, the ground, the water, and in burrows. Crocodiles, caimans, lizards, iguanas, salamanders, geckos, turtles, and tortoises are all part of this ecosystem. The crested basilisk lizard, a type of iguana, lives on stream banks. This lizard is known in Spanish as Jesus Cristo, because it appears to have the ability to walk on water, which it does by skimming over the surface very fast.

Indiscriminate hunting and logging have caused many of these animals to become rare or endangered. Many Central American nations have passed or are in the process of passing legislation that will prohibit hunting and protect forested areas. It will take concentrated, unified, and consistent effort by these national governments and others to save the unique species of Central America.

—Donald J. Nash

See also: Armadillos, anteaters, and sloths; Bats; Birds; Cats; Chaparral; Cranes; Crocodiles; Deer; Dogs, wolves, and coyotes; Ecosystems; Endangered species; Fauna: Caribbean; Fauna: North America; Fauna: South America; Fish; Forests, deciduous; Frogs and toads; Habitats and biomes; Jaguars; Lakes and rivers; Lizards; Mammals; Manatees; Marine animals; Marsupials; Monkeys; Mountains; Opossums; Otters; Parrots; Porcupines; Raccoons and related mammals; Rain forests; Reptiles; Skunks; Snakes; Tidepools and beaches; Wildlife management.

Bibliography

Beletsky, Les. *Belize and Northern Guatemala: The Ecotraveller's Wildlife Guide*. San Diego, Calif.: Academic Press, 1999. A wonderfully illustrated guidebook to the fauna and flora of these countries.

_______. *Costa Rica: The Ecotraveller's Wildlife Guide*. San Diego, Calif.: Academic Press, 1998. Another guidebook by Beletsky.

De la Rosa, Carlos Leonardo, and Claudia C. Nocke. *A Guide to the Carnivores of Central America: Natural History, Ecology, and Conservation*. Austin: University of Texas Press, 2000. Illustrated with maps for easy geographical reference. Very informative.

Reid, Fiona. *A Field Guide to the Mammals of Central America and Southeast Mexico*. New York: Oxford University Press, 1997. This book has many color illustrations and also contains useful geographical maps.

Skutch, Alexander Frank. *Birds of Tropical America*. Austin: University of Texas Press, 1983. Illustrated text of birds of the Central American region.

_______. *Life Histories of Central American Birds*. Berkeley, Calif.: Cooper Ornithological Society, 1954-69. Some colorful illustrations with discussions on Central American bird behavior.

FAUNA: EUROPE

Types of animal science: Ecology, geography
Fields of study: Conservation biology, ecology, environmental science, wildlife ecology, zoology

Millennia of urban life and intensive agriculture have made European wildlife especially endangered. Large mammals are at the greatest risk, with many historical species already extinct. Since the 1990's, increasing efforts have been made to restore endangered populations through breeding programs and advanced reproductive technologies.

Principal Terms

ENDANGERED SPECIES: a species of animal or plant that is threatened with extinction

STEPPE: arid, loess-rich land, usually experiencing extremes of temperature

TUNDRA: a rolling arctic or subarctic plain, located too far north for trees to grow

UPPER PALEOLITHIC: the era from 30,000 B.C.E. to 3500 B.C.E., when humans first began to affect European wildlife populations

The number of wildlife species in decline across Europe increased during the second half of the 1990's. Eleven of Europe's twelve most urgent environmental problems, including waste, climate change, and stratospheric ozone depletion, remained static or worsened over the latter half of the 1990's. The threat to Europe's wildlife continues to be severe, with more than a third of bird species in decline, most severely in northwestern and central Europe. Up to half the known vertebrate species were under threat in many countries.

In the countries of the European Union (EU), intensive and subsidized agriculture has devastated the wildlife. Italy is a good example. The extent of animal life in Italy has been reduced greatly by the long presence of human beings. The primary locales in Italy where wildlife survive are the Italian Alps, the Abruzzi (east central Italy on the Adriatic Sea), and Sardinia. Only in those areas can one find the Alpine ibex, brown bear, wolves, foxes, fallow deer, mouflon (wild sheep), and wild boar.

Portugal and Spain are somewhat different from the rest of Europe in that their wildlife is a mixture of European and North African species. In Portugal, about two-thirds of the wildlife overall is Mediterranean, but the farther south one is in Portugal, the more North African the fauna becomes. Birdlife is rich in both countries of Iberia because the peninsula lies on the winter migration route of western and central European species. Nevertheless, major species in Portugal and Spain are endangered.

Despite its reputation for pollution, the eastern portion of Europe is the stronghold of Europe's wildlife. Agriculture in Eastern Europe is starting to intensify, however. One notable exception to the situation of Europe's fauna is Latvia. Although a small country, Latvia contains a number of species and ecosystems rare in other European nations. As many as four thousand Eurasian otters can still be found in Latvian rivers.

Less fortunate European countries struggle to re-establish their beavers, but Latvia has a population of forty-five to fifty thousand beavers, which were reintroduced in 1952 from Russia. There are two hundred to four hundred wolves, as well as three hundred to four hundred lynx. Latvia has more than two hundred breeding species of birds, some of which are rare elsewhere. These include the white-backed woodpecker, lesser-spotted eagle, and black stork. Estonia, which borders on

Habitats of Selected Vertebrates of Europe

Latvia, is similar in some respects. The moose is the largest animal, along with roe deer and red deer. In the forests of northeast Estonia, bears and lynx are found. Along the riverbanks, mink and nutria are common.

The Effects of Humans

Wild animals have been in retreat since Upper Paleolithic times (30,000 B.C.E. to 3500 B.C.E.), when small human groups held their own against such big game as aurochs, a type of bison, and mammoths, now extinct. In more recent centuries, settlers won the land for crops and domesticated animals. As the population increased in industrializing Europe, humans inevitably destroyed, or changed drastically, the wild vegetation cover and the animal life. With difficulty, and largely due to human tolerance, animals have nevertheless survived.

In January, 2000, a dike holding millions of gallons of cyanide-laced wastewater gave way at a gold-extraction operation in northwestern Romania. The collapse of the dike sent a deadly waterborne plume across the Hungarian border and down the nation's second-largest river. Two hundred tons of dead fish floated to the surface of the blighted waters or washed up on the Tisza River's banks. The toxic brew also killed legions of microbes and threatened endangered otters and eagles that ate the tainted fish.

Wildlife Preserves

Even when there are parks and preserves, human activities endanger the fauna and environment. Doñana National Park, an area of wetlands and sand dunes on the Guadalquivir River delta in southern Spain, is one of the most important wildlife sanctuaries in Western Europe. More than 250 bird species, over half of Europe's total, are found there. It is also home to the rare Iberian lynx. The park covers 190,830 acres (77,260 hectares) and is a World Heritage Site. In the 1990's, the park was threatened by falling water levels caused by thirty dams along the Guadalquivir River.

In 1998, the park was further threatened by a flood of toxic waste from a breached dam that was holding back a reservoir used for dumping mining waste. The waste was successfully diverted, but not before wetlands surrounding the park were contaminated by heavy metals. Those heavy metals may yet poison many of the birds that fly there to feed. The aquifer that supplies the park was contaminated, and thousands of fish and amphibians died. These had to be removed to prevent carrion-eating birds from being poisoned.

Mammals

Larger mammals are mostly gone from Western Europe, with the possible exceptions of Spain and Portugal. In the tundra of European Russia and of Lapland in Finland, Norway, and Sweden, caribou or reindeer thrive. In the short summer of the tundra, arctic fox, bear, ermine, and the wolverine

Efforts to Save Europe's Large Carnivores

Habitat destruction and the loss of prey species have contributed to the decline of large carnivores in Europe. In the year 2000, large carnivores occupied fragmented landscapes dominated by humans. The Iberian lynx population is confined to about ten isolated pockets of Spain and Portugal and numbers less than eight hundred. Elsewhere, relic brown bear populations are dangerously small and highly fragmented in southern, central, and western Europe. Like wolves, brown bears face a hostile reception whenever they move into new areas. Wolverines have been reduced to a few hundred in remote areas of Scandinavia. The Eurasian lynx has disappeared from much of its original habitat; where populations are starting to recover, conflict with humans remains a major stumbling block.

The World Wildlife Fund's (WWF) Campaign for Europe's Carnivores aims to challenge ancient prejudices and to help fund projects supporting peaceful coexistence between people and predators. Wolves are beginning to return to old haunts in France, Switzerland, and even Germany. This fact, along with continuing human conflicts with the wolverine, lynx, and brown bear in other areas, emphasizes the need to secure public support for carnivores. WWF's strong European network is working across international boundaries to help secure pan-European cooperation for carnivore populations and to gain public acceptance of their important place in Europe's natural heritage.

may appear. In the deep forests of Poland, Belarus, and European Russia, one still finds the moose, reindeer, roebuck, and brown bear. The brown bear no longer inhabits Scandinavian forests, but the moose is common in Norway, Sweden, and Finland.

Finland is the home of the only species of freshwater seal in the world—an animal depicted on one of the country's coins. The lynx is mostly gone, but wolves, fox, marten, badgers, polecats, and white weasels survive. The sable, which is much hunted for its valuable fur, only just survives in the northeastern forests of European Russia. In Romania, there can still be found bears, wolves, wild goats, and even the European bison.

In much of the rest of Europe, wild fauna are limited to foxes, squirrels, marmots, and other rodents. In higher elevations in France, Italy, and Austria, one can still find the chamois and ibex. Italy's renowned Gran Paradiso National Park in the Valle d'Aosta saved the alpine ibex from extinction. Alpine marmots and chamois can be seen in the Bavarian Alps near Berchtesgaden, Germany. In the steppes of the Ukraine and European Russia, large wildlife is gone except in the semidesert areas north and northwest of the Caspian Sea. There one can still find two types of antelope, the saiga and the jaran, along with rodent sand marmots, desert jerboas, and the sand badger.

Reptiles, Amphibians, and Fish

Europe's reptiles are most common in Mediterranean and semidesert areas. In the coastal Mediterranean, vipers and similar snakes, lizards, and turtles are found frequently. In the semidesert areas north and northwest of the Caspian, cobras and steppe boas are found, along with lizards and tortoises.

In Italian waters, freshwater fish are the brown trout, sturgeon, and the eel. Off the coast of Italy, one finds the white shark, bluefin tuna, and swordfish. There is an abundance of red coral and commercial sponge on the rocks of the warm southern seas. Spanish waters contain a diversity of fish and shellfish, especially in the southeast where Atlantic and Mediterranean waters mix. Species include red mullet, mackerel, tuna, octopus, swordfish, pilchard, and anchovy. Bottom-dwelling species include hake and whiting. The striped dolphin and the long-finned whale are found off southeastern Spain, and the bottlenosed dolphin off the delta of the Ebro River. In Scandinavian waters, salmon trout, and the much-esteemed *siika* (whitefish) are relatively abundant in the northern rivers. Baltic herring and cod are the most common sea fish.

Wildlife Drowning Under the High Seas

Some of Great Britain's best-known nature reserves are expected to be covered by the sea before the middle of the twenty-first century. The British government forecast that the sea level around southeast England would rise by two feet (sixty centimeters) by mid-century. A wildlife-rich chain of low-lying grazing marshes, reedbeds, and lagoons behind the shingle beaches of Suffolk and Norfolk in southeast England are at risk. In February, 1996, a combination of storms and exceptionally high tides flooded the freshwater Cley and Salthouse Marshes in Norfolk with saltwater to a depth of six feet (two meters) along a two-mile (three-kilometer) stretch of coast. The marshes there are one of the few British haunts of the bittern, a large heronlike bird that is one of the most critically endangered species in Britain. No bitterns bred there in 1996 because the salt stunted the reeds that give the bitterns cover and killed the freshwater fish they eat.

Birds

In the short summer of the tundra, seabirds and immigrant birds such as swans, ducks, and snipes can be found. In the great forests of Eastern Europe, black grouse, snipe, hazel hen, white partridge, owls, and blackbirds are common. The steppes have a more abundant selection of fowl. There are eagles, falcons, hawks, and kites and water and marsh birds such as the crane, bittern, and heron.

Insects

Different kinds of locusts and beetles are common in the steppes of the Ukraine and European Russia. In the summer months of 2000, much of the Eurasian steppes was hit by a plague of locusts, devastating Russia's vital grain crops. In Mediterranean and semidesert areas, scorpions, the karakurt spiders, and the palangid are insects dangerous to humans.

Conservation Efforts

In a remote part of northwestern Greece, the lakes of Mikri Prespa and Megali Prespa combine to form rich wetlands. The area has more than fifteen hundred plant species, forty mammals, including the brown bear, otter, and wolf, eleven species of amphibians, twenty-two species of reptiles, and seventeen fish species. Since 1974, Prespa has been a national park, but there was no management plan. In 1991, the World Wildlife Fund (WWF) helped to establish the Society for the Protection of Prespa (SPP) to intervene at a local level and develop the necessary strategies to maintain the area.

Scientists trying to save the Iberian lynx, the most endangered cat in the world, from extinction are using the latest deoxyribonucleic acid (DNA) technology to track the cat by its droppings. The scientists want to put several healthy cats in a captive-breeding center in the Doñana National Park. Once the population is located, DNA testing will reveal whether inbreeding is further endangering the dwindling species by producing unhealthy offspring. The idea is to capture several cats from different regions and bring them to the breeding grounds. In 2000, there was no stock to breed. There are no more than six hundred lynx in Spain and fifty in Portugal. If the rate of decline continues at the current trend, the animal could be extinct by 2010.

The Mediterranean monk seal is one of the ten most endangered species in the world. At the beginning of the twentieth century, it was prevalent throughout the Mediterranean. It is now a rare sight, although still spotted, particularly in the eastern Mediterranean and out in the Atlantic off the Madeira Islands of Portugal. The monk seal is a shy animal with low reproduction rates. It is thus highly sensitive to changes in its habitat and external disturbances. Increasing pollution from industrial waste, plastics, insecticides, and heavy metals have also affected its habitat, food, and most probably its mating ability. More important, since the seals are perceived by commercial fishermen to be in competition for limited fish stocks, they have been deliberately killed.

The WWF began a project in 1993 in the eastern Mediterranean to monitor the presence of the seal and to determine its interaction with the local population. The WWF has mounted information campaigns and government lobbying in Greece, Turkey, and Cyprus in order to pass protective measures to achieve a balance between increasing human use of decreasing marine resources and the basic survival needs of the monk seal.

—Dana P. McDermott

See also: Bears; Birds; Chameleons; Chaparral; Chickens, turkeys, pheasant, and quail; Cranes; Deer; Dogs, wolves, and coyotes; Ecosystems; Elk; Endangered species; Fauna: Arctic; Fauna: Asia; Fish; Forests, coniferous; Forests, deciduous; Foxes; Geese; Goats; Habitats and biomes; Lakes and rivers; Mammals; Moose; Mountains; Otters; Owls; Pigs and hogs; Reindeer; Seals and walruses; Sheep; Shrews; Storks; Tundra; Weasels and related mammals; Wildlife management; Woodpeckers.

Bibliography

Arnold, Edwin Nicholas, and John S. Burton. *Reptiles and Amphibians of Britain and Europe*. New York: HarperResource, 2000. Comprehensive field guide to the reptiles and amphibians of both Western and Eastern Europe.

Arnold, H. R. *Atlas of Mammals in Britain*. London: Stationery Office Books, 1993. Summarizes over 115,000 sightings of sixty-five species in Great Britain between 1959 and 1991, with distribution maps and tables.

Griffiths, Richard A., ed. *The Newts and Salamanders of Europe*. Fort Worth, Tex.: Harcourt Brace, 1996. Well-illustrated field guide.

Hayward, Peter J., and John S. Ryland, eds. *Handbook of the Marine Fauna of North-West Europe*. New York: Oxford University Press, 1995. A classic reference work, with many illustrations and descriptions.

MacDonald, David, and Priscilla Barrett. *Mammals of Britain and Europe*. New York: HarperCollins, 1998. Another comprehensive field guide, covering 230 species, with photographs and illustrations.

Morrison, Paul. *Mammals, Reptiles, and Amphibians of Britain and Europe*. London: Pan Books, 1994. Covers 180 species, with over five hundred color photographs.

FAUNA: GALÁPAGOS ISLANDS

Types of animal science: Classification, ecology, evolution, geography
Fields of study: Ecology, evolutionary science, herpetology, ornithology, wildlife ecology, zoology

The Galápagos Islands, clustered around the equator six hundred miles west of the coast of Ecuador, are renowned for their remarkable fauna, including the giant Galápagos tortoise, the world's only marine iguana, and Darwin's finches, whose variety and ecological diversity inspired Charles Darwin's revolutionary theory of evolution.

Principal Terms

CARAPACE: hard case covering the back of an animal
DISHARMONIC: ecologically unbalanced
ENDEMIC: belonging to or native to a particular place
NOCTURNAL: active at night and dormant or asleep during the day
VESTIGIAL: less developed or degenerated from an original useful form

The Galápagos are oceanic islands, created by submarine volcanoes that emerged from the ocean devoid of life. Because of the great distance between an oceanic island and the nearest continent, only animals and plants that can disperse easily through the air or water are likely to reach such an island. Even fewer will survive its particular environmental conditions and thrive. For this reason, oceanic islands tend to be disharmonic.

For animals to reach the Galápagos before humans arrived, there were only three options: by sea, whether by swimming, floating, or drifting on natural rafts of logs or other vegetation; by air, either by flying or drifting in wind currents; or by hitching a ride on or in the body of another animal. As a result of such limited, haphazard possibilities, freshwater fishes and amphibians are totally absent from the islands, and endemic mammals are few. Only birds and reptiles well represent the vertebrates, and only beetles and butterflies the invertebrates.

Reptiles

Reptiles dominate the Galápagos. Seventeen out of twenty-seven species are endemic. They include geckos, snakes, lava lizards, iguanas, and giant tortoises. Geckos are small, mostly nocturnal lizards. The one endemic snake is a small constrictor. Lava lizards are so named for their habit of basking on lava rocks, and range in size up to twenty-five centimeters (almost ten inches).

The islands have both land and marine species of iguanas. Land iguanas grow up to 1.2 meters (almost four feet) long, have a large dorsal crest, and can live up to seventy years. They prefer fruits, flowers, cactus pads, and other vegetarian fare but also consume insects and carrion. The marine iguana is the only sea-going iguana in the world and can drink seawater, utilizing special salt glands to excrete the excess salt. It spends most of the day basking but dives to depths of twenty meters (over sixty feet) to feed almost exclusively on marine algae. It can reach a length of 1.5 meters (five feet) and live up to thirty years.

The most famous reptile on the islands is the Galápagos tortoise, *Geochelone elephantopus*. Males can weigh up to 250 kilograms (550 pounds); females are considerably smaller. They drink from pools at any opportunity and browse on plants, but they can survive a year or more without eating or drinking and can live 170 years or more. There are two distinct types of tortoise with one intermediate type. One type has a dome-shaped carapace and lives in moist, lush areas where vegetation is abundant and easy to reach. The other inhabits more arid regions and has a saddle-shaped carapace, allowing extension of its long neck to reach leaves and cactus pads high overhead. It is this tortoise's distinctive shell that gave the islands their name: "Galápago" refers to a light saddle or sidesaddle in Spanish.

Birds

Although sea birds are abundant on the islands, most are nonnative migrants. Only a quarter of sea birds species are endemic: the lava and swallowtail gulls, the waved albatross, the flightless cormorant, and the Galápagos penguin. The Galápagos penguin is the northernmost penguin species and one of the smallest, standing forty centimeters (sixteen inches) tall. The flightless cormorant has a strong, hooked beak and vestigial wings. Like the penguin, it swims underwater to catch fish, propelled by its webbed feet.

Twenty-two out of twenty-nine land birds are endemic. There are four species of mockingbird, one dove, one flycatcher, one hawk, one martin, one rail, and thirteen species of Darwin's finches. Darwin's finches all originated from a species native to the Caribbean, *Melanospiza richardsonii*, evolving over time into thirteen species, with different bill sizes and shapes and varying habits.

Mammals

Not including the sea lions that gather in colonies on the beaches, only eight species of mammals are endemic to the Galápagos, six species of omnivorous rice rats and two species of bats. Unfortunately, many nonnative mammals, feral and domestic, have moved in and decimated many native populations of both flora and fauna. Lush vegetation has been converted to grazing land for goats, cattle, and horses. Donkeys trample tortoise nests while pigs root up and devour the eggs of giant tortoises, sea turtles, and birds. Dogs, cats, and nonnative rats prey on any reptile or bird they can catch.

Arthropods

About one thousand insect species inhabit the Galápagos. Because of the arid environment, most are nocturnal. Although there are relatively few species, insects are plentiful, particularly during the rainy season.

Of the arachnids, two endemic species of scorpions are found on the islands. Although there are over fifty spider species, only a few are endemic. The least popular animal on the island is an endemic centipede that can grow as long as thirty centimeters (approximately one foot) and administers a very painful bite.

—*Sue Tarjan*

See also: Birds; Ecosystems; Endangered species; Fish; Habitats and biomes; Lizards; Marine animals; Mollusks; Penguins; Reptiles; Sparrows and finches; Tidepools and beaches; Turtles and tortoises; Wildlife management.

Bibliography

Constant, Pierre. *The Galápagos Islands*. 4th ed. New York: Odyssey, 2000. Detailed description of fauna, natural history, and geology of the islands.

De Roy, Tui. *Galápagos: Islands Born of Fire*. Los Angeles: Warwick, 1998. Natural history mixed with personal narrative by lifelong islands resident and world-famous wildlife photographer. Stunning illustrations.

Pearson, David L., and Les Beletsky. *Ecuador and its Galápagos Islands: The Ecotravellers' Wildlife Guide*. San Diego, Calif.: Academic Press, 2000. Comprehensive field guide to Ecuador's floras and faunas, including a chapter on the fauna of the Galápagos.

Schafer, Susan. *The Galápagos Tortoise*. New York: Dillon Press, 1992. Well-illustrated account of the tortoise's natural history for young readers.

Steadman, David W., and Steven Zousmer. *Galápagos: Discovery on Darwin's Islands*. Washington, D.C.: Smithsonian Institution Press, 1988. Focuses on evolutionary history and the development of Darwin's theory.

Thornton, Ian. *Darwin's Islands: A Natural History of the Galápagos*. Garden City, N.Y.: Natural History Press, 1971. Older but excellent, with comprehensive coverage of arthropods.

FAUNA: MADAGASCAR

Types of animal science: Classification, evolution
Fields of study: Conservation biology, population biology, wildlife ecology, zoology

Madagascar is home to a rich variety of unique and remarkable animal species, most of which are found nowhere else in the world. These animals have evolved in isolation after the island split off from mainland Africa over 120 million years ago.

Principal Terms

ARBOREAL: living in or among trees
CAMOUFLAGE: to blend in with one's surroundings
ENDANGERED: at risk of becoming extinct
ENDEMIC: existing only in a particular locality
NOCTURNAL: active at night
OMNIVOROUS: feeds on both plants and animals
PROSIMIANS: a group of primates that retain some primitive characteristics absent in higher primates

Madagascar is a large island located in the Indian Ocean about 250 miles east of the African continent. Its varied geography and the resulting differences in temperature and rainfall have led to the identification of five distinct climatic regions, each having its own specialized floras and faunas. Because of the island's isolation, up to 90 percent of its estimated 200,000 evolved animal species so far identified are endemic. Biologists are discovering additional species every year. This unique biodiversity makes it a source of fascination for amateur and professional naturalists alike, as well as a hotspot for bioconservation efforts.

Mammals

There are relatively few mammal species on Madagascar. The ancestors of these unusual, diminutive creatures first arrived on the island by floating across the water on mats of vegetation or by hopping on temporary landmasses. Large mammals are thus conspicuously absent. Excluding bats, Madagascar's native mammals are 100 percent endemic.

Lemurs (order Primates, infraorder Lemuriformes) are Madagascar's trademark animals. These curious prosimians occupy a singular place in primate evolution. Lacking some of the characteristics of the higher primates, such as monkeys and apes, they are nevertheless intelligent and display social organization. Lemurs have excellent night vision, an acute sense of smell, distinctive bright eyes, pointed snouts, long tails, and are about the size of a cat. Most are arboreal and have the ability to jump great distances from tree to tree. They are primarily vegetarians, although some species also eat insects. There are thirty-three living species in five families. Of the true lemurs (family Lemuridae), the ringtail lemur is best known, the most sociable, and the only species that spends most of its time on the ground. Others species include the bamboo, mongoose, brown, black, and ruffed lemurs. The indri (family Indriidae) is the largest lemur, growing up to three feet tall and weighing up to fifteen pounds. It is famous for its loud, wailing, territorial call that can be heard at great distances. Woolly and sifaka lemurs are also in this family. The nocturnal mouse and dwarf lemurs (family Cheirogaleidae) are the smallest primates in the world, travel on all fours, and are the most numerous group on Madagascar. The aye-aye (family Daubentoniidae) is a single species that is quite different from all others. This solitary lemur has scruffy black fur, huge ears, rodentlike teeth, a thick tail, and a long, skeletal middle finger. It hunts for insects by listening for sounds underneath bark, then gnaws a hole and inserts its long finger into the hole to scoop out its prey. Lepilemurs (family Megaladapidae) are medium-sized, nocturnal lemurs.

Madagascan carnivores consist of civets (family Viverridae) and mongooses (family Herpestidae). The fossa is the largest civet and predator on the island, with a body length of about 2.5 feet. It is a catlike creature, with powerful legs and a long tail, and is an agile tree climber. It preys on reptiles, birds, and mammals, including lemurs. Other civets are smaller and more foxlike. The five species of mongoose are even smaller.

Tenrecs (order Insectivora, family Tenrecidae) are among the most primitive of all living mammals and were probably the first to arrive on the island. These small creatures resemble hedgehogs or shrews. Tenrecs feed primarily on insects, but many are omnivorous.

There are twenty-nine species of bats. Only 60 percent are endemic, due to their ability to migrate off the island. The flying fox, a fruit bat, is the largest.

Only eleven species of rodents exist on Madagascar. The giant jumping rat is the largest, about the size of a rabbit and weighing about 2.5 pounds.

Reptiles and Amphibians

Madagascar's amphibian population consists of about 150 species of frogs. The true frogs (family Ranidae) are most abundant, followed by the narrow-mouthed frogs (family Microphyidae) and just a few species of the Hyperoliidae and Rhacophoridae families. Most are arboreal forest dwellers.

The reptiles are better represented, with approximately 260 species identified. The lizards (subclass Lepidosauria, order Squamata, suborder Sauria) comprise the largest diversity, and include Madagascar's famous chameleons, whose color-changing ability, multidirectional vision, sharpshooting tongues, and V-shaped feet make them fascinating subjects. Madagascar is home to the majority of the world's chameleon species. Fifty-four species inhabit the island. The parsonii are the world's largest chameleons, and the pardalis is the most common. They eat insects as well as small birds and lizards. The sixty-three species of geckos include those with perfect camouflage as well as brightly colored ones. Iguanas inhabit only the dry regions. Skinks and girdle-tailed lizards are widespread.

There are no venomous snakes on Madagascar. The largest group are harmless snakes from the family Colubridae, the giant hog-nosed snake being the most common. The rest are blind snakes (family Typhlopidae) and boas (family Boidae). Madagascar is also home to a few species of sea turtles and tortoises. The endangered radiated tortoise is one of the rarest animals on earth. The Nile crocodile is also rare on Madagascar.

Birds

Madagascar has a relative paucity of bird species. Only 250 species have been recorded, and they are generally less colorful than those in other tropical areas. Of these, 106 species are endemic, and in fact there are five endemic families: the mesites, ground-rollers, cuckoo-rollers, asities, and vangas. The couas, a subfamily of the cuckoo, are among the most attractive, as are the kingfishers, flycatchers, sunbirds, weavers, lovebirds, and red fody. Of the sixteen species of birds of prey, eleven are endemic, and include kestrels, fish-eagles, and owls. There are nine species of partridges, quails, and grouse. Most of Madagascar's seabirds and wading birds are also found elsewhere in the world.

Invertebrates

A huge variety of invertebrates exist on Madagascar, due to the abundance of different ecological niches. The largest group is the insects, and most are endemic, many even to a particular locality. These include dragonflies and damselflies, grasshoppers, termites, flies, ants, fifty-two species of praying mantis, and over one hundred species of hissing cockroach. The more than eighty species of stick insects are amazing masters of camouflage. There are over one thousand species of true bugs, many strikingly colored. These include stinkbugs, water bugs, cotton-stainer bugs, and assassin bugs. At least twenty thousand species of beetles exist. These are colorful and varied in appearance, with descriptive names such as jewel-beetles, longhorn beetles, tiger beetles, darkling beetles, scarabs, and the bizarre giraffe-necked weevils. Madagascar has hundreds of moth and butterfly species, including many varieties of swallowtails and the attractive pansy butterflies. Hundreds of species of wasps and bees live in Madagascar, and several have bright metallic green coloring. Of the remaining invertebrates, there are many worms, leeches, and flatworms, and a huge variety of endemic land snails. There are about 430 species of spiders, 12 species of scorpions (only two have a painful sting), 64 species of centipedes, and also many millipedes, some reaching giant sizes of six inches or more.

Conservation

Humans arrived on Madagascar about two thousand years ago, and since then wildlife populations have been steadily decreasing due to hunting and habitat destruction. Several large species have already become extinct. The elephant bird, which stood about ten feet tall and weighed about one thousand pounds, died out about five hundred years ago, and its huge eggs are still found. Several species of the lemur, one the size of a gorilla, have also become extinct, as well as the giant tortoise and pigmy hippopotamus. The Malagasy people of Madagascar use the tavy method of slash and burn rice farming, which destroys the forests and erodes the soil. Only about 10 percent of the original forest cover remains. Cattle grazing, the production of charcoal, and the growing human population add to the problem. Many of Madagascar's unique animal species are severely threatened.

There are now a number of national parks and special reserves throughout the island, created to protect ecosystems and their endangered plant and animal species. A program of education to teach the Malagasy about the importance of their environment and wildlife is also under way.

—*Barbara C. Beattie*

See also: Amphibians; Bats; Bees; Beetles; Butterflies and moths; Centipedes and millipedes; Chameleons; Ecosystems; Endangered species; Fauna: Africa; Fish; Frogs and toads; Habitats and biomes; Insects; Lemurs; Lizards; Mammals; Marine animals; Mollusks; Parrots; Primates; Rain forests; Reptiles; Scorpions; Snails; Snakes; Spiders; Tidepools and beaches; Wasps and hornets; Wildlife management.

Bibliography

Blauer, Ettagale, and Jason Lauré. *Madagascar*. New York: Children's Press, 2000. Written for primary and secondary students, this book gives an overall picture of the island, its people, and its wildlife.

Darling, Kathy. *Chameleons on Location*. New York: Lothrop, Lee & Shepard, 1997. A description of the main characteristics of some of Madagascar's chameleons for the juvenile audience.

Garbutt, Nick. *Mammals of Madagascar*. New Haven, Conn.: Yale University Press, 1999. A detailed and scientific inventory of the island's mammals.

Preston-Mafham, Ken. *Madagascar: A Natural History*. New York: Facts on File, 1991. This comprehensive book describes the geographic and climatic regions of Madagascar, and its floras and faunas.

Tyson, Peter. *The Eighth Continent: Life, Death, and Discovery in the Lost World of Madagascar*. New York: William Morrow, 2000. A fascinating travelogue filled with authoritative scientific observations, historical insight, and personal anecdote.

FAUNA: NORTH AMERICA

Types of animal science: Ecology, geography
Fields of study: Conservation biology, ecology, environmental science, wildlife ecology, zoology

North America, stretching from the Arctic Circle in Canada to the tropics of southern Mexico, encompasses every type of climate and habitat, and as a result has a correspondingly diverse animal population.

Principal Terms

GRAZING SPECIES: animals that eat grass; some are wild, but many grazers have also been domesticated

HERBIVORE: an animal that only eats plants

MIGRATORY SPECIES: a species of animal that lives part of the year in one habitat and then moves to another habitat for another part of the year

PREDATOR: an animal that obtains food by hunting other animals

The wildlife of North America can be grouped within two large regions: the Nearctic realm, which covers most of North America from the Arctic to northern Mexico; and the Neotropical realm, which covers southern Mexico and all of Central America. Species in the Nearctic region are similar to those of Eurasia and North Africa, and originally reached North America from Eurasia by passing over the Bering Strait land bridge that once connected Siberia and Alaska, about 60 million years ago. Species in the Neotropical zone are distinctly different from Nearctic wildlife, and reached Central America and Mexico by gradual movement up the isthmus of Panama from South America.

North America's fauna also can be grouped by regions that reflect such climatic influences as latitude, the position of mountain ranges and oceans, and the plants and trees that grow in that area (grasslands, desert, forest, or tundra). Generally, arctic animals are found to the far north, and on the highest slopes of mountains. As one goes farther south, or farther down the sides of mountains, the animals to be found will be those of the forest, grasslands, or desert environments.

Arctic Tundra

Animals of the far north are similar to those found in Eurasia and are well adapted to their cold, treeless environment. Many of these animals evolved from Ice Age species as the glaciers that once covered North America slowly retreated northward. They are large in size and thickly furred, allowing them to maximize conservation of body heat. Large herbivores such as musk oxen and caribou graze on grasses, lichens, and mosses and are, in turn, a food source for polar bears and arctic wolves. Smaller predators such as the arctic white fox feed on arctic hares and small rodents such as voles, lemmings, or the arctic ground squirrels that subsist upon the small shrubs, berries, and grass seeds of the tundra.

Seals and whales proliferate in the Arctic seas. Birdlife in the Arctic tundra is nearly absent in the winter months, with the exception of willow ptarmigans and snowy owls. In the three to four months of summer, several bird species use the region as a breeding ground where young can be hatched and fed with the abundant insect life that emerges during the long and warmer days of summer. Among these migratory species are many varieties of waterfowl, including Canada geese, snow geese, whooping swans, trumpeter swans, phalaropes, plovers, and arctic terns.

Forests

Farther south, below the tree line, coniferous northern forests of fir, spruce, cedar, hemlock, and pine provide shelter and food for the moose, mule

Habitats and Selected Vertebrates of North America

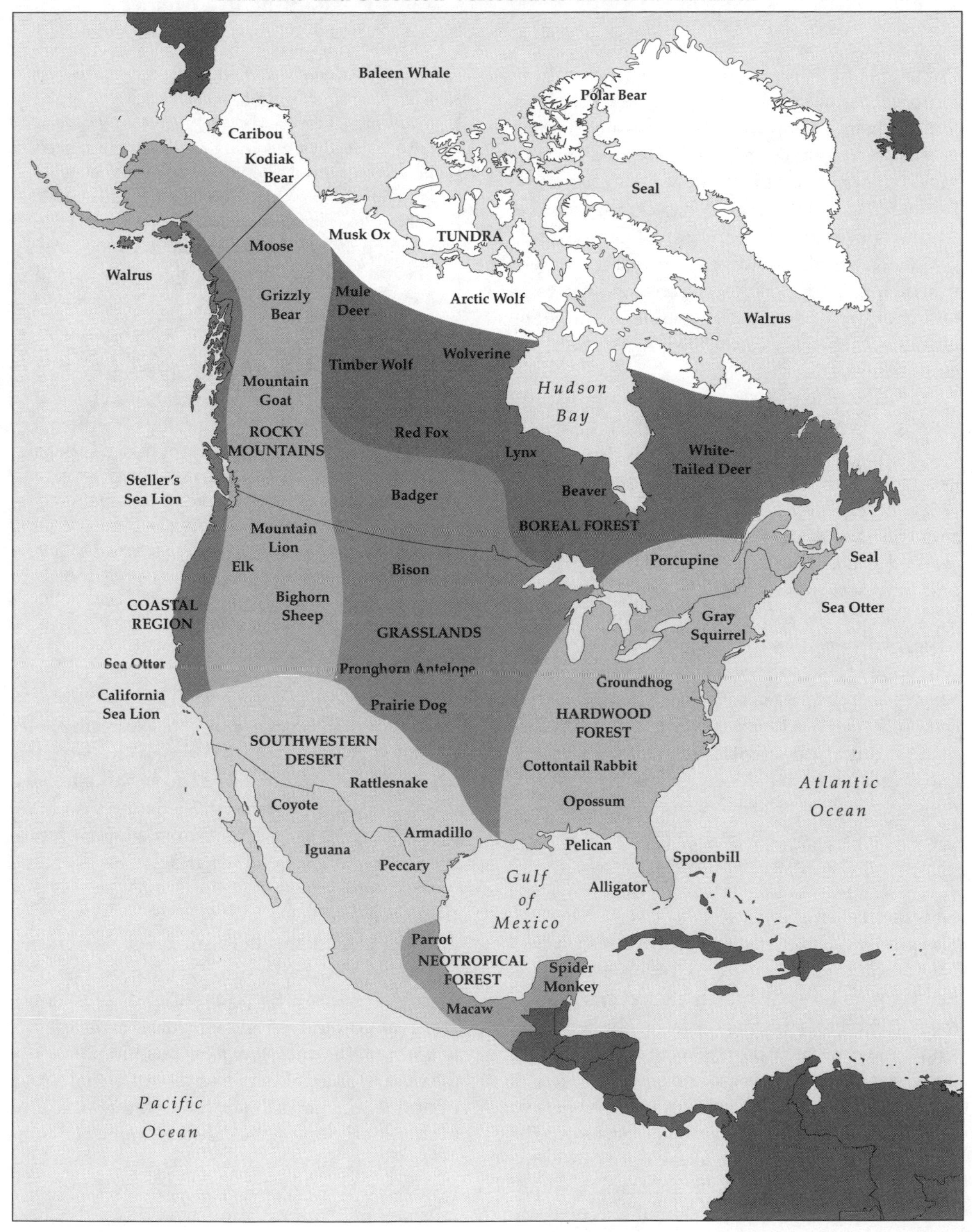

deer, and snowshoe hares that browse on the needles of these trees. Squirrels, chipmunks, and porcupines also thrive in the forests of the far north. Predators of this region include martens, fishers, lynx, wolves, weasels, red foxes, and wolverines. Ponds, rivers, marshes, and swamps are common in this habitat, and provide homes for beavers, muskrats, river otters, mink, and such common fish species as whitefish, perch, pickerel, and pike.

Bird species native to the boreal forest include jays (Canada, blue, and gray), thrushes, finches, nuthatches, loons, osprey, ravens, and crows. A wide variety of songbirds (warblers) and some hummingbirds nest in the boreal woods in the summer months.

The eastern half of North America, from southern Canada to Florida, was once thickly covered by deciduous forests of maple, oak, beech, ash, sycamore, hickory, and other trees that shed their leaves in the winter. Where this forest remains, it provides food for a wide variety of mammals, including black bears, red foxes, raccoons, red and gray squirrels, mink, muskrats, jackrabbits, cottontail rabbits, groundhogs, chipmunks, mice, moles, bobcats, skunks, ermine, opossums, and porcupines. Fish species common to the rivers of eastern and southeastern North America include catfish, suckers, and gar. Several species of salamanders and turtles live in the marshes, streams, and rivers of the deciduous forests, especially in the Appalachians. Waterfowl such as kingfishers, herons, ducks, and grebes also live along the waterways of the hardwood forest region.

Adapting to Extremes

The extremes of climate in North America have caused many animals to adapt in interesting ways. In the far north, animals such as polar bears, musk oxen, and caribou tend to be large in size, so that their external skin area is minimized in relation to their internal portions, allowing the least possible heat loss to the cold air. Many arctic animals have both inner and outer layers of fur, trapping their body heat in the insulating layer. The feet of arctic animals are thickly furred and quite broad, allowing the animals to move over snow without sinking into it.

Animals of the southern deserts have equally effective ways of adapting to the extreme dryness and heat of their environment. They collect and store food when it is abundant after a rare rainfall and then survive long periods of drought on the surplus. They obtain water from the vegetation they eat, rather than from lakes or streams. Many desert animals are pale in color, reflecting back the heat of the sun and blending in with the surrounding landscape. Desert animals are largely nocturnal, moving about only in the cool of the night, after the moist dew falls.

Grasslands

The prairies in the center of North America provide abundant grasses and other herbaceous plants for many small mammals and grazing animals. Jackrabbits, badgers, prairie chickens, and small rodents such as pocket gophers, prairie dogs, and Richardson's ground squirrels, feed on grass and roots, as do such larger herbivores as pronghorn antelope and the American bison. The numerous small rodents and the openness of the grasslands provides optimal habitat conditions for such raptors as owls, hawks, and falcons. Waterfowl nest in the many seasonal watering holes (sloughs) that dot the prairies, although farm drainage and extended periods of drought have greatly reduced this habitat. Seeds and insects are plentiful in the grasslands, supporting such bird species as grouse, quail, partridge, and finches.

Rocky Mountains

The high mountains that run along the western side of North America are inhabited by a number of unique species. Bighorn sheep, Rocky mountain goats, mule deer, and elk graze on the grasses of the foothills and slopes of the mountains. Kodiak bears, grizzly bears, and mountain lions prey upon the grazing animals, and bald and golden eagles subsist upon the ground squirrels, marmots, voles, shrews, and pikas that live in the grasses and scattered forests of the lower mountain slopes. Dipper birds feed from the fast-running mountain streams and are found nowhere else on the continent.

Southwestern Deserts

A large number of animals have adapted to the lack of vegetation and water that exists in much of the southwestern United States and Mexico. Kangaroo rats, pocket mice, jackrabbits, armadillos, peccaries, ring-tailed cats, and ground squirrels all survive in that hostile environment. Predators include bobcats, desert foxes, badgers, and coyotes.

The most prolific forms of wildlife in the arid deserts of southernmost North America are the reptiles, including many different species of lizards, rattlesnakes, toads, and iguanas. Because they can only be active when the outside air temperature provides warmth for basic body functions, reptiles and amphibians are quite rare in the most northern parts of the continent but thrive in the arid and hot southern parts of North America. Roadrunners are a major bird predator of these reptiles, along with eagles and hawks.

Neotropical Forests

The tropical rain forests of Central America and southern Mexico have an astonishing variety of wildlife. The forest canopy has abundant bird life in the form of macaws, parrots, turkey vultures, and flycatchers. Monkeys of many varieties (spider, howler, squirrel, and capuchin) as well as sloths and tamarinds also live in the canopy, feeding on the many fruiting trees. On the floor of the forest, many varieties of ants, spiders, beetles, and chiggers provide food for smaller predators such as anteaters and various species of bats. A number of animals indigenous to South America have adapted to and thrive in Central America, including tapirs, capybaras, pacas, jaguars, ocelots, and agoutis.

Coastal Regions

The beaches, shores, lagoons, and marshes that line the North American continent are home to many kinds of animals that feed upon the ocean life or the intertidal plants and animals that live in the midzone between fresh and salt water. Many different kinds of migratory waterfowl exploit the small crustaceans and mollusks that live at the water's edge, including sandpipers, stilts, curlews, and flamingos. Seals, sea otters, and walruses are found on both coasts in the north of the continent, while Steller's sea lions and California sea lions are found only on the Pacific coast. The lagoons of the southern Atlantic coast and the Gulf of Mexico are home to alligators, pelicans, egrets, and spoonbills.

—*Helen Salmon*

See also: American pronghorns; Antelope; Armadillos, anteaters, and sloths; Bears; Beavers; Birds; Cats; Cattle, buffalo, and bison; Chaparral; Chickens, turkeys, pheasant, and quail; Cranes; Crocodiles; Deer; Deserts; Dogs, wolves, and coyotes; Donkeys and mules; Ducks; Eagles; Ecosystems; Elephant seals; Elk; Endangered species; Fauna: Arctic; Fauna: Caribbean; Fauna: Central America, Fish; Forests, coniferous; Forests, deciduous; Foxes; Frogs and toads; Geese; Goats; Gophers; Grasslands and prairies; Grizzly bears; Habitats and biomes; Horses and zebras; Lakes and rivers; Lizards; Mammals; Manatees; Marine animals; Moles; Moose; Mountain lions; Mountains; Opossums; Otters; Owls; Parrots; Porcupines; Rabbits, hares, and pikas; Raccoons and related mammals; Rain forests; Reptiles; Salmon and trout; Scorpions; Seals and walruses; Sheep; Shrews; Skunks; Snakes; Squirrels; Swamps and marshes; Tidepools and beaches; Tundra; Turtles and tortoises; Vultures; Weasels and related mammals; Wildlife management; Woodpeckers.

Bibliography

Allen, Thomas B., ed. *Wild Animals of North America*. Washington, D.C.: National Geographic Society, 1995. An authoritative reference book on North American animals from Mexico to Canada. Abundant color photographs, many maps.

Brooke, Michael, and Tim Burkhead, eds. *The Cambridge Encyclopedia of Ornithology.*

New York: Cambridge University Press, 1991. A definitive reference book on all birds worldwide, including those of North America.

Jones, David. *North American Wildlife*. Vancouver, British Columbia: Whitecap Books, 1999. A literally huge book, covering North American wildlife in depth, with many photographs.

Lewis, Thomas. *Wildlife of North America*. Southport, Conn.: Hugh Lauter Levin Associates, 1998. Another lavishly illustrated wildlife book, organized by habitat (forest, mountain, prairie, coast, sea, tundra, and desert).

Sherrow, Victoria, and Sandee Cohen. *Endangered Mammals of North America*. Brookfield, Conn.: Twenty-first Century Books, 1995. Written for a juvenile audience. Covers Caribbean manatees, bowhead whales, Mount Graham squirrels, long-nosed bats, gray and red wolves, black-footed ferrets and black-tailed prairie dogs, and Florida panthers.

Whitaker, John O., ed. *National Audubon Society Field Guide to North American Mammals*. New York: Alfred A. Knopf, 1997. Highly detailed guide to North American animals, with photographs and descriptions of each animal's physical features, range and habitat, breeding characteristics, and behavior, as well as listing other animals with which it might be confused.

FAUNA: PACIFIC ISLANDS

Types of animal science: Ecology, geography
Fields of study: Conservation biology, ecology, environmental science, wildlife ecology, zoology

The thousands of islands that dot the Pacific Ocean contain a wide variety of animal species. The restricted conditions of island life result in interesting examples of fauna that have developed in relative isolation, but also increased threat to species that have nowhere to go to escape human encroachment on their habitats.

Principal Terms

ENDEMIC: native to a specific environment and found nowhere else on earth

EXOTICS: nonnative species introduced into an ecosystem

FERAL: an animal, formerly domesticated, which is now living wild

MARSUPIAL: animal with a pouch in which immature young complete their fetal development

The vast region of the Pacific, collectively called Oceania, comprises thousands of islands. Oceania spreads across the Pacific Ocean from 20 degrees north latitude to 50 degrees south latitude and from longitude 125 degrees east to 130 degrees west. The major groupings are Melanesia, Micronesia, Polynesia, and New Zealand. Melanesia ("black islands") is a group of large islands immediately north and east of Australia, from New Guinea to New Caledonia. Micronesia ("little islands") is made up of hundreds of tiny atolls in the western Pacific. Polynesia ("many islands") covers a huge region in the central Pacific. New Zealand lies east and south of Australia. Climates range from tropical to sub-Antarctic, dry to very rainy. Types include volcanic (Fiji, Guam, and Hawaii), tectonic (New Zealand and New Guinea), and low coral atolls (nearly all of Micronesia's islands).

Organisms have a hard time reaching the islands across the broad expanses of the Pacific Ocean. This isolation leads to trends in the number of species found on any given island: Bigger islands have more species; those farthest from continents have fewer species. To reach the islands, animals must fly (birds and bats), float on logs, or be carried in by humans. Birds are usually the first visitors, bringing with them hitchhiking insects. Bats are the only mammals to reach many islands without human help.

Island plants and animals evolve together, affected by difficult conditions; soil is often poor and food limited. Harsh environments and isolation contribute to the formation of new and unique species; some of the strangest creatures on earth are endemic to particular islands. Most of the world's flightless birds developed on islands, where there originally were no large land predators. Island ecosystems are sensitive to disturbances, whether from natural causes such as severe storms, or human activities such as construction, agriculture, logging, and introduced species.

Introduced species (exotics), both accidental and deliberate, are a serious problem. Rats and feral animals can devastate island ecologies. Pigs, cats, rats, and goats are particularly devastating—goats devour vegetation, cats eat birds and small animals, and rats and pigs eat anything.

Exotics also bring diseases to which native animals have no resistance. Humans have tried to deal with exotics, with mixed results. For example, the mongoose was intentionally brought to Fiji to control accidentally introduced rats. No one considered that rats are active at night and mongooses during the day. The mongoose did not control the rats, but they did eat seven native Fijian bird species to extinction.

Another strange example occurred in Hawaii. A Hawaiian bee crawls headfirst into native, barrel-shaped flowers, gathers the nectar, and then backs out. A plant that was introduced by landscapers attracted bees, but the flowers were smaller than the native plants. Once a bee crawled in, it became stuck like a cork in a bottle. There are thousands of these plants, each with hundreds of flowers stoppered with dead bees.

Many tropical and temperate islands have coastal wetlands and mangrove swamps growing at the edge of the sea. Mangroves are low-growing, salt-tolerant trees that form dense tangles that are virtually impenetrable to humans. Wetlands and mangrove swamps are important breeding grounds for many types of fish and crabs, and also trap sediment, stabilize shorelines, and protect coastlines from storms. Humans often fill in the wetlands and cut down the mangroves, causing coastal erosion and the loss of food fish.

Easter Island

Easter Island (Rapa Nui) lies in the South Pacific 2,400 miles (3,862 kilometers) from Chile, South America. It is best known for its huge, mysterious stone statues. Archaeological evidence proves that Easter Island once was densely forested and supported many animal and bird species. The inhabitants burned or cut down the forests before Europeans discovered the island in the eighteenth century. The only plants now living on the island are sparse grasses, a few shrubs, and two species of trees. There are almost no land animals larger than insects. The few thousand people who still live on the island must import nearly all of their food. The devastation of Easter Island is an extreme example of what can happen to island ecologies when human development exceeds natural capacity.

Fiji

The Fiji Islands are mostly volcanic in origin and lie in the South Pacific Ocean about 1,300 miles (2,100 km.) north of Auckland, New Zealand. Some parts of the islands receive up to 13 feet (4 meters) of rain per year, while other parts remain dry. A range of volcanic peaks divides the islands; the highest, Mount Tomanivi (formerly Mount Victoria), is 1,322 meters (4,341 feet). These differences in weather and elevation create a variety of habitats—dense rain forests, grassy savanna, and mangrove swamps—and a large diversity of species.

Human disruption on Fiji has been moderate. About half the total area is still forested, and less than 25 percent of the land is suitable for agriculture. Endemic animals include birds such as lorikeets (parrots that eat flowers or nectar), the Fiji goshawk, spectacularly colored pigeons and parrots, and a pigeon that barks like a dog. The Vanikoro broadbill, which holds its mossy nest together with spider webs, is one of the few native birds that has adapted to forest clearing. It is commonly seen in town gardens and suburbs.

Other creatures native to Fiji are snakes, including a rare, tiny cobra, two species of frog, and several species of geckos and skinks (small lizards). The crested iguana was not discovered until 1978 and lives only on the island of Yaduataba. Yaduataba is now a preserve, and the crested iguana is more likely to survive since feral goats were removed from the island.

There are twelve reserve areas in the Fijian islands, but several are being logged and provide little sanctuary to native plants and animals. The University of the South Pacific is located in Fiji and is a source of serious research into South Pacific species. Tourism is important to Fiji's economy and, with management, could be a source of income to Fijians while still preserving native wildlife.

New Guinea

The world's largest tropical island, New Guinea is located north of Australia and just south of the equator. It is tectonic in origin, with large changes in elevation and many different habitats. Because of its size and varied terrain, New Guinea has a greater variety of habitats than any similar-sized land area in the world. In fact, New Guinea is so rugged that it is one of the least explored or developed places on earth. It provides the best remaining example of the types of organisms that can develop in island isolation.

New Guinea habitats include cold tundra, tropical rain forests, grassy savannas, coastal zones, montane rain forests, cloud forests, and bogs. There are at least twenty thousand species of flowering plants, including more than twenty-five hundred species of orchids, and hundreds of birds and animals. Many New Guinea species are unusual.

Native birds range from the beautiful to the bizarre. There are several species of birds of paradise, some of the most beautiful birds in the world. They have brilliant colors or long, wiry feathers with metallic-looking feather disks that they wave and tremble to attract mates. The bower bird builds large, complicated structures and decorates them with colorful flowers, feathers, or trash, and even uses berry juice as paint. The megapode tunnels into the earth near volcanic hot springs with huge, powerful feet. It lays its eggs in the warm tunnel so that it does not have to sit on them.

New Guinea has several parrots, including the endangered Pesquet's parrot, whose face is completely bald so it can stick its head into fruit without getting sticky feathers. Lorikeets are colorful, nectar-eating parrots. New Guinea is also home to the flightless cassowary, a large bird up to 6.6 feet (2 meters) tall and weighing up to 130 pounds (60 kilograms). Other bird species include feathery crowned pigeons, kingfishers (twenty-two species), ducks, herons, hawks, and egrets.

As in Australia, primates and large mammals never arrived in New Guinea. Marsupials took their ecological places. There are several species of tree-living kangaroos. The basic kangaroo shape is modified in the tree species; they have larger forearms and smaller hind legs. Other marsupials include striped and feather-tailed possums, ringtails, and land wallabies that look like small, dainty kangaroos.

Other odd New Guinea animals are the echidnas, perhaps the most primitive mammal in the world. Echidnas lay eggs and are related to the duck-billed platypus. There are several echidna species; the largest is the giant spiny anteater, which has long spines, dense fur, thick claws, and a long, slender snout. Despite its name, it eats mostly earthworms, which it reels into its mouth using spines on its tongue.

New Guinea has many types of fruit bat. One large species, the flying fox, roosts in colonies. The largest roost is in a tree directly in front of the main police station in the town of Madang. Thousands of bats roosting overhead cause significant problems in street cleaning and for pedestrians without hats.

Reptiles and amphibians on New Guinea consist of snakes, lizards, frogs, and toads. The Salvadori monitor lizard is the longest in the world, growing up to sixteen feet (five meters) long, most of which is tail. Crocodiles live in many rivers and are endangered from hunting and demand for their hides. Some locals have begun crocodile ranches, where they breed and raise the crocodiles for meat and hides without damaging

Biological Mass Murder

The brown tree snake vividly illustrates the dangers of introducing an exotic species to an island ecosystem. The brown tree snake, a native of Australia, was accidentally introduced to Guam after World War II. It reproduced in incredible numbers and reached the highest density of any snake population on Earth, with up to sixteen to twenty thousand snakes per square mile. In the late 1970's, biologists noticed that birds were disappearing from Guam. Of the fourteen bird species endemic to the island, at least nine had become extinct by the 1990's. Many small animal species, as well as chickens raised by the island's inhabitants, also disappeared. The snakes harm humans as well; they bite viciously and cause frequent power outages by climbing on power lines. Millions of dollars have been spent on unsuccessful attempts to control the brown tree snake.

wild populations. Snakes include adders and pythons (eight species), and the deadly taipan and Papuan black snakes. There are many species of frogs, including several odd rain forest species that no longer have a tadpole stage.

New Guinea's insects include the largest moth in the world and the largest butterfly, the Queen Alexandra's birdwing. This species is endangered due to collecting and loss of habitat, but some New Guineans run butterfly ranches. They raise the insects, sell them to collectors and museums, and still preserve the species. There are also giant millipedes, stick insects twelve inches (thirty centimeters) long, and a fly with its eyes on stalks. Several species of flies have antlers on their heads, which they use to fight in defense of egg-laying territory.

Many forests host "ant-plants," warty looking epiphytes that have hollow mazes inside their tissues. Ants live in the maze, safe from predators. The ants provide nutrition for the plant in the form of droppings, scraps of food, and dead ants.

Even though New Guinea is rugged and isolated, human impact is increasing. It has proven difficult to develop New Guinea economically without destroying the unique life of the island. It is hoped that lessons learned on other islands, such as Guam and New Zealand, may be applied to New Guinea. The government has tried incentives to keep wild areas wild, such as encouraging ecologically friendly businesses like crocodile and butterfly farms and ecotourism. The National Park reserve that includes Mount Jaya is the only place in the world where it is possible to visit a glacier and a coral reef in the same park.

New Zealand

Located off the eastern edge of Australia, New Zealand has a fairly moderate climate that comes from conflicting warm, humid Pacific and colder Antarctic weather. It is similar to New Guinea, with rugged terrain, high mountains, and habitats from grassy open plains to dense forests, wet areas to near-deserts. Unlike New Guinea, however, New Zealand has been occupied and developed by humans for hundreds of years. Before large-scale agriculture, about half of New Zealand was covered with forests and one-third with grassland communities. Now, half is pasture for grazing and a quarter is forest, mostly introduced species. Much of the remaining native forest is maintained as national parks and reserves. Pasture land usually consists of a single species of grass and does not support the wide variety of bird and animal life of the original grassland communities.

It is sometimes said that the dominant mammal in New Zealand is the sheep. Sheep are vital to the economy but have caused problems for native animals; more than six hundred native species are threatened and several are extinct, including 43 percent of New Zealand's frogs and more than 40 percent of its native birds.

Like New Guinea and Australia, New Zealand did not originally have any large mammals or carnivores. Fourteen flightless or weak-flying birds developed there. The native Maori people hunted the large, flightless moa to extinction before the Europeans arrived, one of the rare instances in recent history when humans not from a Western culture were responsible for a species' extinction.

Flightless birds that still survive in New Zealand are the kiwi and the kakapo parrot. The introduction of cats, dogs, rats, and pigs has severely endangered both species, and at the end of the twentieth century, there were fewer than seventy kakapo left. All had been moved to predator-free, offshore islands. The kiwi is New Zealand's state bird, and efforts are being made to save it as well. Two species of another flightless bird, the penguin, breed on the south and east coasts. There are more than two hundred species of flying birds, at least forty of them introduced, including tropic birds, gulls, hawks, harriers, skuas, spoonbills, and pheasant.

The only original endemic mammals were two species of bats. In the last two hundred years, many exotic marsupials and mammals have appeared, including rabbits, rats, mice, weasels, otters, cats, pigs, cattle, deer, goats, and sheep. Domestic cattle, sheep, and pigs are economically important.

Amphibians and reptiles are interesting but

not numerous. Among them is a frog that retains its tail as an adult and one that bypasses the tadpole stage. Some frog species have been introduced but have had little impact. Reptiles include geckos, skinks, and the extremely primitive, lizardlike tuatara.

Invertebrate species include such oddities as a giant carnivorous land snail. There are more than four thousand species of beetles, two thousand species of flies, and fifteen hundred species of butterflies and moths. In a reversal of the usual ecological concerns, some native insects are destroying introduced pasture grasses.

Micronesia

The Federated States of Micronesia consists mainly of small atolls. Coral atolls are found only in tropical latitudes because coral (small, colonial animals) grows only in warm water. Coral reefs support a tremendous variety of fish, crabs, and mollusks. Atolls tend to have porous, infertile soil and to be very low in elevation; the inhabited state of Tokelau, comprising three small islands, has a maximum elevation of sixteen feet (five meters). Fauna usually consists of lizards, rodents, crabs, and other small creatures. Pigs, ducks, and chickens are raised for food, and dogs and cats are kept as pets.

Human disturbances on coral atolls often have been particularly violent; several nuclear test bombs were exploded on Bikini Atoll and other islands in the 1940's and 1950's. Kwajalein, the largest atoll in the world, is used by the U.S. military for intercontinental ballistic missile target practice. Johnston Atoll, about 820 miles (1,320 km.) southwest of Honolulu, Hawaii, is a U.S. military base and storage facility for radioactive and toxic substances. It is also designated as a protected area and bird-breeding ground.

The Prospect for Island Ecologies

Island ecologies are unique and fragile. Some, like those in Guam and New Zealand, can never be returned to their original state, but with extensive wildlife management, many native species can be saved from extinction. Ironically, people in the developed world who visit zoos see more native Pacific island species than those who live on those islands. Some endemic species have been wiped out or had their habitat destroyed, so zoos are the last refuge for many creatures.

New Guinea, Fiji, and many smaller islands are in earlier stages of modern development and wildlife destruction can still be controlled. Conservationists sometimes do not realize that islands are not small, geographical zoos; people live there and want to improve their lives. Development cannot be stopped, but it can be managed so that the humans can improve their standard of living as they wish, and the original, amazing island dwellers can still survive.

—*Kelly Howard*

See also: Amphibians; Bats; Birds; Butterflies and moths; Centipedes and millipedes; Coral; Crocodiles; Ecosystems; Endangered species; Fauna: Asia; Fauna: Australia; Fish; Flies; Frogs and toads; Grasslands and prairies; Habitats and biomes; Insects; Lizards; Marine animals; Marsupials; Mollusks; Mountains; Ostriches and related birds; Parrots; Penguins; Rain forests; Reefs; Snails; Snakes; Swamps and marshes; Tidepools and beaches; Wildlife management.

Bibliography

Adams, Douglas, and Mark Carwardine. *Last Chance to See*. New York: Ballantine Books, 1990. An account of a year-long journey to seek out endangered species of the world, including several in Oceania.

Eliot, John L. "Bikini's Nuclear Graveyard." *National Geographic* 6 (June, 1992): 70-83. Discusses the ecology of Bikini Atoll in the aftermath of nuclear testing.

Flannery, Tim. *Mammals of the Southwest Pacific and Moluccan Islands*. Ithaca, N.Y.: Cornell University Press, 1995. Identifies, classifies, and describes every living native mammalian species of the region as well as introduced and prehistorically extinct species, the result of a five-year field survey. Color and black-and-white photos, glossary.

Jaffe, Mark. *And No Birds Sing: The Story of an Ecological Disaster in a Tropical Paradise*. New York: Simon & Schuster, 1994. An investigation, written by an environmental journalist for the *Philadelphia Inquirer*, into the sudden disappearance of the birds of Guam. A cautionary tale about the danger of introduced species.

Malloy, Les, and Gerald Cubitt. *Wild New Zealand*. Cambridge, Mass.: MIT Press, 1995. A beautifully photographed book on the flora and fauna of New Zealand. Includes maps and a list of major conservation areas.

FAUNA: SOUTH AMERICA

Types of animal science: Ecology, geography
Fields of study: Conservation biology, ecology, environmental science, wildlife ecology, zoology

South America, stretching from the tropics of southern Mexico and Central America to the Antarctic, encompasses every type of climate and habitat, and as a result has a correspondingly diverse animal population.

Principal Terms

GRAZER: an animal that eats grass; some are wild, but many grazers have also been domesticated

HERBIVORE: an animal that only eats plants

NOCTURNAL: active at night and dormant or asleep during the day

PREDATOR: an animal that obtains food by hunting other animals

A wide range of animals, both ordinary and exotic, inhabit the continent of South America. The types of animals found in any geographical area are determined by the climate and the terrain. In the Andes, animals such as the llama have adapted to the terrain and climate of the high, steep mountains. North of Antarctica, in the archipelago known as Tierra del Fuego, there are many penguins, whose layers of feathers help them to survive the frigid sea waters in that area.

Camelids of the Andes

Four members of the camel family live in the Andes Mountains of Peru, Bolivia, Ecuador, and Chile: alpacas, vicuñas, guanacos, and llamas. These camelids all have commercial value in the Andes, as pack animals and for their meat and fur. Many people of the Andes also raise sheep on the mountains for both wool and meat.

More than four thousand years ago, alpacas, which are raised for meat and for their fine cashmere fur, were reserved for the exclusive use of the Incas, who prized their coats. Alpacas live from fifteen to twenty-five years. The average adult is about three feet (one meter) high at the shoulder and weighs up to 180 pounds. In 1999, there were approximately three million alpacas in South America, mostly in Bolivia, Chile, and Peru.

Closely related to the alpaca is the llama, domesticated as a work animal more than three thousand years ago. Llamas, used primarily in Peru and Bolivia, have historically been the beasts of burden in the Andes Mountains. A single animal can carry about two hundred pounds (ninety kilograms) for twelve hours a day. However, they cannot be ridden, and when they tire, they often simply lie down and refuse to move. They even spit at their drivers when they no longer want to work. Reaching heights of nearly 4.5 feet (1.5 meters) llamas are generally larger than alpacas. A llama's fur is usually white with black and brown, but some are pure white and others pure black. Although llamas are used for work, they are also kept as pets in many homes of South America.

Vicuña fur has historical importance in South America. It was used by Incas to make cloth, and only members of Inca royalty could wear clothing made from this cloth. Anyone else found with such clothing was executed. By 1979, only four thousand vicuñas were left in South America. They had been hunted for their fleeces by poachers who killed the animals. Representatives of the governments of Bolivia, Chile, Ecuador, and Peru signed a treaty for protection of the vicuña. Twenty years later, there had been a resurgence of vicuñas in Peru, Chile, and Argentina: in 1999, there were 103,000 in Peru, 30,000 in Argentina, 16,000 in Chile, and a small transplanted herd in Ecuador. Even though *el chacu*, the communal hunting of vicuña, continued after 1979, laws lim-

Habitats and Selected Vertebrates of South America

ited these hunts to local people. Those local hunters sold the fiber from the animals as an important source of income for their families.

Vicuñas weigh ninety pounds and are a little less than three feet (one meter) in height at the shoulder. They have long necks, slender legs, padded cloven feet, large round eyes, and a fine, dense, tawny coat. Aside from their economic value, the vicuña is valuable for scientific study. They are highly communicative animals, signaling each other with body postures and ear and tail positions. They emit soft humming sounds as symbols of bonding and greeting.

The fourth member of the camel family found in South America is the guanaco. The guanaco is more adaptable than the other three camelids. It is found throughout the Andes, in the dry Atacama Desert of Chile, and in Tierra del Fuego, where it rains year-round. This animal, from which the llama was domesticated, began life in the semiarid desert and has developed physiological mechanisms for coping with both heat and dehydration. It is similar in structure to the other camelids, but is the largest member of the camel family living in South America.

Other Andean Mammals

Throughout the Andes, from Argentina to Colombia, and on into Central America, pumas roam. This reddish brown feline can reach lengths of about 6.5 feet (2 meters), not including its long tail. In some areas of South America, the puma is endangered. It is a carnivore whose natural prey are elk, deer, and small wild animals; however, it also eats sheep and cattle. Thus, ranchers have retaliated by killing the predators. Another member of the cat family is the elusive Andean mountain cat. Rarely sighted by humans, it is the least-known New World cat. The Andean wild cat is considered sacred by the native people of the *altiplano* of Bolivia, the Aymara.

In the forests of the Andes lives the spectacled bear. Its range extends as far north as Ecuador. This bear, which is endangered because of overhunting and destruction of its environment, has a shaggy brown coat with yellow facial markings and a cream-colored muzzle, throat, and chest.

Several rodents are native to the Andes. Chinchillas were found living in crevices in the mountains when early Spanish explorers first arrived there. Living off berries and fruits in Peru, Chile, and Argentina, these rodents belonged to Inca royalty, who used their fur to make chinchilla stoles. In the latter part of the twentieth century, they were nearly extinct in the wild but existed in captivity. Related to the chinchilla is the viscacha, which is prey for such animals as the Andean mountain cat. Mountain viscachas have long, rabbitlike ears, and long, squirrel-like tails. East of the mountains lives the plains viscacha, which has shorter ears and a blunter head. The cavy, the South American guinea pig, lives in the crevices of the Andes.

Andean Birds

Various exotic birds also live in the Andes, many of which also are found in the Amazon Basin to the east. Among these birds are the Andean cock-of-the-rock, the scarlet macaw, the quetzal, the Andean condor, and the James flamingo. The cock-of-the-rock is a huge dancing bird found in the mountain forests. The scarlet macaw, a brilliantly plumed member of the parrot family, is an endangered species. The quetzal had religious significance to early Andeans; even at the beginning of the twentieth-first century, it is regarded as a symbol of the Andes Mountains. The Andean condor is found in the high plains area of Bolivia and Chile. With a wingspan of twelve feet (four meters), it is the largest flying bird in the world. It can soar to a distance of 26,000 feet (7,925 meters) above sea level.

In the southern Andes lives the rhea, a flightless bird related to the ostrich, which often is called the South American ostrich. The rhea is much smaller than an ostrich and has three toes on each foot, whereas an ostrich has two toes on each. Rheas live in flocks of twenty to thirty in Brazil's southern plains and in Argentina, Paraguay, and Uruguay.

Animals of the Amazon Basin

Representatives of almost one-fourth of all known varieties of animals live in South America, mostly in the Amazon Basin. The basin includes the rain forests, plateaus, rivers, and swamps southeast of the Andes Mountains.

The tapir, found in the Andes and in the forests east of the Andes, is South America's largest animal. With its short, hairy body, it resembles a pig, but is actually related to the horse and the rhinoceros. The number of tapirs in South America lessened over the twentieth century because they were hunted for their flesh and their thick hides, and also because the cutting of forest has reduced the land available for their habitat.

The tapir's natural enemy, the jaguar, also is found in the eastern Andes and in the forests east of the high mountains. This feline, which was worshiped by pre-Columbian civilizations as a god, lives in the area between the southern United States and northern Argentina and is especially prevalent in Brazil. Strong swimmers, jaguars like to live near rivers and other streams. At the end of the twentieth century, they were on the list of threatened animals in South America. They were threatened because farmers were farming lands that previously were their natural habitat, and also because farmers, claiming that the cats killed their cattle and sheep, were killing the jaguars.

Several types of foxes are indigenous to the Amazon Basin and roam through lower mountain areas to the east and the west of the Amazon. Among them are the gray fox and the crab-eating zorro. The gray fox roams over the plains, the Pampas, the desert, and the low mountains, but its number is decreasing because farmers are cultivating lands that previously were its habitat. In Argentina, where it has been hunted for its skin, the gray fox has been placed on the endangered species list. In Chile it is protected by law, but enforcement of the law has been lax. The crab-eating zorro is found in Colombia, Venezuela, Suriname, eastern Peru, Bolivia, Paraguay, Uruguay, Brazil, and northern Argentina. An omnivore, it eats not only crabs but also insects, rodents, fruit, reptiles, and birds.

Also living within the Amazon Basin are the giant anteater, the sloth, and the peba. The giant anteater has value to the environment and to farmers because it consumes up to thirty thousand insects per day. The peba, a nine-banded armadillo found widely throughout South America, also contributes to agriculture by consuming insects and worms. This nocturnal animal is protected from its predators by a horn and bony plates covering its body. The sloth is the world's slowest-moving large mammal.

Within the Amazon Basin lives one of the world's most interesting rodents, the huge capybara. It generally is as large as a big dog, but can reach 4 feet (1.2 meters) in length and can weigh as much as one hundred pounds (forty-five kilograms). Humans have little to fear from the amphibious capybara, however, since it is a vegetarian.

The coatimundi inhabits areas from Arizona to northern Argentina. A member of the raccoon family, it is brown or rust-colored. It eats snails, fish, berries, insects, spiders, lizards, birds, eggs, and mice, and is often kept as a pet by South Americans.

The rain forest is also inhabited by many bats, squirrels, and parrots, which eat the fruits and nuts of the upper and lower canopies. In the lower canopy live lemurs, flying squirrels, and marmosets, small monkeys found mostly in eastern Brazil. These animals use their sweat glands for communication. Animals of the lower canopy eat the fruits, nuts, and insects that are found there. Within the rain forests also live vampire bats; these fascinating animals have to have two tablespoons of blood per day in order to survive.

Other residents of the Amazon Basin include the yapok, a member of the opossum family that has webbed feet for swamp travel; the sapajou monkey, a small New World primate; and the octodont, an eight-toothed rodent also known as a spiny rat or a spiny hedgehog because of the sharp spines embedded in its fur.

Birds and Reptiles of the Amazon

Many types of birds live in the Amazon Basin, some deep within the rain forest, others closer to the mountains. Among the birds in the basin are hummingbirds, parrots, ospreys, macaws, boat-billed herons, great egrets, white-necked herons, least bitterns, and blue and yellow macaws. Tou-

The Anaconda

Another name for the anaconda is the water boa, an appropriate name for a snake that is almost always found near water. Anacondas live in the Amazon and Orinoco basins of tropical South America, and their habitat extends to Trinidad. Like the crocodile, the anaconda has nostrils high on its snout so that it can swim with its head above water to breathe. The anaconda lies near the shore, waiting for its prey. When a deer, bird, or other prey comes to the water to drink, the anaconda quickly strikes, dragging its victim underwater to drown it. The anaconda then eats the unfortunate animal whole. A good meal can last an anaconda for several weeks, during which it will lie in the water, digesting its food.

cans, which also live at high elevations up to 10,000 feet (3,050 meters), can be found deep within the Amazon rain forest.

Both the land iguana and the lava lizard live in the Amazon Basin. Land iguanas can live up to twenty-five years and weigh up to 15 pounds (6.8 kilograms). They eat low-growing plants, shrubs, fallen fruits, and cactus tree pads. The lava lizard, about 1 foot (30.5 centimeters) long, is smaller than the land iguana. Lava lizards are beneficial to agriculture because they eat beetles, spiders, and ants.

The boa constrictor also lives in the jungles of this basin. This snake, which is usually six to nine feet (two to three meters) in length, but can reach thirteen feet (four meters), kills its prey by squeezing it to death, using its body coils to suffocate its victims. After killing its meal, the boa constrictor stretches its jaws wide apart and pulls the entire victim into its mouth. Using this method of killing, a boa is able to eat animals that are much larger than its head.

Another large snake native to the Amazon Basin is the anaconda. Most anacondas weigh several hundred pounds (100 kilograms) but can reach weights of 550 pounds (250 kilograms) and can reach thirty-six feet (eleven meters) in length. The anaconda is found in the Guyanas and throughout tropical South America, east of the Andes. With eyes high on its head, the anaconda can submerge its body in water and watch for the approach of unsuspecting prey.

River Animals

Many animals spend all or part of their lives within the Amazon. The black caiman, an alligator that is nearly extinct, is one such animal. It can weigh as much as a ton. It will eat all vertebrates, including humans, if that is the only food available.

The semiaquatic brown water lizard also is found in the jungle area around the Amazon. Within the Amazon are manatees as well as Boto River dolphins, also known as an Amazon River dolphin. This endangered animal is the only dolphin to have a neck. Giant otters live in the waters of the Amazon as do many types of fish, including piranhas, which, with their sharp teeth, can quickly strip the flesh from their prey.

Animals of the Eastern Highlands

The Eastern Highlands of South America host many unique animals, including the bush dog, woolly tree porcupine, maned wolf, peccary, bushmaster, and coypu rat, and many birds such as flamingos.

The bush dog is a wild dog, but, with its webbed feet, it resembles an otter more than a dog. Bush dogs live in packs and hunt small deer and rodents. Bush dogs can be found from the rain forests into the grasslands, in Colombia, Venezuela, the Guyanas, Brazil, Paraguay, northeastern Argentina, eastern Bolivia, and eastern Peru.

The maned wolf is one of South America's most beautiful and revered animals. Weighing on average one hundred pounds (forty-five kilograms), it is South America's largest canid. It is found mostly in Argentina and Brazil and has no natural enemies. An omnivore, it will eat nearly anything, including fruits, insects, and small vertebrates. At the end of the twentieth century, the Smithsonian Institution estimated that fewer than ten thousand maned wolves existed in the wild, living mostly in Argentina and Brazil. Because of its beautiful red and gold fur, the maned wolf is a tourist attraction. Many South Americans regard it as an important part of their cultural heritage. The rural people of the Sierra de Canastra of Brazil believe that the maned wolf has medicinal and supernatural powers.

The peccary and the bushmaster are also found in the Eastern Highlands. The peccary resembles the tapir, but is much smaller. It has a big head, sharp teeth, and prickly fur. It eats smaller animals and plants such as cactus flowers. The bushmaster, the largest poisonous snake in the Americas, is a type of pit viper related to the rattlesnake. Like the rattlesnake, it shakes its tail before striking, but it has no rattles. Gray and brown with a diamond pattern, it averages 8 to 12 feet (2.5 to 5.5 meters) in length.

The coypu rat is also known as a swamp beaver. This relative of the muskrat is found in southern Brazil, Bolivia, and Colombia. The agouti, a rodent nearly two feet (sixty centimeters) long,

Baron Alexander von Humboldt

Born: September 14, 1769; Berlin, Germany
Died: May 6, 1859; Berlin, Germany
Fields of study: Environmental science, wildlife ecology, zoology
Contribution: Von Humboldt was a great German naturalist and explorer who founded the field of physical geography. He also studied the relationship between regional geography and its flora and fauna.

Alexander von Humboldt was born in Berlin, Germany, the son of an army officer. After his father's death in 1779, his mother raised him in the family castle at Potsdam. Von Humboldt was privately tutored to prepare him for a life in public service. He began studying economics at the University of Frankfurt, then switched to engineering. At that time, von Humboldt became interested in botany.

Study at the University of Göttingen turned von Humboldt's interest to geology, and he studied at the Frieberg School of Mines for two years. This led to an appointment to the Prussian Department of Mining. In this position, he reorganized numerous mines and supervised mining activities. In 1797, von Humboldt resigned this post to study meteorology and geomagnetism. He obtained permission from Spain to visit its Central and South American colonies.

In 1799, his self-financed expedition set off in a small ship, the *Pizarro*. He was accompanied by botanist Aimé Bonpland. Over the next five years, the expedition thoroughly explored Central and South America. First, it proved that the Casiquiare River connected the Amazon and Orinoco rivers. Next came exploration of the Andes, including von Humboldt's climb to near the top of Mount Chimborazo, a world record that stood for many years.

In 1803, the *Pizarro* sailed to Mexico for a study of that country. Then it returned to France with a huge collection of plants and data on physical geography. From 1804 to 1827, von Humboldt processed the data. In Paris, he published thirty volumes of findings. Meteorological data included weather maps which helped to found comparative climatology, studies of the relationship between regional geography and its flora and fauna, and proof that the earth's surface was not formed by sedimentation from the liquid state.

In 1827, von Humboldt returned to Berlin and the royal court, where he tutored the crown prince, was on the privy council, and was court chamberlain. In 1829, he was invited to Russia to examine gold mines and advise on mining techniques. This experience produced important geographical, geological, and meteorological observations on Central Asia. Von Humboldt, also hugely interested in fluctuations of Earth's geomagnetic field (magnetic storms), saw that a world net of observatories was needed to identify their origin. In 1836, he convinced the British Royal Society to establish such stations. Data obtained proved that the storms were due to sunspot activity.

Amid these other activities, over twenty-five years von Humboldt wrote four volumes of *Kosmos*, a work which gave a clear account of the structure of the known universe. A great success, it was translated into every European language. At age eighty-nine, while working on the fifth volume, von Humboldt died in Berlin.

—Sanford S. Singer

also lives in the Eastern Highlands. Farmers detest the agouti because it eats sugar and banana plants. A cousin of the jaguar, the ocelot, also makes its home in this area. This slender cat is camouflaged in the forests and deserts of the highlands.

Birds that live in the Eastern Highlands include the James flamingo, which lives on Bolivia's frigid salt lakes, and the giant antshrike, more than a foot (30 centimeters) in length.

Animals of Tierra del Fuego

In the islands that make up Tierra del Fuego, many unusual animals are found, including penguins and many other types of birds. Penguins cannot fly; they use their wings for swimming in the icy waters near their home. Penguins are insulated from the frigid ocean waters by three layers of short feathers and an underlying layer of fat. Other birds common to Tierra del Fuego are Magellanic cormorants, imperial cormorants, albatrosses, and various petrels. Sea lions also live on these islands.

—Annita Marie Ward

See also: Armadillos, anteaters, and sloths; Bats; Bears; Birds; Camels; Cats; Crocodiles; Deer; Deserts; Dogs, wolves, and coyotes; Dolphins, porpoises, and toothed whales; Donkeys and mules; Ecosystems; Endangered species; Fauna: Antarctica; Fauna: Caribbean; Fauna: Central America; Fish; Flamingos; Foxes; Frogs and toads; Habitats and biomes; Jaguars; Lakes and rivers; Lizards; Mammals; Manatees; Marine animals; Marsupials; Monkeys; Mountains; Opossums; Ostriches and related birds; Parrots; Penguins; Porcupines; Raccoons and related mammals; Rain forests; Reptiles; Rodents; Shrews; Snakes; Tidepools and beaches; Wildlife management.

Bibliography

Dunphy, Madeline. *Here Is the Tropical Rain Forest*. Westport, Conn.: Hyperion Press, 1997. Written for a juvenile audience, uses a cumulative format to illustrate the connections between animal life and environment.

Eisenberg, John F. *Northern Neotropics: Panama, Colombia, Venezuela, Guyana, Suriname, French Guiana*. Vol. 1 in *Mammals of the Neotropics*. Chicago: University of Chicago Press, 1989. A taxonomically organized guide to South American mammals, covering external measurements, a physical description, geographical distribution, information on life history, notes on ecology and behavior, and distribution maps.

Eisenberg, John F., and Kent H. Redford. *The Central Neotropics : Ecuador, Peru, Bolivia, Brazil*. Vol. 3 in *Mammals of the Neotropics*. Chicago: University of Chicago Press, 2000. A taxonomically organized guide to South American mammals, covering external measurements, a physical description, geographical distribution, information on life history, notes on ecology and behavior, and distribution maps.

Kalman, Bobbie. *Rainforest Birds*. New York: Crabtree, 1998. A guide to Amazonian birds for children, well illustrated by Barbara Bedell.

Matthews, Downs. *Beneath the Canopy: Wildlife of the Latin American Rain Forest*. San Francisco: Chronicle Books, 1999. Primarily a photo book with captions, but shows many animals very rarely caught on film.

Redford, Kent H., and John F. Eisenberg. *The Southern Cone: Chile, Argentina, Uruguay, Paraguay*. Vol. 2 in *Mammals of the Neotropics*. Chicago: University of Chicago Press, 1992. A taxonomically organized guide to South American mammals, covering external measurements, a physical description, geographical distribution, information on life history, notes on ecology and behavior, and distribution maps.

Savage, Steven. *Animals of the Rain Forest*. Austin, Tex.: Steck Vaughn, 1995. Maps and photographs illustrate a simple introduction to rain forest life.

FEATHERS

Types of animal science: Anatomy, evolution
Fields of study: Anatomy, archaeology, ornithology

Feathers are a unique physical characteristic of birds. There are many types of feathers serving a variety of purposes, including flight, insulation, communication, breeding, and camouflage.

Principal Terms

KERATIN: fibrous proteins that are the chemical basis for feathers
MOLT: the process of replacing feathers
PIGMENTS: a variety of colored substances which impart color to feathers
SHAFT: long, central spine of the feather
VANE: flat, broad web emerging from opposing sides of the feather shaft

In 1860, a quarryman from southern Germany discovered a fossil preserved in limestone revealing a single, 2.5-inch-long feather. This feather was asymmetric, with the vane on one side of the quill being twice the width of the vane on the other side, and similar to the flight feathers of modern birds. A year later, a fossil was discovered in a nearby quarry which demonstrated a feathered creature named *Archaeopteryx*, which means "ancient wing." *Archaeopteryx* is generally regarded as the first bird because of its feathers, and it flew 150 million years ago.

Structure and Color of Feathers

Feathers continue to define birds. They are the only animals that have feathers, and all birds have some type of feathers. Feathers result from a modification of the outer layer of the bird's skin, the epidermis, and probably evolved from scales as found on reptiles, such as snakes and lizards. Feathers are composed largely of keratin, an inert substance that is light, strong, and long-lasting. The developing bird inside the egg is covered with bumps called papillae, and the epidermis folds inward around each of the papillae to form follicles. From these structures feathers are grown and regrown. The number of feathers on birds varies greatly, from less than 1,000 on a ruby-throated hummingbird to over 25,000 on a tundra swan.

A typical feather has a central shaft with two vanes arising from opposite sides. On flight feathers, the vane from the leading edge of the feather is narrower and more rigid, thereby maintaining the wing structure and producing the lift necessary for flight. Amazingly, this basic feather structure has remained unchanged since *Archaeopteryx* flew above the dinosaurs of the late Jurassic period. The portion of the feather shaft within the follicle and beneath the skin is called the calamus. The portion of the shaft above the skin is called the rachis, from which extend numerous opposing parallel branches, called barbs, that collectively form the vanes. Hooked structures called barbules join the barbs together, producing a smooth, sheetlike surface.

Color is imparted to feathers by colored substances called pigments and by variations in feather structure, which alter the manner in which light is reflected. There are three types of pigments. The most common pigment is melanin, which produces black, a spectrum of brownish shades, and light yellow. Examples include the black of a crow or yellow of a baby chicken. Birds synthesize melanin from dietary proteins. Carotenoids are produced by plants and are ingested by birds. They produce red, orange, and yellow feathers, such as the red of the northern cardinal. Porphyrins are metabolic breakdown products produced by birds from hemoglobin, the oxygen-carrying component of red blood cells, when red blood cells become old and are broken down by the liver. Porphyrins produce a variety of colors, including red, brown, green, or pink. Porphyrins produce the brown feathers of many owls. Structural properties of the feathers, such as layers, often containing pigments, and microscopic air bubbles, produce iridescent colors, such as the colored throats of hummingbirds, or noniridescent colors, such as the blue of a bluejay. Often the multiple colors of a bird's feathers result from some combination of both pigment and structure.

A fossil dromaeosaur that lived 130 million years ago, discovered in northeastern China in 2000, shows evidence of downy fluff and primitive feathers. (AP/Wide World Photos)

Functions of Feathers and Molting

While flight is certainly the most outstanding function of feathers, they serve many important functions, enabling birds to exist and thrive in environments all over the globe. Feathers provide insulation by trapping air and thereby helping to control body temperature. The color and form of feathers help birds communicate and signal breeding, as well as keep them camouflaged from predators. Adults sitting on a nest or fledgling birds often depend greatly on their feather coats to make them less visible against the background of the environment. Feathers also help birds remain waterproof while swimming or diving. Some feathers help birds remain clean, and others help the birds to support themselves, as can be seen when the stiff tail feathers of a woodpecker act as a strut against the trunk of a tree. Specially modified feathers serve as sense organs able to detect feather position and movement essential for the complexities of flying.

A fully developed feather is a dead structure and when it becomes worn or broken it must be replaced, as repair is not possible. Molting is the process of replacing all (complete molt) or a portion (partial molt) of the feather coat. This process varies widely among birds in timing as well as completeness and annual number. However, most adult birds undergo a partial molt prior to breeding season. This partial molt frequently yields bright colorful feathers and is called the alternate plumage. A good example is the bright yellow and black of the American goldfinch, seen in the spring. A complete molt occurs after the breeding season and results in the basic plumage. The basic plumage of the American goldfinch is a drab olive. This complete molt provides some migrating birds with new feathers for the long flights to their home territories. Other birds get new feathers to help insulate them from the winter cold.

The beauty of feathers and the wonder of flight have attracted the attention of humans throughout history. Feathers have adorned priests and warriors as well as the hats of fashionable women. In 1905, Guy Bradley, a warden hired to patrol wading-bird rookeries in southern Florida, was shot and killed by plume hunters. He is buried in Everglades National Park and his death resulted, in 1913, in stricter laws to protect birds in the United States and a change in public opinion that abolished the use of feathers in hats sold in America. It is currently illegal to possess wild bird feathers in the United States.

—H. Bradford Hawley

See also: *Archaeopteryx*; Birds; Evolution: Animal life; Flight; Fur and hair; Insects; Locomotion; Molting and shedding; Scales; Velociraptors; Wings.

Bibliography

Clark, George A., Jr. "Form and Function: The External Bird." In *Home Study Course in Bird Biology*. 2d ed. Ithaca, N.Y.: Cornell Laboratory of Ornithology, 2000. An excellent chapter in the newly revised text produced for a popular college-level ornithology course, which is taken at home with mail-in tests.

Feduccia, Alan. *The Origin and Evolution of Birds*. 2d ed. New Haven, Conn.: Yale University Press, 1999. This outstanding, extensive review of avian evolution, including feathers, advocates feathered reptiles as the ancestors of birds. The author believes that early birds climbed trees and began to fly by gliding from tree to tree and to the ground.

Gill, Frank B. *Ornithology*. 2d ed. New York: W. H. Freeman, 1995. This is the gold standard of basic ornitholgy textbooks.

FERTILIZATION

Types of animal science: Development, reproduction
Fields of study: Cell biology, developmental biology, embryology, reproduction science

Fertilization occurs when the genetic information in a haploid sperm combines with the genetic information of the haploid ovum to form the diploid zygote.

Principal Terms

CORONA RADIATA: the layers of follicle cells that still surround the mammalian egg after ovulation
VITELLINE ENVELOPE: the protective layers that form around the egg while it is still in the ovary
ZONA PELLUCIDA: mammalian protective layer analogous to the vitelline envelope

For fertilization to occur, several things must happen: Sperm and eggs must be in close proximity, the gametes need to be compatible, the sperm must be able to penetrate the egg, and the haploid egg nucleus must combine with the haploid sperm nucleus. If any one of these is missing, fertilization will not occur.

Assuring Eggs and Sperm Are in Proximity

Animals have many mechanisms to ensure that sperm and eggs are in close proximity. This can be a major concern for aquatic organisms with external fertilization, and many release gametes in the millions or even billions to assure that at least some sperm reach the appropriate eggs. To increase the chances of a meeting between same-species gametes, animals often have specialized mating behaviors. Corals are among those animals that release their gametes into the water and depend on currents to bring egg and sperm together. This is not, however, as random as it may seem. As the first coral releases its gametes, it also releases hormones that induce nearby corals of the same species to release their gametes. These also release the same chemicals with their gametes, and soon there are clouds of eggs and sperm, and the chances of a proper meeting are increased dramatically. One species of polychaete annelid, *Eunice viridis*, or the palolo worm, has another method of assuring male and female gametes are in the same place. In this species, sexually mature worms called epitokes swarm together at the ocean's surface in response to the lunar cycle. Females then secrete a hormone that induces males to release sperm, and the sperm induce the females to shed eggs. Many fish go through elaborate courtship rituals, during which males and females release gametes at a specific point, thus assuring that egg and sperm are together. Other fish build nests where females lay eggs and males deposit sperm. Frogs and toads usually breed in the water, but the female will only release her eggs when the male is clasped to her back in amplexus. Thus, sperm are deposited on the eggs as they are being laid.

Males of other species place sperm directly in the female's reproductive tract. The male octopus has a special tentacle that is used to place one of his sperm packets in the mantle cavity of the female. Some salamander males deposit their sperm packets on the substrate during a squat dance courtship ritual. Females also do the squat dance and pick up the packet with the lips of their cloacae. In some species of water mites, females mount a special saddle-shaped extension of the males' abdomen. The male squats to deposit a sperm packet, moves ahead slightly and then squats again when the opening of the female's reproductive system is over the packet, forcing the packet into her reproductive system. Another interesting way to assure fertilization is seen in the sea horse. In these animals, a female deposits her eggs into a pouch on the male's abdomen and the male releases sperm into the pouch at the same time. The most common way to introduce sperm into a female's reproductive tract is through copulation, where the male ejaculates sperm directly into the female's reproductive tract. The motile sperm then travel to the egg. For a sperm to gain full motility, it usually must undergo a little-understood process called capacitation.

Penetration

Once eggs and sperm are in close proximity, the sperm must begin to penetrate the egg's protective layers. All eggs have at least one protective layer outside the cell membrane. Called the vitelline envelope in most organisms, it is synthesized in the ovary and composed primarily of polysaccharides and glycoproteins. The oviducts and uterus often secrete other protective layers around the egg. In some instances, the sperm must also penetrate these layers, for example, the jelly layers that surround sea urchin and frog eggs. In other instances, the egg is fertilized before these layers are added, as is the case with the many protective layers that surround bird and reptile eggs. A protective layer made up of cells is seen in most mammals, since the egg is released by the ovary with cells of the cumulus oophorus still attached. For the sperm to penetrate these layers, its acrosome must contain the appropriate enzymes to lyse (disintegrate) the chemicals that block its way. The acrosomal reaction must also take place in order to expose the digestive enzymes of the acrosome. This reaction depends on changes in membrane permeability to ions and subsequent changes in pH.

Once through the protective layers, the sperm makes contact with the egg's plasma membrane. If the sperm and egg are of the same species, sperm receptor molecules on the egg membrane attach to complementary molecules, called bindins, on the sperm membrane and the two membranes fuse. If the bindins on the sperm do not complement the receptors on the egg, there is no fusion and fertilization does not continue, thus preventing most interspecies crosses. However, closely related species often have bindins and receptors sufficiently alike to allow some fertilization to proceed. The products of these interspecific matings are hybrids, such as the mule.

Once the first sperm fuses with the egg, mechanisms to prevent polyspermy, the fertilization of an egg by more than one sperm, are put into place. The first block to polyspermy is common to most animals studied: a very quick and only temporary depolarization of the plasma membrane. In sea urchins, the resting membrane potential of the egg plasma membrane is approximately –70 millivolts, the inside being more negative than the outside. Fusion of the sperm plasma membrane with the egg cell membrane causes a rapid influx of sodium ions. The positive charges neutralize negative charges in the egg until the membrane potential is raised to +10 millivolts. All this happens in less than five seconds, and lasts for about one minute before the egg cell has actively transported enough sodium out of the cell to repolarize it. While the cell is depolarized, no further sperm membranes can fuse with the egg membrane. This is often referred to as the fast or temporary block to polyspermy and seems to occur in all animals thus far studied. The fast block also sets into motion the slow or permanent block to polyspermy. The changed membrane potential of the fast block and the release of nitrous oxide by the sperm allows cells to release calcium ions from storage. The initial calcium ion release causes the egg to release nitrous oxide, which then increases the egg's release of calcium ions. The release of calcium ions induces the cortical reaction by which cortical granules move to the surface of the cell, fuse with the cell membrane, and empty their contents into the space between the cell membrane and the vitelline envelope. In sea urchins, the first acrosomal enzymes released break the bonds between the cell membrane and the vitelline envelope. In the presence of water, other chemicals released by the cortical granules swell, lifting the vitelline envelope away from the cell membrane. Finally, other enzymes released by the cortical granules alter the vitelline envelope, knocking off any attached sperm and causing the release of peroxide ions, which harden the envelope, making it impermeable to sperm. This impermeable barrier is renamed the fertilization membrane. The released peroxide may also provide another benefit. Any sperm that had penetrated the vitelline envelope before it hardened would be killed by the peroxide and would thus not lead to polyspermy. In other animals studied, although cortical granules do empty their contents into the perivitelline space, the permanent block to polyspermy does not seem to involve the same extensive changes to the vitelline envelope (or *zona pellucida* in mammals) that are seen in the sea urchin. In large, yolky eggs, some polyspermy does occur, but the extra sperm remain in the yolk and never reach the egg nucleus for fusion.

Cell Metabolism and Meiosis

Concomitant with the cortical reactions is an increase of metabolism in the egg, which will be necessary for nuclear fusion and cleavage. In species where the egg has not completed meiosis, it does so at this time. Which parts of the sperm enter the egg is dependent on the species. In many mammals, the entire sperm enters, while all but the tail enters in echinoderms. In other organisms, the head with the nucleus and centrioles seem to be the only things that enter. There is no evidence that any parts of the sperm other than the nucleus and centrioles are used by the zygote, and other parts that enter most probably degenerate and their components are recycled.

Studies on the mitochondria of sperm indicate that soon after entering the egg, the sperm's mitochondria are tagged by ubiquitin, the first step in breakdown and recycling. After entry, the sperm nucleus imbibes water and is converted into the male pronucleus. At the same time, the egg nucleus becomes the female pronucleus. In most animals, the male pronucleus and the female pronucleus fuse to form the diploid zygote nucleus. In some nematodes, mollusks, and annelids, however, the pronuclei remain separate until after the first cleavage division. In a few others, like the copepod *Cyclops*, the pronuclei divide separately for several cleavage divisions.

The fusion of the sperm with the egg nucleus affects many other cellular processes. One of the most interesting is the displacement of some cytoplasmic constituents. These constituents of the egg determine the fate of cells derived from the parts of the egg in which they were located and probably determine the plane of bilateral symmetry. Sperm attachment and entry often causes shifts in the position of the viscous cortical and subcortical cytoplasm, where many of the fate-determining chemicals are located.

—*Richard W. Cheney, Jr.*

In Vitro Fertilization of Endangered Species

Techniques of in vitro fertilization were first developed to aid couples who had not been able to conceive through normal sexual relations. In this technique, eggs were surgically removed from the mother and mixed with the father's sperm in the laboratory. If fertilization took place, one or more embryos were introduced into the mother's uterus in the hope that an embryo would implant and develop into a full-term infant. Almost immediately, these techniques were used in other animals, especially endangered species. It offered many advantages over natural reproduction. In pairs that showed little sexual interest in each other, eggs and sperm could be extracted, mixed in the laboratory, and viable embryos could be introduced into the female's uterus. Also, if there was little genetic diversity in a zoo population, sperm from a donor at another location could be sent and used. By the end of the twentieth century, in vitro fertilization was being coupled with surrogate motherhood. Here, after the embryos are formed, they are introduced into the uteri of females of similar, but not endangered, species. This increases the number of uteri available for the endangered species' reproduction.

See also: Asexual reproduction; Breeding programs; Cleavage, gastrulation, and neurulation; Cloning of extinct or endangered species; Copulation; Courtship; Determination and differentiation; Development: Evolutionary perspective; Estrus; Gametogenesis; Hermaphrodites; Hydrostatic skeletons; Mating; Parthenogenesis; Pregnancy and prenatal development; Reproduction; Reproductive strategies; Reproductive system of female mammals; Reproductive system of male mammals; Sexual development.

Bibliography

Balinsky, B. *An Introduction to Embryology*. 5th ed. Philadelphia: Saunders College Publishing, 1981. A thorough chapter on "Fertilization and the Beginning of Embryogenesis."

Bronson, F. H. *Mammalian Reproductive Biology*. Chicago: University of Chicago Press, 1989. A very complete look at mammalian development. One chapter gives a brief overview of development for each mammalian order.

Carlson, B. *Patten's Foundations of Embryology*. 6th ed. New York: McGraw-Hill, 1996. A comprehensive chapter on fertilization with emphasis on vertebrates.

Kumé, Matazo, and Katsuma Dan. *Invertebrate Embryology*. Translated by Jean C. Dan. Belgrade, Yugoslavia: NOLIT Publishing House for the U.S. Department of Health and Human Services, 1968. Extensive compendium of invertebrate development by phylum.

FINS AND FLIPPERS

Type of animal science: Anatomy
Fields of study: Anatomy, behavior

Vertebrate animals first evolved in an aquatic environment. The necessity to move in a stable and directed manner resulted in the evolution of fins, which impart both stability and lift while swimming. As fishes have evolved, fins and their related structures have become more complex and may play a direct role in starting, stopping, and turning. When certain fish groups moved from an aquatic to a terrestrial environment, the paired fins of fishes eventually gave rise to forelimbs and hindlimbs. When, much later, certain mammalian groups (including whales and dolphins) returned to the oceans, their forelimbs evolved into finlike structures called flippers, which retained the internal bone structure of the limbs from which they are derived.

Principal Terms

FIN-FOLD THEORY: theory that fins initially evolved as long folds of tissue extending around the body

FLIPPER: finlike structures of marine mammals that have evolved from the forelimbs of their terrestrial ancestors

FOUR-FIN SYSTEM: the combined activity of paired fins in some bony fishes that makes them highly maneuverable

HETEROCERCAL: a tail in which the spine extends into the upper lobe, giving a distinctly sharklike impression

HOMOCERCAL: a type of tail at which the spine ends at the base of the tail, which consists of two equal lobes

LEPIDOTRICHIA: modified scales that form the supporting rays of the fins of bony fishes

PECTORAL AND PELVIC GIRDLES: skeletal structures that form a structural base for attachment of the paired fins in fishes, connecting them to the rest of the body's skeleton

One of the characteristic features of aquatic vertebrate animals is the presence of single and paired appendages used for locomotion. In fishes these structures are known as fins and in marine mammals they are known as flippers or flukes (on the tail). Although these structures bear superficial resemblances to each other, there are significant differences in their anatomies, attesting their different evolutionary histories. However, at a deeper level, they are related structures, having evolved from the same basic structures found in the earliest vertebrates. This is a classic example of evolutionary parallelism: Finlike structures have evolved independently in both fishes and marine mammals from a common ancestral structure.

The Evolution of Fins

The earliest vertebrates were elongated aquatic animals, and locomotion in these animals was probably accomplished by an eel-like undulation of the body. The efficiency of this form of locomotion is decreased by unwanted motion resulting in an up-and-down (pitch) or side-to-side (yaw) seesawing and rolling around the long axis of the body. Fins first evolved as stabilizers to resist these motions and increase swimming efficiency. Since their appearance, however, they have also evolved other functions.

There are two theories as to how fins first evolved. The fin-fold theory suggests that the early vertebrates had two paired folds of tissue extending along the side of the body. These folds fused just behind the anus to form a single fin which extended around the tail and up onto the midline of the upper body (dorsal) surface. The theory states that several regions of these fin folds have persisted, resulting in the paired and unpaired fins in modern fishes, whereas the rest of the early fin folds have been lost. As evidence for this theory, the sand launce, *Brachiostoma* (amphioxus) is used as an example. Amphioxus is an animal that is closely related to the vertebrates, and the earliest protovertebrates are believed to have resembled it. It has extensive folds that closely resemble the theoretical fin-folds of the earliest vertebrates. A different theory, the body-spine theory, states that the earliest vertebrates possessed two or more pairs of spines extending from both sides of the lower portion of the body. Fins were then formed when membranes extended from the tip of the spine to the side of the body, rather like a sail extending from the mast of a boat. Internal support structures within the fin (the endoskeleton) developed at a later date. At the current time there is no evidence to conclusively support one theory or the other.

The Uses of Fins

Fins in fishes are either paired, meaning there are equivalent fins on either side of the body trunk, or unpaired, meaning that there is a single fin located on the midline of the upper or lower body. The paired fins are the pectoral fins, generally located just behind the gills, and the pelvic fins, which are located behind and below the pectoral fins. The unpaired fins comprise one or more dorsal fins, located on the midline of the back, the anal fin, located on the midline of the bottom (ventral) surface of the body, behind the anus, and the caudal fin, or tail.

Dorsal fins, as on this killer whale, help maintain stability and straightforward movement while swimming through the water. (Digital Stock)

Other Uses of Fins

Although the principal roles of fins are in locomotion, there are numerous other ways in which they are employed. Many fishes possess fins with stout or serrated spines that serve in defense; often such spines are associated with venom glands. In the elasmobranch fishes (sharks and rays) and the related ratfishes, the pelvic fins of males are modified into claspers, the male organs of reproduction. In the common aquarium swordtail (platy), the unpaired anal fin performs a similar role.

The dorsal fin of the deep-sea anglerfishes has an elongated fin ray that possesses a fleshy lobe at its tip. A fish pursuing this "lure" soon finds itself engulfed in the sharp-toothed maw of the anglerfish.

The remoras and sharksuckers are well known for attaching themselves to larger fishes such as sharks and tunas. Despite their name, these fishes do not attach themselves to larger fishes by "sucking" with their mouths, but rather via a modified first dorsal fin. This fin has been modified into a series of flattened plates that can be retracted by the remora, creating a suction that leaves the fish attached to its host.

The deep-sea tripodfishes (family Bathyperidae) have pelvic fins with extremely long fin rays, set well forward on the body, and long fin rays on the lower lobe of the caudal fin. When extended, these form a three-point landing gear, upon which the fish rests above the oozy surface of the ocean floor.

Finally, the first three pectoral fin rays of the sea robin are covered with taste buds and can be moved independently. As this fish "walks" across the bottom, it is actually tasting the substrate, searching for food.

In fishes, the tail, or caudal fin, provides a large portion of the forward thrust required for moving through the water and has evolved as a specialized portion of the rearward end of the body trunk and spine. The spine itself is composed of repeating skeletal structures, the vertebrae, which together form a flexible column extending through the long axis of the animal. Sharks, sturgeon, and paddlefish possess a heterocercal tail, in which the vertebral column extends into the upper portion, or lobe, of the tail, giving it a distinctly asymmetric and sharklike appearance. The majority of bony fishes possess a homocercal tail, in which the spine ends at the base of the tail and in which the tail is composed of equal-sized upper and lower lobes. A few extinct species of fishes possessed hypocercal tails, in which the spine extended into the ventral lobe of the tail. These animals therefore possessed caudal fins in which the lower lobe was larger than the upper, the opposite of what is seen today in sharks.

The paired and unpaired fins of sharks are solid, broad-based, and relatively inflexible. Like all fins they possess an internal skeleton, which provides structural support as well as attachment for muscles that allow the fin to be moved. These consist of a series of cartilages, known as basals and radials, located in the base of the fins. Long rods of cartilage called ceratotrichia extend from the radials out to the edges of the fin, providing support. The principal role of these fins is to resist the yawing, pitching, and rolling motions generated during swimming. Because of the asymmetric heterocercal tail, the thrust generated in forward swimming extends downward from the upper lobe of the tail through the shark's center of gravity, pushing the animal downward. An important function of the large pectoral fins in sharks is to generate lift, in much the same manner as an airplane's wing. The lift generated by the pectoral fins counters the downward thrust and moves the animal forward through the water. The pelvic fins are specialized in male sharks and rays to form claspers, which serve as the male organs of reproduction in internal fertilization.

The fins of bony fishes are distinctly different from those seen in sharks, although the endoskeleton of the fins also consists of basal and radial bones. The fins of the majority of bony fishes, particularly the paired fins, are much narrower at their base, more flexible, and may play a more direct role in locomotion in addition to stabilizing the fish during swimming. The radials of the endoskeleton are reduced to very small structures located within the muscles at the base of the fin. These articulate with modified scales called lepidotrichia, or fin rays, which extend out to form the main structural elements of the fin. Membranous connective tissue extends between the fin rays, giving them their typical weblike appearance. The vast majority of bony fishes possess this type of fin structure, and they are known collectively as the ray-finned fishes, or Actinopterygii.

The Fins of Bony Fishes

A key development in the evolution of bony fishes was the appearance of a swim bladder, a gas-filled sac in the abdominal cavity that counters the fish's tendency to sink in the water and thus provides it with neutral buoyancy. With the evolution of this structure, it was no longer necessary for the pectoral fins to generate lift, and they could then be employed as brakes to stop forward motion. The pectoral fins of most bony fishes have a large surface area and a narrow base that is inserted into the body wall almost vertically, as opposed to horizontally in sharks. They are therefore admirably suited to act as brakes both singly and together, and greatly increase the ability of the fish to stop or change direction rapidly. The pelvic fins act to counterbalance any pitching or rolling motion generated by the pectorals. The combined activity of the paired pectoral and pelvic fins, often called the four fin system, provides enhanced maneuverability and control. This system is best observed in many types of coral reef butterfly fishes (Chaetodontidae), in which the pelvic fins are inserted almost directly below the pectoral fins. These fishes are highly maneuverable, able to move, turn, and stop quickly and accurately within the complex and constrained multidimensional environment of the coral reef.

A relatively small group of bony fishes possesses paired fins of a different sort. This group of fish, which includes the lungfishes and the coelacanth, is characterized by pelvic and pectoral fins that possess fleshy lobes at their base. The lobes contain muscles and skeletal elements; however, only a single basal bone articulates with the rest of the skeleton of the fish. These lobe-finned fishes (the Crossopterygii) are important because they represent descendants of the lineage that moved from the aquatic environment onto land. The pectoral and pelvic fins of these fishes gave rise to the forelimbs and hindlimbs of terrestrial vertebrates, and the single bone articulating with the body skeleton is the forebear of the humerus, the bone of the upper arm.

In all fishes, the paired fins are connected to the rest of the body's support framework, the skeleton, via structures known as the pectoral and pelvic girdles. The pectoral girdle in sharks is relatively simple and consists of a large bar of cartilage that extends across between the two pectoral fins, known as the coracoid bar. Scapular processes extend above the base of the pectoral fin and connect the pectoral girdle to the skeleton. The pelvic girdle of sharks has fused into a single bar of cartilage, the puboischiac bar.

The pectoral girdle of bony fishes contains a large number of small bones including the cleithrum, supracleithrum, and the clavicle. A posttemporal bone attaches the pectoral girdle to the rear of the skull. The pelvic girdle in most bony fishes is composed of a pair of bones that act as extensions of the basal bones of the fins.

Flippers

During the Tertiary period (approximately thirty-five to fifty-seven million years ago) several groups of land-dwelling mammals returned partly or completely to the aquatic environment. The best known of these are the cetaceans (whales and dolphins), but other groups include the Sirenidae (manatees) and the pinnipeds (seals and sea lions). These animals faced the same problems of locomotion in water as did their fish ancestors, and in the course of evolution they have evolved finlike structures called flippers and flukes. These perform the same stabilizing functions as in fishes, and bear a superficial resemblance to fins. However, closer examination of whales and dolphins demonstrates that the pectoral flippers of cetaceans are in fact modified mammalian forelimbs and contain the same bones present in the forelimbs (arms) of terrestrial mammals. Thus, the bones of the pectoral fins, which gave rise to the forelimbs of terrestrial vertebrates, have evolved back into finlike structures in these mammals as an adaptation to an aquatic lifestyle. The hindlimbs, along with their bones and supporting pelvic girdle (the hip bones in land mammals) have been lost in modern dolphins and whales, although they are still present in fossil forms such as the extinct fossil whale *Basilosaurus*. These mammals also possess fleshy dorsal fins to help stabilize them during swimming and large horizontal flukes on the tail, which provide the main forward propulsive thrust to these animals as they move through the water.

—*John G. New*

See also: Anatomy; Bone and cartilage; Convergent and divergent evolution; Dolphins, porpoises, and other toothed whales; Fish; Locomotion; Manatees; Marine animals; Physiology; Seals and walruses; Tails; Whales, baleen.

Bibliography

Bailey, Jill. *Animal Life: Form and Function in the Animal Kingdom*. New York: Oxford University Press, 1994. A well-illustrated volume dealing with the relationship between the form of animal's bodies and their function.

Bone, Q., N. B. Marshall, and J. H. S. Blaxter. *Biology of Fishes*. 2d ed. New York: Blackie, 1995. A solid introductory text in the biology of fishes.

Moyle, Peter B., and Joseph J. Cech. *Fishes: An Introduction to Ichthyology*. 4th ed. Upper Saddle River, N.J.: Prentice Hall, 1999. An upper-division college-level textbook. Covers fish anatomy, structure, form, and evolution.

Riedman, Marianne. *The Pinnipeds: Seals, Sea Lions, and Walruses*. Reprint. Berkeley: University of California Press, 1991. Detailed, comprehensive coverage of these flippered marine mammals.

FISH

Type of animal science: Systematics (taxonomy)
Fields of study: Anatomy, liminology, marine biology, oceanography, zoology

Bony fishes constitute one of the seven living classes of vertebrates. There are approximately twenty thousand recognized living species, nearly as many as all other vertebrates combined.

Principal Terms

CTENOID SCALES: thin, flat, bony scales with tiny spines on the exposed rear edge, found on sunfish, perch, sea bass, and other advanced teleosts

CYCLOID SCALES: thin, flat bony scales with a smooth surface; rounded in shape, found on herrings, minnows, trout, and other primitive teleosts

GANOID SCALES: thick, diamond-shaped, bony scales that are covered with ganoine, a hard inorganic substance; found on bichirs, gars, and other primitive bony fishes

OSTEICHTHYES: the taxonomic class in which the bony fishes are placed; contains species related to the ancestors of higher vertebrates

PECTORAL FINS: paired fins found near the head end of the fish body; related to the forelimbs of higher vertebrates

PELVIC FINS: paired fins found either near the tail end of the fish body or below the pectoral fins; related to the hindlimbs of higher vertebrates

SWIM BLADDER: the hydrostatic (buoyancy) organ of teleost fishes derived from the lung of more primitive bony fishes

TELEOSTS: members of the infraclass Teleostei, the most advanced of the ray-finned fishes; they compose the vast majority of living bony fish species

The Osteichthyes, or bony fishes, constitute the largest and most diverse of the classes of vertebrates. Like the jawless and cartilaginous fishes, they are characterized by gills, fins, and a dependence on water as a medium in which to live. Unlike those fishes, however, they typically possess a skeleton made of bone. Additional features characteristic of most bony fishes include a lateral line system, scales, osmoregulation (salt balance) by means of salt retention or secretion, and a bony operculum (gill cover) over the gill openings.

The fossil record of bony fishes begins nearly 400 million years ago in the early Devonian geological period, mostly in freshwater deposits. Thus there is reason to believe that bony fishes originated in freshwater habitats. Living bony fish species inhabit both freshwater habitats (58 percent of species) and marine habitats (41 percent), and some (1 percent) move between the two environments on a regular basis. This distribution does not reflect the relative proportions of these environments, since 97 percent of the earth's water is in the oceans and only 0.001 percent is in freshwater lakes, rivers, and streams (the rest is ice, groundwater, and atmospheric water). Rather, the high diversity of freshwater species is a reflection of the ease with which freshwater populations become isolated and evolve into new species.

Fish Subclasses

There are four subclasses of bony fishes: Dipneusti (or Dipnoi), Crossopterygii, Brachiopterygii, and Actinopterygii. The first three of these include a total of only eighteen primitive living species.

Subclass Actinopterygii includes all the rest. The Dipneusti, or lungfish, are named for their possession of lungs, an ancestral characteristic suggesting that the earliest steps of bony fish evolution took place in tropical freshwaters subject to stagnation. Modern lungfishes (six species) are able to cope with such conditions by swallowing air and exchanging respiratory gases (oxygen and carbon dioxide) in the lung. Once considered closely related to terrestrial vertebrates, they are now believed to share certain similarities merely because of convergence (independent evolution of characteristics that appear similar).

The Crossopterygii, or fringe-finned fishes, were the dominant freshwater predators of the Devonian period. One fossil subgroup, the rhipidistians, had many features intermediate between fishes and ancestral amphibians, including tooth structure, lobed fins, and a jaw connected directly to the skull. Therefore, they are believed to represent a link between fishes and higher vertebrates. The other subgroup, the coelacanths, was also believed to be extinct (for 70 million years) until a coelacanth was taken from deep water off South Africa in 1938. This species, *Latimeria chalumnae*, is the only known living crossopterygian fish, and is of great interest as a kind of "living fossil."

The Brachiopterygii, or bichirs, include eleven living species known from swamps and rivers in tropical Africa. Though they share some characteristics with the other bony fish subclasses, they have some distinct features that warrant placing them in a separate subclass. One such feature is a dorsal fin consisting of many separate finlets, each supported by a single spine.

The Actinopterygii, or ray-finned fishes, comprise three major infraclasses: the Chondrostei, Holostei, and Teleostei. Chondrosteans, which have reverted to a largely cartilaginous skeleton, include the sturgeons (twenty-three species) and paddlefishes (two species). One species, the beluga sturgeon (*Huso huso*), source of the famous Russian caviar, may be the largest living Osteichthyes species. It is known to achieve a length of 8.5 meters and a weight of nearly 1,300 kilograms.

Holosteans include eight species: seven gar and one bowfin species, all known from North America. These freshwater piscivors (fish predators) are characterized by a skeleton made entirely of bone, but they do have certain other features in common with their more primitive ancestors, such as the ability to breathe air (with the swim bladder) and ganoid scales (also found in sturgeons).

The vast majority of ray-finned fishes, hence, of bony fishes—and indeed nearly half of all living vertebrates—belong to the infraclass Teleostei. It includes nineteen to twenty thousand living species. Among the features characteristic of teleosts are cycloid or stenoid scales (though some are scaleless), a swim bladder (lost in many bottom fishes), highly maneuverable fins, and a homocercal tail (meaning that its upper and lower lobes are symmetrical). Teleosts are represented by an amazing range of body sizes and shapes. A large number of species are quite small, enabling them to occupy niches (ways of living) unavailable to other fishes. The smallest known fish, in fact the smallest vertebrate of any kind, is a goby from the Indian Ocean, *Trimmatom nanus*, which matures at 8 to 10 millimeters in length.

Fish Shapes and Habits

There are several common body shape categories among teleosts which relate strongly to the fishes' habits. "Rover-predators" have the fusiform (streamlined) body shape that is perhaps most typically fishlike. Fins are distributed evenly around the body, and the mouth is terminal (at the end of the snout). This category includes minnows, basses, tunas, and others that typically are constantly moving—searching for and pursuing prey. "Lie-in-wait predators" tend to be more elongated, with the unpaired fins far back on the body, favoring a sudden lunge for their prey. The pike, barracuda, and needlefish typify this category.

"Bottom fishes" include a wide variety of shapes. Some are flattened for lying in close contact with the bottom (as are flatfishes such as flounder), some have flattened heads and sensory barbels (filaments with taste buds) near the mouth (as do catfishes), and some have fleshy lips for

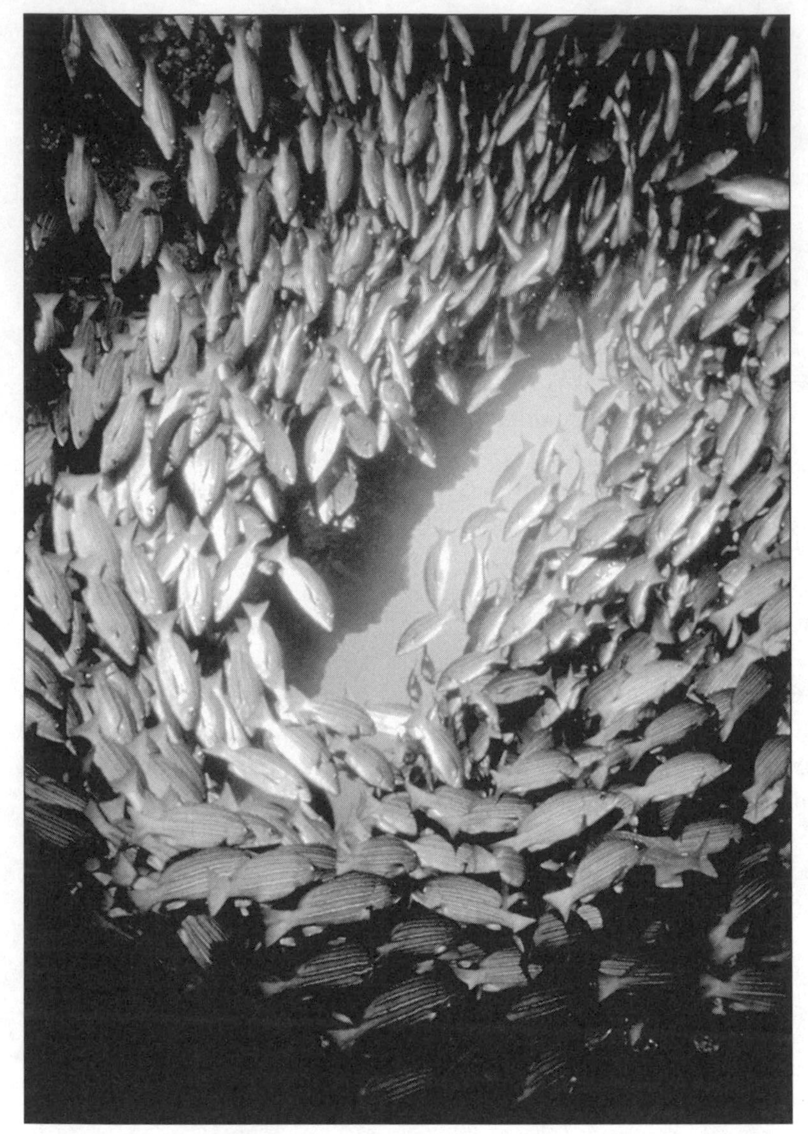

Many species of marine fish live in large schools, such as these four-line snappers. (Digital Stock)

tened and have pectoral fins high on the body, with pelvic fins immediately below. This arrangement favors maneuverability in tight quarters such as coral reefs, thick plant beds, or dense schools of their own species. Examples include angelfishes, surgeonfishes, and freshwater sunfishes. "Eel-like fishes" have highly elongated bodies, tapering or rounded tails, and small, embedded scales (or no scales at all). They are adapted for maneuvering through crevices and holes in reefs and rocks and for burrowing in sediments. Eels, loaches, and gunnels typify this category.

Teleosts occupy habitats ranging from torrential streams high in the Himalayas to the bottom of the deepest oceanic trenches. They are found in the world's highest large lake (Lake Titicaca) and deepest lake (Lake Baikal). Some blind species live in the total darkness of underground caves. One *Tilapia* species lives in hot soda lakes in Africa at 44 degrees Celsius, while the Antarctic icefish *Trematomus* lives at –2 degrees Celsius.

The vast majority of teleost fishes, both marine and freshwater, are tropical. Southeast Asia contains the greatest number of freshwater fish species, but the Amazon and its tributaries contain almost as many (and perhaps many hundreds more, still undiscovered). Marine teleosts are most diverse in the Indo-Pacific region, especially in the area from New Guinea to Queensland, Australia. A single collection made in the Great Barrier Reef off northeastern Australia may contain one hundred or more species. Some marine teleost species have a nearly worldwide distribution, while certain freshwater spe-

sucking food from the bottom sediment (as do suckers). A number of bottom-fish species have structures, usually modified pelvic fins, that enable them to cling to the bottom in areas with strong currents (sculpins and clingfishes have these).

"Surface-oriented fishes" tend to be small, with upward-pointing mouths, heads flattened from top to bottom, and large eyes. The mosquitofish, killifish, and flying fish belong to this category. "Deep-bodied fishes" are laterally flat-

cies have highly restricted ranges. The Devil's Hole pupfish (*Cyprinodon diabolis*), for example, is found only in one small spring in Nevada.

Many teleosts have highly specialized associations, called symbioses, with other organisms. Some live among the stinging tentacles of sea anemones (the clownfish *Amphiprion*), within the gut of sea cucumbers (the pearlfish *Carapus*), within the mantle cavity of giant snails (the conchfish *Astrapogon*), or among the stinging tentacles of the Portuguese man-of-war jellyfish (the man-of-war fish *Nomeus*).

Reproduction

Reproduction among teleosts is incredibly varied. Most species are egg-layers, often producing an enormous number of eggs. The female ocean sunfish *Mola mola* may produce up to 300 million eggs, making it the most fecund vertebrate of all. Some species are livebearers, such as the platyfishes, swordtails, and surfperches. Some species are oral brooders, incubating the eggs in the mouth of the male (as in many cardinal fishes) or of the female (as in many cichlids). In one South American cichlid species, *Symphysodon discus*, the female "nurses" its young with a whitish milklike substance secreted by the skin.

Many teleost species are hermaphroditic. A few of these are synchronous hermaphrodites (functioning as male and female at the same time), such as the hamlet *Hypoplectrus*, but many more are sequential hermaphrodites (first one sex, then the other), such as the sea bass *Serranus*. In some coral reef fishes in the wrasse family, a dominant male mates with a harem of females. If this male is removed, the largest female becomes male and takes over the missing male's behavioral and reproductive function.

In a few species, all individuals are female, as in the Amazon molly *Poecilia formosa*. It has been shown that this species is a "sexual parasite" of two related "host" species. Sperm from host males are required to activate development of Amazon molly eggs, but male and female chromosomes (genetic material) do not join, and the offspring are all genetically uniform females.

Ichthyology

The scientific study of bony fishes dates back to Aristotle, who was the first to note, for example, that the sea bass is hermaphroditic. The "father of ichthyology" in more recent times was Peter Artedi (1705-1735), whose classification system was used by Carolus Linnaeus (1707-1778) in his *Systema Naturae*, which became the basis for all future classification systems.

Bony fish classification depends on the study of taxonomic features, or characters, which vary from one species, or group of species, to another. Useful characters include countable features (meristic characters), such as the number of fin supports (rays) or the number of scales in the lateral line, and measurable features (morphometric characters), such as the relative lengths of body parts. Such studies are typically done on museum specimens that are preserved in alcohol solution after fixation in formaldehyde solution. Dissecting tools, microscopes, and even X-ray machines are used for revealing meristic and morphometric characters. For studying bones, dry skeletons are sometimes prepared, or (especially for small species) specimens are "cleared and stained." This latter technique involves clearing the flesh with potassium hydroxide and staining the bones with Alizarin red stain.

Other techniques use samples of living tissue for finding taxonomic characters. Karyotyping (analysis of the chromosomes) and enzyme electrophoresis (using an electric field to separate similar proteins) are also important sources of taxonomic information.

Specimens for taxonomic studies are collected by means of netting, trapping, catching with hook and line, and spearing. Specialized techniques include electrofishing (use of an electric shocking device) for stream fishes, and ichthyocide (fish poison such as rotenone) for coral reef fishes.

Understanding the evolutionary history and classification of the Osteichthyes also depends on paleontological studies (the study of fossils). Bony fishes are well represented in the fossil record because of the superior fossilizing nature of their bony skeletons. Many fish biologists are concerned with matters other than taxonomy. Because of the economic importance of both marine and freshwater bony fishes, the science of fisheries biology (concerned with the management and exploitation of fish populations) is of great significance. Fish populations are often studied with "age and growth" techniques. Age (determined by scale analysis), length, and weight data can be used to calculate growth and mortality rates, age at maturity, and life span. Other techniques for studying fish populations involve tagging individuals (useful for making estimates of population size) and even using tiny radio transmitters that can be followed by aircraft (useful for studying fish migrations).

Ecologists and ethologists (behavioral biologists) are also active in fish studies, particularly since the invention of scuba diving, which allows direct observation of fishes in their natural habitat. An example of an important discovery made possible by scuba diving is cleaning symbiosis, common in coral reef areas. This symbiosis (an association involving members of two different species) involves a "cleaner" species (often a goby or wrasse), which feeds on the external parasites and diseased tissue of a host ("cleanee") species, which visits the cleaner for this service.

Questions Still to Be Answered

Bony fishes are by far the most numerous of all vertebrates. They are also arguably the most diverse in terms of body form, reproductive habits, symbiotic relationships, and other characteristics. Yet much remains to be learned. Virtually every ichthyological expedition into the Amazon region, for example, returns with specimens of previously unknown species. Some ichthyologists estimate that perhaps five or ten thousand undiscovered teleosts remain in unexplored streams and remote coral reefs.

Many biological mysteries remain about even some of the most familiar species. A good example is the American eel, *Anguilla rostrata*. This predatory species spends most of its life in the rivers, streams, and lakes of eastern North America, where it is often one of the dominant species. After six to twelve years in these habitats, the adult eels swim to the ocean and apparently migrate more than five thousand kilometers to spawn in deep water in the Sargasso Sea (an area in the western Atlantic south of Bermuda).

This general location of eel spawning has been inferred from the appearance there of the tiniest eel larvae (called leptocephali, these were once considered a separate species). The larvae become larger and larger as they drift in the Gulf Stream toward the North American coast. This much has been known since 1922. The adult migration has never actually been followed, however, and no one knows exactly where, at what depth, or how they mate and spawn, nor is it known what then happens to the adults.

Despite many advances in scientific knowledge, much remains to be learned about the interrelationships, ecology, behavior, and fishery potential of the world's bony fish species.

—George Dale

See also: Cold-blooded animals; Deep-sea animals; Eels; Fins and flippers; Lakes and rivers; Lampreys and hagfish; Lungfish; Lungs, gills, and tracheas; Marine animals; Marine biology; Reefs; Salmon and trout; Scales; Seahorses; Sharks and rays; Tidepools and beaches; Vertebrates; Whale sharks; White sharks.

Bibliography

Bond, C. E. *Biology of Fishes*. 2d ed. Philadelphia: W. B. Saunders, 1996. Intended as an introduction to the study of fishes for the general reader and for the college student. Includes a concise summary of bony fish diversity, and ten chapters on general topics in fish ecology, behavior, and physiology. Illustrated with black and white drawings. Each chapter has a reference list.

Cailliet, G. M., M. S. Love, and A. W. Ebeling. *Fishes: A Field and Laboratory Manual on Their Structure, Identification, and Natural History*. Belmont, Calif.: Wadsworth, 1986.

This manual, used in college-level ichthyology courses, is the best source of information on the techniques used in the field and laboratory study of fishes. Concise, well illustrated, and thoroughly referenced. Spells out, in step-by-step detail, the actual methods used by modern ichthyologists.

Chiasson, Robert B. *Laboratory Anatomy of the Perch*. 4th ed. Boston: McGraw-Hill, 1991. The North American yellow perch *Perca fluviatilis* has long served as a "typical bony fish" for college zoology classes. In fact, anatomical structure is highly similar throughout the six or seven thousand perchlike species in the teleost order Perciformes. This short (67-page) dissection guide is an excellent introduction to any of them, and to fish anatomy in general.

McClane, A. J. *McClane's Field Guide to Freshwater Fishes of North America*. New York: Henry Holt, 1978. Probably the best angler's guide to North American freshwater gamefishes available. "Tackle-box-size," nontechnical, and packed with the kind of information sport fishermen would find most interesting and useful. Arranged by family, with accurate full-color illustrations and useful glossary.

_______. *McClane's Field Guide to Saltwater Fishes of North America*. New York: Henry Holt, 1978. Companion edition to McClane's freshwater fish guide, and contains similar features. A handy, popular guide to the saltwater gamefishes of North American coastal waters. Arranged taxonomically, with color illustrations and a glossary. Highly recommended for the saltwater sport fisherman.

Mills, D., and G. Vevers. *The Golden Encyclopedia of Freshwater Tropical Aquarium Fishes*. New York: Golden Press, 1982. An attractive popular introduction to tropical freshwater aquarium fishes in two parts. Part 1 is a practical guide to starting and maintaining an aquarium, illustrated with full color drawings and photographs. Part 2 is a detailed guide to two hundred of the most popular and attractive species, illustrated with stunning photographs.

Moyle, P. B., and J. J. Cech. *Fishes: An Introduction to Ichthyology*. 4th ed. Englewood Cliffs, N.J.: Prentice Hall, 2000. One of the best college-level textbooks of general ichthyology available. Covers all aspects of fish biology and surveys the diversity of bony fishes in eight detailed chapters (16 to 23). Large, exhaustive bibliography.

Paxton, John R., and William N. Eschmeyer. *Encyclopedia of Fishes*. 2d ed. San Diego, Calif.: Academic Press, 1998. Beautifully illustrated encyclopedia, with information on levels of extinction threat. Covers twenty-four thousand fish species, including habitats and adaptations and behavior. Photographs, illustrations, diagrams, and maps.

Pough, F. H., J. B. Heiser, and W. N. McFarland. *Vertebrate Life*. 5th ed. Upper Saddle River, N.J.: Prentice Hall, 1999. This large college text, a superlative introduction to vertebrate biology, is revised to incorporate changes in the philosophy and methodology of vertebrate taxonomy. Chapters 1 to 5 provide an excellent general background to vertebrate evolution, and chapter 8 covers the bony fishes in detail.

FLAMINGOS

Types of animal science: Anatomy, classification
Fields of study: Anatomy, ornithology, zoology

Eight flamingo types exist, all beautiful red, pink, or white water birds. Hunted to near extinction in the United States, they inhabit Europe, Africa, Asia, South America, and the West Indies. A great threat to flamingos everywhere is pollution and destruction of their habitats

Principal Terms

GASTRONOMIC: pertaining to the art of fine dining
MANDIBLES: the beaks of birds
PLUMAGE: the feathers of birds

Eight kinds of flamingos make up the avian family Phoenicopteridae. Flamingos are beautiful water birds with long legs and luxuriant deep red, light red, pink, or white plumage. They inhabit Europe, Africa, Asia, South America and the West Indies. European flamingos migrate to Africa in the winter.

The birds usually live in tropical and temperate regions along oceans and lakes or in marshes. Flamingos are also found in the Andes mountains. It is thought that flamingos are pink to red because they eat varied amounts of blue-green algae and other organisms, which contain the substances that make carrots orange and tomatoes red. Flamingos also eat diatoms, shrimp, and small mollusks.

There are four flamingo (Phoenicoparrus) species: the American and Caribbean, the Andean, the James', and the lesser flamingos. Several Phoenicoparrus species have subspecies. For example, American and Caribbean flamingos (*Phoenicopterus ruber*) have three subspecies: greater (*P. ruber roseus*), Chilean (*P. ruber chilensis*), Galápagos (*P. ruber ruber*) flamingos. Regrettably, few of these birds are seen in the United States, as feather hunters made them almost extinct. A Chilean flamingo is a bit smaller than a greater flamingo. It is pink, with red streaks on its back, and nests in Andean mountain lakes and southern South American lowlands. Two smaller species, are the Andean (*Phoenicoparrus andinus*) and James' flamingo (*Phoenicoparrus jamesi*). The smallest, most abundant species, numbering in the millions, is the lesser flamingo (*Phoeniconaias minor*) of Africa and India. It has a subspecies (*Phoeniconaias minor jamesi*).

Physical Characteristics of Flamingos

All flamingos have very long legs, webbed feet, and long, flexible necks. The long legs and webbed feet allow them to wade into fairly deep waters and stir up the muddy bottoms of lagoons and lakes, causing food to rise up closer to them. Their bills bend sharply about halfway from their ends. The upper mandible (beak half) is narrow and, when closed, fits tightly into the lower mandible. To feed, flamingos dip their heads into water upside down and scoop backward, taking in food-containing water. Then they press their mandibles together and push their tongues upward. As mandible edges have small ridges, the tongue pressure pushes water out and the strainerlike action retains the small animals and vegetation they feed on.

Different species and subspecies are colored differently and have different sizes. Overall, adult flamingos attain heights and weights of 2.5 to 5.5 feet and weigh four to seven pounds, depending on species. Females are usually shorter and weigh less than males. Flamingo plumage is white, various pinks, or crimson red. Their legs, webbed feet, bills, and faces are red, pink, orange, or yellow.

The Flamingo Life Cycle

Most flamingos live in colonies which number tens to hundreds of thousands. The colonies are usually located in or around lagoons and lakes. A well-known, very populous example is Kenya's Lake Nakuru, where millions of flamingos congregate. During breeding season, a male and female mate. It is believed that once mated, pairs of flamingos are monogamous.

The female lays one 3.5-ounce white egg in a depression atop a nest which is a conical mound of mud, one foot tall, built by the breeding pair. The pair then incubate the egg for about a month, until it hatches. On hatching, the baby flamingo stays in the nest for about three months. At first, it has gray, downy feathers and its legs and bill are pink. Its feathers turn pink and its bill curves into the adult shape as it grows. Both parents feed the young bird. It is given regurgitated food for as long as it remains in the nest, though it can feed itself thirty days after hatching. In the wild, flamingos may live for forty to fifty years.

Flamingo Conservation

Flamingos live in isolated habitats and have few natural predators except for humans. In the distant past, the ancient Romans hunted flamingos for their tasty tongues, thought to be a gastronomic delicacy. Regrettably, the American greater flamingo (*P. ruber roseus*), once common in the South, is now seen only rarely in the United States. They were hunted for their beautiful plumage faster than they could reproduce. This is unfortunate, because few sights are more beautiful than a flock of pink, rose, or scarlet flamingos standing

Flamingos are wading birds, living at the edges of lakes or oceans, or in marshes. (PhotoDisc)

together or flying in the sun of the United States tropics. They are still plentiful in the West Indies and South America.

It might be thought advantageous to restock the wild with zoo-bred flamingos. However, this has not been possible because flamingos captured for zoos often die in transit, and those in zoos rarely breed successfully. It is hoped that with time and with the cessation of feather hunting, flamingos will reestablish themselves in the United States. A great threat to this prospect, and to flamingos elsewhere, is pollution and destruction of their habitats

—Sanford S. Singer

See also: Beaks and bills; Birds; Feathers; Flight; Molting and shedding; Nesting; Respiration in birds; Wildlife management; Wings.

Flamingo Facts

Classification:
Kingdom: Animalia
Phylum: Chordata
Subphylum: Vertebrata
Class: Aves
Order: Ciconiiformes
Family: Phoenicopteridae (flamingos, three genera, seven species)
Geographical location: Europe, Africa, Asia, South America, the West Indies
Habitat: Lagoons, lakes, and marshes
Gestational period: One month of incubation
Life span: Forty to fifty years in the wild
Special anatomy: Long, storklike legs; bills that strain shellfish and other food out of water

Bibliography

Arnold, Caroline. *Flamingo.* New York: Morrow Junior Books, 1991. This book, for juvenile readers, examines different flamingo types, their physical characteristics, habitats, and behavior.

McMillan, Bruce. *Wild Flamingos.* Boston: Houghton Mifflin, 1997. This photoessay describes flamingo physical characteristics, natural habitat, and behavior.

Ogilvie, Malcolm, and Carol Ogilvie. *Flamingos.* Gloucester, England: A. Sutton, 1986. This book describes flamingo natural history, reproduction, and distribution.

Wade, Nicholas, ed. *The Science Times Book of Birds.* New York: Lyons Press, 1997. This illustrated book includes information on flamingos, their lives, and their habitats.

FLATWORMS

Type of animal science: Classification
Fields of study: Anatomy, invertebrate biology, physiology

Flatworms are wormlike animals with a single major opening to the gut. Many are parasitic for at least part of their life cycle.

Principal Terms

DEFINITIVE HOST: the host in which a symbiont (the organism living within the host) matures and reproduces

ECTOPARASITE: a parasitic organism that attaches to the host on the exterior of the body

ENDOPARASITE: a parasitic organism that attaches to an interior portion of the host's body

FREE-LIVING: an organism that does not have to spend a portion of its life cycle attached to another organism

HERMAPHRODITIC: a situation in which both functional ovaries and testes are present in the same organism

PROGLOTTID: a body segment of a tapeworm that contains a set of reproductive organs, usually both ovaries and testes

SNAILING: the process in which the free-swimming larva (miracidium) of the flukes utilizes the tissue of a snail as an intermediate host

The flatworms or Platyhelminthes are wormlike animals with a single major opening to the gut. This opening functions as both a mouth and an anus. Between the gastrodermis (lining of the gut) and the epidermis, the body is filled with tissues, including layers of muscle, connective tissue, and reproductive organs. Included in the flatworms are free-living forms (class Turbellaria), and two major groups of animal parasites (class Trematoda—flukes—and class Cestoda—tapeworms). Most tubellarians are bottom dwellers in marine water or freshwater or live in moist terrestrial environments, but a few species are symbiotic or parasitic. The majority of the larger species are found on the underside of rocks or other hard objects in freshwater streams or in the littoral zones of the ocean. All of the cestodes and trematodes exist as endoparasites and most exhibit indirect life cycles with more than one host. The initial host is usually an invertebrate, and the final host is most often a vertebrate. A number of species utilize humans as a final host.

General Characteristics of Flatworms

The free-living flatworms generally range in size from five to fifty millimeters. The epidermis is covered with cilia, and locomotion is achieved through a combination of ciliary movements and the contraction and relaxation of layer of circular muscles that go around the body, and a layer of longitudinal muscles that extent down the length of the body. The most commonly studied turbellarians are the planarians. The digestive system consists of a mouth on the ventral surface, a pharynx, and an intestine. Planarians are mostly carnivorous, feeding mainly on nematodes, rotifers, and insects. In contrast to the parasitic species, the turbellarians have simple life cycle. Some can reproduce by asexual fission, but most reproduce sexually. While the turbellarians are hermaphroditic, they generally crossbreed. Planarians demonstrate a remarkable ability to regenerate. If a section is excised from the middle of the worm, it will regenerate both a new tail and a new head.

Flukes primarily differ from the turbellarians in their adaptions for parasitism, including organs

Flatworm Facts

Classification:
Kingdom: Animalia
Subkingdom: Bilateria
Phylum: Platyhelminthes
Classes: Turbellaria (free-living), Monogenea (ectoparasitic flukes with a one-host life cycle), Trematoda (endoparasitic flukes), Cestoda (tapeworms)
Subclasses: Cestoda—Cestodaria (body not segmented), Eucestoda (body segmented into proglottids)
Orders: Turbellaria—Acoela (no gut cavity), Rhabdocoela (simple tubular gut), Alloeocoela (gut with one main branch and small side branches), Tricladia (gut with three branches), Polycladia (gut with many main branches); Trematoda—Aspidogastrea (endoparasitic with a one-host life cycle), Digenea (endoparasitic with at least a two-host life cycle); Eucestoda—Bothriocephaloidea (fish tapeworms), Taenioidea (pig and beef tapeworms)
Geographical location: All over the world
Habitat: Turbellaria—generally found in ponds, lakes, streams, and oceans; Monogenea, Trematoda, and Cestoda—larvae may be found in streams, but adults live within the body of a host
Gestational period: Varies among species, but most species lay eggs within a few days after fertilization; eggs usually hatch within a few days to a few weeks after being deposited
Life span: Varies among species; can be as short as a year in some turbellarians and up to thirty years in some flukes
Special anatomy: Elongated, bilateral invertebrates without appendages, have neither a true body cavity nor a circulatory system; parasitic species have specially adapted mouth parts for attaching to the tissues of the host

for adhesion such as suckers or hooks and an increased reproductive capacity. They are generally leaf-shaped, varying in size from ten to twenty millimeters. Most flukes, such as those in the class Digenea, have complex life cycles. The eggs produced by the mature fluke pass from the definitive host and hatch in water to form free-swimming larvae called miracidia. The miracidium enters the tissue of an intermediate host, snails, and transforms into a sporocyst, which reproduces asexually to form rediae. The rediae reproduce asexually to form cercariae, which leave the snail and penetrate a second intermediate host, such as fish, or encyst on vegetation, where they become metacercariae, juvenile flukes. When the metacercariae are eaten by the definitive host, they develop into mature flukes.

Tapeworms are also keenly adapted for parasitism, but unlike the flukes, they have long, slender bodies that can reach lengths of several meters and lack a digestive system. They obtain digested nutrients directly from the gut of the host. The tapeworm body consists of a linear series of proglottids. The tapeworm grows lengthwise by adding new proglottids. Mature proglottids contain fertilized eggs and break off the end of the tapeworm to be excreted out of the host. Almost all cestodes require at least two hosts, and the adult is a parasite in the digestive tract of vertebrates. One of the intermediate hosts is usually an invertebrate. Almost all species of vertebrates are subject to tapeworm infection.

—D. R. Gossett

See also: Asexual reproduction; Hermaphrodites; Invertebrates; Regeneration; Roundworms; Symbiosis; Worms, segmented.

Bibliography

Ehlers, U. "Phylogenetic Relationships Within the Platyhelminthes." In *The Origins and Relationships of Lower Invertebrates*, S. Morris, J. D. George, R. Gibson, and H. M. Platt, eds. Oxford, England: Clarendon Press, 1985. This is an excellent presentation on the relationships of the groups normally assigned to the class Turbellaria.

Hickman, Cleveland P., Larry S. Roberts, and Allan Larson. *Integrated Principles of Zoology*. 11th ed. Boston: McGraw-Hill, 2001. An introductory zoology text that gives a very good review of the characteristics of flatworms.

Schell, Stewart C. *Handbook of Trematodes of North America North of Mexico*. Moscow: University Press of Idaho, 1985. A very good reference book for the identification of trematodes.

Schmidt, Gerald D. *CRC Handbook of Tapeworm Identification*. Boca Raton, Fla.: CRC Press, 1986. A very good reference for the identification of cestodes.

Strickland, G. Thomas. *Hunter's Tropical Medicine and Emerging Infectious Disease*. 8th ed. Philadelphia: W. B. Saunders, 2000. An excellent source of information on the parasites of medical importance.

FLIES

Types of animal science: Anatomy, classification, reproduction
Fields of study: Anatomy, entomology

Flies are members of the order Diptera, two-winged insects, of which there are 95,000 species. Flies are related to the other dipterans, the mosquitoes.

Principal Terms

IRIDESCENT: showing the colors of the rainbow depending on light reflection
MOLT: shed an insect shell
SYMBIOSIS: beneficial relationship between two organisms

Flies belong to the fourth largest animal order, Diptera, which comprises 95,000 species of two-winged insects. The Diptera evolved from four-winged insects, and their vestigial rear wings are stalks that act as gyroscope balance organs, enabling the forewings to move the fly forward without causing nose dives, and keeps it on a steady course.

Dipterans occur worldwide, including Antarctica, but are most plentiful in moist, warm climates. The main dipteran suborders are Nematocera, Brachycera, and Cyclorrapha. Nematocera are mosquitoes, slender, with long antennae. The other orders have stout bodies and short antennae; among them are houseflies and tsetse flies.

Physical Characteristics of Flies

A fly body has three parts: a head, thorax or middle part that holds six legs, and an abdomen or rear end. Two compound eyes containing up to four thousand facets cover most of the head. The eyes see light changes and sudden movement from many different directions at once. This is why flies are hard to catch

Atop a fly's head, paired antennae provide the senses of touch and smell. Dipterans have a mouth part called a proboscis. It is funnel-shaped, with its wide part at the bottom. The proboscis is like a straw, sucking up fluid via a pump in the head. The proboscis of a housefly is soft, so it cannot bite. Bee flies have a long proboscis that enters flowers for nectar. Biting flies, such as horseflies, have hard, sharp proboscises that pierce the skin of their victim.

Each foot on a fly is tipped by claws that grip rough surfaces. Under the claws are pads called pulvilli. On smooth surfaces, they flatten and grip tightly, allowing a fly to walk upside down on ceilings without falling. Flies breathe through openings called spiracles, located on each side of thorax and abdomen.

Fly Life

The life of a fly begins when a female lays hundreds to thousands of eggs in manure, garbage, vegetable waste, fruit, plant stems, or stagnant water. Houseflies have telescoping ovipositors that place eggs in soft matter. Other species have stiff ovipositors that penetrate plant stems or fruit. Flies, like mosquitoes, also lay eggs on water.

The heat given off in these environments incubates the eggs. In one or two days they hatch as white, legless larvae (maggots), which eat the material surrounding them. Maggots rapidly grow too large for their skins, which split and allow molting. After molting twice, the larvae find sheltered places to form pupariums, where they molt a final time and become pupae, and then become winged flies. The process usually takes four to ten days. Adult flies emerge from the pupae full-sized. Most live for thirty days. In winter most die,

Houseflies often lay their eggs on manure or other decaying materials, and may thus transmit bacteria onto foodstuffs. (Jeff J. Daly/Photo Agora)

but larvae and pupae live to become adults in the spring. The development of a fly from egg, to larva, to pupa, to adult is called metamorphosis.

Tsetse Flies, Horseflies, and Blowflies

Tsetse flies are twenty species of genus *Glossina*. Five of these species carry sleeping sickness. They live in grasslands, forests, and river and lake shores in Africa, between the Sahara and Kalahari deserts. Tsetse flies suck blood with their sharp proboscises; the blood is digested with the assistance of a membrane that is secreted around the blood meal in the foregut. Tsetse flies mate year round. Females have one offspring per mating. After fertilization, an egg hatches in the mother and is later born as a full-sized maggot, which burrows into the soil, pupates, and becomes adult in a month. The flies carry trypanosomes that can infect people with sleeping sickness. When the tsetse fly bites a victim, the trypanosomes enter, multiply, make them very tired, and quickly kill them.

The 3,500 Brachycera species of horseflies occur worldwide, most in tropical and temperate fields and forest areas near water. They have inch-long, stout bodies, large heads, short antennae, and iridescent compound eyes. Their mouths pierce the surfaces of plants and animals to suck fluids. Males eat nectar and plant juices, while females suck blood. The time of mating varies according to species and climate. Females must eat blood before laying eggs or the eggs do not develop. A female lays one thousand eggs in damp sites such as rotten wood. Larvae hatch in two to three weeks, although some hibernate all winter. Before becoming adults, larvae pupate for about three weeks. Adults live for six weeks before mating and dying. Horsefly bites are painful and may cause anthrax or tularemia.

Fly Facts

Classification:
Kingdom: Animalia
Subkingdom: Bilateria
Phylum: Arthopoda
Subphylum: Uniramia
Class: Insecta
Order: Diptera
Suborder: Brachycera, thirty-five families
Geographical location: Every continent
Habitat: Grasslands, forests, near rivers and lakes
Gestational period: Eggs hatch in two to seven days; adulthood occurs in one to six weeks
Life span: One to six months
Special anatomy: Six legs; compound eyes; one pair of wings, antennae, proboscises; pulvilli

Blowflies are metallic blue-green and are larger than houseflies. Females lay eggs on meat or in the open wounds of animals. Eggs or larvae that are in food swallowed by animals and humans cause pain and sickness when the maggots eat into the wound or tissue where they were laid. When the larvae enter an animal's skin, puss-filled sores form. The screwworm, a blowfly larva, harms livestock. If screwworms are not controlled, animals die.

Flies and Disease

A housefly can carry pathogenic bacteria disease because it lives in manure and garbage. Thousands of related species transmit germs to whatever surface they land on, spreading disease. Often it is best to kill flies and stop their reproduction. Flies destroy crops; parasitize animals; and carry typhoid and cholera.

However, flies are also useful. Some flies, such as hoverflies, pollinate plants. Larvae also eat aphids, which kill crops. Flies speed the decomposition of animal carcasses and manure. In addition, flies consume other harmful insects, controlling their numbers. Finally, flies are the food source for numerous insectivores higher up the food chain.

—Sanford S. Singer

See also: Antennae; Flight; Insects; Metamorphosis; Molting and shedding; Mosquitoes; Wings.

Bibliography

Chapman, R. F. *The Insects: Structure and Function*. 4th ed. New York: Cambridge University Press, 1998. A clear, comprehensive, systematic entomology text.

Fischer-Nagel, Heiderose, and Andreas Fischer-Nagel. *The Housefly*. Minneapolis: Carolrhoda Books, 1990. Text and illustrations describe flies' physical characteristics, habits, environment, and housefly-human interactions.

Lawrence, Peter. *The Making of a Fly: The Genetics of Animal Design*. Boston: Blackwell Science, 1992. Traces the development of the common fruitfly, *Drosophila melanogaster,* from egg formation through the differentiation of all the tissues, organs, and body parts. Illustrated.

Miller, Sara Swan. *Flies: From Flower Flies to Mosquitoes*. New York: Franklin Watts, 1998. A brief book describing fourteen fly species and recommendations for finding, identifying, and observing them.

FLIGHT

Types of animal science: Anatomy, evolution
Fields of study: Anatomy, biophysics, evolutionary science, invertebrate biology, ornithology, paleontology, zoology

Flight has evolved in a number of groups of organisms. Seed pods, insects, pterosaurs, birds, and bats are all familiar examples of organisms that have evolved some flight capability. Depending on the definition of flight, certain organisms inhabiting the oceans and freshwater realms, such as fish, ocean turtles, penguins, sea snakes, and even crocodiles can fly underwater.

Principal Terms

DRAG: a force that acts in the opposite direction of the movement of a body through a fluid medium; sources of drag vary but include friction and pressure suction

FLUID: a substance, either liquid or gas, that flows or conforms to the outline of its container

INERTIA: the property of an object with kinetic energy to move in a straight line unless acted upon by an outside force

LIFT: an aerodynamic force created through differential flow above and below a structure

REYNOLDS NUMBERS: the results of a formula that takes into account the velocity of an object, its characteristic length divided by the dynamic viscosity of the fluid

TURBULENCE: flow that is chaotic and may create stall conditions through the loss of lift

VISCOSITY: the stickiness of a fluid created by internal forces as molecular attractions

There are two types of movement through a fluid environment: rowing and flying. Organisms that row use the viscosity (stickiness) of the fluid to propel themselves along. These organisms generally have limbs used as oars with which to row; these oars push up against the fluid (usually water) and the fluid exerts a force back onto the oars that drives the animal forward. Organisms that fly use the inertial qualities of the fluid (usually air); essentially, they will use differential flow rates over the body surfaces to create thrust and lift. Flying is necessary when animals are large, and the drag on the body becomes too high for rowing to be effective. The choice between rowing and flying has to do with the organism's Reynolds number, which is a ratio between length and inertial viscosity forces. Viscosity dominates at low Reynolds numbers, whereas inertial forces dominate at high Reynolds numbers. Drag can be defined as any force that tends to retard the forward movement of an object in a fluid environment. Drag can be due to friction on the body surface; pressure drag is due to the fluid not adhering to the shape of the object and peeling away as a wake. The size of an organism capable of achieving airborne flight by means of rowing is quite small (hypothesized for insects such as mymarid wasps, which are smaller than fruit flies). In aquatic environments, larger forms can row due to the greater viscosity of water.

How Flight Works

Regardless of the organism, flow must take on different velocities above and below the body or wing, according to Bernoulli's principle, whereby flow is understood in terms of conservation of energy. What this means is that in an ideal fluid where there is no friction (that is, no viscosity), if the fluid accelerates, its pressure goes down. Conversely, when the velocity is reduced, the pressure goes up. While this may be counterintuitive, it nonetheless provides a basis for understanding flight. Thus, wings are not necessary to fly, but rather an organism needs to create a surface that is longer on top than it is on the bottom. A concave body form might possibly provide lift by its shape alone in an air streamline. Wings can assume such a shape by twisting themselves, as in insect wings, or creating a curvature with feathers, as in birds. In this situation, air flows faster above than below the structure, and lift is produced perpendicular to the upper surface.

This only works because fluids follow the law of continuity: if a streamline of air is split by an airfoil, the streamline will flow both over and under the structure. Both the upper and lower streamline must theoretically meet at the end of the structure. Thus, a curved structure or a structure that is inclined at an angle to the oncoming flow will create a longer upper surface than lower surface, hence the differential velocity as the streamlines race to the rear edge of the structure. If the aerodynamic structure is inclined downward, the lifting force can now be separated into a force that drives the structure forward and upward.

This picture of flow is very basic, and it is more complicated in practice. The flow over wings circulates around the wing in circles. Bernoulli's principle may still be applied to the circulating flow to understand how lift is generated. This circulation is shed off the wings at the tips in particular patterns based on the movements of the

The feathers on bird wings create sufficient curvature to produce lift and allow the bird to fly. (Corbis)

From Dinosaurs to Birds

Paleontologists have suggested that ground-dwelling bipedal dinosaurs gave rise to birds, and thus avian flight developed from the ground up. This viewpoint is flawed, since the dinosaurian proavis would have had to overcome gravity by running, which would not be energy-efficient. Moreover, bipedal dinosaurs, by virtue of their well-developed hindlimbs, had their center of mass in the pelvis, a situation which would create an unbalanced torque that would flip the animal attempting to fly. Any lift generation in the chest would act vertically at the forelimb area. The pelvic center of mass would act as a downward force in the rear of the animal, resulting in a rotation of the body. Flying birds have their center of mass and center of lift in roughly the same place, and thus rotations of this kind do not occur. It is likely that birds originated from arboreal reptilian ancestors that were quadrupeds. Jumping or falling from a tree would create a cost-free air flow about the body surface to create lift and thus extend a glide between trees. Quadrupeds have a center of mass close to the pectoral area, unlike bipeds, and thus have less problems with unbalanced rotational forces, such as torques. In fact, those land vertebrates that glide today are all quadrupeds.

wing or (in the case of most fishes) the tail. This circulation of fluid around the wing must be generated for lift and thrust to be generated. These circulating rings of fluid are shed off the wings or oscillating tail as vortex rings. This why aircraft are spaced apart during landings, as these vortex rings may cause such turbulence that the aircraft flying behind may experience loss of lift.

Laminar flow is a flow whose constituent pattern is in one direction, an even flow. Turbulent flow occurs when areas within a flow become chaotic, causing heat production and a loss of velocity. Worse than this is a condition when the turbulence becomes so great that the flow no longer adheres to the aerodynamic surface and departs from the structure, causing a loss of lift. This can happen for a variety of reasons. If the angle of attack of a wing becomes too steep, the flow may become turbulent and separate off the structure. If there is no circulation on the wing, there is no lift, a stall develops, and gravity takes over, with dire consequences. In addition, if the velocity is too great, the flow will become turbulent and unable to stay on the surface.

Shape also may contribute to the production of turbulence. Surface roughness may cause turbulence to occur through collisions with the streamlines. In addition, convexities and concavities can cause turbulence to form and prevent the adherence of the fluid on the surface of the object. While this has practical applications in design, it points to the fact that aerodynamic structures are streamlined, having tapered ends that preclude the necessity of fluids having to adhere to abrupt curvatures and creating potentially turbulent flows.

What has been described here is large-scale turbulence, where separation of flow from the object's surface occurs. Small-scale turbulence can actually benefit the lift on the aerodynamic structure, since the longer the flow adheres to the airfoil, the more lift and thrust the structure can generate. Small-scale turbulence may actually maintain the flow on a surface, and thus the lift produced far outweighs the drag force produced by the turbulence.

Flight in Animals

The evolution of flight in birds is centered mainly on the evolution of feathers. Feathers are complex epidermal structures that may have been derived via a developmental program alteration of scales. Feathers are the most complex epidermal structure found in vertebrates. They have a central strut, called a rachis, and a series of barbs emanating from both sides of the rachis in pinnate fashion to form the feather vane. These barbs are hooked together by a series of hooklets that act very much like Velcro. Feathers form the major aerodynamic surface in a bird. They are lightweight, but very strong. Contour feathers maintain a uniform surface for the bird so that there are no abrupt curvatures that may create turbulence. The long flight feathers, known as primaries, secondaries, and tertiaries, form the wing. Long, narrow wings, termed high aspect-ratio wings, consume less power when flapped than broad or low aspect-ratio wings. The bird wing is cambered and so creates lift in an airflow. Shore birds, such as gulls, are seen lifting out of the water just by holding out their wings in a breeze without the need of flapping. Flapping the wings faster than the velocity of the air increases the lift and thrust imparted to the air. Birds are able to fly because they can generate more lift than their body weight, and it has taken them considerable evolutionary time to perfect the weight reduction necessary for flapping flight. Penguins use their wings to fly underwater. They are extremely maneuverable and use their agility and speed to catch fish. The wing beat does not proceed through the same distance as aerial flying birds because of the greater viscosity of water.

Bats are also evolved from a quadrupedal ancestor. In their case, a skin membrane is stretched between the elongated fingers. In this way the camber and aspect ratio of the wing can change quickly. Many bats retain a membrane between the legs, the uropatgium, that increases the lifting surface of the body. Bats are agile and rapid flyers and, equipped with ultrasonic pulses, can locate a variety of prey from insects to frogs. Mammals have evolved gliding forms in much the same way across other taxa. A wing membrane is stretched between the fore and hindlimbs in marsupials such as sugar gliders, and among placental mammals such as flying squirrels (rodents) and the so-called flying lemur (*Cynocephalus*).

Pterosaurs also used a skin membrane stretched between the elongated fourth digit and the hindlimb. This design worked well for over a hundred million years, but a tear in the membrane would destroy the lift-generating ability of the wing. While most pterosaurs were small, some giant forms evolved with wing spans of over forty feet. Lizards have never evolved powered flight, but have generated gliders and parachutists. In the Triassic and Cretaceous periods as well as in the present day, lizards evolved a flight membrane supported by the ribs. The ribs can be folded against the body when the lizard is climbing among the branches in search of insect prey. Other lizards, such as geckos, have evolved body fringes of skin and the ability to flatten themselves to create an aerodynamic surface. Some species of snakes launch themselves into the air, and some frogs use their expanded webbed hands and feet for gliding.

Although fishes essentially fly with the tail, and many flap the pectoral fins like bird wings, some fish have taken to the air. Flying fish will gain speed underwater using their tail and then leap out of the water spreading their large pectoral fins. These fishes can glide for some distance before reentering the water. The flight strategy is associated with escape behavior from predaceous fish.

There is controversy surrounding the origin of insect flight. Some postulate that insect wings developed for temperature control, as wing beating would heat the organism. Moreover, wings could be used to collect or dissipate heat depending upon exposure to the sun or wind. Others contend that insect wings evolved from lateral extensions of the exoskeletal tergites to aid in the stabilization of jumps. However it began, insects are the only invertebrate to evolve aerial powered flight. Normally insects have two pairs of wings, the first pair essentially acting as covers for the posterior wings involved in lift and thrust generation. Some insects have two functional pairs of wings, such as dragonflies. The wings are operated by muscles that have short contraction lengths to obtain rapid wing movements. As is the case in birds, small insects have the highest wing beat frequencies, reaching over one thousand beats per second. Muscles attach either directly to the wing bases, or indirectly to a specialized part of the dorsal carapace termed the notum. The deflection of the notum moves joints that connect to the wings, allowing them to beat in a twisted fashion and create the aerodynamically shaped surfaces.

—Samuel F. Tarsitano

See also: *Archaeopteryx*; Bats; Birds; Feathers; Insects; Locomotion; Mammals; Pterosaurs; Wings.

Bibliography

Ellington, C. P. "Aerodynamics and the Origin of Insect Flight." *Philosophical Transactions of the Royal Society of London*. B305 (1991): 1-181. This paper is a review of the basic principles of flight and their application to the origin and mechanics of insect flight. Theories of the origin of insect flight are discussed along with the evolution of powered flight.

Feduccia, Alan. *The Origin and Evolution of Birds*. 2d ed. New Haven, Conn.: Yale University Press, 1999. An extensive review of the controversies surrounding the evolution of flight in birds. Discusses the ecology and adaptations of the major groups of birds.

Pennycuick, C. J. *Animal Flight*. London: Edward Arnold, 1972. A general but extensive review of aerodynamic principles of flight. Gliding, flapping, and hovering flight are explained in understandable terms using living animals as examples.

Tarsitano, S. F., A. P. Russell, F. Horne, C. Plummer, and K. Millerchip. "On the Evolution of Feathers from an Aerodynamic and Constructional Viewpoint." *American Zoologist* 40 (2000): 676-686. This paper deals with the aerodynamic basis for the evolution of feathers. Results from water tunnel tests on protofeathers are used to explain why feathers may have been selected first for flight rather than for insulation.

Vogel, S. *Life in Moving Fluids*. 2d ed. Princeton, N.J.: Princeton University Press, 1994. The best overall source laying down the basis of flow around objects and flying organisms. The flight of bats, birds, insects, and even tree seeds is described and discussed.

FOOD CHAINS AND FOOD WEBS

Types of animal science: Behavior, ecology, geography, physiology, scientific methods
Fields of study: Ecology, environmental science, marine biology, physiology, population biology, wildlife ecology

In the biosphere—the portion of earth that sustains life—organisms including plants, animals, and microbes interact and interconnect with one another through organized and complex networks. Understanding this network, also referred to as a food chain or food web, is vital to the study of many subjects in biology.

Principal Terms

BIOLOGICAL MAGNIFICATION: the increasing accumulation of a toxic substance in progressively higher feeding levels

CONSUMER: an organism that eats other organisms

DECOMPOSERS: microbes such as fungi and bacteria that digest food outside their bodies by secreting digestive enzymes into the environment

DETRITUS FEEDERS: an array of small and often unnoticed animals and protists that live off the refuse of other living beings

ENERGY PYRAMID: a graphical representation of the energy contained in succeeding trophic levels, with maximum energy at the base (producers) and steadily diminishing amounts at higher levels

NUTRIENT CYCLE: a description of the pathways of a specific nutrient (such as carbon, nitrogen, or water) through the living and nonliving portions of an ecosystem

PRODUCERS: organisms that produce food for themselves as well as for nearly all other forms of life, including plants, plantlike protists, and cyanobacteria

TROPHIC LEVEL: the categories of organisms in a community, and the position of an organism in a food chain, defined by the organism's source of energy

All activities of life are powered, whether directly or indirectly, by a single source of energy: sunlight. The energy enters the biosphere through primary producers, which trap and convert solar energy into chemical energy, first in form of sugars and ultimately in other complex organic molecules. Energy in its chemical form is then passed along from one type of organism to another through a complex feeding relationship. Two fundamental laws underlie this function: Energy moves through ecosystems in a one-way street, while nutrients cycle and recycle. Each time the energy is used, some of it is lost as heat. Energy needs constant replenishment from an outside source, the sun. In contrast to energy, nutrients constantly cycle and recycle in a circular flow. Nutrients enter the system from soil or water or atmosphere through primary producers (plants), and pass along to herbivores, then to carnivores. The wastes and dead bodies or body parts degraded by detritus feeders return nutrients to the ecosystem.

Feeding Relationships Within a Community

Feeding relationships in a community are often defined or described through food chains and food webs. A linear feeding relationship spanning all trophic levels is called a food chain, whereas many interconnecting food chains in a community make up the food web. Obviously, different ecosystems have drastically different food chains. To illustrate who feeds on whom in a community, it is better first to examine some basic laws and structures that govern a community in general. Energy and nutrients are two common elements that sustain all communities and ecosystems. Energy enters communities through the process of photosynthesis, by which plants and other photosynthetic organisms trap a small portion of sunlight and convert it into sugars. Photosynthetic organisms, from the mighty sequoia to the zucchini and tomato plants in a garden to single-celled diatoms in the ocean, are called autotrophs or producers, because they produce food for themselves. They also produce food for nearly all other organisms, called heterotrophs or consumers.

The amount of life a community can support is determined by how much energy the producers within it can capture. Within the community, energy flows from producers, occupying the first trophic level, through several levels of consumers. The consumers that feed directly and exclusively on producers are herbivores, ranging from caterpillars to buffaloes to wheat aphids. These herbivores, also called primary consumers, form the second trophic level. Carnivores, such as the spider, eagle, fox, and birds that eat caterpillars, are meat eaters, feeding primarily on primary consumers. Carnivores are secondary consumers that form the third or higher trophic level. Some consumers, such as the black bear that eats both blueberries and salmon, occupy more than one trophic level.

A food web, however, represents many interconnecting food chains in a community, describing the actual, complex feeding relationships within a given community. A food web also reflects the feeding nature of organisms that occupy more than one trophic level. Animals such as raccoons, bears, rats, and a variety of birds are omnivores, eating at different consumer levels at different times. In addition to producers and consumers, a functional ecosystem also consists of detritus feeders and decomposers that release nutrients for reuse. The extremely diverse network of detritus feeders is made up of earthworms, mites, protists, centepedes, nematodes, worms, some crustaceans and insects, and even a few verte-

Dodos and Bats: Cases of Disrupted Ecosystems

Within complex ecosystems, animals and plants interact with each other in very intricate and dynamic ways. Plants provide animal species with nutritious fruits. In return, animals help to disperse their seeds. These mutually beneficial relationships sustain both plant and animal populations, and ultimately the ecosystem itself.

A pair of organisms, the dodo bird and the tambalacoque tree, coexisted on the island of Mauritius in the Indian Ocean for a long time. When humans arrived at the island, the slow dodos were easy prey. By 1681, humans had hunted the dodo to extinction. The damage, however, reached far beyond just dodos. It turned out that dodos helped disperse the seeds and promote the germination of the tambalacoque trees of Mauritius. Dodos ate the tambalacoque fruit before it had a chance to rot, thoroughly cleaning the seeds and thereby protecting them from infections by destructive fungi. Dodos also dispersed the seeds all over the island, ensuring that some of them reached fertile soil for germination. With the dodos gone, the fruit rapidly rots, the seeds are destroyed by fungi, and the tambalacoque trees are seriously threatened.

In tropical forests, bats serve as the most important agents for seed dispersal. Bats may eat up to twice their weight in fruit in a day. They also fly more than twenty miles per night, defecating the seeds in flight. After passing through bats' digestive tract, seeds have more than a 95 percent germination rate, as compared to 10 percent in seeds planted directly from fruit. As bat populations (and those of other fruit-eating animals) decline dramatically, seed dispersal has stopped or been significantly reduced. Tropical fruits are rotting on the forest floor or sending up doomed sprouts under the shade of their parents and other canopies. Some types of tropical forests may not survive.

brates such as vultures. Except for vultures, these organisms thrive in the garden, and compost by extracting energy stored within dead organic matter, in turn releasing it in a further decomposed state. The excretory products released serve as food for other detritus feeders and decomposers, which are primarily fungi or bacteria. Fungi and bacteria digest food outside their bodies by releasing digestive enzymes into the environment. They then absorb the nutrients they need, and leave the remaining nutrients for recycling. Without detritus feeders and decomposers, nutrients would soon be locked into organic matters, and the ecosystem will cease to be functional.

Energy Transfer and the Nutrient Cycle

One important principle that governs the flow of energy through the ecosystem is that the energy transfer from one trophic level to the next is never efficient. The net transfer of energy between two trophic levels is roughly 10 percent. During the transfer, 90 percent of the energy is lost as heat or in other forms. Of one thousand calories stored in the producer (plants), the caterpillars that consume all plant tissues will obtain one hundred calories, a bird that eats caterpillars will extract ten calories, and when a hawk catches the bird, that portion of the energy is reduced to a mere one calorie. This inefficient energy transfer between trophic levels is called the 10 percent law, or an energy pyramid. This 10 percent law has profound impacts within an ecosystem. Plants have the most energy available to them; the most abundant animals will be those directly feeding on plants, and carnivores will always be relatively rare, especially those of large size that occupy a higher trophic level.

However, when toxic substances pass through trophic levels, the exact opposite is true. While energy diminishes in the process of flowing from lower to higher trophic levels, toxic substances progressively increase in concentration along the food chain. This phenomenon, called biological magnification, was discovered through the study of the use of the pesticide dichloro-diphenyl-trichloroethane (DDT). Tests of water samples following the use of this pesticide showed a trace amount of DDT. Tissue analyses of predatory birds in the same aquatic ecosystem, however, revealed a DDT concentration a million times greater than that in the water. Fish caught from the same waters also contained much higher DDT levels than the water, but substantially lower levels than that of the birds that consumed those fish. DDT has since been confirmed as the cause for population declines of several predatory birds, especially fisheaters such as bald eagles, brown pelicans, and cormorants. Understanding biological magnification is crucial to the prevention of widespread loss of wildlife.

Unlike energy, no mechanism or source exists to allow a constant replenishment of nutrients. The same pool of nutrients has been supporting life from the beginning. Nutrients are elements and small molecules that form all the building blocks of life. Macronutrients are those acquired by organisms in large quantities, including water, carbon, hydrogen, oxygen, nitrogen, phosphorus, sulfur, and calcium. Micronutrients, including zinc, molybdenum, iron, selenium, and iodine, are acquired in trace quantities. Nutrients cycle from producer, to consumer, to detritus feeder and decomposer, and eventually back to producer. The major reservoirs of nutrients are in the nonliving environment, such as soil, rock, water, and atmosphere. In an undisrupted ecosystem, nutrients have cycled and recycled in a sustainable manner for thousands of years. However, human intervention, either through industry or agriculture, has created enormous problems for sustainable nutrient cycling. One example is the nitrogen cycle. Although plants may obtain nitrogen from soil, the primary source of nitrogen is the atmosphere. Nitrogen gas (mainly N) makes up over 70 percent of the air. N may be extracted by some plants or converted by lightning into useable forms that drop as rainfall. Through industrial production and use of nitrogen fertilizers in agriculture, humans have overcharged the nitrogen cycle, causing acid rains and surface water and groundwater contamination, which pose a serious threat to natural ecosystems.

Why Preserve Biodiversity?

Biodiversity is defined as the total number of species within an ecosystem, and also as the resulting complexity of interactions among them. It measures the "richness" of an ecological community. Among an estimated eight to ten million unique and irreplaceable species existing on Earth, fewer than 1.6 million have been named. A tiny fraction of this number has been studied. Over thousands of years, organisms in a community have been molded by forces of natural selection exerted by other living species, as well as by the nonliving environment that surrounds them. The result is a highly complex web of interdependent species whose interactions sustain one another and provide the basis for the very existence of human life as well. Thus, it is easy to see why biodiversity should be preserved.

Loss of biodiversity poses a serious challenge to the sustenance of many communities and ecosystems. For example, the destruction of tropical rain forests by clearcut logging produces high rates of extinction of many species. Most of these species have never been named, and many never even discovered. As species are eliminated, the communities of which they were a part may change and become unstable and more vulnerable to damage by diseases or adverse environmental conditions. Aside from disrupting the natural food webs, potential sources of medicine, food, and raw materials for industry are also lost. As Harvard professor Edward O. Wilson once said, "The loss of species is the folly our descendants are least likely to forgive us."

Food Chains and Food Webs in a Terrestrial Ecosystem

Since different ecosystems have drastically different food chains and food webs, they can be better illustrated using examples of different communities. The first example is a land community, to provide an overview of terrestrial food chains and food webs. Plants such as maple trees and squash are the producers that occupy the first trophic level. Aphids, caterpillars, grasshoppers, and other animals that forage directly plants or plant tissues are primary consumers that occupy the second trophic level. Birds, spiders, and other insects that feed on primary consumers are secondary consumers occupying the third trophic level. Large birds, such as eagles, owls, and hawks, that eat secondary consumers are tertiary consumers making up the fourth trophic level. This food chain can go on to even higher trophic levels.

However, natural communities rarely contain well-defined groups of primary, secondary, and tertiary consumers in a linear pattern. In reality, a food web showing many interconnecting food chains in a community describes much more accurately the actual feeding relationships within a given land community. This is in part due to the omnivorous, or "eating all" nature of some animals. These animals include but are not limited to bears, rats, and raccoons. They act as primary, secondary, and even tertiary consumers at different times. Many carnivores will eat either herbivores or other carnivores, thus acting as secondary or tertiary consumers, respectively. An owl, for instance, is a secondary consumer when it eats a mouse, which feeds on plants, but a tertiary consumer when it eats a shrew, which feeds on insects. Once a shrew eats carnivorous insects, it is by itself a tertiary consumer, making the owl that feeds on the shrew a quaternary consumer that occupies the fifth trophic level. Since organisms are interlocked in such complex yet organized networks, disruption at any particular point of the food web (damage to one group of organisms) might have far-reaching effects on a whole community or ecosystem.

Food Chains and Food Webs in a Marine Ecosystem

Coral reefs will be used as an example for marine ecosystem. Reefs are created by concerted efforts of producers—algae—and consumers—corals. In warm tropical waters, with just the right combination of bottom depth, wave action, and nutrients, specialized algae and corals build reefs from their own calcium carbonate skeletons. The reef-building corals grow best at depths of less than forty meters, where light can penetrate and allow their algal partners to photosynthesize. Algae and corals are involved in a mutualistic relationship, where algae benefit from the high nitrogen, phosphorus, and carbon dioxide levels in the coral tissues. In return, algae provide food for the coral and help produce calcium carbonate, which forms the coral skeleton.

Coral reefs provide an anchoring place for many other algae, a home for bottom-dwelling animals, and shelter and food for the most diverse collection of invertebrates and fish in the oceans. In essence, algae is the producer that occupies the first trophic level. Corals that feed on algae are primary consumers sitting at the second trophic level. Many fish (such as blue tang) that feed on corals are secondary consumers occupying the third trophic level. Larger fish, such as sharks that eat small

Ecosystem Energetics

The diverse forms of life in all ecosystems are powered by a single energy source, sunlight, which enters the biosphere through a process called photosynthesis. Organisms such as plants and algae capture a small fraction of sunlight's energy, storing it as chemical energy in sugar or other complex organic molecules. The energy then passes through an ecosystem via different feeding (trophic) levels. Each time the energy is passed on, a portion of it is lost as heat. Thus, energy needs constant replenishment from the original source, the Sun.

How does energy flow through communities? It flows through ecosystem energetics. Of the energy that reaches Earth, much is either reflected or absorbed as heat by the atmosphere and Earth's surface, leaving only about 1 percent to power all life. Of this 1 percent, green plants capture 3 percent or less. All life on this planet is therefore supported by less than 0.03 percent of the energy reaching Earth from the Sun. Photosynthetic organisms are called autotrophs, or producers, because they produce food for themselves. Directly or indirectly, they produce food for nearly all other forms of life as well. Organisms that cannot photosynthesize are called heterotrophs, or consumers, because they must acquire energy prepackaged in the molecules of the bodies of other organisms.

There are three basic principles that govern ecosystem energetics. First, the amount of life an ecosystem can support is defined by the energy captured by the producers. This energy, made available to consumers over a given period, is called net primary productivity. It is usually measured in units of energy stored per unit area in a given period, or measured as biomass, the dry weight of the total organic material added to the ecosystem per unit area in a given time span. The net primary productivity is influenced by a variety of environmental factors, including the amount of sunlight, the availability of water, the amount of nutrients available to producers, and the temperature. Among all environmental variables, the most limiting variable is the one that determines net primary productivity, for instance, water in the desert or light in the deep ocean.

Second, within the community, energy is passed from one feeding level to another. Energy flow moves from producers to primary consumers to secondary and tertiary consumers. Primary consumers are normally herbivores that directly consume producers. Secondary and tertiary consumers are typically carnivores that eat meat of other consumers. Certain consumers may occupy more than one feeding level. As energy is passed through feeding levels via food chains or food webs, the transfer is never efficient. This is the third basic principle of ecosystem energetics. Each time the energy is transferred to the next feeding level, the bulk of it is lost as heat. On average, only 10 percent of the energy is transferred from one feeding level to the next. In other words, the higher the feeding level an organism occupies, the less energy is available to it. This so-called 10 percent law or energy pyramid puts a cap on how much life a particular ecosystem can sustain.

—Ming Y. Zheng

fishes, are tertiary consumers at the fourth trophic level. A vast array of zooplanktons, invertebrates such as sponges, the poisonous blue-ringed octopus, and so on, also live in coral reef ecosystems to make extremely complex marine food webs. For example, the Great Barrier Reef in Australia is home to more than two hundred species of coral, and a single reef may harbor three thousand species of fish, invertebrates, and algae.

Similar to terrestrial ecosystems, aquatic ecosystems are also prone to human disturbance. Of all aquatic or marine ecosystems, coral reefs are probably most sensitive to certain types of disturbance, especially silt caused by soil eroding from nearby land. As silt clouds the water, light is diminished and photosynthesis reduced, hampering the growth of the corals. Furthermore, as mud accumulates, reefs may eventually become buried and the entire magnificent community of diverse organisms destroyed. Another hazard is sewage and runoff from agriculture. The dramatic rise in fertilizer in near-shore water causes eutrophication, by which excessive growth of algae blocks sunlight from the corals, deprives corals of nutrients, and suffocates corals and other organisms. A third threat to coral reefs, overfishing, is also strictly a result of human interference. It is estimated that in over eighty countries, an array of species, including mollusks, turtles, fish, crustaceans, and even corals, are being harvested much faster than they can replace themselves. Collectively, these human activities had destroyed over 30 percent of coral reefs worldwide by year 2000. Assuming no effective measure is taken to preserve or restore coral reef ecosystems, another 50 percent of reefs will disappear by year 2030. The message is clear: Once humans disturb an ecosystem, through damaging one or more species in an intricately networked food web, balance and sustainability within the whole system is affected. The price for such disruption is high and far-reaching.

—*Ming Y. Zheng*

See also: Biodiversity; Biogeography; Carnivores; Chaparral; Digestion; Digestive tract; Ecosystems; Forests, coniferous; Forests, deciduous; Grasslands and prairies; Habitats and biomes; Herbivores; Ingestion; Lakes and rivers; Marine biology; Mountains; Nutrient requirements; Omnivores; Predation; Rain forests; Savannas; Symbiosis; Tidepools and beaches; Tundra.

Bibliography

Baskin, Y. "Ecologists Dare to Ask: How Much Does Diversity Matter?" *Science* 264 (April 8, 1994): 202-203. Outlines and discusses the scientific evidence for the importance of biodiversity to ecosystem function.

Bell, R. H. V. "A Grazing Ecosystem in the Serengeti." *Scientific American* 255 (July, 1971): 86-93. The great migrations of a variety of herbivores of the African savanna are illustrated and described.

Earthworks Group. *Fifty Simple Things You Can Do to Save the Earth*. Berkeley, Calif.: Earthworks Press, 1989. Provides compelling, practical advice on how individuals can make a difference in being good stewards to global resources and the environment.

Goreau, T. F., et al. "Corals and Coral Reefs." *Scientific American* 241 (August, 1979): 124-136. Good discussions on the ecology and formation of the great reefs built by mutualistic relation between algae and coral.

Holloway, M. "Sustaining the Amazon." *Scientific American* 269 (July, 1993): 90-99. Describes the threats to the rain forests and possible ways to preserve tropical forests and their richness in biodiversity.

Quamman, David. *The Song of the Dodo: Island Biogeography in an Age of Extinction*. New York: Simon & Schuster, 1997. This beautifully written book takes the reader on a globe-spanning tour of wild places and ideas through interwoven personal observations, scientific theory, and history.

FORESTS, CONIFEROUS

Types of animal science: Ecology, geography
Fields of study: Conservation biology, ecology, environmental science

Coniferous forests are those in which the predominant species of trees carry their seeds in cones. Although conifers are often associated with cooler climates, they have wide global distribution. Conifers form the plant phylum Pinophyta and include such well-known species as sequoias, spruces, and cedars.

Principal Terms

CONIFER: member of the phylum Pinophyta, characterized by carrying seeds in cones and often having needlelike foliage

DECIDUOUS: trees which lose their leaves annually, such as maples and elms

GYMNOSPERM: a plant whose seeds are borne on seed scales arranged in cones

NAVAL STORES: conifers used for production of turpentine, pitch, and other materials once considered necessary for maintenance of navy vessels

TOLERANCE: ability of a tree to grow in the shade of other trees

Forests in which the dominant species of tree are members of the plant phylum Pinophyta are referred to as coniferous forests. Coniferous forests are often associated with cooler climates, such as the boreal forest stretching across northern Europe, Asia, and North America, but conifers are actually found in a wide range of climates and locations. Conifers are sometimes referred to as evergreens, meaning that the trees seemingly never lose their leaves, but not all conifers are evergreen, nor are all evergreens conifers.

Conifers do, in fact, lose their leaves, but with the exception of the larches (also known as tamaracks), which change color uniformly and shed their leaves seasonally, needle drop is a gradual process. Depending on the species, leaves will remain green and on the tree for many years rather than being replaced annually. The new growth replaces the dropped needles gradually enough that often only the thick mulch of dead needles at the base of pines, spruces, and other conifers provide definitive evidence that needle drop occurs. In some coniferous forests, the thick carpeting of needles on the woodland floor chokes out shade-tolerant species of vegetation, while the evergreen canopy blocking the sun discourages intolerant species from sprouting. Over time a pure stand of conifers may emerge, in which few other species of trees grow and even smaller trees of the same species are crowded out. Many conifers are highly shade intolerant, so a coniferous forest will thin itself naturally. The spreading branches of the taller trees will block the light needed by smaller ones. A mature stand of red pine, for example, will eventually be composed of so many trees of similar size and even spacing that natural red pine forests can be mistaken for artificially planted commercial stands.

Physical Characteristics of Conifers

Conifers take their name from the fact that their seeds are carried in cones and from the observation that many of the most widely recognized coniferous trees, such as spruces and pines, present a rather conical appearance, broad-based at the bottom and tapering symmetrically to a pointed top. The trunks of conifers rarely fork, but instead form a straight central trunk with branches radiating from it. In the event a conifer loses its tip, or leader, a branch near the break may shift its orientation to grow upward to replace the lost leader, while other branches will continue to radiate out horizontally. Not all conifers form such a neat appearance, however. The bristlecone pine is noted for its twisted, gnarled appearance, and many junipers and yews tend to sprawl rather than grow upright. The narrow, needlelike leaves of many conifers are another easily identifiable characteristic, although some conifers, such as the cedars, have leaves that are small and scalelike rather than needle-shaped.

The long, straight trunk of many conifers has contributed to the trees being valued as a source of building materials. During past centuries, millions of acres of white pine forest in North America were cut, first for use as masts on sailing ships, and later for lumber for general construction. Species such as western red cedar, redwoods, and bald cypress continue to be prized for their natural resistance to rot and water damage, while Douglas fir and white spruce are important sources of construction lumber.

In addition to being cut for lumber, conifers are harvested for pulp and the production of chemicals. Prior to the introduction of petroleum products, sailing ships required huge quantities of pine tar. Governments would set aside forest reserves as naval stores, which meant that the trees were not to be cut for lumber. The conifers were instead reserved as a source for turpentine and pine tar, which were produced from the sap of conifers such as pitch pine. The bark of the trees were slashed to allow sap to run into pans for collection. The sap was then distilled into turpentine and pine tar.

Conifer Range

Although coniferous forests dominate the cooler regions of the globe, conifers are also found in tropical and desert regions. Cypress trees thrive in warm, swampy areas, such as the bayous of Louisiana, while piñons and other pines constitute the dominant species on the desert plateaus of the American southwest. The thin, waxy needles of conifers conserve moisture while allowing photosynthesis to occur year-round, an adaptation that serves conifers equally well in the harsh conditions of a subalpine mountain slope or during long Canadian winters.

Some mature conifers are also noted for their extremely thick bark, which reduces their vulnerability to fire. As recently as the 1990's, foresters continued to use fire routinely as a management tool in plantations of loblolly and southern yellow

Coniferous forests are composed of very tall trees with relatively little undergrowth. (Corbis)

pines. Controlled burns removed the undesirable broadleaf species, such as rhododendrons, from the understory, while leaving the taller, thick-barked pines unharmed. Other species have evolved so that natural fire is a necessary part of the reproductive cycle. Until humanity began suppressing forest fires, fires triggered by lightning strikes were a common occurrence in many coniferous forests. The litter of dead needles and fallen branches on the ground rarely built up to catastrophic levels, and low intensity fires remained on the ground where they did little long-term harm to the forest. Lodgepole pine, for example, bears seed cones that will remain tightly closed for many years, until exposed to extreme heat. When fire sweeps through an area, it both clears the ground of competing species and opens the lodgepole pine cones, allowing for a new generation of trees to sprout.

Conifer Size and Age

Species of trees found in coniferous forests range from the stunted-looking spruces growing in the arctic taiga, where the boreal forest merges into the tundra, to the giant sequoias (*Sequoiadendron giganteum*) in the mountains of southern California. In far northern regions, spruce trees hundreds of years old may stand barely ten feet tall and be only a few inches in diameter. Sequoias, in contrast, are the most massive trees on earth, as they attain heights of 275 feet and diameters as large as 30 feet. Young sequoias present a pleasing symmetrical appearance, the typical cone shape of many conifers, but as the trees mature they often lose their tops to lightning strikes and become rather ragged and uneven looking. The largest sequoia on record measured 293 feet in height, with a diameter of 37 feet. Once widespread throughout North America, giant sequoias are now found only in isolated groves high in the mountains of California. A closely related species, the redwood (*Sequoia sempervirens*) is the dominant species in the forests along the Pacific coast in northern California and Oregon. Although redwoods grow as tall as sequoias, they do not achieve as a large a diameter, nor are they as long-lived. Redwoods are considered mature at 400 years and may live to 1,500. No sequoia is known to have died of old age, although mature trees have been killed by lightning strikes and fire.

The oldest known living examples of conifers, however, are not the mammoth redwoods and sequoias, but bristlecone pines found in the mountains of the North American West. Scientists have identified individual bristlecone trees over four thousand years old. Unlike the sequoias and the bristlecone pines, which may live for millennia, other conifers, such as balsam fir, are relatively short-lived, reaching maturity in only a few decades and dying off while less than a century old.

Based on the fossil record, scientists believe conifers have existed on earth much longer than the broad-leaved species of trees. Botanists speculate that conifers evolved from the giant fern trees common during the days when dinosaurs roamed the earth. Broad leaf trees, such as maples, are angiosperms, or plants whose flowers contain both male and female parts. Angiosperms are the most common plants on earth, with hundreds of thousands of different species growing in a wide range of environments. Angiosperms encase seeds within fruits, such as acorns, that both protect the seed and provide food for the embryonic plant. In contrast, the flowers of gymnosperms will be either male or female, never both at once, and seeds are set on scales arranged in cones. Although the flowers of gymnosperms are less complex than angiosperms, gymnosperms may have both male and female flowers on the same plant. Just as the needles of many conifers may remain on the tree for five or more years, so too do the cones of some species cling to the tree through many seasons. Sequoias, for example, will hold cones for as long as twenty years, with the cones remaining green and seemingly recently formed the entire time they are on the tree. Gymnosperms such as the conifers number fewer than one thousand species globally and are more limited in range than angiosperms.

Conifers as Habitat

All forests provide wildlife habitat on multiple levels, with some animals living on or under the forest floor, others taking advantage of resources available in the lower sections of the trees, and still others living in the canopy. Coniferous forests are no exception. Small rodents, such as chipmunks and mice, live in underground burrows, subsisting on seeds gathered from fallen cones. While the needles on conifers may not appear palatable to humans, deer, elk, and moose depend on conifers for winter browse. Moose, in particular, rely on conifers such as balsam fir as a primary food. Squirrels often thrive in dense spruce and balsam stands, building their nests high in the trees and rarely setting foot on the ground. A wide variety of small birds and mammals often dwell in coniferous forests, as the evergreen canopy provides a protective cover from large birds of prey such as eagles. Kirtland's warbler, the marbled murrelet, and the spotted owl are only a few examples of birds that call coniferous forests home.

—*Nancy Farm Männikkö*

See also: Chaparral; Ecosystems; Forests, deciduous; Grasslands and prairies; Habitats and biomes; Lakes and rivers; Mountains; Rain forests; Savannas; Tundra.

Bibliography

Maser, Chris. *Forest Primeval: The Natural History of an Ancient Forest*. Reprint. Corvallis: Oregon State University Press, 2000. An in-depth study of the old-growth forests of the Pacific Northwest.

Massa, Renato. *The Coniferous Forest*. Austin, Tex.: Raintree/Steck-Vaughn, 1996. A good overview of coniferous forests. The book is aimed at young adult readers and is written in terms easily understood by anyone.

Nakagoshi, N., and Frank B. Golley, eds. *Coniferous Forest Ecology, from an International Perspective*. The Hague, Netherlands: SPB Academic Publishers, 1991. A collection of essays covering a variety of ecosystems globally, ranging from the rain forests of the Olympic Peninsula and British Columbia to the Siberian taiga.

Preston, Richard J., Jr. *North American Trees*. 4th ed. Ames: Iowa State University Press, 1989. A good general field guide to the trees of North America that includes a glossary of technical terms as well as a discussion of the commercial uses, if any, of various species.

Smith, William K., and Thomas M. Hinckley, eds. *Ecophysiology of Coniferous Forests*. San Diego, Calif.: Academic Press, 1995. Detailed essays discussing the unique characteristics of coniferous forests; may prove to be difficult reading for younger students.

Van Gelderen, D. M. *Conifers: The Illustrated Encyclopedia*. Portland, Oreg.: Timber Press, 1996. This is the definitive work on conifers, and it is accessible to any interested reader.

Walker, Laurence C. *The Southern Forest*. Austin: University of Texas Press, 1991. A thoroughly interesting discussion of forests in the southern United States, including the Mississippi pineries and Florida pitch pine naval stores.

Welch, Humphrey J. *The Conifer Manual*. Boston: Kluwer Academic Publishers, 1991. A good reference guide to conifers, although not quite as complete as Van Gelderen's.

FORESTS, DECIDUOUS

Types of animal science: Classification, ecology, reproduction
Fields of study: Anatomy, ecology, zoology

Deciduous forests contain trees that shed their leaves. Temperate deciduous forests shed their leaves in autumn, and monsoon or tropical deciduous forests do so in the dry season. They hold many animal species, including predators, omnivores, and herbivores.

Principal Terms

CARNIVORE: animal that eats other animals
DECIDUOUS: shedding or losing leaves at the end of a growing season
GESTATION: the term of pregnancy
HERBIVORE: animal that eats plants
OMNIVORE: animal that eats plants and animals

A forest is an aggregate of vegetation, mostly trees and other woody plants, on land that may encompass many square miles. Forests exist relatively unchanged over very long time periods, having in them lasting populations of specific trees, shrubs, grasses, and animals. The interactions of these creatures with each other and with the forest helps to maintain the status quo. The climate, soil, and geography of the region where a forest occurs determine the main tree types it contains and their locations in the aggregate. Within any section of a forest, dominant trees help to control the specific shrubs, herbs, and grasses that are present.

Forests are divided into ten types on the basis of tree leaf types and ambient climate. Two are deciduous temperate and deciduous monsoon or tropical forests. Deciduous temperate forests are exemplified by the typical forest of the eastern and midwestern United States, Balkan forests, and Scandinavian forests such as the ones in Norway. Deciduous temperate forests hold trees which lose their leaves in autumn, when the weather turns cold. Deciduous monsoon or tropical forests include those of Southeast Asia, western coastal Mexico, and Central America. The climate in those areas varies between long periods of heavy daily rainfall and dry periods during which trees shed their leaves.

Temperate Deciduous Forests

Typical temperate deciduous forests are those occurring in the mountains of southeastern Europe's Balkan Peninsula. These mountains begin at the western Romanian border and run south to the Bulgarian border. In Bulgaria, they turn east and run toward the Black Sea. The Balkans contain deciduous forests whose trees are mostly oak, ash, beech, birch, hazel, and maple. Elk, deer, foxes, wolves, badgers, weasels, martens, wild boar, lynx, bear, and roe deer are among larger animals found in them.

In the United States, over 40 percent of Wisconsin's land is forest. The southern part of the state has huge deciduous forests composed largely of maple, basswood, oak, elm, and hickory trees. Many animals inhabit the forests, including white-tailed deer, black bears, badgers, foxes, hares, porcupines, woodchucks, coyotes, skunks, mink, beavers, muskrats, wolves, martens, copperhead snakes, rattlesnakes, and lynx. In addition, wild birds include pheasant, ruffed grouse, partridges, robins, wrens, swallows, owls, chickadees, and eagles.

Forests also cover about one third of Norway. Scandinavian deciduous forests occur near the south and southwest coasts. The main trees in the forests are oak, ash, birch, hazel, elm, and maple. Elk, deer, foxes, otters, wolves, martens, pit vipers, other reptiles, and many birds inhabit the forests.

Fauna of Temperate Deciduous Forests

Among the interesting organisms in temperate deciduous forests are carnivores, including the weasel family (mustelids). These furry mammals weigh from a few ounces to one hundred pounds. They include weasels, martens, otters, skunks, and badgers. All have short legs and paws with five toes and sharp claws for grasping prey and burrowing. Beautiful mustelid fur is used in expensive fur garments. Mustelids have glands that make musks, which produce odors to mark territory or enable self-defense. Mustelid musk is most offensive in skunks, who use it in self-defense. Male and female mustelids live alone, except when mating. Mating seasons, gestation (two to nine months), and litter size (up to ten young) depend upon species. Life spans of those mustelids that reach old age are up to twenty-five years.

Lynx, also carnivores, are cats. They have very long legs and large paws, are found all through temperate and subarctic regions of the northern hemisphere, and can grow to three feet long. Most have hair tufts at each ear tip and are agile climbers, spending much of their time on the limbs of trees, waiting to ambush passing prey. Lynx live in caves and hollow trees. Their litters contain two to four offspring. The lynx of Europe, the United States, and Canada are different species.

Wolves belong to the dog family. The two wolf species are the gray (timber) wolves of Europe, Asia, and the New World and the red wolves of Texas and Louisiana. An adult timber wolf has a maximum body length of 6.5 feet and a weight of 185 pounds. These wolves are usually reddish-yellow to yellowish-gray, with white underbodies. Red wolves are smaller and darker colored. All wolves have powerful teeth and bushy tails. They live on prairies, in forests, and on mountains. In winter, their packs seek small animals, birds, and large mammals as prey. Wolves den in caves, hollow trees, or holes they dig. Breeding season is the spring, and a female has three to nine cubs, which stay with parents until the following winter.

Deer and Elk

Deer and elk inhabit all temperate forests. The name deer denotes hoofed, artiodactyl mammals of the family Cervidae, whose males have bony, branching antlers that are shed and regrown annually. There are forty cervid species in the Americas, Asia, Europe, and North Africa. Most inhabit forests, especially deciduous forests. Deer species range in size from moose, seven feet tall at shoulder height, to tiny species only a foot tall.

Deer and elk eat the twigs, leaves, bark, and buds of bushes and trees, as well as grasses. They feed mostly at twilight. Does (or female moose) birth one or three offspring each year, after six- to ten-month gestations. In the United States, the most common deer is the white-tailed deer, known for a tail with a white underside. The American elk, or wapiti, is called the red deer in Europe. Moose—the largest deer in the world—found throughout the United States, are known as elk in Europe.

Dear and elk have compact bodies and long, powerful legs suited for woodland terrain. They are also fine swimmers. They are herbivorous ruminants (cud chewers). Also, their solid and bony antlers, unlike the hollow, permanent horns of other ruminants, form only in males. The antlers are used to make territorial markings and threaten or combat other males. Male moose have the largest antlers, which can be six feet wide and weigh fifty pounds.

Monsoon or Tropical Deciduous Forests

Deciduous monsoon or tropical forests, with their broadleaf trees, occur in southern Brazil and along highland slopes of the Andes, South America, Central America, and the lowlands of Mexico, a neotropical region. Its animals are quite varied and differ from those of all other continents, including North America north of Mexico's plateau region.

Deciduous forests are usually composed of a wide variety of trees of different sizes, with a lot of undergrowth. (PhotoDisc)

Found throughout tropical deciduous forests, as well as in rainforests, are mammals confined to the neotropical region, such as monkeys different from those seen elsewhere. Ruminants such as cameloid llamas and alpacas graze forest meadows and carnivorous puma or jaguar eat them. Many neotropical birds such as tanagers, macaws, hawks, and condors also live in or seek prey in the deciduous forests and their open meadows. Reptiles include boas, bushmasters, fer-de-lance, and other poisonous snakes. Also present in these forests are peccaries, tapirs, capybaras, and agoutis.

Jaguars are the largest, most powerful American big cats. Found in Mexico and throughout Central and South America, they are most abundant in Central America and Brazil. Jaguars are the only big cats absent in Africa and Asia. They inhabit forests, swamps, and grasslands. Mature jaguars have lengths of seven feet, shoulder heights of two feet, and weights of 275 pounds. Their yellow coats are spotted with black rosettes, each a circle of spots around a central spot. Adept climbers and swimmers, jaguars eat a wide range of animals, such as peccaries, deer, tapirs, monkeys, agoutis, capybaras, birds, fish, and rodents.

Jaguars of either sex live alone except when mating, each maintaining a hunting territory. They have no set breeding season and gestation is about 3.5 months. Mothers give birth to two to four babies. They mate at age two years and have life spans up to twenty-two years. Jaguars are close to the top of the food chain in their environment.

Among the interesting animals that are prey for jaguars and pumas are tapirs. Brazilian tapirs have stocky bodies, short necks, and sturdy legs built to push through forests. They have short trunks with flexible "fingers" at the trunk tips. Trunk fingers bring leaves to their mouths. Nocturnal animals, Brazilian tapirs have short, stiff neck manes which provide some protection from predators. Tapirs have thick brown skins, and can attain lengths of 6 feet, heights of 2.5 feet, and weights of five hundred pounds. These herbivores eat grass, aquatic plants, leaves, buds, twigs, and fruit. They are solitary except when mating or nursing young. Mating occurs year round and gestation is 12.5 months. A mother births one offspring, which cannot mate until age two to three. Tapirs can live for over thirty years.

Peccaries, forest-dwelling pigs, are also prey for big cats. They inhabit high-altitude forests from Mexico to South America. These artiodactyls reach lengths of four feet, heights of two feet, and weights of one hundred pounds. They are gray-black to brown, have slender legs, and can run quite fast. Omnivorous, peccaries eat roots, seeds, fruits, and insects. They have tusks, which they use to dig and cut roots. Peccaries live in herds of up to one hundred. After mating, gestation is six months and produces two or three young, dependent on their mother for six months. Peccaries can mate at a year old and can live for up to ten years in the wild or twenty years in captivity.

Damaging Deciduous Forests

A deciduous forest is a self-sustaining ecosystem. In it, interactions of trees, shrubs, and grasses help to maintain viability and forest identity. Added to this, appropriate interaction of plants and resident herbivores help to assure that plant species do not overgrow and damage the balance or severely change the status quo that gives the forest its identity. In addition, predation by forest carnivores adds to the picture by assuring that herbivores do not become so plentiful that they overgraze the forest and alter its identity.

Any untoward alteration of the population of an animal species in a deciduous forest has a ripple effect which profoundly alters the sum of relationships that, taken together, lead to forest continuity of identity. For example, decreasing numbers of wolves in deciduous forests due to their hunting by humans has damaged food webs in forests, where their predation helped to prevent overpopulation of herbivores such as deer. Deer overpopulation, in turn, caused overgrazing of many forest areas, followed by starvation of the excessive number of deer. The overgrazing also has damaged the lives of many other plant and animal species because of the lack of the ability of such forests to provide for their needs. Thus it is crucial to assure that appropriate numbers of all plants and animals inhabit deciduous forests by suitable monitoring and conservation measures, if it is desired to maintain them in their primal states.

—*Sanford S. Singer*

See also: Chaparral; Ecosystems; Forests, coniferous; Grasslands and prairies; Habitats and biomes; Lakes and rivers; Mountains; Rain forests; Savannas; Tundra.

Bibliography

Allaby, Michael. *Temperate Forests*. New York: Facts on File, 1999. Covers forests, forest ecology, forest health, and forest management.

Edlin, Herbert L., et al. *The Illustrated Encyclopedia of Trees: Timbers and Forests of the World*. New York: Harmony Books, 1978. This illustrated book covers forests, forestry, trees, and tree identification.

Gerlach, Duane, Sally Atwater, and Judith Schnell, eds. *Deer*. Mechanicsburg, Pa.: Stackpole Books, 1994. This book discusses deer habits, habitats, species, and more.

Kobalenko, Jerry, Thomas Kitchin, and Victoria Hurst. *Forest Cats of America*. Willowdale, Ontario, Canada: Firefly Books, 1997. This useful book covers pumas, bobcats, and lynx.

Perry, Richard. *The World of the Jaguar*. New York: Taplinger, 1970. This book contains a wealth of information on jaguars.

Sutton, Ann, and Myron Sutton. *Wildlife of the Forests*. New York: Abrams, 1979. This book describes many animals found in forests

FOSSILS

Types of animal science: Classification, evolution
Fields of study: Evolutionary science, paleontology, systematics (taxonomy)

Fossils are the preserved remains of once-living organisms or the traces left by biological activity. They provide a record of the development of life through time and are used extensively in evolutionary studies as well as in the dating of rocks and the interpretation of past environments.

Principal Terms

BIOSTRATIGRAPHY: the dating of rocks using fossils

CAMBRIAN: the period of time between 544 and 505 million years ago

ICHNOLOGY: the study of trace fossils

PALEONTOLOGISTS: scientists who study fossils

TAPHONOMY: the study of the processes that lead to fossilisation

ZONE FOSSILS: fossils that characterize a period of time and can be used to provide a relative date for a rock

Fossils are now recognized as the remains of once-living organisms; however, in the past the term was not so restricted and included anything dug out of the ground (Latin *fossilis*, "to dig up"). Thus minerals, gems, and archaeological remains were all included as fossils. The record of fossils goes back 3.4 billion years to preserved single-celled organisms such as bacteria and blue-green algae. Complex organisms or metazoans did not appear until about 700 million years ago, and hard parts were not preserved until about 544 million years ago, at the base of the Cambrian when there was a tremendous development of organisms often called the Cambrian Explosion. Since then, a diverse assemblage of fossil organisms is available for study by paleontologists.

Recognition of Fossils

Early ideas about the source of fossils were that they were the remains of mythical animals or that they had grown in the rocks like crystals. Aristotle suggested that the fossils of fish were the remains of sea animals that had swum into cracks in the rocks and become stranded, while others thought that they developed from seeds or grew from fish spawn washed into cracks during Noah's Flood. These ideas remained an important influence until the Renaissance, when Leonardo da Vinci recognized that fossil shells in the Apennines represented ancient marine life, and Nicolaus Steno in the mid 1600's showed that "tongue stones," thought to be the petrified tongues of dragons, were actually ancient shark teeth. At about the same time that Steno was writing, the British scientist Robert Hooke studied and described fossils and used early microscopes to study their cellular structure. However, their ideas were slow to become accepted, and it was not until the mid 1700's that naturalistic concepts of fossils began to prevail; when Linnaeus published his classification of all organisms, the *Systema Naturae* (1735; *A General System of Nature*, 1800-1801), fossils were treated as living organisms. By 1800, Baron Georges Cuvier was able to apply comparative anatomy to fossil organisms, both showing their relation to modern organisms and arguing that they must represent extinct animals, as they had not been found living on even the most remote continents. Prior to this, the fact of extinction was not accepted, as it went against the view that

God would not allow any of his creations to become extinct. Cuvier also recognized that fossils occur in a regular succession and can thus be used to date the sediments in which they are found. At about the same time, this fact was also recognized by William Smith in England, who developed it to the point that he was able to produce geological maps of England and Wales in 1815. This was the basis of dating and mapping rocks, and by the time of the publication of Charles Darwin's *On the Origin of Species* in 1859, the understanding of the fossil record had reached the point where few scholars took biblical ideas about the history and evolution of life literally.

Fossilization

Fossilization is a rare occurrence, and it has been estimated that of the more than one million living species, only 10 percent are likely to be preserved as fossils. A single square meter of seafloor could, during a few million years, produce enough sea shells to swamp the museums of the world if they were all preserved. It must be appreciated, therefore, that the fossil record is selective and thus may be incomplete and biased. This is because some organisms with hard parts, such as shells or skeletons, tend to fossilize readily, and therefore much is known about their past. Others are soft-bodied and rarely if ever fossilize, so their fossil record is minimal. The study of how organisms become fossils, the study of death and decay, is called taphonomy. This is now an important area of paleontology, because to understand and interpret the fossil record it is necessary to understand the processes that have resulted in fossilization. Once an organism dies, its remains are subject to

These fossilized giant oysters were found almost two miles above current sea level in the Peruvian Andes. Discoveries such as this offer clues to the changes that have occurred in the earth's crust over the eons, as well as examples of the changing morphology of animals that still exist today. (AP/Wide World Photos)

decay and perhaps transport and mechanical breakdown before they are buried. The more of that lost information that can be reconstructed, the more reliable the final hypothesis about the original community will be.

Preservation is dependent on a number of factors, of which the most important are the composition and structure of the organism, its abundance, the sedimentary environment in which it lives, and the postdepositional changes that occur. All organisms are composed of delicate tissues, known as soft parts, but many also have more resistant tissues, referred to as hard parts. The hard parts may be mineralized, as in the shell of bivalves, or composed of organic material, such as the chitin that makes up the exoskeleton of arthropods. Although in general the possession of mineralized hard parts is a prerequisite for fossilization, soft-part preservation can occur, and some fossil groups have only organic soft parts.

In a few exceptional cases, organisms are preserved entire, and this results from unusual preservational circumstances. Woolly mammoths have been preserved by freeze-drying, so that their soft tissues and even their last meals are preserved, and humans have been found completely preserved in peat bogs. Preservation in amber occurs when small organisms are trapped by tree resin that subsequently hardens around them, and in some cases the preservation is so good that original biomolecules can be extracted and deoxyribonucleic acid (DNA) sequencing carried out. Although this was the premise of the film *Jurassic Park* (1993), it is clear that the genetic material is so incomplete that it will never be possible to reconstruct a complete organism in this way.

In most cases, the soft tissue has been lost and some change has taken place to the hard tissues during the process of fossilization. Permineralization is the process by which the natural pores in wood, bone, or shell are filled with minerals by groundwater that percolates through them in the sediment. In this case, none of the original material is lost, and even in petrified wood there is still some of the original material, although the logs have been completely permeated by silica. Only part of the original material is preserved in carbonization, a process in which most of the volatile organic materials disperse, leaving a black carbon film. This may result in exceptional preservation in insects, for instance, or ichthyosaurs (marine reptiles) in which the outline of the body is preserved as a carbon film. More frequently, the hard tissues are recrystallized or are replaced by some other mineral. Recrystallization results in the loss of internal detail, and the same is true of replacement, as in both cases none of the detailed structure is left. However, replacement by more resistant minerals, such as the replacement of a calcite shell by silica, can improve the survival potential of a fossil and preserve delicate structures, such as spines, that might otherwise be lost.

Preservation Potential

The preservation potential of an organism is affected by a number of factors that operate at different stages in the process of fossilization. Necrolysis, or the breakup and decay of an organism after death, reduces the potential for preservation of all but the most resistant organisms, and it has been estimated that only 20 to 30 percent of shallow marine organisms are likely to be preserved. The numerical abundance of organisms is generally considered to be important as, all other things being equal, abundant organisms would seem to be more likely to be preserved than rare ones. However, in a number of cases this has been shown not to be true, and this differential preservation is often related to the preservability of the hard tissues. Mobile organisms tend to have lighter and less durable skeletons that would be more easily destroyed than those of static animals such as corals, in which the skeleton is formed of a solid mass of calcite.

A number of agents may operate to destroy even resistant skeletons. Biological agents include scavengers, which may break up shells and bones to extract nutrition, and burrowing organisms, which may use the hard tissues as a substrate and thus weaken them. Mechanical agents such as wind, waves, and currents can also be very effective, particularly in shallow environments where

Lagerstätten

Although generally only a small proportion of organisms become fossilized, there are a number of examples in the fossil record in which there are unusual concentrations of fossils or in which the fossils are remarkably well preserved. These are termed "fossil *Lagerstätten*," a term that derives from the German mining tradition in which *Lagerstätte* are rocks containing constituents of economic interest. These exceptional fossil occurrences are separated into concentration *Lagerstätten*, in which there are uncommon concentrations of fossils, although their preservation may be unremarkable, and conservation *Lagerstätten*, in which the individual quality of preservation is unusual. Concentration deposits often occur in situations in which fossils have been preserved initially in an unconsolidated sediment and then eroded and redeposited together with more recent material. Conservation deposits include some of the most celebrated and famous fossil occurrences, including the Late Jurassic Solnhofen Limestone in Germany, which has yielded fossils of *Archaeopteryx*, the earliest known bird, and the Burgess Shale in the Canadian Rockies, which provides a window on the development of organisms during the Cambrian, 530 million years ago. In the Burgess Shale, the organisms were transported from a carbonate bank area into deep, anoxic waters by submarine landslides, where they were buried in fine muds and thus protected from decay. This has resulted in the preservation of soft-bodied organisms as well as those with hard parts, and has provided evidence for the presence of many unique and sometimes bizarre animals that would otherwise be completely unknown. Although most of these fossils can be assigned to existing phyla, many are considered to be evolutionary experiments that did not survive into later periods, thus demonstrating that there was a great deal of experimentation in body plans during the Cambrian, most of which did not survive.

energy is highest. Experiments in which shells were tumbled with pebbles have shown that thin-shelled organisms break up quickly, while thicker shells last longer. Similar studies for terrestrial vertebrates have shown that the least durable parts of a skeleton are the ribs and vertebrae, while the skull, jaws, and particularly the teeth are the most resistant to breakdown. In general, it appears that the shape, density, and thickness of the bone or shell are the most important factors in determining its survival during mechanical transport.

Many apparently unfossiliferous rocks may have had fossils at one time but lost them during the changes that take place in sediments after deposition. Many snails and bivalves have shells composed of the unstable form of calcite called aragonite, and this may dissolve early, leaving no trace of the fossil, while other organisms composed of more stable calcite will not be dissolved. In these cases, therefore, the representation of the original community will be biased more by the chemical composition of shells than by their mechanical durability or by their relative abundance.

Trace Fossils

Trace fossils (or ichnofossils, from the Greek *ichnos*, "track") are sedimentary structures created by organisms and, therefore, reflect their behavior. They include tracks, trails, burrows, and borings and may occur in sediments that contain no body fossils, thus providing the only record of life and certainly the only record of soft-bodied organisms that are not normally preserved. One important aspect of trace fossils is that they are almost always preserved in place, as reworking would result in their destruction, and thus they can always be directly related to the sediments in which they are found. Because of this, trace fossils are particularly important in understanding and reconstructing paleoenvironments.

Trace fossils are difficult to classify, as they represent behavior rather than the remains of an animal. They are also difficult to relate to the animal that formed them, as traces are rarely found with the organism that made them, different organisms may form similar traces, and the same organism may form a variety of different traces in different circumstances. Because of these problems, traces are organized within behavioral categories, including dwelling structures made by organisms that feed outside their burrows; mining structures that are formed by animals feeding in the sediment; grazing traces made by animals feeding across a surface; locomotion traces formed by animals during travel; and resting traces formed by an animal during a temporary stop. Thus, dinosaur tracks would come within the locomotion category and the burrow of a filter-feeding worm would be recognized as a dwelling structure.

Understanding the behavioral genesis of trace fossils has enabled their use in reconstructing past environments. They are grouped in recurrent ichnofacies (traces reflecting a particular environment), and these are in turn linked to a set of environmental parameters, such as salinity, light level, and temperature, that are generally correlated with depth. This approach has been particularly valuable in recognizing shallow marine facies and has the advantage of being independent of time, as particular environments and, therefore, behavior persist through time, although the trace formers themselves may change.

The study of marine traces is relatively recent; however, terrestrial traces have been studied for much longer, as the first dinosaur tracks were reported in 1802, although they were misidentified as bird tracks at that time. Terrestrial trackways can give a considerable amount of information about the locomotion and behavior of the trace former, and can lead to quite precise information about speed, as well as insights into posture, activity levels, and metabolism. Relative speed is related to the length of the leg and the stride length (the distance between one foot impression and the next by the same foot). Hip height (calculated from foot length) is divided into the stride length (measured from the trackway) to give a proportional estimate of the speed. At 2.0 an animal moves from walking to trotting, and above 2.9 it is running, information that has been used in studies of dinosaur locomotion. In addition, the fact that trackways show dinosaurs moving together in herds has been used to suggest a level of social organization greater than that demonstrated by any modern reptiles.

Fossils and Time

In geology, both relative time and measured time are recognized. Measured time is obtained from the decay rates of certain radioactive minerals in rocks and gives a date in years. Relative time refers only to the sequence of strata, showing that some beds are older or younger than others. Although the relative age of rocks can be obtained from their sequence, as those lower in the sequence are older and those higher in the sequence are younger, the lithology of the rocks cannot be carried laterally, because sedimentary environments change over distance. This problem was resolved in the late 1700's in England by William Smith, who recognized that there was a sequence to the fossils in the sedimentary record and that this could be used to characterize the rocks and correlate them to each other over a distance. This recognition of "faunal succession," which represents the successive preservation of organisms as they evolved through time, has been extremely important in the development of the study of biostratigraphy, in which fossils are used to provide dates for rocks. In this system, the basic unit is the zone, which is represented by zone or index fossils that characterize a period of time. The potential of different organisms as zone fossils will vary, and to be most useful they should embody certain characteristics. They must have a high preservation potential, be relatively common, and also be distinctive, so that accurate identification is possible. They must also have a wide lateral distribution but a short vertical range, so that they characterize a short time period over a wide area. It is also helpful if they are independent of environment. Because of this and because most sediments are marine in origin, the organisms used are often planktonic, that is, they float near the surface of the ocean but will be preserved in various sedimentary environments when they die. Microfossils, such as spores and pollen, are very useful because they are wind-borne and thus may be found in both terrestrial and aquatic environments.

—*David K. Elliott*

See also: *Allosaurus*; *Apatosaurus*; *Archaeopteryx*; Dinosaurs; Evolution: Animal life; Extinction; Hadrosaurs, Ichthyosaurs; Paleoecology; Paleontology; Prehistoric animals; Pterosaurs; Sauropods; Stegosaurs; *Triceratops*; *Tyrannosaurus*; Velociraptors.

Bibliography

Ausich, William I., and N. Gary Lane. *Life of the Past*. 4th ed. Upper Saddle River, N.J.: Prentice Hall, 1999. Explores the origin and development of life through time and includes chapters on fossils and fossilization.

Briggs, Derek E. G., and Peter R. Crowther, eds. *Palaeobiology: A Synthesis*. Boston: Blackwell Scientific Publications, 1990. An encyclopedic treatment, consisting of 120 concise articles that provide an up-to-date review of the state of the art in aspects of paleontology other than descriptive morphology.

Gillette, David D., and Martin G. Lockley, eds. *Dinosaur Tracks and Traces*. New York: Cambridge University Press, 1989. A series of papers that provides a good coverage of the methods used in developing information on locomotion and behavior from trackways.

Prothero, Donald R. *Bringing Fossils to Life*. Boston: WCB/McGraw-Hill, 1998. An undergraduate text that provides a full coverage of the fossil record both from the perspective of the taxonomy of the organisms and their paleobiology.

Rudwick, Martin J. S. *The Meaning of Fossils*. Chicago: University of Chicago Press, 1985. A fascinating and readable account of the history of study of fossils from the fifteenth century on.

FOXES

Type of animal science: Classification

Fields of study: Anatomy, ecology, evolutionary science, physiology, reproduction science, systematics (taxonomy), wildlife ecology, zoology

There are approximately twenty-one species of fox, in the order Carnivora. Foxes constitute only 0.5 percent of the four thousand mammalian species, but play a very important role in the environment.

Principal Terms

CANID: doglike in appearance

ESTRUS: the period of maximum sexual receptivity

TEMPERATE: moderate with respect to temperature

TERRITORY: the area of land that an animal defends against intruders

VIXEN: a female fox

The fox is a canid, related to coyotes, jackals, and wolves. The fox is the smallest member of the dog family. It is also the most numerous and widespread carnivore in the world. Although primarily carnivorous, foxes will also eat fruits and grains. Foxes hold territories, the size of which depends on habitat. A dog fox, a vixen, and their cubs occupy each territory. A differing number of chromosomes eliminate the possibility of foxes interbreeding with other canids.

Fox populations fluctuate along with the population cycles of their prey. In many cases, availability of dens and secure sites for hiding will also play an important role in population size. Foxes have few natural predators. However, in mountainous regions, coyotes and cougars will sometimes attack foxes. In arid lands, wild dogs, large cats, and hyenas are problematic, and will often compete with foxes for food. In arctic regions, the wolf can also be a dangerous competitor.

Foxes appear to have characteristics of both dogs and cats. Some of their catlike characteristics include vertical slit eyes, partially retractable claws, light body weight, and a stalking and pouncing style of hunting. Foxes are also good tree climbers. Fox coloration varies among black, red, silver, silver-gray and white.

The Life Cycle of Foxes

During December and January, vixens leave scent marks that tell the male fox she is in estrus. The dog will follow the marks until he finds her. The courtship is very short, since the vixen is only fertile for about one week each year. Litter sizes are normally one to six. The dog fox helps the vixen

Fox Facts

Classification:
Kingdom: Animalia
Subkingdom: Deuterostomia
Phylum: Chordata
Subphylum: Vertebrata
Class: Mammalia
Subclass: Eutheria
Order: Carnivora
Family: Canidae
Genus: Vulpes, with twelve species

Geographical location: All continents except Antarctica

Habitat: Temperate climates, although some live in desert and Arctic regions; generally next to water, meadows, and forests

Gestational period: Fifty to sixty days

Life span: Up to six years in the wild; up to thirteen years in captivity

Special anatomy: Pointy muzzles, large ears, long thin bodies, long legs, long bushy tails

Foxes are territorial and will drive off outsiders who intrude on their range. (Corbis)

raise the cubs during the first two months. For the first month or so, the vixen depends entirely on the dog fox to bring her food. In the first two weeks, the cubs are unable to see or hear anything. They are fed by the mother's milk. The fur is short and black. After three weeks, the fur will slowly turn dark brown on the head, and gray on the body. The teeth break through the gum, and the cub starts to chew on anything. Soon the cub starts to show the first signs of social behavior, and the first fights are fought between the siblings. After four weeks, the cubs start exploring the world outside the den for the first time. Later, they start to follow their mother when she goes hunting. In the summer, the young foxes get more and more independent from the mother fox. After about three months, the dog fox will go back to his solitary lifestyle, but he will try to find the same vixen again when the mating season comes back. In the fall, the young foxes come into puberty and become sexually mature. It is then that the vixen begins to force the cubs out of the den.

Destructive and Beneficial Foxes

Foxes may cause serious problems for farmers. Losses may be heavy in small farm flocks of chickens, ducks, and geese. Damage by foxes can be difficult to detect because the prey is usually carried from the kill site to a den, or uneaten parts are buried. Foxes will also scavenge carcasses, making the actual cause of death difficult to determine.

Like other scavengers, foxes will clean up carcasses, or kill dying or weak animals for food. In this manner they are a major contributor to the processes of natural selection.

—*Jason A. Hubbart*

See also: Carnivores; Cats; Dogs, wolves, and coyotes; Predation; Scavengers; Urban and suburban wildlife.

Bibliography

Ewer, R. F. *The Carnivores*. Reprint. Ithaca, N.Y.: Comstock, 1998. A refreshingly descriptive book, describing many of the mammalian carnivores. It is thorough and easy to understand.

Feldhamer, George A., Lee C. Drickamer, Stephen H. Vessey, and Joseph F. Merritt. *Mammalogy: Adaptation, Diversity, and Ecology*. Boston: WCB/McGraw-Hill, 1999. In standard style, these authors do a superior job systematically describing and documenting mammals.

Grambo, Rebecca L. *The World of the Fox*. San Francisco: Sierra Club Books, 1995. Abundantly photographed guide to foxes around the world, covering the arctic fox, swift fox, kit fox, gray fox, and red fox.

Henry, J. David. *Red Foxes: The Catlike Canine*. Washington, D.C.: Smithsonian Institution Press, 1996. Well-illustrated guide to fox ecology.

Vaughan, T. A., James M. Ryan, and Nicholas Czaplewski. *Mammalogy*. 4th ed. Fort Worth, Tex.: Saunders College Publishing, 2000. Thorough and easily understood. A great place to start for those unfamiliar with mammalian diversity.

Wilson, D. E., and D. M. Reeder. *Mammal Species of the World: A Taxonomic and Geographic Reference*. 2d ed. Washington, D.C.: Smithsonian Institution Press, 1993. A detailed account of the mammalian species. Nicely organized and easy to follow.

FROGS AND TOADS

Types of animal science: Anatomy, classification
Fields of study: Anatomy, conservation biology, herpetology, systematics (taxonomy)

The more than four thousand living species of frogs and toads are classified in the order Anura. Most species have a biphasic life cycle, in which a larval stage, the tadpole, develops in water (or land, in some species) and undergoes an abrupt metamorphosis into the adult body form.

Principal Terms

EXTERNAL FERTILIZATION: the union of eggs and sperm in the environment, rather than in the female's body

GRANULAR GLANDS: one of many kinds of glands in the skin of frogs and toads; granular glands secrete toxins for defense from predators

METAMORPHOSIS: an abrupt change from a larval body form, accompanied by many physiological changes, into an adult body form

REPRODUCTIVE MODE: a combination of life-history characteristics, including egg-deposition site and type of parental care

TADPOLE: the larval stage of frogs and toads

Frogs and toads, the Anura, are amphibians, a group of vertebrates that first evolved in the Devonian era, about 400 million years ago. Two other groups of amphibians are the Caudata, salamanders and newts, and the Gymnophiona, caecilians. More than four thousand species of frogs and toads currently inhabit the planet. In contrast, about 400 species of salamanders and newts and 165 species of caecilians are extant. Frogs and toads thus account for approximately 88 percent of all living amphibians.

Amphibians have many unique characteristics that separate them from other vertebrates. In terms of skeletal features, the skull articulates with the vertebral column by a specialized vertebra called the atlas, and ribs are lacking in most frogs and toads. Many amphibians have a two-phase life history, an aquatic larval stage and a terrestrial adult stage, although some species may be entirely terrestrial or entirely aquatic. Amphibian skin is unlike that of other vertebrates in that it is glandular and lacks scales, feathers, or hair. Respiration is by means of the skin, which is porous, freely permitting the exchange of oxygen and carbon dioxide, with gills in aquatic larvae and adults, and lungs in terrestrial species.

Scientists classify frogs and toads into twenty-five families. Although the terms "frog" and "toad" are commonly used, toads are simply a specialized group of frogs. The true toads are in the family Bufonidae. They are distinguished from other families by a number of osteological (bony) features, and they also typically have thick, glandular skin with wartlike pustules. Many species have large parotoid glands that produce toxic secretions located on the back of the head. In contrast, most frogs have smooth skin and no parotoid glands. Toads in the family Bufonidae can generally withstand drier conditions than frogs. Other kinds of frogs have common names that include the word "toad," such as spadefoot toads, which are in the family Pelobatidae, or narrow-mouth toads, which are in the family Microhylidae.

Anatomy of Adult Frogs and Toads

Frogs and toads are tailless amphibians. In general, they are small animals; the largest has a body length of about three hundred millimeters, whereas the smallest is under ten millimeters.

They have a truncated body with a greatly reduced number of vertebrae and elongated hindlimbs. The vertebrae in the lower part of the body are fused into a rodlike structure, which provides rigidity for their jumping mode of locomotion.

Two types of glands are present in the skin of all adult amphibians: mucous and granular. Mucous glands continuously secrete mucopolysaccharides, which function to keep the skin moist so that cutaneous respiration can occur. All amphibians, including frogs and toads, have granular or poison glands in their skin, and these glands produce a variety of toxic or noxious chemicals used for defense. The chemicals are broadly classified into four major groups, of which alkaloids are one. The great majority of these chemicals are not harmful or deadly to humans, but they can be irritating. For example, if a secretion from one of the tree frogs found in the eastern United States is rubbed into the eyes, an intense burning sensation will be felt for a short time. If a frog or toad is grabbed by a predator, such as a bird or a snake, it may be quickly released because of the bad-tasting secretion. Some predators have evolved an immunity to the secretions, however, so frogs and toads are not free from predation.

Anatomy of Tadpoles

Frogs and toads are unique among amphibians in that the young, called tadpoles, are very different from adults in morphology and physiology. Tadpoles typically have a round, globular body, with a long tapering tail. After a period of time that varies from one to two weeks to several years, depending on species, the larval forms undergo metamorphosis and change into adult frogs. Many radical changes, initiated by the hormone thyroxin, occur during this period. The most obvious of these is the loss of the tail, which is gradually absorbed, and the growth of the limbs. The hindlimbs begin to develop early in the tadpole stage, but are held flat against the body and are not used until they are quite large. The front limbs develop at the same rate, but they are hidden beneath the operculum, a thin layer of skin that covers the gills. At metamorphosis, as the tail is absorbed, the hindlimbs enlarge and the forelimbs break through the operculum.

The tadpoles of most species of frogs are herbivorous or detritivorous, and the digestive system consists of a long, coiled intestine. At transformation, the digestive system undergoes a radical modification,

Frog and Toad Facts

Classification:
Kingdom: Animalia
Phylum: Vertebrata
Class: Amphibia
Subclass: Lepospondyli
Order: Anura
Families: Twenty-five, including Ascaphidae (tailed frogs), Bufonidae (true toads), Centrolenidae (glass frogs), Dendrobatidae (poison frogs), Hylidae (treefrogs), Leptodactylidae (tropical frogs), Microhylidae (narrowmouth toads), Pelobatidae (spadefoot toads), Ranidae (true frogs)

Geographical location: Throughout the world, except for the Arctic, Antarctica, some oceanic islands, and some very dry deserts

Habitat: Many habitats, including forests, savannas, prairies, freshwater ponds and streams, and ephemeral pools

Gestational period: Varies among species; some toads and spadefoot toads have a tadpole stage that lasts only for ten days to two weeks; in other species, such as bullfrogs, the tadpole stage may last for three years; in many species, the tadpole stage extends for about two months

Life span: Varies among species; less than a year in many species to as much as six to thirty-six years in some species, particularly for individuals kept in captivity

Special anatomy: Truncated head and body with no neck; long back legs that facilitate jumping; tadpole anatomy, globose body with long tail, very different from adult body shape

Declines in frog populations may be linked to pollutants in the water where they live. (Adobe)

and the long, coiled intestine becomes greatly shortened and divided into a distinct esophagus, stomach, small intestine, and large intestine. The internal gills of the tadpole are lost, and the lungs become fully functional. In many species, the lungs begin to develop and become partially functional long before metamorphosis. Tadpoles of these species are frequently seen rising to the surface of the water to gulp air.

Tadpoles have unusual mouthparts, consisting of a fleshy disc surrounded by small papillae; the disc has rows of small "teeth" that are not the same as the teeth of the adult. The actual opening of the mouth is surrounded by beaklike structures that have sharp, serrated edges. The rows of teeth serve to anchor the disc, and the beak is used to scrape algae and other detritus from the substrate. The food is captured by strands of mucus in the mouth that carry it into the digestive tract. At metamorphosis, the tadpole's mouthparts are completely shed and the adult mouth develops.

Life History of Frogs and Toads

The life history of frogs and toads includes four stages: egg, larva, juvenile, and adult. Eggs consist of an egg cell supplied with yolk surrounded by a varying number of mucoid capsules. Because the eggs have no shells and are susceptible to drying out, they must remain in water or in a humid environment.

In temperate zones (nontropical areas), frogs and toads generally breed in the spring and summer when temperature and moisture conditions are suitable. Some, such as chorus frogs, spring peepers, and wood frogs, begin breeding very early, usually in February, while ponds are still covered with ice. Others, such as American toads and leopard frogs, breed later in the spring, and

Reproductive Modes of Frogs

Gladiator frogs are large treefrogs that construct nests in which eggs are deposited. Nests are made by males, who pivot around in soft sand or mud at the edge of a pond or stream to make a depression that fills with water. Eggs are deposited as a surface film in the nest.

Glass frogs deposit a small clutch of eggs on a leaf overhanging a small stream or river. When the eggs hatch, the tadpoles fall into the water below.

Leptodactylus frogs construct foam nests for their eggs. An amplexing male and female rotate their legs rapidly and secretions emitted with the eggs and sperm are whipped into a meringuelike froth. Eggs and tadpoles develop entirely within the foam; most predators cannot enter the sticky foam.

Poison frogs deposit their small clutches of eggs on land. The male (but sometimes the female) guards them. After the eggs hatch, the parent transports the tadpoles on his or her back to a small pool or stream where the tadpoles undergo the rest of their development.

Eleutherodactylus frogs deposit small clutches of eggs on land. Development of the egg to a small froglet occurs entirely within the egg capsule.

Gastric-brooding frogs swallow their eggs, and development of the tadpoles occurs in the mother's stomach. A chemical secreted by the developing tadpoles inhibits the production of gastric fluids so that the tadpoles are not digested.

still others, such as bullfrogs and cricket frogs, breed primarily in the summer months.

Most temperate-zone frogs migrate to ponds during rains to breed. Males arrive first and begin calling to attract females. Each species of frog and toad has a unique call, so it can be identified only by sound. Frogs make more than one kind of call. The most common calls, termed advertisement calls, are of three kinds: the courtship call, which is used to attract females; the territorial call, used in response to the call of a nearby male that may be intruding on the caller's territory; and the encounter call, used when two males in a breeding chorus are very close. The latter call may be accompanied by grappling or fighting between the two males.

Although female frogs do not make advertisement calls, those of some species (and males, too) can produce a very dramatic distress call when grabbed by a predator. This call is produced with the mouth wide open and is a very loud piercing scream. Some types of predators are startled by this call and will release the frog. Snakes, however, have no ears and do not seem to be affected by the distress call.

Frogs and toads have external fertilization. When a female approaches and nudges a calling male, he will climb atop her and clasp her with his arms behind her arms; this position is called amplexus. The pair floats in the water and simultaneously expels eggs and sperm in the water; thus, fertilization of the eggs occurs outside the female's body.

Eggs generally require several days to one week to hatch, depending on temperature and the species of frog or toad. After hatching, tadpoles of most species disperse throughout the pond to feed and grow. Tadpoles of some species, however, congregate together in schools, similar to fish. Many species of toads, for example, have small, black tadpoles that form tight schools in shallow water at the edges of ponds. Most tadpoles eat algae and other kinds of detritus that they scrape from the substrate of ponds or streams in which they live. A few kinds of tadpoles are predaceous and eat other tadpoles or mosquito larvae and other small aquatic insects. Tadpoles undergo a period of development and then metamorphose into the adult body shape.

Frogs and toads have evolved numerous ways of removing their eggs or tadpoles from aquatic situations, where, presumably, density of egg and tadpole predators is high. Scientists have identified thirty kinds of reproductive modes in frogs and toads, mostly in tropical species. Because tropical regions have high rainfall and warm temperatures, frogs can deposit their eggs on land without risk of desiccation.

Declining Amphibian Populations

Populations of frogs and toads in many parts of the world, even in pristine (undisturbed) locations, are declining and disappearing. The causes of these declines are numerous, and many scientists are now conducting research to determine the causes of the declines. Some of the causes are obvious. For example, habitat destruction is a major cause of loss of many animals and plants, not just amphibians. Most recently, frogs and toads of many species in rain forests in Central America and Australia are reported to have been killed by a type of fungus called a chytrid. The fungus normally infects decaying matter, so why it is now infecting frogs is not yet understood. One theory is that frogs are being stressed by some other environmental problem not yet identified; this stress may reduce the ability of their immune systems to fight disease.

Also being investigated as possible causes of frog population declines are the use of agricultural chemicals and pesticides, acid rain, ozone depletion leading to a rise in ultraviolet (UV-B) radiation, environmental estrogens, and introduction of fish that prey on eggs and tadpoles. It is possible that at any one location, one or several of these factors may be interacting.

—*Janalee P. Caldwell*

See also: Amphibians; Chameleons; Communication; Defense mechanisms; Ecosystems; Metamorphosis; Salamanders and newts; Vocalizations.

Bibliography

Duellman, W. E., ed. *Patterns of Distribution of Amphibians: A Global Perspective*. Baltimore: The Johns Hopkins University Press, 1999. Covers all aspects of distribution of amphibians with respect to climate, geography, vegetation, and evolutionary history.

Duellman, W. E., and Linda Trueb. *Biology of Amphibians*. New York: McGraw-Hill, 1986. A comprehensive book that covers all aspects of amphibian biology, including paleontology, morphology, reproduction, and biogeography.

Heatwole, H., and B. K. Sullivan. *Amphibian Biology*. Vol. 2 in *Social Behaviour*. Chipping Norton, New South Wales, Australia: Surrey Beatty and Sons, 1995. A thorough treatment of amphibian social behavior, including territoriality, mating behavior, and parental care.

McDiarmid, R. W., and R. Altig, eds. *Tadpoles: The Biology of Anuran Larvae*. Chicago: University of Chicago Press, 1999. A comprehensive book that summarizes current knowledge about the biology of tadpoles.

Phillips, K. *Tracking the Vanishing Frogs: An Ecological Mystery*. New York: St. Martin's Press, 1994. Written for a general audience, this book describes some of the causes of amphibian declines. The beauty and diversity of frogs and toads are highlighted.

Vitt, L. J., J. P. Caldwell, and G. Zug. *Herpetology*. San Diego, Calif.: Academic Press, 2001. A general introduction to the biology of frogs and toads, as well as other reptiles and amphibians. Color photographs of examples from all families. Aimed at advanced high school or beginning college students.

FUR AND HAIR

Type of animal science: Anatomy
Fields of study: Physiology, zoology

Fur is the hairy covering of the skin of a mammal. This covering is called fur when its individual hairs are fine and spaced closely together; when the hair is soft, kinky, and matted together, it is called wool, which grows in a fleece, and when it consists of coarse, stiff hairs, it is called bristles, or when pointed, spines or quills. Its primary role is as protection, insulation, and ornamentation.

Principal Terms

CORTEX: the main part of a hair, made of pigment-containing cells, surrounding a central medulla

CUTICLE: the outermost layer of a hair, made of scales

EPIDERMIS: the dead, outermost portion of the skin

FOLLICLE: the saclike organ from which a hair grows; its blood vessels nourish the hair

KERATIN: a tough fibrous protein, seen in large quantities in epidermal structures such as hair

MEDULLA: the innermost layer of a hair

SHAFT: the main hair part, made of dead cells arranged in a complex fashion

The term fur is used in several ways. All are related to its being the hairy covering of the skin of a mammal. In its most common usage, fur is the dense, hairy body covering of a mammal. However, even the sparse hair covering on the arms and legs of humans may be viewed as fur, since it is the hairy covering of the skin of a mammal.

Each individual hair in a mammal is a thread-like epidermis outgrowth. Collectively, hairs form the body coverings (or pelage) of all mammals. Each hair is made mostly of a fibrous, sulfur-containing protein called keratin. The pelage is most correctly called fur when its individual hairs are fine and spaced closely together. In cases where the hair is soft, kinky, and matted together, the pelage is wool, which grows in a fleece. Coarse, stiff hairs are bristles, and when pointed (as in porcupines) they are spines or quills. Hair grows on most parts of the bodies of mammals. As their body covering, its primary role is as protection, insulation, and ornamentation, like bird feathers and reptile scales.

The Structure of a Hair

Any hair consists of two main parts, the root and the shaft. The shaft is the hair part outside of the skin. It contains the hair's unattached end. Hair shaft cross-sections range from round to flattened. Due to their composition, round and flat hairs are straight and curly, respectively. Hair shafts consist of dead epithelial cells, arranged in columns which surround a central medulla (or core), covered with flat scales. The scales—a hair's cuticle—overlap like roof shingles, but their free ends face upward, away from the hair's root. Beneath the cuticle is a second layer of dead cells, the cortex, which surrounds a central core made of yet other dead cells (the medulla).

The cortex, which makes up most of each hair, consists of many dead, longitudinally arranged, keratin-rich, spindle-shaped cells which are tightly attached to each other. Hair color results from pigments in cortex cells and light reflected from the medulla. The medulla is less dense than the cortex and its cells are only loosely attached to one another.

Wool and Woolens

Wool denotes soft, curly fleece fibers from domesticated sheep. It differs from hair in that its scales are more numerous, smaller, and pointy. Curliness makes wool very resilient. This high tensile strength and elasticity help wool fabrics to retain their shape. These properties, wool's lightness, and its fine insulating ability make wool fabrics desirable.

Wild sheep have sparse, woolly undercoats and coarse hair, useless in fabrics. The hair has been bred out in domestic breeds. Sheep fleeces are usually shorn annually, from late spring to early summer. The wool is cut off very close to the skin and removed in one piece, weighing nine to eleven pounds. Wool from different fleece parts varies in its fiber length, fineness, and structure. The shoulder and sides yield the best fibers. Merino sheep yield the best overall wool. It makes up 40 percent of all wool produced commercially. Crossbreeds of merino sheep and strains that produce longer, coarser wool yield most of the rest. Some apparel wool comes from alpacas, goats, and llamas.

Woolen cloth manufacture begins by pulling fleeces apart and choosing the best fibers for given uses. Next, fibers are cleaned to remove lanolin and dried-on sweat. The clean fibers are disentangled and drawn straight by carding, which entails passing them between rotating cylinders to yield a thin film or web. Web processing varies, depending on whether it will be used in tweed or worsted yarn. Tweeds, woven from bulky yarn made of short, randomly arranged fibers, are thick and fuzzy. Worsted fabrics such as gabardines are woven from web made of longer, thinner fibers, tightly twisted for smoothness.

At every hair's base is a saclike hair follicle. The hair grows from the bottom of the follicle, nourished by blood vessels in a structure called a papilla. The papilla extends into the follicle and into the hair's root. A tiny muscle is attached to each hair follicle. Action of the nervous system can cause the muscle to contract to make hair "stand on end."

Hair Growth and Replacement

A hair forms from cells that grow from the surface of the papilla, which means that it grows from the root, not the free end. As new cells develop, they push forward old ones, which become part of the shaft. Hair growth continues as long as follicle and papilla are functional. The lifetime of a hair from start of growth until it is shed depends upon the organism which produces it. When an old hair falls out a new one takes its place.

Hair follicles produce hairs in cycles of hair growth, in which the hair follicle and the shaft pass through a complex series of morphological changes. During hair growth, the follicle penetrates into the dermis, and cells of the shaft are joined together. In addition, the follicle's melanocyte cells deposit pigment into shaft cells. Once a hair shaft attains its characteristic length, the follicle contracts and a "dead" hair protrudes from it. The growth period of a single hair ranges from three years in humans to around two weeks in rodents.

Hairs are continually replaced, or shed, throughout the life of a mammal. However, their development and loss occur asynchronously, so mammals are never completely naked of pelage. In rodents, replacement is in waves, across the body. In primates, each follicle passes through the growth cycle, independent of those around it. Hormones control hair growth; however, there are other, as yet unknown components that must also affect the process, because hormones are carried in the blood, and if they acted alone they would simultaneously affect all follicles in the body. This would be disastrous, because if all follicles grew in together, at the same rate, all hairs would be shed at the same time. Then the mammal involved would have naked periods where it was deprived of hair's protection and insulation. Continuous growth of hairs can also be hazardous. For example, in merino sheep, long growth phases produce long-stranded wool. However, if these sheep strayed, were not minded well, or were sheared irregularly, they might starve to death from becoming entangled in underbrush.

Hair Origin and Function

Hair in the pelage of contemporary mammals acts mainly to insulate against temperature variation. It has been proposed that mammal hair evolved into pelage from "prehair" which had the same functions. One theory of hair origin is that it evolved from epidermal mechanoreceptors, a concept supported to some extent by the existence of sinus hairs in mammals. These hairs, whiskers (vibrissae) in mammals such as felines and rodents, have blood-filled sinuses in the skin around the follicle. This tissue, together with associated nerve fibers, engenders mechanoreception, which facilitates nocturnal movement. However, most mammal body hairs lack nerves and only insulate and protect. The basis for hair evolution from vibrissae is therefore unclear.

Certainly, mammalian hair has other functions today. For instance, it can serve the unusual protective function of the quills of porcupines, and perhaps it may more generally serve to attract mates. It seems possible that the change of prehairs into attractive pelage drove development into these main contemporary forms.

Protecting Endangered Fur-Bearers

The desire on the part of humans for fur garments has led to atrocities committed on many mammal species which have gorgeous pelage. One of many examples is clubbing young fur seals to death. Beyond that, many species have been hounded to near extinction by hunters. Classic examples of such endangered species are the big cats, such as

Domesticated sheep have been bred to produce the type of very soft, kinky hair called wool. (PhotoDisc)

The Unusual Porcupine

Porcupines, whose name derives from a French word meaning "spiny pig," are animals that possess very unusual hair coverings. Included in these coverings are quills, which are used for defense against predators. A coat of regular hairs serves the porcupine as thermal insulation.

An adult porcupine possesses between twenty-five thousand and thirty-five thousand quills. Each of the quills is actually a group of long, stiff hairs that have grown together very tightly. The quills range in length from three to eighteen inches, depending upon the porcupine species and the body positions of the quills. Quills have sharp barbs on their ends.

When a porcupine is attacked or otherwise disturbed, it arches its back and makes the quills stand up straight. Attacking predators are very likely to impale themselves painfully on the quills, which pull out of the porcupine's body quite easily. A porcupine can also swing its quill-filled tail at a predator. When the tail connects with the aggressor, quills enter the flesh of that animal deeply and are bound to do damage. When a quill hits an attacker's eye, it may cause blindness. If quills enter its jaw, they may prevent eating and lead to the predator's death by starvation. Porcupines grow new quills, as needed.

tigers and leopards, and rodents, such as the beaver. Many other mammals are threatened species, likely to become endangered in the foreseeable future.

Fortunately, several organizations have sought to protect these fur-bearing mammals. Efforts of organizations such as the World Wildlife Fund have focused public opinion and led to animal conservation legislation. Preeminent are the Endangered Species Act of 1973, and a 1977 Convention, signed by eighty nations, including the United States. These actions have led to agreement that furs will not move interstate or between signatory countries without proof that the species from which they were harvested is not threatened or endangered. This bodes well for the future of hair.

—*Sanford S. Singer*

See also: Claws, nails, and hooves; Exoskeletons; Feathers; Horns and antlers; Molting and shedding; Scales; Shells; Skin; Teeth, fangs, and tusks.

Bibliography

Deems, Eugene F., Jr., and Duane Pursley. *North American Furbearers: A Contemporary Reference*. Baltimore: International Association of Fish and Wildlife Agencies, 1983. Presents a great deal of information on fur and fur-bearing animals.

Robbins, Clarence R. *Chemical and Physical Behavior of Human Hair.* 3d ed. New York: Springer-Verlag, 1994. The first chapter of this text covers the morphological and macromolecular structure of hair.

Robertson, James, ed. *Forensic Examination of Hair.* London: Taylor & Francis, 2000. This book on the forensics of human hair has two solid, basic chapters on hair physiology and growth and on its microscopic examination.

Spearman, R. I. C., and P. A. Riley, eds. *The Skin of Vertebrates*. New York: Academic Press, 1980. The proceedings of an international congress, including sound coverage of skin and hair.

Stanford Environmental Law Society. *The Endangered Species Act*. Stanford, Calif.: Stanford University Press, 2001. Describes endangered species law and legislation, with a solid bibliography.

GAMETOGENESIS

Types of animal science: Development, reproduction
Fields of study: Cell biology, developmental biology, embryology, genetics, reproduction science

Gametogenesis is the process of sex cell formation. It includes the events that lead to a reduction in chromosome number so that the sex cells will have one-half the chromosomes that are found in normal body cells.

Principal Terms

DIPLOID: the number of chromosomes or the amount of genetic material normally found in the nucleus of body cells; this number is constant for a particular species of animal

GAMETE: a sex cell; the egg or ovum in the female and the sperm in the male

HAPLOID: one-half of the diploid number; the number of chromosomes or the amount of genetic material found in a gamete

MEIOSIS: reduction division of the genetic material in the nucleus to the haploid condition; it is the process used by animal cells to form the gametes

OOGENESIS: gamete formation in the female; it occurs in the female gonads, or ovaries

SPERMATOGENESIS: gamete formation in the male; it occurs in the male gonads, or testes

SPERMIOGENESIS: the structural and functional changes of a spermatid that lead to the formation of a mature sperm cell

Sexual reproduction is the predominant mode of reproduction in animals. Sexual reproduction involves the production of gametes: the eggs and sperm. In most animals, these gametes are produced in specialized organs called gonads (ovaries and testes). The sex cells in most animals are separate—that is, each individual animal contains either testes or ovaries, but not both. Such animals are said to be dioecious. In dioecious animals, the sex cells from two different individuals (one male and one female) will fuse together in a process known as fertilization to form the offspring. The advantage of sexual reproduction seems to be in its potential to produce variability in the gametes and therefore in the new organism.

Gametes are highly specialized cells that are adapted for reproduction. These egg and sperm cells develop by a process of gametogenesis, or gamete formation. Sperm cells are relatively small cells that are specialized for motility (movement); egg cells are larger, nonmotile cells that, in many species, contain considerable amounts of stored materials that are used in the early development of the zygote (fertilized egg).

In animals, gametogenesis consists of two major events. One involves the structural and functional changes in the formation of the gamete. The other involves the process of meiosis. Animal body cells normally contain the diploid amount of genetic material. Each species of animal has a characteristic diploid number that remains the same from generation to generation. Because fertilization involves the fusion of the egg and the sperm, bringing together each cell's set of genetic material, some mechanism must reduce the amount of genetic information in the gamete, or it would double every generation. Meiosis is a special nuclear division whereby the genetic material is reassorted and reduced to form haploid cells. Therefore, gametes are haploid, and gamete fusion during fertilization reestablishes the diploid content in the zygote.

Sperm

Sperm are highly motile cells that have reduced much of their cellular contents and are little more than a nucleus. Sperm are produced in the testes from a population of stem cells called spermatogonia. Spermatogonia are large diploid cells that reproduce by an equal division process called meiosis. Spermatogenesis is the process by which these relatively unspecialized diploid cells will become haploid cells; it is a continuous process that occurs throughout the sexually mature male's life. When a spermatogonium is ready to become sperm, it will stop dividing mitotically, enlarge, and begin the reduction division process of meiosis. These large diploid cells that begin to divide meiotically are known as primary spermatocytes.

The first step in the division process involves each primary spermatocyte dividing to form two secondary spermatocytes. Each secondary spermatocyte continues to divide, and each forms two spermatids. These spermatids are haploid cells. For each primary spermatocyte that undergoes spermatogenesis, four spermatids are formed. The spermatids are fairly ordinary cells; they must go through a process that will form them into functional sperm. The transformation process of a spermatid into a sperm is called spermiogenesis and involves several changes within the cell. The genetic material present in the nucleus begins to condense, while much of the cytoplasm and its subcellular structures are lost. The major exception to this latter event is the retention of mitochondria, cytoplasmic structures involved in energy production. The mature sperm has three main structural subdivisions: the head, the neck (or midpiece), and the tail. All are contained within the cell's membrane. The oval head has two main parts, the haploid nucleus and the acrosome. The acrosome comes in various shapes but generally forms a cap over the sperm nucleus. The acrosome functions differently in various animals, but generally its functions are associated with the fertilization process (union and subsequent fusion of egg nucleus and sperm nucleus). Acrosomes contain powerful digestive enzymes (organic substances that speed the breakdown of specific structures and substances) that allow the sperm to reach the egg's membrane. The midpiece of the sperm contains numerous mitochondria, which provide the energy for the sperm's movement. The tail, which has the same general organization as flagella or cilia (subcellular structures used for locomotion or movement of materials), uses a whiplike action to propel the sperm forward during locomotion. The structural changes that occur during spermiogenesis are meant to streamline and pare down the sperm cell for action of a special sort and of a limited duration. The sperm's function is to "swim" to the egg, to fuse with the egg's surface, and to introduce its haploid nucleus into the egg's interior.

Eggs

The female gamete, the egg or ovum, is produced by a process known as oogenesis. This process occurs in the female gonads, the ovaries. At first glance, oogenesis and spermatogenesis appear to be very similar, but there are some striking differences. The major similarity is that both processes form gametes, which contain genetic material that has been reduced to the haploid condition. To understand oogenesis, one must consider that its goal is to produce a cell that is capable of development. The mature egg in all animal cells is large in comparison with other cells, particularly with the sperm. There are two important features of the egg that must be considered: the presence of a blueprint for development and the means to construct an embryo from that blueprint. In other words, the egg must be programmed and packaged during oogenesis. The programming refers to the information that is coded within the structure of the egg. This information includes the genetic material as well as the cytoplasmic information. Together, the nucleus and cytoplasm provide the egg with the potential to transform a simple cell into a complex preadult form. Since it is within the egg that this transformation occurs, the programming must be within the organization of the egg, and the directions for development must be within that organization. The packaging refers to the presence of all the material necessary to build embryonic structures, to nourish this developing embryo, and to provide its energy until it can obtain nourishment on its own.

As happens in spermatogenesis, the potential eggs are formed from unspecialized stem cells, in this case called oogonia. Oogonia contain the diploid amount of genetic material and divide by the process of mitosis. At some point in their life, oogonia stop dividing mitotically, enlarge, and prepare to become eggs—that is, they begin meiosis. The cell that begins this reduction division process is called the primary oocyte. Each primary oocyte divides into two cells—one large cell, the secondary oocyte, and a very small cell, the first polar body. The secondary oocyte continues the final reduction phase of meiosis and forms two cells, one large one (the ovum) and one very small one (the second polar body). The first and second polar bodies are nonfunctional by-products of meiosis. The one functional cell, the mature egg, contains most of the cytoplasm of the primary oocyte and one-half of its genetic material. In many animals (primarily the vertebrates), all oogonia present in the ovaries enter meiosis at the same time; the initial events of oogenesis are synchronous within the animal. Oogenesis in many animals is not a continuous process, as is spermatogenesis in the male. Rather, the primary oocytes in the first stages of reduction division may remain inactivated for a long time—in some cases, for several decades. Therefore, in female animals with this format of oogenesis, a primary oocyte population is maintained, and eggs will mature as they are needed.

Thus far it appears that the egg's formation differs from the sperm's in three ways. First, in many female animals, there is a limited number of primary oocytes capable of going on to form eggs; second, this egg formation is not necessarily a continuous process; third, one primary oocyte yields one mature egg at the end of meiosis. Although these are three very important differences, there are other distinctly egg events that deal with the developmental programming and the packaging of materials in this potential gamete.

Eggs and RNA

Little is known about the egg's storage of developmental directions or the actual programming of information, but developmental and molecular biologists are beginning to elucidate events that occur during oogenesis that are concerned with function of the egg. One such event, fairly widespread among the animal kingdom, is the formation of so-called lampbrush chromosomes during oogenesis. The chromosome's backbone unravels at many sites so that regions, composed of specific genes, loop outward from the backbone. These loops give the chromosome its distinctive lampbrushlike appearance. Large amounts of a nucleic acid known as messenger ribonucleic acid (mRNA) are being made on each loop. This mRNA is then processed and sent into the developing eggs's cytoplasm, where most of it will be stored for use during early development. After fertilization, these maternal (egg-derived) mRNAs can be used to make specific proteins necessary for the embryo.

Another event present in some developing eggs is the mass production of another type of RNA known as ribosomal RNA (rRNA). Most of this rRNA will also be stored until fertilization. After fertilization, these rRNA particles will help form cytoplasmic structures called ribosomes (the sites of protein synthesis). In addition to these egg products, many animal eggs must become filled with yolk. Yolk is the general term that covers the major storage of material in the egg.

Because the maternal proteins (yolk and other protein components) and nucleic acids (various RNAs) form the bulk of the egg cytoplasm, they have profound influences on the development of the embryo. In particular, the positions of maternal mRNAs, ribosomes, and proteins affect the organization of the embryo. It is evident, then, that the maternal genetic information and the arrangement of the products of this information provide crucial developmental information that will control much of the course of embryonic development. Therefore, the egg contributes considerably more than a haploid nucleus to the zygote.

Studying Gametogenesis

There are several approaches to the study of gametogenesis. Early biologists employed cytological techniques (methods of preparing cells for the study of their structure and function) and microscopy to study gamete formation. These early studies were, in fact, observations of the actual events themselves. Although these early descriptive approaches gave much information about the cells involved at each stage of gamete formation, they did not provide any information about the control mechanisms for this process. Biochemical studies have contributed to the understanding of certain regulatory substances and how they function in gametogenesis. By enhancing or inhibiting the presence of these regulatory substances in the organism, investigators have been able to elucidate many of the normal events of gametogenesis.

Beginning at puberty, the hormones (substances released from endocrine glands, generally functioning to regulate specific body activity) of the hypothalamus, the pituitary gland, and the gonads interact to establish and regulate gametogenesis in the organism. Gonadotropin-releasing hormone (GnRH) from the hypothalamus stimulates the release of follicle-stimulating hormone (FSH) and luteinizing hormone (LH) from the anterior portion of the pituitary gland. All three of these hormones are necessary for spermatogenesis and oogenesis.

Surgical removal of the mammalian pituitary gland (hypophysectomy) in the male leads to degeneration of the testes. Testicular function can be restored in these hypophysectomized animals by administering the hormones FSH and LH. These studies suggest that FSH and LH are necessary for normal functioning of the testes. LH appears to stimulate the release of testosterone (male hormone) by certain cells (Leydig cells) of the testes. Both testosterone and FSH are necessary for spermatogenesis, but the exact role that each of these hormones plays in male sexual physiology has yet to be determined.

Oogenesis in the female has been the subject of intense investigation. At the beginning of each ovarian cycle, from puberty to menopause, one primary oocyte present in the female's ovaries is activated to continue the process of gamete formation. Release of GnRH from the hypothalamus at the beginning of each cycle stimulates the anterior portion of the pituitary gland to release FSH. FSH, in turn, affects the ovaries: It stimulates a primary oocyte to mature to the point that it can be released from the ovary, as a secondary oocyte, and it causes certain cells (follicle cells) in the ovary to produce estrogens, female hormones. High estrogen levels will cause the pituitary to inhibit FSH release, a negative feedback mechanism, and stimulate LH release. These estrogen-mediated events occur at approximately the middle of the ovarian cycle. LH also affects the ovaries. LH, however, is responsible for ovulation (the release of the oocyte from the ovaries) and for the formation of a cellular structure called the corpus luteum. LH also stimulates the corpus luteum to produce progesterone, another female hormone. Eventually, high levels of progesterone will inhibit LH release from the pituitary gland, and the cycle begins anew.

—Geri Seitchik

See also: Aging; Asexual reproduction; Cell types; Cleavage, gastrulation, and neurulation; Determination and differentiation; Embryology; Fertilization; Growth; Parthenogenesis; Pregnancy and prenatal development; Regeneration; Reproduction; Reproductive system of female mammals; Reproductive system of male mammals; Sex differences: Evolutionary origins.

Bibliography

Epel, D. "The Program of Fertilization." *Scientific American* 237 (November, 1977): 128-140. A summary of the events of fertilization and an excellent account of the world of the egg and the sperm. It explains the events that occur after the formation of the gametes. Written for the college-level reader with some background in general biology.

Kinne, Rolf K. H., ed. *Oogenesis, Spermatogenesis, and Reproduction*. New York: Karger, 1991. Covers gametogenesis in fish.

Sadler, R. M. *The Reproduction of Vertebrates*. New York: Academic Press, 1973. A survey of reproductive patterns and mechanisms of reproductive control in the major vertebrate groups. Useful to the student who wants a comparative presentation of all aspects of reproductive events.

Van Blerkom, Jonathan, and Pietro M. Motta, eds. *Ultrastructure of Reproduction: Gametogenesis, Fertilization, and Embryogenesis*. Boston: Kluwer, 1984. A collection of essays focusing on the use of electron microscopy to understand reproductive processes such as gametogenesis.

GAS EXCHANGE

Type of animal science: Physiology
Fields of study: Biochemistry, biophysics, cell biology

Gas exchange refers to the processes used by animals to take up oxygen and eliminate carbon dioxide. These processes are the basis for understanding how animals breathe, particularly in diverse conditions.

Principal Terms

DIFFUSION: the passive movement of a gas across a membrane from a region of high pressure to one of low pressure

EPITHELIUM: a thin layer of cells that lines a body surface, such as the lining of the lungs or the intestines

PARTIAL PRESSURE: that part of the atmospheric pressure caused by only a single gas of many in a mixture; it is determined by how much of the gas is present in the mixture

PERMEABILITY: the tendency, in this case of a membrane, to permit the movement of a gas across that membrane

RESPIRATORY MEDIUM: the water or air that contains the oxygen used by an animal to carry out biochemical reactions

VENTILATION: the movement of the respiratory medium to and across the site of gas exchange

Gas exchange is the uptake of oxygen and the loss (or elimination or excretion) of carbon dioxide. It refers to two major steps in the overall oxygen consumption (or carbon dioxide excretion) by the whole animal. These two steps are the movement of the respiratory medium (containing oxygen) past the site of gas exchange, known as ventilation, and the diffusion of oxygen across the gas-exchange surface into the animal. The final step in diffusion of oxygen into an animal always involves diffusion from a liquid into a liquid, even in air breathers. Both ventilation and diffusion depend on the design of the structures as well as the way in which the systems and structures work. The additional steps in the whole respiratory system are the internal counterparts to ventilation and diffusion. These are perfusion, or blood flow, and diffusion from the blood to the tissues.

Basically, gas exchange takes place at a respiratory surface where the source of oxygen, the respiratory medium, is brought into contact with the surface. Oxygen diffuses into the animal, carbon dioxide diffuses out, and the spent or used respiratory medium is removed. The movement of the respiratory medium is termed ventilation. Once oxygen is in the animal, it is transported to the site of oxygen utilization, the tissues. Carbon dioxide, on the other hand, must be transported from the tissues to the site of gas exchange for excretion in gaseous form.

Gas-Exchange Organs

Animals have three basic types of gas-exchange organs: skin, invaginations (inpocketings of the epithelium), and evaginations (outpocketings of the epithelium). All three show modifications to improve the conditions of gas exchange. Skin always permits gas exchange unless it is coated with some material that limits diffusion. The skin of a snake or a turtle is so coated and permits very little gas exchange. The skin of a worm or an octopus, on the other hand, is quite thin and permits gas exchange quite freely. Invaginations of the external epithelium are basically what lungs and insect tracheas are, but in a highly modified condition. Evaginations of the skin are represented by the gills of aquatic animals; even when inside a cavity, as are fish and crab gills, they are still evaginations.

There is only one way that animals take up oxygen from the external medium, regardless of whether that medium is air or water. Gas must passively diffuse across the membrane that separates the animal from its environment. That membrane is a type of tissue called epithelium and is similar in nature and structure to the tissue that lines other body surfaces. The different types of epithelia are classified according to their locations and functions; those lining gills, lungs, and certain other organs of gas exchange are all known as respiratory epithelia. The respiratory epithelium not only separates the internal and external fluids but also represents a barrier to the movement of materials such as gas.

Gas Diffusion

Diffusion of a gas across a membrane occurs according to the laws of physics. The driving force for gas diffusion is the difference in the partial pressure of the gas across the membrane. A high external partial pressure and low internal partial pressure will provide a large difference and will enhance diffusion. Oxygen makes up 20.9 percent of the air, so that 20.9 percent of the atmospheric pressure at sea level (14.72 pounds per square inch) is attributable to oxygen (3.08 pounds per square inch). At higher altitudes, atmospheric pressure and partial pressure of oxygen are reduced.

The other factors that determine the diffusion of a gas are the thickness of the membrane across which it diffuses, the membrane's total surface area, and the nature or composition of the membrane. Obviously, a thick membrane will retard diffusion of gas because the gas must move across a greater distance. The distance the gas must diffuse is known as the diffusion distance. Additionally, the total surface area of the membrane available for diffusion has a direct effect on the rate of diffusion from one place to another. The greater the surface area, the greater the quantity of gas that can diffuse in a given time. Finally, the composition of the membrane is of critical importance in determining the diffusion of a gas. The nature of the membrane is referred to as the permeability of the membrane to the gas in question. The greater the permeability, the more easily gas diffuses. A membrane with a layer of minerals (calcium, for example) on the cells will not be as permeable as one without such a layer.

A very important point to note is that gases diffuse according to the difference in the partial pressure of the gas and not according to the concentration of the gas in the liquid. Several scientists have proved this by constructing artificial systems with two dissimilar fluids separated by a membrane. The movement of oxygen is always from high partial pressure to low and not from a high concentration to low. The reason is that pressure is a measure of molecular energy, but concentration of a gas in a liquid depends on the amount of that gas that can dissolve in the liquid—its solubility.

A Fluid-Fluid Boundary

All gas exchange occurs across a fluid-fluid boundary—that is, from one liquid to another—even in air breathers. The explanation for this is that all respiratory epithelia are moist and are kept so by the cells that line the surface. If the surface were to dry out, the permeability to gases (and other materials) would be substantially reduced. Thus, in air breathers, oxygen must first dissolve in the thin fluid layer lining the surface before diffusion across the epithelium takes place.

Organs of gas exchange work as do radiators, except that a gas is exchanged instead of heat. In this system, there are two liquids, one the source and the other the sink for the transferred material, the gas. The source is the external supply, and the sink is the blood or other internal fluid. Both fluids are contained in vessels or tubes that channel and direct it, with a thin layer of epithelium between the two. In the most efficient transfer systems, both the source and the sink flow, and they flow in opposite directions. If they did not flow, then the two fluids would simply come to an equilibrium, with oxygen partial pressure the same in both of the fluids. By moving in opposite directions, each is renewed, and the difference between them is always maximized. This type of exchange system is a "countercurrent system," because the two fluids flow in opposite directions. The most efficient of the gas-exchange organs function this way.

Two other types of gas-exchange systems are based on fluids flowing in directions other than perfectly opposite to each other. Birds have a respiratory system in which blood and air do not flow in opposite directions; rather, the blood flows perpendicularly to the direction of the air flow. This is referred to as crosscurrent flow. While not as efficient as countercurrent flow, it provides for an acceptably high level of efficiency. A system in which the respiratory medium is not channeled, but the blood is instead in vessels, is known as a mixed volume system. The mammalian lung functions in this way; the air is pumped into sacs that are lined with tiny blood vessels, but the air does not flow.

Ventilation

Ventilation of gas-exchange organs is accomplished by a pumping mechanism that brings the respiratory medium to and across the gas-exchange surface. Both water and air breathers use ventilatory pumps, but water breathers must move a much heavier and denser medium than air breathers. Pumping mechanisms may be located at the inflowing end of the system or at the outflowing end. The former are positive pressure pumps that push respiratory medium, and the latter are negative pressure pumps that pull respiratory medium into the cavity. Negative pressure is used in mammalian lungs, insect tracheas, and crab gills, while positive pressure is used in some fish that push water from the mouth into the gill chamber.

There are two basic patterns of flow of the respiratory medium through gas-exchange organs: one-way and tidal. One-way flow is found in fish, crabs, clams, and a number of other aquatic animals. Interestingly enough, the bird respiratory system also uses one-way flow. In one-way flow systems, the medium is always moving and passes over the gas exchange surface only once. In tidal-flow systems, the respiratory medium moves in and out (like the tide) through the same passages and tubes. The mammalian and insect respiratory systems both utilize tidal ventilation. The respiratory medium is not always moving, and when it is exhaled, there is some amount remaining in the cavity. The remaining respiratory medium will contain more carbon dioxide and less oxygen than fresh respiratory medium, with which it will mix upon inhalation.

Measuring Gas Exchange

The total amount of oxygen used by an animal is a gross measure of gas exchange known as oxygen uptake or oxygen consumption. Its counterpart for carbon dioxide is carbon dioxide excretion. Oxygen uptake is expressed as the amount of oxygen used per minute per kilogram of animal mass. Carbon dioxide excretion is expressed in the same terms. In theory, oxygen uptake and carbon dioxide excretion will be numerically the same, but in live animals there are several circumstances that cause the two to differ. Measurement of both rates is accomplished in similar ways. One method is the use of a respirometer and involves placing an animal in a sealed container and measuring the rate at which oxygen is depleted or carbon dioxide produced by an animal. Alternatively, respiratory medium, either air or water, is pumped through the respirometer and the oxygen or carbon dioxide measured in the inflowing and outflowing medium; the difference will be the amount used by the animal. The flow rate of the air or water must also be known for the calculations. The use of a respirometer is preferable but may not be practical for large animals, such as a horse.

In the case of animals too large to use a respirometer, oxygen uptake or carbon dioxide excretion is determined by measuring the rate of flow of the respiratory medium through the gas exchange organ and measuring the oxygen in the inspired and expired air or water. The result is the ventilation rate and the amount of oxygen extraction, the product of which is the oxygen consumption rate. This measurement is straightforward in animals with one opening for inspired and another for expired respiratory medium, such as a fish or a crab. In animals that inhale and exhale through the same organ, however, there are complications that make the measurements more difficult. Still, it is possible to measure the flow of air in and out of a lung and to collect at least some of the gas and measure either the oxygen or carbon dioxide in that air.

There is an additional advantage to the latter technique, measuring ventilation and the oxygen in the water or air. That advantage is that another measure of gas exchange is provided in these measurements. The difference between the amount of oxygen in inspired and expired water or air is the extraction (the amount taken out) and assesses the efficiency of the gas exchange organ. The efficiency is usually given as the percentage of oxygen taken out of the respiratory medium (the amount removed divided by the amount in inspired air or water). There are numerous factors that affect extraction, and measuring the efficiency provides one piece of information.

Studies of gas exchange encompass all levels of organization of animals, from the cellular to the whole animal. One of the most important levels concerns the structure of the organ and the parts of the organ. For this, it is necessary to see the spatial relationships among the parts, measure distances and areas, and count structures. The surface area, the volume, the number of structures or substructures, and the diffusion distances must all be measured. The results describe the morphology and morphometrics of the organ. Both whole, intact animals, and preserved specimens are used to make these measurements. The techniques are those used in surgery and dissection, and the results are critical to an understanding of the basic function of the respiratory organ. The electron microscope has been a powerful tool in this regard, permitting the accurate measurement of cellular-level distances, such as the diffusion distance.

Measuring Partial Pressure Difference

Measuring the partial pressure difference between the inside and outside of the animal is critical because of the role that this pressure difference has in gas exchange. Partial pressure of oxygen may be measured in two ways: in the intact animal or in a sample removed from the animal and injected into an instrument. The instrument most commonly used for measuring oxygen or carbon dioxide partial pressure is an electrode that changes electrical output when oxygen diffuses across an artificial membrane into a salt solution. Some of these electrons have been miniaturized and are only four millimeters across, and they will fit in a syringe needle. Still, it is difficult to use one of these in an intact mammal. The other way to measure partial pressure of oxygen or carbon dioxide on either side of the respiratory epithelium is to withdraw a sample of the air or water on the outside or the blood from the vessels on the inside. This procedure may be routine (in animals such as fish and crabs) or somewhat difficult (as in a mammal). A small tube is threaded into the lung to withdraw the air sample.

Measuring Ventilation

The movement of the respiratory medium, ventilation, is an important measure in determining the rate at which oxygen is brought to the respiratory surface. The blood flow (perfusion) on the inside is the counterpart to ventilation and is equally important. Ventilation can be measured either indirectly (meaning it is calculated) or directly. Indirect determinations require measuring other functions and then calculating ventilation based on known equations. If the rate of oxygen uptake and the extraction are measured, for example, then ventilation can be calculated.

Direct measures of ventilation use an electronic sensing device to determine the flow of water or air at the site of intake or outflow of respiratory medium on the animal. A human subject can simply breathe into such an electronic or mechanical device. Nonhuman mammals are more difficult and frequently require indirect techniques.

Direct measures may be the flow rate of the respiratory medium, the frequency of breathing, the hydrostatic pressure in the respiratory chamber, or a change in shape and size of the respiratory chamber. Any of these measures can be used to monitor routine respiratory function, but all are needed to assess gas exchange completely and accurately.

It is also necessary to know the general pattern of water or air movement at the respiratory surface. To do so often requires some invasive technique and the use of an indicator, such as a dye in the respiratory medium. The movement of the medium can then be visualized to determine the pattern. In some animals, video cameras can be used to photograph flow patterns of dyed medium, particularly water.

Uses of Gas Exchange Study

Gas exchange is studied by researchers and health practitioners both to assess basic function and to determine the source and nature of limitations of the systems of the body. These two areas may seem quite different at first; one is applied research, and the other is considered basic research. Both, however, have the same bases and use the same equations and principles. Only the animals or conditions differ.

One of the clinical applications, or contexts, in which gas exchange is studied is in respiratory distress or pulmonary (lung) disease. In these cases, the respiratory epithelium may become inflamed and thickened. This will increase the diffusion distance and retard, or limit, oxygen uptake and carbon dioxide release at the lung. Secretion of mucus by a respiratory epithelium may have a similar result for the same reasons. Mucus secretion occurs in several diseases and also takes place in fish gills when irritated by noxious chemicals in the water.

Gas exchange is also studied in diverse animals to understand evolutionary trends and pressures. Animals that live at high altitudes, for example, are constantly faced with low oxygen pressure in the air, and therefore some adjustment must be made by the animal. Scientists study the respiratory systems of these animals to determine if one of the other factors that affects gas exchange, such as diffusion distance or total surface area, is altered to compensate for the lower pressure difference.

All animals have similar basic physiological needs, including a need for oxygen to fuel the conversion of food materials into energy and other substances. Many animals have unique or specific forms or structures enabling them to survive in a particular habitat. Some of these forms affect the respiratory system—as in the differences between the respiratory surface in land animals compared with similar species that live in water. Scientists have compared the gas-exchange systems in aquatic and terrestrial species to learn more about evolutionary processes.

—Peter L. deFur

See also: Cell types; Circulatory systems of invertebrates; Circulatory systems of vertebrates; Hibernation; Lungs, gills, and tracheas; Metabolic rates; Respiration: Adaptation to low oxygen; Respiration in birds; Respiratory system.

Bibliography

Boutilier, R. G., ed. *Vertebrate Gas Exchange: From Environment to Cell*. New York: Springer-Verlag, 1990. Covers the comparative physiology of respiration in vertebrates.

Dejours, Pierre, ed. *Principles of Comparative Respiratory Physiology*. 2d ed. New York: Elsevier, 1981. An excellent book for professionals and for those who have mastered other readings in the field. Dejours gives the whole story, from basic concepts to more complex research, but the primary focus is clearly on the latter. He presents such a strong foundation that the information is not overwhelming. Uses equations freely and provides complete documentation for the graphs he has drawn from other scientists' work.

Hill, R. W., and G. A. Wyse. *Animal Physiology*. New York: Harper & Row, 1989. Provides a relatively complete discussion of the topic of gas exchange from first principles to application. The illustrations are of very high quality, and the explanations are neither too complex nor filled with jargon; they are understandable to entering college students. Additional readings are listed.

Prange, Henry D. *Respiratory Physiology: Understanding Gas Exchange*. New York: Chapman & Hall, 1996. Covers the biological, chemical, and physical principles of respiratory physiology. Takes a practical approach to explaining not only how gas exchange works, but why.

Rahn, H., A. Ar, and C. V. Paganelli. "How Bird Eggs Breathe." *Scientific American* 240 (February, 1979): 46-55. This application of basic diffusional-based gas exchange provides the fundamental concepts and an example of how these concepts are used to understand evolutionary processes.

Raven, P. H., and G. B. Johnson. *Biology*. 4th ed. Boston: McGraw-Hill, 1996. The chapter on respiration is not very long but has good illustrations and simple explanations of gas exchange. Others on various animal groups do provide additional comparative information in one text. Written at the introductory college level.

West, John B. *High Life: A History of High-Altitude Physiology and Medicine*. New York: Oxford University Press, 1998. An introduction to the field of high-altitude respiratory physiology, taking a historical approach to topics such as the understanding of acute high-altitude disease, the physiological effects of long-term high-altitude residence, and the origins of the field in the eighteenth and nineteenth century scientific investigation. Written for specialists.

GEESE

Type of animal science: Classification
Fields of study: Anatomy, ornithology, physiology, zoology

Geese are migratory water birds. Over two dozen species of geese have been identified; seven live in North America.

Principal Terms

BANDING: attachment of identification tags to individual birds
CLUTCH: number of eggs in the nest
COVERTS: feathers covering the bases of the large feathers of the wings and tail
FLEDGLING: young bird still unable to fly
LAMELLAE: toothlike structures in the beak, forming a strainer that permits birds to retain food particles while still enabling water to flow from the closed mouth
MANDIBLES: upper and lower beak parts
MOLTING: the loss of feathers that will be succeeded by new growth

Relatively difficult to identify, geese are perhaps best known for their migration, being harbingers of both summer and winter. Geese live comfortably at the intersection of water and land, eating a vegetarian diet of grasses, seeds, roots, berries, and aquatic plants. Unlike many birds, geese pair-bond and mate for life, building a simple nest in the ground and lining it with down. Depending on their species, geese lay a clutch of four to nine eggs, and fledglings remain grounded for forty to seventy-three days. Geese molt once a year after the breeding season; during this period the bird is unable to fly for four to five weeks.

Physical Characteristics of Geese

Geese are medium- to large-sized birds, between the size of ducks and swans. They have long necks and short legs in the center of their bodies. Because of this leg placement, geese do not waddle as much as other waterfowl when moving about on land. Their feet are webbed, which enables them to swim rapidly.

The bills of geese have serrations on the mandibles called lamellae; these are useful, because most geese gather some of their food while submerged. Because the bill is equipped with a horny covering at the tip of the upper mandible and because of the curved shape of the bill, geese can easily clip grasses, grains, or other foods.

Most species of geese are blackish in color, although many species have considerable white

Geese Facts

Classification:
Kingdom: Animalia
Subkingdom: Bilateria
Phylum: Chordata
Class: Aves
Order: Anseriformes
Family: Anatidae
Subfamily: Anserinae
Tribe: Anserini (swans and geese, four genera, thirty-six species)
Geographical location: Primarily the Northern Hemisphere
Habitat: Land (fields, prairies, tundra, forests) and water (ponds, lakes, rivers, streams, inlets, tidal areas, oceans)
Gestational period: From three to four weeks
Life span: Most geese live about twenty years; can (rarely) reach fifty-five years
Special anatomy: Webbed feet, long necks, large bills

Migrating geese make stopovers en route, forming huge, seasonal flocks on lakes along their flyways. (PhotoDisc)

markings in their plumage, often restricted to their bellies or lower tail coverts. Even though each species has distinguishing physical characteristics, it is often difficult to identify a particular species in the field. Even Canada geese, which are probably the easiest to identify with their white chin straps, can be a source of frustration for bird watchers attempting to distinguish between the dozen or so subspecies of Canada geese. Male and female geese of each species resemble each other.

Goose Migration

Many species of birds migrate, but geese, because they are so visible, epitomize this phenomenon to most people. Geese migrate for survival: Their northern breeding grounds become inhospitable when water areas freeze over, making food unattainable, and their southern grounds cannot sustain large populations of birds year around. In the United States, geese migrate northward from February to March and southward from September to November.

Geese, like many birds, often return to the same summer and winter areas year after year. Clearly, geese make use of visual landmarks when migrating; even at night some landmarks, such as large bodies of water, can be seen from the air. Young geese learn about these landmarks, as well as about traditional migratory stopping places, from their elders. Landscape alone, however, is not all that helps geese navigate. The position of the sun and stars can help them, as can the wind and the earth's magnetic field.

Most species of geese fly in some sort of V-formation. Flocks of snow geese, for example, form U's, checkmarks, or irregular masses. Only Canada geese fly in an almost perfect V. These formations have less to do with aesthetics than aerodynamics. Just as bicycle racers will draft lead bikers by riding slightly behind and to the side of them, so a goose will draft the lead bird, the one doing most of the work by breaking the wind's resistance for the birds that follow. Lead birds change during a flight. Geese normally fly at about forty miles per hour, a speed that can change if the bird is in a hurry or being chased.

By plotting the origins and destinations of banded birds, biologists have discovered four main flyways used by geese in the United States: the Atlantic Flyway, the Mississippi Flyway, the Central Flyway and the Pacific Flyway. Banding geese in order to study them is more than an academic exercise. Knowing migration paths has allowed experts to make decisions on hunting quotas and on the location of wildlife refuges.

—*Cassandra Kircher*

See also: Beaks and bills; Birds; Domestication; Feathers; Flight; Migration; Molting and shedding; Nesting; Pair-bonding; Respiration in birds; Wildlife management; Wings.

Bibliography

Bellrose, Frank C. *Ducks, Geese, and Swans of North America*. 3d ed. Harrisburg, Pa.: Stackpole Books, 1980. This book is the definitive word on geese as well as on ducks and swans. Bellrose has also written countless technical articles on geese.

Fegely, Thomas. *Wonders of Geese and Swans*. New York: Dodd, Mead, 1976. Amply illustrated, this book is an accessible introduction to geese.

Furtman, Michael. *On the Wings of a North Wind*. Harrisburg, Pa.: Stackpole Books, 1991. A disciple of Frank Bellrose, Furtman chronicles his travels with waterfowl. This is not a reference book.

Lishman, William. *Father Goose*. New York: Crown, 1996. This story of a man who flew with geese and taught them to navigate, teaches and entertains. Although this is not a standard reference text, it is essential reading for anyone studying geese.

Short, Lester. *The Lives of Birds*. New York: Henry Holt, 1993. Focusing on bird behavior, this book answers questions about the world's birds. Two dozen pages focus on geese.

GENE FLOW

Type of animal science: Evolutionary science
Fields of study: Anthropology, ecology, genetics

In biology, "migration" refers to the movement of a member or many members of a species of animal or plant from one geographic location to another. If other members of the same species interbreed with the migrating species, either along the way or in the new location, an exchange of genes occurs. Biologists call this exchange "gene flow" and consider it to be fundamental to the evolutionary process.

Principal Terms

ALLELE: one of a group of genes that occurs alternately at a given locus

DEME: a local population of closely related living organisms

FOSSIL: a remnant, impression, or trace of an animal or plant of a past geologic age that has been preserved in the earth's crust

GENE POOL: the whole body of genes in an interbreeding population that includes each gene at a certain frequency in relation to other genes

MUTATION: a relatively permanent change in hereditary material involving either a physical change in chromosome relations or a biochemical change in the codons that make up genes

POPULATION: a grouping of interacting individuals of the same species

SPECIATION: the process whereby some members of a species become incapable of breeding with the majority and thus form a new species

SPECIES: a category of biological classification ranking immediately below the genus or subgenus, comprising related organisms or populations capable of interbreeding

Prior to the nineteenth century, religious dogmatism retarded the activities of most scientists investigating the origins and nature of life by insisting on the immutability of species created by God. Despite mounting fossil evidence that many species of flora and fauna that once inhabited the earth had disappeared and that many extant species could not be found in the fossil record, pre-nineteenth century naturalists could find no viable explanation (other than divine intervention) for the disappearance of life-forms and their replacement by other forms. Then, in 1859, Charles Darwin published his epochal *On the Origin of Species*, which proposed the theory that all contemporary life-forms have evolved from simpler forms through a process he called "natural selection."

Many individuals before Darwin had proposed theories of evolution, but Darwin's became the first to be widely accepted by the scientific community. His success resulted from the careful and objective presentation of an overwhelming amount of evidence showing that species can and do change, and his concurrent promulgation of a convincing explanation of the mechanism that produces that change—natural selection. Since Darwin, scientists have modified and added new concepts to his theory, especially concerning the ways in which species change (evolve) over time. One of those new concepts, which was only dimly understood in Darwin's lifetime, is the importance of genetics in evolution, especially the concepts of migration and gene flow.

Genes and Gene Exchange

Genes are elements within the germ plasm of a living organism that control the transmission of a hereditary characteristic by specifying the structure of a particular protein or by controlling the function of other genetic material. Within any breeding population of a species, the exchange of genes is constant among its members, ensuring genetic homogeneity. If a new gene or combination of genes appears in the population, it is rapidly dispersed among all members of the population through inbreeding. New alleles may be introduced into the gene pool of a breeding population (thus contributing to the evolution of that species) in two ways: mutation and migration. Gene flow is integral to both processes.

A mutation is the appearance of a new gene or the almost total alteration of an old one. The exact causes of mutations are not completely understood, but scientists have demonstrated that they can be caused by radiation. Mutations occur constantly in every generation of every species. Most of them, however, are either minor or detrimental to the survival of the individual and thus are of little consequence. A very few mutations may prove valuable to the survival of a species and are spread to all of its members by migration and gene flow.

When immigrants from one population interbreed with members of another, an exchange of genes between the populations ensues. If the exchange is recurrent, biologists call it "gene flow." In nature, gene flow occurs on a more or less regular basis between demes, geographically isolated populations, and even closely related species. Gene flow is more common among the adjacent demes of one species. The amount of migration between such demes is high, thus ensuring that their gene pools will be similar. This sort of gene flow contributes little to the evolutionary process, since it does little to alter gene frequencies or to contribute to variation within the species. Much more significant for the evolutionary process is gene flow between two populations of a species that have not interbred for a prolonged period of time.

Populations of a species separated by geographical barriers often develop very dissimilar gene combinations through the process of natural selection. In isolated populations, dissimilar alleles become fixed or are present in much different frequencies. When circumstances do permit gene flow to occur between two such populations, it results in the breakdown of gene complexes and the alteration of allele frequencies, thereby reducing genetic differences in both. The degree of this homogenization process depends on the continuation of interbreeding between members of the two populations over extended periods of time.

Hybridization

The migration of a few individuals from one breeding population to another may, in some instances, also be a significant source of genetic variation in the host population. Such migration becomes more important in the evolutionary process in direct proportion to the differences in gene frequencies—for example, the differences between distinct species. Biologists call interbreeding between members of separate species "hybridization." Hybridization usually does not lead to gene exchange or gene flow, because hybrids are not often well adapted for survival and because most are sterile. Nevertheless, hybrids are occasionally able to breed (and produce fertile offspring) with members of one or sometimes both the parent species, resulting in the exchange of a few genes or blocks of genes between two distinct species. Biologists refer to this process as "introgressive hybridization." Usually, few genes are exchanged between species in this process, and it might be more properly referred to as "gene trickle" rather than gene flow.

Introgressive hybridization may, however, add new genes and new gene combinations, or even whole chromosomes, to the genetic architecture of some species. It may thus play a role in the evolutionary process. Introgression requires the production of hybrids, a rare occurrence among highly differentiated animal species. Areas where hybridization takes place are known as contact zones or hybrid zones. These zones exist where populations overlap; in some cases of hybridization, the line between what constitutes different species and what constitutes different populations of the same species becomes difficult to draw. The significance of introgression and hybrid zones in the evolutionary process remains an area of some contention among life scientists.

Biologists often explain, at least in part, the poorly understood phenomenon of speciation through migration and gene flow—or rather, by a lack thereof. If some members of a species become geographically isolated from the rest of the species, migration and gene flow cease. The isolated population will not share in any mutations, favorable or unfavorable, nor will any mutations that occur among its own members be transmitted to the general population of the species. Over long periods of time, this genetic isolation will result in the isolated population becoming so genetically different from the parent species that its members can no longer produce fertile progeny should one of them breed with a member of the parent population. The isolated members will have become a new species, and the differences between them and the parent species will continue to grow as more ages pass. Scientists, beginning with Darwin himself, have demonstrated that this sort of speciation has occurred on the various islands of the world's oceans and seas.

Studying Gene Flow

Scientists from many disciplines are currently studying migration and gene flow in a variety of ways. For decades, ornithologists and marine biologists have been placing identifying tags or markers on members of different species of birds, fishes, and marine mammals to determine the range of their migratory habits in order to understand the role of migration and subsequent gene flow in the biology of their subjects. These studies have led, and will continue to lead, to important discoveries. Most studies of migration and gene flow, however, relate to human beings.

Many of the important discoveries concerning the role of gene flow in the evolution of life come from the continuing study of the nature of genes. A gene, in cooperation with such molecules as transfer ribonucleic acid (tRNA) and related enzymes, controls the nature of an organism by specifying amino acid sequences in specific functional proteins. In recent decades, scientists have discovered that what they previously believed to be single pure enzymes are actually groups of closely related enzymes, which they have named "isoenzymes" or "isozymes." Current theory holds that isozymes can serve the needs of a cell or of an entire organism more efficiently and over a wider range of environmental extremes than can a single enzyme. Biologists theorize that isozymes developed through gene flow between populations from climatic extremes and enhance the possibility of adaptation among members of the species when the occasion arises. The combination and recombination of isozymes passed from parent to offspring are apparently determined by deoxyribonucleic acid (DNA). Investigation into the role of DNA in evolution is one of the most promising avenues to an understanding of the nature of life.

A classic example of the importance of understanding migration and gene flow in the animal kingdom is the spread of the so-called killer bees. In the 1950's, a species of ill-tempered African bee was accidentally released in South America. The African bees mated with the more docile wild bees in the area; through migration and gene flow, they transmitted their violent propensity to attack anything approaching their nests. As the African genes slowly migrated northward, they proved to be dominant.

Further research into migration and gene flow promises to provide information indispensable to the attempt to unravel the mysteries of life. Coupled with the concept of mutation, gene flow is a crucial component of evolution.

—Paul Madden

See also: Adaptive radiation; Demographics; Ecological niches; Ecology; Evolution: Animal life; Evolution: Historical perspective; Genetics; Habitats and biomes; Hardy-Weinberg law of genetic equilibrium; Natural selection; Population analysis; Population genetics; Punctuated equilibrium and continuous evolution.

Bibliography

Ammerman, A. J., and L. L. Cavalli-Sforza. *The Neolithic Transition and the Genetics of Populations in Europe*. Princeton, N.J.: Princeton University Press, 1984. This book is rather technical in its language, but it presents a thoughtful discussion of the influence of migration and gene flow on the Neolithic revolution and the complex workings of gene flow on human evolution.

Bailey, Jill. *Evolution and Genetics: The Molecules of Inheritance*. New York: Oxford University Press, 1995. An encyclopedia of the current understanding of genetics and evolution.

Cavalli-Sforza, Luigi Luca. *Genes, Peoples, and Languages*. Berkeley: University of California Press, 2000. Although this book focuses on humans and their languages, Cavalli-Sforza's introductory sections on genetics and gene flow are exceptionally clear and well written, accessible to nonspecialists, and applicable to all animal species.

Crow, J. F., and Motoo Kimura. *An Introduction to Population Genetics*. New York: Harper & Row, 1970. An excellent starting place for those whose knowledge of migration and gene flow and their influence on evolution is limited and who wish to learn more about the subject. The book is relatively free of the technical jargon that can make some biological texts difficult for the nonbiologist.

Endler, John A. *Geographic Variation, Speciation, and Clines*. Princeton, N.J.: Princeton University Press, 1977. Endler's book is valuable primarily because of an excellent chapter on gene flow and its influence on the evolutionary process. Endler sees evolution as a very slow and gradual process in which gene flow and small mutations cause massive change over long periods of time.

Hoffmann, Ary A., and Peter A. Parsons. *Evolutionary Genetics and Environmental Stress*. New York: Oxford University Press, 1991. Another excellent starting place for those whose knowledge of migration and gene flow and their influence on evolution is limited and who wish to learn more about the subject. The book is relatively free of the technical jargon that can make some biological texts difficult for the nonbiologist.

Raup, D. M., and D. Jablonski, eds. *Patterns and Processes in the History of Life*. New York: Springer-Verlag, 1986. Raup and Jablonski's book is a compilation of articles from the Dahlem workshop, "Patterns and Processes in the History of Life," held in Berlin in 1985. The articles in the book discuss the evidence concerning the role of migration and gene flow in the evolutionary process. Thought-provoking.

GENETICS

Types of animal science: Evolution, fields of study, reproduction
Fields of study: Biochemisty, cell biology, developmental biology, embryology, evolutionary science, reproduction science

Genetics is the study of inheritance of characteristics from one generation to the next. Humans have been studying genetics since prehistoric times with the first selective breeding of wolves for companion animals. In the 1800's, an Austrian monk, Gregor Mendel, described the basic laws that govern the inheritance of genetic traits. In the twentieth century, the field of molecular genetics was created as biologists determined the actual chemical makeup of genes.

Principal Terms

ALLELE: alternative forms of a single gene
CHROMOSOME: a long strand of DNA with supporting proteins, that contains many genes
DEOXYRIBONUCLEIC ACID (DNA): the chemical polymer that is the genetic material of multicellular organisms
GENE: factors in cells that are responsible for an observable characteristic of an organism
GENOME: all of the genetic material of an organism
GENOTYPE: the actual genetic makeup of an organism
MUTATION: any heritable change in the genetic material
PHENOTYPE: the observable characteristics of an organism (for example, black fur color in a cat)

Before any recorded history, ancient man chose alert pups from a litter of wolves for breeding. This practice of selectively breeding the wolves that were good companions eventually gave rise to the domesticated dog. The oldest undisputed dog bones known, excavated from a twenty-thousand-year-old Alaskan settlement, demonstrate that prehistoric humans knew that traits could be passed from one generation to the next, and that selectively breeding animals (or plants) could produce an organism that possessed desired characteristics. This practice of deliberate breeding is known as artificial selection.

Humans have practiced artificial selection on numerous animals, including pigs, cattle, goats, and sheep. Homer and other Greek poets wrote about selective breeding, and part of the wealth of the ancient city of Troy was attributed to its expertise in horse breeding. Although humans had some control over the traits of domesticated animals through selective breeding, the results of matings were not always predictable, and nothing was known about the mechanism through which traits were passed from one generation to another until the mid-1800's.

Mendelian Genetics

Gregor Mendel, an Austrian monk, is the undisputed father of the science of genetics. Working with garden peas, Mendel analyzed thousands of breeding experiments to describe laws that governed the inheritance of traits. Though Mendel studied a plant, his laws for the inheritance of traits apply to all sexually reproducing organisms, including humans.

Mendel chose seven distinct traits to study in his garden peas: flower color, plant height, seed shape, seed color, pod shape, pod color, and flower position. He concluded that each of these

Gregor Mendel's experiments with garden peas laid the foundation for the study of genetics. (National Library of Medicine)

traits was determined by a single, discrete factor called a gene. For instance, there was a gene for flower color and a gene for seed shape. Each gene had several variations, or alleles. The gene for flower color had a white allele that produced white flowers and a purple allele that produced purple flowers.

Mendel's experiments revealed that organisms have two copies of any gene for a trait. Those two copies can be identical, two purple alleles of the flower color gene, for instance; or those two alleles can be different. A pea plant could have one purple allele of the flower color gene and one white allele of the flower color gene. When an organism has two identical alleles of a gene, it is homozygous for that gene. When an organism has two different alleles of a gene, it is heterozygous for that gene.

An organism inherits one allele, or copy of a gene, from one parent and one allele from the other parent. An organism, or cell, that has two copies of all of its genetic information is called diploid. In most sexually reproducing animals, the offspring are formed when a sperm cell from the male parent fertilizes an egg from the female parent. The sperm and the egg only contain half of all the genetic information. They are said to be haploid. However, the new organism they create is diploid because it gets one copy of the genetic information from the sperm and a second copy from the egg.

Mendel's first law, or the law of segregation, states that the two copies of each gene separate during the formation of gametes (eggs and sperm), and that fertilization of the egg by the sperm is a random event. Any sperm containing any allele of a gene can fertilize any egg of the same species, regardless of the allele carried by that egg.

Mendel noted that certain alleles seemed to dominate over others. For instance, when a plant had a purple allele for flower color and a white allele for flower color, the plant always had purple flowers. Mendel called the allele that was seen in the heterozygote, in this case the purple allele, the dominant allele. The allele that was hidden or masked, he called the recessive allele. In order to show a recessive allele, an organism has to have two identical copies of a gene, both containing the same recessive allele. This is known as the homozygous recessive condition. Garden peas that have white flowers are homozygous recessive for the white allele of the flower color gene.

Homozygous recessive describes the organism's genotype, or its genetic makeup. It has two copies of the recessive allele of the gene. The observable characteristic of the organism, having white flowers, is called its phenotype.

Mendel also demonstrated that the segregation of alleles of any one gene is not dependent on the segregation of alleles of any other gene. For instance, a gamete could receive a dominant allele for an eye color gene and a recessive allele for height, or that gamete could receive the recessive alleles for both genes or the dominant alleles of both genes. This is Mendel's second law, the law of independent assortment, and it applies to any genes that are located on separate chromosomes.

Mendel's work was far ahead of its time. Although Mendel published his research in the 1800's, it was not until after his death that his work gained recognition in the scientific community. In 1900, three other scientists, each working separately on inheritance, came across Mendel's work in the course of their research. They gave him credit for his insights, and Mendel's research provided the foundation for the new discipline of genetics.

Genes and Chromosomes

Although Mendel described the gene as the factor that was responsible for a particular trait, nothing was known about the physical makeup of a gene. One of the first questions scientists needed to answer was where genes are found in cells. Early studies in frogs and sea urchins indicated that the nucleus of the sperm and the nucleus of the egg combined with each other during fertilization. This observation suggested that the genetic material that determined how the fertilized egg would develop might reside in the nucleus.

As microscopes improved, scientists were able to distinguish structures within the nuclei of cells. These long, threadlike structures stained blue and were called chromosomes (Greek *chroma*, "color"). Several scientists observed that when animal and plant cells divided, the chromosomes duplicated, then separated, and each daughter cell inherited a complete set of chromosomes. The one exception to this was the cell division that produced the gametes (eggs and sperm). When an egg or a sperm cell was produced, it only contained half the number of chromosomes as the cell that produced it. If genetic information was carried on chromosomes, scientists reasoned that a sperm and an egg could each contribute half of the genetic information to the new organism at fertilization.

Some of the first evidence that chromosomes were linked to observable traits came from the studies of American graduate student Walter S. Sutton. Sutton studied grasshoppers, and his observations indicated that male grasshoppers always had an X and a Y chromosome, whereas female grasshoppers contained two X chromosomes. Several other scientists observed similar things in other organisms, such as fruit flies, and concluded that the physical characteristic of sex was determined by the kind of chromosomes an organism possessed.

Since chromosomes determined the trait of sex, it was possible that chromosomes contained the genes that Mendel had shown to determine physical characteristics. The first scientist to demonstrate that genes were located on chromosomes was Thomas Hunt Morgan, who showed that an eye-color gene in the fruit fly, *Drosophila melanogaster*, was located on the X chromosome.

Next, scientists wanted to know what kind of chemical molecule actually carried the genetic information. Chromosomes contain two kinds of molecules, protein and a weak acid called deoxyribonucleic acid (DNA). Experiments in the early 1930's first demonstrated that DNA is the genetic material. Oswald Avery, Colin MacLeod, and Maclyn McCarty showed that adding DNA to these bacterial cells could change their physical traits. In their experiments, they mixed a harmless strain of bacteria with DNA from bacteria that caused disease in mice. When they did this, the previously harmless bacteria changed (or transformed) into disease-causing bacteria. Two other scientists, Alfred Hershey and Martha Chase, later obtained similar results by studying a virus that infects *E. coli*.

Molecular Genetics

By the 1940's, scientists knew that genetic information was carried by genes made of DNA molecules inside cell nuclei. However, scientists did not know how the genetic information was copied accurately from one generation to the next—from one cell division to the next. Nor did scientists know how the DNA could account for the appearance of inherited changes or mutations. In order to answer these questions, scientists needed to know the precise chemical structure of DNA.

Many scientists contributed to the understanding of the structure of DNA. Erwin Chargaff obtained data that indicated that specific molecular components of the DNA molecule were always present in equal parts. These components were nitrogen-containing molecules (or nitrogenous bases). Chargaff determined that the nitrogen-containing bases adenosine and thymine were always present in a one to one ratio, and the bases guanine and cytosine were always present in a one to one ratio, no matter what species' DNA was analyzed.

Simultaneously, two scientists at Kings College in London, Rosalind Franklin and Maurice Wilkins, were attempting to make X-ray pictures of DNA molecules. Rosalind Franklin obtained an X-ray film that indicated that DNA was a helical molecule. Just previous to Franklin's work, an American chemist, Linus Pauling, had made a breakthrough in solving the structure of the protein alpha helix using a model-building approach.

Two scientists working at Cambridge University in England, James Watson and Francis Crick, decided to use Pauling's method of model building to attempt to solve the structure of the DNA molecule. Combining the data from a variety of sources including the data of Chargaff, Wilkins, and the crucial X-ray crystallography data of Rosalind Frankin, Watson and Crick solved the structure of the DNA molecule.

Watson and Crick created a model of DNA: a double helix, like a twisted ladder. The DNA molecule was a long polymer of repeating nucleotides. Each nucleotide contained three chemical parts: a sugar, a phosphate group, and a nitrogen-containing base. The sides of the double helix ladder were formed by alternating sugars and phosphates, and the rungs were formed on the inside of the helix by specific pairings of the nitrogen-containing bases. Adenine paired with thymine to form one kind of rung. Guanine paired with cytosine to form a second kind of rung.

The order of the bases provided the information within DNA. Certain combinations of bases could form "words" that stood for parts of proteins or other molecules encoded by the DNA. The double helix could unzip like a zipper, each strand serving as a template to guide the construction of a new strand. This provided an accurate means for copying the DNA molecules from a parent cell to a daughter cell.

Animal Model Genetic Organisms

ROUNDWORM (*Caenorhabditis elegans*): This millimeter-long worm allowed scientists to test the concepts of gene therapy, to develop methods for sequencing large amounts of DNA, and provided information about the biology of human diseases such as Alzheimer's disease and cancer. Research on this worm has also enabled scientists to develop effective control measures for plant and animal parasitic roundworms.

FRUIT FLY (*Drosophila melanogaster*): Studies in this organism allowed scientists to determine that genes reside on chromosomes, and gave insight into the nature of mutations. Studies of the development of complex structures such as the eye continue to provide insight into some of the ways cell specialization is regulated and directed by DNA.

ZEBRA FISH (*Danio rerio*): The zebra fish, with its transparent embryos, provides an excellent system in which to study the genes that regulate vertebrate development. Additionally, zebra fish have been used in studies investigating bioaccumulation of organic compounds in the environment.

COMMON MOUSE (*Mus musculus*): Mice are useful for genetic study because of the availability of hundreds of single gene mutations. Studies in mice demonstrated that Gregor Mendel's laws of inheritance were as applicable to mammals as to plants. Transgenic genetic analysis of mice has allowed the creation of mouse strains that mimic human genetic diseases.

Genetic Engineering

The details of how DNA is passed from one generation to the next, of how mutations arise, and of how the information of DNA is actually translated into the activities of cells forms the basis of genetic research at the beginning of the twenty-first century.

One of the most important scientific discoveries that led to modern genetic technology was the discovery of a particular kind of protein, a restriction enzyme, from bacteria that cuts DNA molecules at specific sequences of bases. These restriction enzymes gave scientists the tool they needed to break DNA down into smaller pieces, eventually allowing the isolation of individual genes from the huge amount of DNA inside the nucleus of the cell.

Herbert Boyer and Stanley Cohen combined their knowledge of restriction enzymes and bacterial transformation (getting bacteria to take up DNA from the environment) to clone genes. Gene cloning involves isolating a gene of interest by using a restriction enzyme to cut it away from other DNA, and placing it in a piece of DNA called a vector that can be taken up by bacterial cells. One of the first applications of this technology was the production of human insulin. Scientists isolated the gene that encodes the information for making insulin from human DNA, cloned it into a bacterial vector, and placed the vector with the insulin gene in *E. coli*. The *E. coli* cells were able to produce large quantities of insulin. This new insulin was considerably cheaper and safer than insulin purified from human tissue.

Variations on this technique of taking a piece of DNA from one species and inserting it into the cells of another species are involved in genetic engineering of multicellular organisms. In multicellular organisms such as plants or monkeys, the DNA vector is usually a modified virus. These techniques are the basis of human gene therapy.

In the last decade of the twentieth century, entire organisms have been cloned. In Scotland, Ian Wilmut and colleagues reported the first mammalian cloning of a sheep named Dolly. In Wisconsin and Japan, scientists have cloned cattle. When an organism is cloned, all of its DNA, usually contained within an intact nucleus from a cell of the adult animal, is transferred to an egg cell from which all the genetic information has been removed. The egg is then allowed to develop into a new organism. Although the new organism is young, it has the same DNA as the parent from which the nucleus was obtained.

Scientists have also developed techniques for sequencing DNA, determining the exact order and number of nitrogenous bases within the DNA of an organism's genome. In 2000, the Human Genome Project announced that the entire genome of the human had been sequenced. Many other genomes have been sequenced, including the roundworm, *C. elegans*, several plants, and even baker's yeast. The sequence of an organism gives scientists another tool in answering questions about how DNA regulates and determines the activities of cells.

The ethical consequences of genetic engineering are not clear. DNA forensic evidence is now used to convict or exonerate criminal suspects on a routine basis. The genetic engineering of food crops that are pest resistant or contain additional nutrients is fairly routine. With the cloning of entire organisms now possible, the cloning of a human is not science fiction. Parents can have an embryo tested for devastating genetic diseases before it is born. While many of these advances are clearly positive, many of them are double-edged swords, begging for informed public debate.

—Michele Arduengo

See also: Asexual reproduction; Breeding programs; Cleavage, gastrulation, and neurulation; Cloning of extinct or endangered species; Copulation; Courtship; Determination and differentiation; Development: Evolutionary perspective; Estrus; Fertilization; Gametogenesis; Hermaphrodites; Hydrostatic skeletons; Mating; Parthenogenesis; Pregnancy and prenatal development; Reproduction; Reproductive strategies; Reproductive system of female mammals; Reproductive system of male mammals; Sexual development.

Bibliography

Hartwell, L., L. Hood, M. Goldberg, A. Reynolds, L. Silver, and R. Veres. *Genetics: From Genes to Genomes*. Boston: McGraw-Hill, 2000. Up-to-date, definitive genetic textbook covering all major fields of investigation in genetics: Mendelian, bacterial and viral, human, population, and genetic engineering.

Marshall, Elizabeth L. *The Human Genome Project: Cracking the Code Within Us*. New York: Franklin Watts. 1996. Discusses the goals and structure of the Human Genome Project and its implications. Contains a chapter on the contributions of animal model systems to this project.

Mousseau, T. A., B. Sinervo, J. A. Endler, eds. *Adaptive Genetic Variation in the Wild*. New York: Oxford University Press, 1999. Provides excellent background and current information for readers interested in population genetics and evolution in natural systems.

Sayre, Anne. *Rosalind Franklin and DNA*. New York: W. W. Norton, 1975. Details the invaluable contribution of Dr. Rosalind Franklin to the discovery of the structure of the DNA molecule.

Sponenberg, D. P. *Equine Color Genetics*. Ames: Iowa State University Press, 1996. A fascinating, in-depth look at the genetic mechanisms that cover coat color in horses and donkeys.

Watson, James D. *The Double Helix: A Personal Account of the Discovery of the Structure of DNA*. Edited by Gunther S. Stent. New York: W. W. Norton, 1980. Watson's personal story of the characters and events of the race to solve the structure of the double helix. This edition provides viewpoints and commentary from other players in the story as well as reprints of the original research articles.

GIRAFFES

Type of animal science: Classification
Fields of study: Anatomy, behavior, ecology, physiology

Giraffes are even-toed, herbivorous mammals that are found exclusively in Africa. They have long necks and legs, and distinctive brown patches all over their bodies.

Principal Terms

BROWSER: feeds on shoots and leaves
DIURNAL: active during daylight hours
FLEHMEN: behavior involving curling and wrinkling of lips and nostrils, with the activation of the Jacobson's organ
HIERARCHY: a social structure in which animals are dominated by those higher on the linear ladder
JACOBSON'S ORGAN: a sense organ between the roof of the mouth and nasal passages, which detects chemical signals associated with reproduction

Because of their long necks and legs, giraffes are the tallest animals in the world. Patch coloration and shape can vary within their extensive habitat range. Females are distinguished from males in that the females have shorter, inward-curving horns. In both sexes, a long mane of stiff, brushlike hairs extends from the back of the head to the shoulders. Giraffes exhibit a unique, fluid gait. When walking, the fore and hind legs on the same side appear to move almost in unison. Swift and fleeting, giraffes can gallop up to thirty-three miles per hour. Their gallop can be described as a motion in which the front legs move together and their hind legs move forward and outward, enveloping the forelegs in a unique rhythmic pattern. Long and graceful, their sleek necks swing back and forth rhythmically with their legs. The neck has remarkable range of motion. A system of blood vessels and valves in the neck protects the brain and reduces blood pressure when the animal lowers its head.

Diet

Although some may feed at night, giraffes are classified as herbivorous diurnal eaters. They are browsers, and competition for food is greatly reduced because of the height at which they feed. The male feeds at greater heights, with his head stretched upward, whereas the female feeds at lower heights, often bending her head and neck to reach the leaves. Giraffes feed mainly on the highly nutritious leaves, fruit, and flowers of acacia trees. Their long, dexterous tongues strip leaves from the acacia twigs. Giraffes often consume soil and bones to balance the phosphorus and calcium in their blood. When feeding on sprouted vegetation and when drinking water, giraffes splay their front legs and bend their knees. In such a position, they are vulnerable to predators, especially lions. Giraffes can go without water for days.

Protection, Defense, and Communication

Generally docile creatures, giraffes may kill other animals with a kick of a fore or hind hoof. Their heads, used like a knight's mace, land formidable blows on the body or legs of opponents. The effective use of their heads as weapons is enhanced by the physical structure of horns and knobs. The skull bone is solid and thick, so that its force can result in a fatal blow. The giraffe's own head is protected by extensive sinuses, which absorb shock. Giraffes snort, grunt, bleat, bellow, and moo to communicate. Because of their height, giraffes' sense of smell is not as keen as other animals. However, their sensitivity to sound and their visual acuity more than compensate for their underdeveloped sense of smell.

Reproduction and Birth

Mixed herds of variable numbers have been recorded in the field. Old males are often solitary. About a day before mating, the female becomes sexually attractive. At the onset of mating, the male licks the female's genitals and catches her urine on his tongue. Chemical signals in her urine are detected by flehmen as the male's Jacobson's organ becomes activated. The male remains with the female in heat unless he is displaced by a higher-ranking challenger. The female gives birth in cover, with her back legs bent to reduce the height from which the calf falls. Initially, the mother is alone with her calf, but later she may form a nursery group assisted by other females. The calf stands within five minutes of birth and suckles within the hour. The calf becomes independent of the mother around the sixteenth month. Play behavior, called nose-to-nose sniffing, between the young cements social bonds between them.

—*William P. Carew*

See also: Defense mechanisms; Fauna: Africa; Herbivores; Horns and antlers.

Giraffes are the tallest animals in the world, with long necks that allow them to eat leaves from the tops of trees unreachable by other herbivores. (Corbis)

Giraffe Facts

Classification:
Kingdom: Animalia
Subkingdom: Bilateria
Phylum: Chordata
Subphylum: Vertebrata
Class: Mammalia
Order: Artiodactyla
Family: Giraffidae
Genus and species: Giraffa camelopardalis, with nine subspecies

Geographical location: Africa, south of the Sahara

Habitat: Dry, open savanna biomes, covered with bush and acacia trees

Gestational period: About 450 days

Life span: Up to twenty-five years in the wild; up to twenty-eight years in captivity

Special anatomy: A long neck with seven vertebrae (characteristic of all mammals); adult height of between fifteen and eighteen feet; chestnut brown patches of various sizes and shapes; sloping back ending in a long, tufted tail; a pair of horns on top of the head

Bibliography

Apps, Peter. *Creatures of Habit: Understanding African Animal Behavior.* Cape Town, South Africa: Struik, 2000. This book is a good introduction to African wildlife, including giraffes. It is user-friendly and enriched with beautiful photographs.

Benyus, Janine M. *The Secret Language and Remarkable Behavior of Animals*. New York: Black Dog & Leventhal, 1998. All kinds of unusual giraffe behaviors are documented in this book. The descriptions are enhanced with excellent diagrams.

Nowak, Ronald M. *Walker's Mammals of the World*. 6th ed. Baltimore: The Johns Hopkins University Press, 1999. This is a very well-known text on mammals. It is a detailed and valuable reference work.

Vaughan, Terry A., James M. Ryan, and Nicholas J. Czaplewski. *Mammalogy*. 4th ed. Fort Worth, Tex.: Saunders College Publications, 2000. This is a standard textbook for students specializing in the study of mammals.

GOATS

Types of animal science: Anatomy, classification, reproduction
Fields of study: Anatomy, zoology

Goats, artiodactyls similar to sheep, can live on hills and mountains where sheep would starve. They were domesticated to provide meat, milk, and leather.

Principal Terms

ARTIODACTYLS: hoofed mammals with an even numbers of toes
GESTATION: pregnancy
HERBIVORE: an animal that eats only plants
KERATIN: a tough, fibrous major component of hair, nails, hooves, horns
RUMINANT: a herbivore that chews and swallows plants, which enter its stomach for partial digestion, are regurgitated and chewed again, and reenter the stomach for more digestion.
UNGULATE: any hoofed mammal

Goats are herbivorous artiodactyls—even-toed ungulates—of the family Bovidae, genus *Capra*, which usually have hooves and hollow horns. They are also ruminants, chewing and swallowing food, regurgitating it, and chewing and swallowing it again. This cud chewing allows them to get the most nutrients possible from the low-quality foods they eat.

Wild goats are mountain dwellers, adept at leaping between rocks, sure-footed due to their hoofs. The hoofs have a hard outer layer and a softer, inner layer that wears away quickly and leaves hard edges useful for climbing. Domesticated goats, raised for milk, meat, and leather, retain many of these characteristics. Swiss goats, the most common domesticated variety, have pointy ears and horns, while Nubians are hornless.

Physical Characteristics of Goats

Most adult goats weigh up to 125 pounds. They are not as large as sheep, which they resemble. Their horns are twisted flat and turn backward. Their hoofs are divided in two (cloven). Males are called rams or billies, while females are called does or nannies. Males emit strong odors during mating season. Males and most females have chin beards, leading to the name "goatee" for the similar style of facial hair in men. Goats are normally covered with straight hair, but some grow wool, such as angora goats. Their coats are red, brown, tan, or white. Goats find enough to eat on poor, dry land where horses, cows, and sheep would starve. Adult female goats reach lengths of 2.5 to 3.5 feet and are approximately 4 feet at the shoulder; they weigh between 100 and 120 pounds. Males are 20 percent larger and heavier than females and have longer horns.

Domesticated goats derive from ten wild goat species. They live on hills and mountains and are either goat antelope or true goats. All except the Rocky Mountain goat inhabit Europe and the Asian Himalayas. Rocky Mountain goats and chamois are goat antelope, having physical characteristics of both goats and antelope.

Types of Goats

Rocky Mountain goats are goat antelope inhabiting mountains from the American Northwest to Alaska. They live in snowy, craggy habitats and are excellent climbers, due to hoofs having soft pads with hard rims that work well on snow, ice, or rock. Rocky Mountain goats are about three

Goat Facts

Classification:
Kingdom: Animalia
Subkingdom: Bilateria
Phylum: Chordata
Subphylum: Vertebrata
Class: Mammalia
Order: Artiodactyla
Family: Bovidae
Subfamily: Caprinae
Tribes: Rupicaprini (four genera, five species); Ovibonini (two genera, two species); Caprini (five genera, seventeen species)
Geographical location: Asia, Europe, North America, Africa, New Zealand, Australia
Habitat: Dry mountains or hills
Gestational period: Around 5.5 months
Life span: Fourteen to twenty years
Special anatomy: Permanent, hollow horns, ruminant stomach, very tough hooves

feet tall at the shoulder, with black horns, white, shaggy pelage, and goatees. They eat any plants available and are solitary, except during mating.

Chamois goat antelope have light brown pelage in the summer, which turns dark in winter. The males' deadly curved horns are eight inches long. Chamois are about the same size as Rocky Mountain goats and have similar hoofs. They live above mountain tree lines in Europe, Asia, and New Zealand and prefer eating grass and lichens. During winter, they eat pine needles and bark. Females form herds of up to one hundred, while males live alone except to mate. Chamois battle for mates, and defeated males often die after their belly and throat are ripped open by the victor's horns.

Angora goats, true goats, have body shapes like domesticated Swiss goats. They inhabit Turkey, South Africa, the United States, Argentina, Australia, and New Zealand. Horned and bearded, their long, thick, hair is used in mohair cloth. Angora goats grow approximately 80 percent of the size of Rocky Mountain goats. They live on hillsides, eating woody vegetation and grass.

Nubian goats, hornless true goats, have short hair and droopy ears, and the males have goatees. Adult males weigh up to 180 pounds, while females weigh up to 140 pounds. Males are about five feet tall at the shoulder. Their coats are black, tan, or red, sometimes with white spots. They eat briars, thistles, and brambles, as needed, with their tough mouths and teeth. Their splayed hooves aid climbing hills. Nubians live in large herds and males fight for mates.

Goat Life

Goats prefer eating leaves and grass, but can eat thistles, briars, and brambles with their tough mouths and teeth. Thus, they can survive where soft vegetation is scarce and other herbivores starve. Goats are sociable and live in groups of from a dozen to thousands of individuals.

Mountain goats are adept at leaping from rock to rock, balancing on their hooves. (PhotoDisc)

In the wild, males fight for mates and may die in such battles. Mating usually occurs in the spring among domesticated goats, while wild goats generally mate in the fall. The goat gestation period is about 5.5 months and yields one to four young, nursed for six months. Young goats are born with hair, eyes open, and can run and jump within twelve hours of birth. Goats can live for fourteen to twenty years.

A number of goat breeds are raised for meat, milk, leather. Goats are fine milk producers, and the milk is often used to make cheese. Their milk is sweet, nourishing, and easy to digest. It also has more fat and protein than cow milk and is helpful to persons with digestive troubles. More people throughout the world use goat milk than cow milk. Toggenburg and Saanen Swiss goat imports are popular milk goats in the United States. Toggenburgs are brown, with light side stripes, while Saanens are short-haired and white or cream-colored; both breeds produce around five quarts of milk per day. Goatskins also make high-grade leather; for instance, "Morocco leather" is chamois skin. Angora and cashmere goats are raised for their coats, which are used to make wool.

—Sanford S. Singer

See also: Domestication; Mountains; Ruminants; Sheep; Ungulates.

Bibliography

Hetherington, Lois, and John G. Matthews. *All About Goats*. 3d ed. Ipswich, England: Farming Press, 1992. Contains a lot of information on the natural history of goats.

Lavine, Sigmund A. *Wonders of Goats*. New York: Dodd, Mead, 1980. Covers goat lore, relationship to humans, and characteristics.

Porter, Val, and Jake Tebbit. *Goats of the World*. Ipswich, England: Farming Press, 1996. Describes many goat types and breeds in detail.

Staub, Frank J. *Mountain Goats*. Minneapolis: Lerner, 1994. Describes the natural history of mountain goats. Written for children; lots of photographs.

GOPHERS

Types of animal science: Anatomy, classification, reproduction
Fields of study: Anatomy, zoology

Gophers are small, burrowing rodents with fur-lined cheek pockets. When they eat bulbs and roots they become agricultural pests, damaging crops.

Principal Terms

GESTATION: time in which mammalian offspring develop in the uterus
HERBIVORE: an animal that only eats plants
INCISOR: a cutting tooth that acts like a scissors or chisel
RODENT: any gnawing animal

Gophers are small, herbivorous rodents. They are rat-sized, but somewhat more rotund than rats. They form the families Geomyidae and Thomomyini, which have eight genera, thirty-five species, and hundreds of subspecies. Gophers burrow in the ground and do not leave their burrows during daylight hours. Different species are found in deserts, shrublands, and grasslands across much of North and Central America, from central and southwestwern Canada through the western and southeastern United States, Mexico and south to the Panama-Colombia border.

Striped gophers (*Citellu tridecemlineatus*, prairie squirrels), which have thirteen body stripes, live from the western plains of the United States to Panama. Camas rats (*Thomomys bulbivorus*) are the largest gophers, reaching body lengths of over one foot. Plains pocket gophers (*Geomys bursarius*) are dark brown, and common in the Mississippi Valley.

French settlers in North America first noticed that gopher burrows honeycombed the soil and named them *gaufres* (French for "honeycomb"). These animals all have pouches (pockets) in their cheeks, and are more correctly called pocket gophers (pouched rats). Pocket gophers are divided into twenty-six Geomyini species and nine Thomomyini species. The main gopher types include eastern pocket gophers, yellow pocket gophers, and western pocket gophers.

Physical Characteristics of Gophers

Gophers are plump, ratlike rodents, up to 1.25 feet long and covered with soft, short, black and red-

Gopher Facts

Classification:
Kingdom: Animalia
Subkingdom: Bilateria
Phylum: Chordata
Subphylum: Vertebrata
Class: Mammalia
Order: Rodentia (rodents)
Family: Geomyidae (gophers)
Tribes: Geomyini (four genera, twenty-five species); Thomomyini (one genus, nine species)
Geographical location: North and Central America, from central and southwestern Canada, western and southeastern United States, Mexico, and south to the Panama-Colombia border
Habitat: Deserts, shrubby land, grasslands, agricultural areas, and tropical lowlands
Gestational period: About three weeks
Life span: Up to four years in the wild; up to six years in captivity
Special anatomy: Fur-lined pockets in cheeks, whiskers, strong incisor teeth, front claws for digging, tactile tails

brown to gray fur. All have whiskers, to help them navigate underground and at night, and two large, fur-lined pockets, one in each cheek. These pockets, used to carry food, lead to the name pocket gopher. The pockets are lined with fur, and a gopher can turn them inside out to clean them.

Gophers have wide, blunt heads, with underdeveloped ears and eyes. Their incisors are large and well designed for gnawing. Gophers have short limbs and feet with powerful claws, longer on the forefeet. They dig tunnels with the claws of the front feet. Their thick, almost hairless tails, about three inches long, are sensitive tactile organs, used to help them find their way around their tunnels when moving backward. Gophers are able to run backward about as quickly as forward. Their body lengths range from 4.5 to 17 inches, and they weigh between 0.75 and 2 pounds.

Gopher Life

Gophers eat leaves, grass, roots, nuts, tubers, buds, and farm vegetables. Their main foods, garnered in their tunnels, are roots and tubers. The other foods are gathered on nocturnal surface forays. They need water to survive, but when water is scarce in arid regions, they eat cacti to obtain it. The gophers store most of their food in the tunnels and chambers of their burrows. They carry the food to their burrows in their cheek pouches.

Gophers are solitary and males are territorial. They come together only to breed. Female gophers can have one or several litters each year. They birth two to ten young, depending on the species. Gestation is about three weeks. The young are totally dependent on their mothers at birth, weaned after a month, and have their own tunnels by age three to four months. Pocket gophers have life spans of one to seven years.

Gophers have fur-lined pockets in their cheeks that are used to carry food. (Joerg Boetel/Photo Agora)

Western Pocket Gophers

Western pocket gophers—like other gophers—have small, round bodies. They are tan to gray and live in deserts, meadows, and farmlands in Canada, the southwest United States, Mexico, and Central America. They like areas where the soil is easy to dig. These gophers eat plant leaves and underground roots and tubers. In dry areas, they eat cacti to obtain water. They travel on the surface at night, seeking and cutting through underground roots with sharp incisors. Western pocket gophers carry their food in their cheek pouches. They are territorial, and a given male's territory may be up to one thousand square feet.

Western pocket gophers can breed up to four times a year, once during each season of the year. Gestation is approximately three weeks, and young, born completely dependent on their mothers, develop their own burrows and territories within three months. These gophers live for four years in the wild and seven years in captivity.

Gophers are very destructive when they tunnel in farmland such as meadows, farms, and orchards. The tunnels are identified by the mounds of earth left along their courses. Very voracious, gophers eat all vegetation they find underground. They destroy food trees, tuber crops, and flower crops by gnawing roots, tubers, and bulbs.

—Sanford S. Singer

See also: Fauna: Central America; Fauna: North America; Home building; Mice and rats; Nesting; Rodents.

Bibliography

Alderton, David. *Rodents of the World*. New York: Facts on File, 1996. Lots of information and illustrations of many rodents, as well as useful bibliographical references.

Crouch, Glenn L., and Larry R. Frank. *Poisoning and Trapping Pocket Gophers to Protect Conifers in Northeastern Oregon*. Portland, Oreg.: Department of Agriculture, Forest Service, 1979. Describes pocket gophers and how to deal with them where they are pests.

Gould, Edwin, and George McKay, eds. *Encyclopedia of Mammals*. 2d ed. San Diego, Calif.: Academic Press, 1998. This book contains a nice chapter on rodents.

Lacey, Eileen A., James L. Patton, and Guy N. Cameron, eds. *Life Underground: The Biology of Subterranean Rodents*. Chicago: University of Chicago Press, 2000. This book describes the natural history, behavior, and burrowing styles of many rodents, including gophers.

GORILLAS

Type of animal science: Classification
Field of study: Anatomy

Gorillas, the largest, rarest, most powerful apes, are shy vegetarians living in bands in the wild. They have language and are an endangered species.

Principal Terms

BAND: A gorilla group of five to twenty apes, including males, females, and young

SILVERBACK: An older, male band leader, whose back hair has turned gray

Gorillas, the largest, strongest, rarest apes, look the most human of all other primates. They live in equatorial West African forests, from lowlands near the Cameroon Coast to highland altitudes of 10,000 feet in Rwanda, Zaire, and Uganda. All gorillas are believed to belong to one species, *Gorilla gorilla*.

Physical Characteristics of Gorillas

Males are 5.75 feet tall when on all fours, over six feet when standing erect, and thus are taller than the average man. They are much heavier, weighing three hundred to six hundred pounds. Unsurprisingly, their bone structure is heavier than that of humans. Females, a foot shorter than males, weigh half as much. Gorilla skin is dark and, except for the face and hands, covered with long, coarse, dark brown to black hair. This hair turns gray on the backs of older males, who are then called silverbacks.

Gorillas, especially males, look fierce, although they are timid unless mating, cornered, or threatened. Their fierce looks are due to hulking bodies and somber faces, with dark, hairless, strong jaws and long, powerful teeth. In addition, gorillas have brutish brow ridges which jut out above small eyes. Finally, the face of the male is dominated by a large, flat nose with coarse nostrils. Combined, these features produce what humans perceive as a somber and threatening expression.

Gorilla brain cases and brain sizes are smaller than in humans, promising lower intelligence. As to skeletal structure, the gorilla is similar to humans, but its bones are thicker. Its arms are much longer and its legs are much shorter. Gorilla spines lack the structures needed for a continual erect posture. Therefore, while gorillas can stand upright and walk erect, most often they walk on all fours, using the knuckles of their hands as supports.

Life and Sexual Cycles of Gorillas

Contrary to their legendary savagery, reinforced by their appearance, gorillas are shy, friendly creatures. Once they become used to unthreatening intruders, as Dian Fossey discovered, these individuals are accepted. Initially, males of gorilla groups charge intruders, growling and beating their chests. Intruders who run are often killed. Those who stand their ground and behave in an unthreatening manner are not harmed.

Gorillas live in families and extended families (bands) of five to twenty individuals. A band has a silverback leader, up to three subordinate males who help protect it, several mature females, and numerous young. The silverback drives most young gorilla males away at maturity. At these times all males are fierce and use their strength to attain and maintain supremacy. Males driven off form bachelor groups or join other bands. Silverbacks, challenged and defeated, live alone.

Gorilla bands each have a territory they allow

Gorilla Facts

Classification:
Kingdom: Animalia
Subkingdom: Bilateria
Phylum: Chordata
Subphylum: Vertebrata
Class: Mammalia
Order: Primates (apes, monkeys, and humans)
Family: Pongidae
Genus and species: Gorilla gorilla
Geographical location: Equatorial West Africa
Habitat: Lowland and highland forests
Gestational period: Nine months
Life span: Thirty to forty years in the wild, up to fifty years in captivity
Special anatomy: Long arms used to walk on all fours

others to enter. A band lacks permanent dwellings. Rather, its members build temporary shelters each night after a long day of travel to forage for the honey, eggs, plants, berries, bark, and leaves that make up the gorilla diet. In the wild or in captivity gorillas will eat meat, but do not seek it or kill other animals, except in self defense. When the terrain permits, females and young sleep on tree platforms of branches and leaves. Mature males nest at the bases of trees occupied by other band members.

Female gorillas, like women, menstruate monthly and mate successfully at any time. Pregnancy lasts 9.5 months and yields one or two young. Baby gorillas are suckled for a year. They are adults ten to eleven years later. Wild gorillas live for thirty to forty years. A few captive gorillas have attained more than fifty years.

The gorilla's spine lacks the support structures that would allow it to stand upright, so it usually uses its arms for support, resting on its knuckles. (Corbis)

Dian Fossey

Born: January 16, 1932; San Francisco, California
Died: December 26, 1985; Karisoke Research Center, Rwanda
Fields of study: Anthropology, zoology
Contribution: Fossey, an American zoologist, was a world authority on mountain gorillas.

Dian Fossey was born in San Francisco. After completing elementary and high school, she studied occupational therapy at San Jose State College, graduating in 1954. After this, she worked as a therapist in a children's hospital in Louisville, Kentucky.

In 1963, Fossey traveled to East Africa. There she met the famous anthropologist Louis Leakey and had her first exposure to mountain gorillas. Fossey returned to the United States and to her job, staying there until 1966. Then the persuasive Leakey convinced her to return to East Africa to study mountain gorillas in the wild, from the perspective of a trained therapist. Fossey began to work with these rare, elusive apes, which fascinated her and became her lifework. In 1967, she formalized this fascination by establishing the Karisoke Research Center in the solitude of Rwanda's Virunga Mountains. Initially she carried out her work alone, with no professional staff.

After expending a great deal of effort on the task, Fossey was able to accustom the gorillas to her presence in their midst and to study them quite extensively. The data that Fossey was able to accumulate greatly enlarged the scientific understanding of the mountain gorilla's lifestyle, habits, and methods of communication in the wild. To expand her scientific credentials, Fossey left Rwanda for England in 1970 to work on a doctorate in zoology at venerable Cambridge University. She received her degree in 1974, with a dissertation titled "The Behavior of the Mountain Gorilla."

Fossey soon returned to Rwanda and to her beloved apes. Several volunteer student aides accompanied her. This staff of educated assistants made possible much more extensive research on the mountain gorillas. During the course of her endeavors, one of Fossey's best and most-liked study subjects, the gorilla Digit, was killed by poachers. This motivated her to engage in a pitched battle against the poachers. In 1978, Fossey's efforts to this end began the international media coverage of the issue.

In 1980, she left Rwanda again, accepting a visiting associate professorship at Cornell University in Ithaca, New York. While teaching at Cornell, Fossey completed her well-known book *Gorillas in the Mist* (1983), which was made into a film of the same name (1988). Then she returned to her beloved Rwanda to resume the study of the mountain gorilla. Fossey also increased her campaign against the poachers. Sadly, in 1985 Dr. Fossey's dead body was found in the bush, presumably a murder victim of poachers whose activities she had sought so bravely to stop.

—Sanford S. Singer

Dian Fossey, in 1970, shows a photograph of one of the gorillas she studied. (AP/ Wide World Photos)

Gorilla Language and Intelligence

Gorilla language is composed of several sounds. Hooting signals alarm or indicates unusual events. Hooting by a silverback gains immediate attention from all band members. Other language sounds include sharp grunts that discipline young gorillas and low growls signifying pleasure. All gorillas beat their chests. In males, this is a symbol of power and intimidates other creatures.

The mental capacity of gorillas was long thought inferior to chimpanzees. Their intelligence is still being explored and testing is changing experimenters' opinions. Different techniques are needed to train gorillas, who are not as curious as chimpanzees. Trained correctly, gorillas have better memories and problem-solving skills than chimpanzees. They also discriminate between geometric shapes better. In addition, it is reported that they are most likely to perform tasks associated with intelligence out of interest, not for rewards. In exploration of communication via American Sign Language (ASL), some gorillas have mastered over one hundred words.

Gorillas' Endangered Status

Gorillas are close to extinction because of intrusion on their habitat of farmers, animal herders, and hunters. In the early 1970's, the estimated gorilla population was one thousand. About 25 percent were in Zaire's Kahuzi-Biega National Park (KBP) and 40 percent in its Mount Virunga area (MVA). The rest were scattered but relatively numerous in Rwanda's Parc des Volcans and Uganda's Gorilla Game Reserve.

In 1980, the number of gorillas in KBP was unchanged and those in MVA had dropped 40 percent. The implied decline, which may be continuing, is explained by increased farming and cattle herding in gorilla habitats. Also, killing and poaching of elephants and buffalo causes these animals to move into gorilla habitats, cutting down food sources for the great apes. The Rwandan Mountain Gorilla Project has slowed the decline in Rwanda. Similar efforts are in place elsewhere.

—Sanford S. Singer

See also: Apes to hominids; Baboons; Cannibalism; Chimpanzees; Communication; Communities; Evolution: Animal life; Evolution: Historical perspective; Fauna: Africa; Groups; Hominids; *Homo sapiens* and human diversification; Human evolution analysis; Infanticide; Intelligence; Language; Learning; Lemurs; Mammalian social systems; Monkeys; Orangutans; Primates.

Bibliography

Dixson, A. F. *The Natural History of the Gorilla*. New York: Columbia University Press, 1981. This excellent book gives information on gorilla anatomy, habitat, and reproductive processes. It also contains a good bibliography for interested readers.

Mowat, Farley. *Woman in the Mists: The Story of Dian Fossey and the Mountain Gorillas of Africa*. New York: Warner Books, 1987. This biography of Dian Fossey is very well illustrated and describes observations on the lives of gorillas.

Nichols, Farley. *The Great Apes: Between Two Worlds*. Washington, D.C.: National Geographic Society, 1993. This well-illustrated book contains photos, maps, and expert sections on gorillas and the other apes contributed by Jane Goodall and George B. Schaller.

Verrengia, Joe. *Cenzoo: The Story of a Baby Gorilla*. Boulder, Colo.: Roberts Rinehart, 1997. This charming juvenile book describes the life of a baby gorilla in captivity in a U.S. zoo. It holds a lot of interesting information.

GRASSHOPPERS

Types of animal science: Anatomy, classification, reproduction
Fields of study: Anatomy, entomology, invertebrate biology

Grasshoppers, leaping insects of the order Orthoptera, include long-horned and short-horned species. They eat grasses and shrubs, as well as agricultural crops, and some human societies eat them.

Principal Terms

MANDIBLES: insect jaws
MOLTING: shedding an insect shell, to enable continued growth
NYMPH: a grasshopper larva

Grasshoppers, leaping insects of the order Orthoptera, include all locusts. However, not all grasshoppers are locusts. The main difference between the locusts and other grasshoppers is the length of their horns (antennae). Locusts (Acrididae) have shorter antennae than other grasshoppers (Tettigoniidae). The amazing leaps of grasshoppers are due to long, slender hind legs with large thighs. These leaps are each many times the grasshopper's body length.

Most grasshoppers also have large, straight, delicate hindwings, which enable flight. When a grasshopper is at rest, these wings are folded up and protected by tough front wings that cover them entirely. Grasshoppers are found in most areas of the world except for northern Canada, Greenland, northern Asia, northwest Africa, West Australia, and the Arctic and Antarctic regions.

Long-horned grasshoppers are herbivores and found wherever vegetation grows. Their thread-like antennae are longer than their bodies. They are related to katydids. When endangered, they spit out brown liquid called "tobacco juice," and take huge, vigorous leaps to escape. The green color of these grasshoppers conceals them in grass, where they eat pieces of grass leaves and stems.

Long-horned grasshoppers do not usually eat crop plants. Short-horned grasshoppers, locusts, are called true grasshoppers because they live only in grasses and leaves. They are well known for traveling in huge swarms that lay bare whole farms or whole regions of countries. The huge populations of swarms and the destruction they have caused are mentioned in the Bible. *Schistocera perigrina*, a North African locust, may have been the species described in the biblical account of the plagues of Egypt.

Physical Characteristics of Grasshoppers

Grasshoppers are one to eight inches long when fully grown. Some species undergo seasonal color changes, being green at some times and red, olive, or brown at others. Like other insects, the grasshopper body is divided into three parts: head; thorax, or mid-section; and abdomen, or hind-section. A grasshopper's antennae, which have tactile functions, are found on its head.

Each grasshopper has a pair of compound eyes with many lenses, located on the front of its head. Grasshoppers also have three pairs of legs, extending from the thorax. The last pair is much larger and longer than the others and enables jumping. Grasshoppers eat leaves, roots, and stems of grasses, herbs, and shrubs, chewing with strong mandibles (jaws), moving these jaws from side to side to break apart their food.

Most grasshoppers have two pairs of wings along the back of the abdomen. Two hard forewings serve as protection and two membranous hindwings are used to fly. When a grasshopper is not flying, its hindwings fold up and are covered and protected by its forewings. All long-horned

Grasshoppers' long, slender hind legs and strong thighs allow them to make amazing leaps many times the length of their bodies. (Digital Stock)

grasshoppers "sing" by rubbing the bases of their forewings together. Some male locusts make calls to females by rubbing their hind legs against their wings, and others do so by rubbing their hind legs or forewings against other parts of their bodies.

The hearing organs of long-horned grasshoppers are small growths just beneath the knee joints of their front legs. In short-horned grasshoppers, these ears are clear, circular areas on the abdomen at points just behind the junction of hind legs and body. In females, growths shaped like sickles are located at the rear of the abdomen. These ovipositors drill holes in grass, twigs, or the ground, where eggs are deposited. Ovipositors of short-horned grasshoppers are specially designed to deposit pouches of eggs in the ground.

Grasshopper Life Cycles

In the spring, grasshoppers hatch from eggs as pale, wingless nymphs (larvae). Then, within ninety days, they develop into full-grown locusts, molting four to five times in the process. Mature insects mate, and about a week later, females lay the eggs for the next generation. They die a few weeks after this.

Locusts, such as the Rocky Mountain locust (*Melanoplus spretus*), lay their eggs in holes in the ground in the fall. The eggs hatch in the spring, and young reach maturity in July or August. Those of long-horned grasshoppers, such as meadow grasshoppers (*Orchelimum vulgare*) are laid in low bushes or crevices in tree bark.

Not only do locusts eat human crops, but in turn, humans have eaten them for centuries. For example, Talmudic law exempts locusts and other grasshoppers

Grasshopper Facts

Classification:
Kingdom: Animalia
Subkingdom: Bilateria
Phylum: Arthropoda
Class: Insecta
Order: Orthoptera (grasshoppers and related species)
Families: Include Acrididae (short-horned grasshoppers, locusts), Tettigoniidae (long-horned grasshoppers)
Geographical location: Most world locations, except Arctic and Antarctic regions, northern Canada, Greenland, northern Asia, northwest Africa, and West Australia
Habitat: Grass and shrubbery
Gestational period: Natural gestation time is uncertain, as eggs laid in the fall do not hatch until spring
Life span: A year in temperate to cool climates, indeterminate in the tropics
Special anatomy: Six legs, three on each side of the thorax; two pairs of wings, two hard forewings that serve as protection for two membranous hindwings used in flying; ears on or near legs; compound eyes

from the taboo on eating flying or creeping creatures "going on all fours." Also, Shakespeare's play *Othello* mentions food "luscious as locusts." Candied locusts are eaten throughout China and the Philippines. In North Africa, locusts dried and ground into powder are mixed into flour used to bake bread.

—Sanford S. Singer

See also: Antennae; Ears; Insects; Molting and shedding; Wings.

Bibliography

Bailey, Jill, and Carolyn Scrace. *Grasshoppers*. New York: Bookwright Press, 1990. A brief book describing the life cycle of a grasshopper.

Brown, Valerie K. *Grasshoppers*. New York: Cambridge University Press, 1983. A brief text, holding useful information on grasshopper natural history.

Chapman, R. F. *The Insects: Structure and Function*. 4th ed. New York: Cambridge University Press, 1998. A clear, comprehensive, systematic text on entomology.

Chapman, R. F., and A. Joern, eds. *Biology of Grasshoppers*. New York: John Wiley & Sons. 1990. A solid work describing a wide variety of grasshoppers clearly and completely.

GRASSLANDS AND PRAIRIES

Type of animal science: Ecology
Fields of study: Conservation biology, ecology, environmental science, wildlife ecology

Grasslands, including the prairies of North America, are a biome characterized by the presence of low plants, mostly grasses, and are distinguished from woodlands, deserts, and tundra. They support a great variety of plants and animals. At present, the remaining grasslands provide grazing for livestock and wildlife.

Principal Terms

BIOMASS: the dry weight of the living material in an organism, population, or community

BIOME: region with a particular climate and a characteristic community of living organisms

CARNIVORE: animal dependent on animal material in its diet

CARRYING CAPACITY: the maximum number of animals that a given area can support indefinitely

DENUDE: strip the covering from, as vegetation from overgrazed grassland

FERAL: once domesticated but no longer under the control of humans

HERBIVORE: animal dependent on plant material in its diet

RUMINANT: animal with four stomach compartments that regurgitates its food and rechews it

Grasslands occupied vast areas of the world over ten thousand years ago, before the development of agriculture, industrialization, and the subsequent explosive growth of the human population. They are characterized by the presence of low plants, mostly grasses, and are distinguished from woodlands, deserts, and tundra. They experience sparse to moderate rainfall and are found in both temperate and tropical climatic zones. The main grasslands of the planet include the prairies of North America, the pampas of South America, the steppes of Eurasia, and the savannas of Africa. Grasslands are intermediate between deserts and woodlands in terms of precipitation and biomass. The warmer tropical savannas average 60 to 150 centimeters (25 to 60 inches) of rain. The temperate grasslands range between twenty-five to seventy-five centimeters (ten to thirty inches) of precipitation, some of which may be in the form of snow. The biomass of grasslands, predominantly grasses, is quantitatively intermediate between that of deserts and woodlands, which produce 10 to 15 percent and 200 to 300 percent, respectively, of the amount of plant material. It should be recognized that the grassland biomes can be subdivided in terms of climate, plant species, and animal species. It should also be noted that grasslands do not always shift abruptly to deserts or woodlands, leading to gradations between them. In addition, grasslands do have scattered trees, often along streams or lakes, and low-lying brush.

Grasses have extensive root systems and the ability to become dormant. These permit them to survive low rainfall, including periodic droughts, or the winter cold typical of temperate regions. Furthermore, grasslands have always been subjected to periodic fires, but the deep root systems of grassland plants also permit them to regrow after fire. Grasses coevolved over millions of years with the grazing animals that depend on them for food. Ten thousand years ago, wild ancestors of cattle and horses, as well as antelope and deer, were on the Eurasian steppes; bison and pronghorn prospered on the North American prairies; wildebeest, gazelle, zebra, and buffalo dominated African savannas; and the kangaroo was the predominant grazer in Australia. Grazing is a symbiotic relationship, whereby animals gain their nourishment from plants, which in turn benefit from the activity. It removes vegetative matter, which is necessary in order for grasses to grow, facilitates seed dispersal, and disrupts mature plants, permitting young plants to take hold. Urine and feces from grazing animals recycle nutrients to the plants. The grassland ecosystem also includes other animals, including worms, insects, birds, reptiles, rodents, and predators. The grasses, grazing animals, and grassland carnivores, such as wolves or large cat species, constitute a food chain.

Humans have been an increasing presence in grassland areas, where over 90 percent of contemporary crop production now occurs and much urbanization and industrialization has taken place. Remaining grassland areas are not used for crops, habitation, or industry because of inadequate water supplies or unsuitable terrain but instead are used for grazing domesticated or wild herbivores. In addition, many woodland areas around the world have been cleared and converted to grasslands for crops, livestock, living, or working.

The Prairies of North America

Originally stretching east from the Rocky Mountains to Indiana and Ohio, and from Alberta, Canada to Texas, the prairies were the major grassland of North America. The short-grass prairie extended about two hundred miles (three hundred kilometers) east of the mountains, and the long-grass prairie bordered the deciduous forest along the eastern edge, while the mixed-grass prairie was between the two. Going from west to east, the amount of precipitation increases, causing changes in plant populations. The short-grass prairie receives only about twenty-five centimeters (ten inches) of precipitation each year, mostly as summer rain, and, as its name suggests, has short grass, less than sixty centimeters (two feet) tall. Today, it is used primarily for grazing because the soil is shallow and unsuited for farming without irrigation. The mixed-grass prairie receives moderate precipitation, ranging from 35 to 60 centimeters (14 to 24 inches) and has medium-height grasses, ranging from 60 to 120 centimeters (2 to 4 feet) tall. Much of it is now used for growing wheat. The tall-grass prairie receives more than 60 centimeters (24 inches) of precipitation, mostly in the summer, and had grasses that grow to over 150 centimeters (5 feet) tall. It has rich soil and has been mostly converted to very productive cropland, primarily for corn and soybeans. The prairies experience very cold winters (down to –45 degrees Celsius, –50 degrees Fahrenheit) and very hot summers (up to 45 degrees Celsius, 110 degrees Fahrenheit). They are often windy and experience severe storms, blizzards in winter, thunderstorms and tornadoes in summer.

Like other biomes, the prairies have a characteristic assortment of animals, herbivores that eat the plants and carnivores that prey on the herbivores. Before 1500 C.E., two ruminants, the bison (commonly but inaccurately called buffalo) and the pronghorn (not a true antelope), were the major grazers on the prairies. The prairie dog, a herbivorous rodent that burrows, lived in large communities on the prairies. The major predators were the wolf and coyote for bison and pronghorn, and the black-footed ferret and fox for prairie dogs. A variety of birds, herbivorous and carnivorous, reptiles, and insects also made their home on the prairie.

Overgrazing Grasslands

While grazing is of mutual benefit to plant and animal, overgrazing is ultimately detrimental to both the plant and animal populations, as well as the environment. Continued heavy grazing leads to deleterious consequences. Removal of leaf tips, even repeated, will not affect regeneration of grasses provided that the basal zone of the plant remains intact. While the upper half of the grass shoot can generally be eaten without deleterious consequences, ingesting the lower half, which sustains the roots and fuels regrowth, will eventually kill the plants. Overgrazing leads to denuding the land, to invasion by less nutritious plant species, to erosion due to decreased absorption of rainwater, and to starvation of the animal species. Because the loss of plant cover changes the reflectance of the land, climate changes can follow and make it virtually impossible for plants to return, with desertification an ultimate consequence. It is not just the number of animals, but the timing of the grazing that can be detrimental. Grasses require time to regenerate, and continuous grazing will inevitably kill them. Consumption too early in the spring can stunt their development.

Semiarid regions are particularly prone to overgrazing because of low and often unpredictable rainfall; regrettably, these are the areas of the world where much grazing has been relegated, because the moister grassland areas have been converted to cropland. Overgrazing has contributed to environmental devastation worldwide. Excessive grazing by cattle, sheep, goats, and camels is partly responsible for the desert of the Middle East, ironically the site of domestication for many animals and plants. Uncontrolled livestock grazing in the late 1800's and early 1900's negatively affected many areas of the American West, where sagebrush and juniper trees have invaded the grasslands. Livestock overgrazing has similarly devastated areas of Africa and Asia. In the early twenty-first century, feral horses in the American West and the Australian outback are damaging those environments. Overgrazing by wildlife can also be deleterious. The 1924 Kaibab Plateau deer disaster in the Grand Canyon National Park and Game Preserve is one such example, where removal of natural predators led to overpopulation, overgrazing, starvation, and large die-offs.

Riparian zones, the strips of land on either side of a river or stream, are particularly susceptible to overgrazing. Because animals naturally congregate in these areas with water, lush vegetation, and shade, they can seriously damage them by preventing grasses from regrowing and young trees from taking root, as well as trampling and compacting the soil and fouling the water course. The ecosystem can be devastated, threatening survival of plant and animal species and leading to serious erosion. While herding and fencing can be used to control animals in these areas, a less expensive method is to disperse the location of water supplies and salt blocks to encourage movement away from rivers or streams. If deprived of salt, grassland animals crave it and will seek it out.

Grassland Management

Grassland areas need not deteriorate if properly managed, whether for livestock, wildlife, or both. Managing grasslands involves controlling the number of animals and enhancing their habitat. Carrying capacity, which is the number of healthy animals that can be grazed indefinitely on a given unit of land, must not be exceeded. Because of year-to-year changes in weather conditions and hence food availability, determining carrying capacity is not simple; worst-case estimates are preferred in order to minimize the chances of exceeding it. The goal should be a healthy grassland achieved by optimizing, not maximizing, the number of animals. For private land, optimizing livestock numbers is in the long-term interest of the landowner, although not always seen as such. For land that is publicly held, managed in common, or with unclear or disputed ownership, restricting animals to the optimum level is particularly difficult to achieve. Personal short-term benefit often leads to long-term disaster, described as the "tragedy of the commons" by biologist Garrett Hardin.

Appropriate management of grasslands involves controlling animal numbers and enhancing grassland plants. Restricting cattle and sheep is physically easy through herding and fencing, although it can be politically difficult and expensive. Much more problematic is controlling charismatic feral animals, such as horses, or wildlife, when natural predators have been eliminated and hunting is severely restricted. As for habitat improvement, the use of chemical, fire, mechanical, and biological approaches can increase carrying capacity for either domesticated or wild herbivores. Removing woody vegetation by burning or mechanical means will increase grass cover, fertilizing can stimulate grass growth, and reseeding with desirable species can enhance the habitat. Plants native to a particular region can be best for preserving that environment. Effective grassland management requires matching animals with the grasses on which they graze.

—James L. Robinson

See also: Chaparral; Ecosystems; Forests, coniferous; Forests, deciduous; Habitats and biomes; Lakes and rivers; Rain forests; Savannas; Tundra.

Bibliography

Collinson, Alan. *Grasslands*. New York: Dillon Press, 1992. Written for adolescents, this accurate and well-photographed book examines the grasslands around the world and the life they support, as well as the effects of human activities on them.

Demarais, Stephen, and Paul R. Krausman, eds. *Ecology and Management of Large Mammals in North America.* Upper Saddle River, N.J.: Prentice Hall, 2000. This text for students, professionals and interested laypersons includes chapters on carrying capacity, a crucial feature in managing grasslands, and on various prairie species, including bison, pronghorn antelope, and wolves.

Humphreys, L. R. *The Evolving Science of Grassland Improvement*. New York: Cambridge University Press, 1997. A review of grasslands, their ecology, and management by a leading scholar in the field. It includes a discussion of various grazing systems.

Owen, Oliver S., Daniel D. Chiras, and John P. Reganold. *Natural Resource Conservation*. 7th ed. Upper Saddle River, N.J.: Prentice Hall, 1998. This undergraduate text includes an excellent description of grasslands and a chapter on "Range Management," which applies primarily to grasslands.

Pearson, C. J., and R. L. Ison. *Agronomy of Grassland Systems*. 2d ed. New York: Cambridge University Press, 1997. An undergraduate text that emphasizes a holistic approach to the biology and management of grassland systems.

Steele, Philip. *Grasslands*. Minneapolis: Carolrhoda Books, 1997. This well-illustrated book for adolescents describes the various grassland ecosystems (prairies, pampas, and savannas), including their animals.

GRIZZLY BEARS

Type of animal science: Classification
Fields of study: Anatomy, conservation biology, physiology, reproduction science, wildlife ecology

Grizzly bears are a kind of brown bear found in inland North America. Brown bears den during the winter and their reproduction includes a period of delayed implantation. While many brown bears are found in the world, their numbers have been decreasing.

Principal Terms

DELAYED IMPLANTATION: an extended period after fertilization, when an embryo stops developing, and before it attaches to the uterine wall and resumes development

DENNING: the period of winter sleep during which a bear does not eat, drink, urinate, or defecate

DIGITIGRADE: walking on the toes, with the heel raised

PLANTIGRADE: walking on the sole, with the heel touching the ground

Grizzly bear is a loose term used to describe a subspecies of brown bears found in inland North America. Around the world, brown bears vary in color from black to very light brown. Grizzly bears have a brown coat with silver-tipped hairs, which gives them a "grizzled" coloration. Grizzlies vary in weight, usually between four hundred and eight hundred pounds, but are generally smaller than brown bears found on the North American coast. In contrast, the largest brown bears in the world, found in coastal Alaska, sometimes weigh over one thousand pounds. Weight differences in brown bears are probably due to the availability of dietary protein. All brown bears have small, round ears, and large, round, dish-shaped faces with a large brow. They also have a characteristic hump over their shoulder that contains fat, and powerful digging muscles. Five long, nonretractable claws aid their digging. Unlike dogs and cats, which walk on their toes (digitigrade), bears walk flat on their whole foot (plantigrade), as humans do.

Denning

One of the most amazing characteristics of brown bears is their ability to den through the winter. During this time, bears appear to be asleep. Their heart and breathing rates slow dramatically, and their energy use is cut in half. They do not eat, drink, urinate, or defecate for three to five months. Although this is often referred to as hibernation, bears are not true hibernators. Their body temperature only falls a few degrees (to about 90 degrees Fahrenheit) and they are easily awakened.

In order to survive such a long period without eating and drinking, bears break down their fat stores. During the winter, they lose 15 to 30 percent of their body weight. In order to build up the fat needed to make it through the winter, brown bears must eat around ninety pounds of food per day during the fall. Most of this food consists of plant material, such as berries, grasses, nuts, and roots, which they unearth with their powerful digging muscles. They also eat some animal material, as available. This may include fish, deer, or elk, and small mammals, such as squirrels or insects. The denning period allows bears to survive winter, a time of food shortage, by using their own stores. Captive bears that are fed through the winter do not den.

Some scientists believe that understanding bears' denning abilities could aid human medicine. A person who must remain bedridden for several months will suffer from bone and muscle loss. Yet brown bears do this every winter with no ill consequences. They are able to go several months without urinating by reabsorbing water from the bladder and converting wastes back into proteins. Understanding this process could help humans who suffer from kidney diseases.

Reproduction and Distribution

Brown bears mate in the late spring or early summer each year. In the days after mating, the fertilized egg divides and grows into a small cluster of cells. Then, the embryo stops growing and remains free-floating in the uterus until the beginning of winter. This unusual process is known as delayed implantation. In winter, the embryo attaches to the uterine wall, and after a sixty-day pregnancy, cubs are born. Brown bears usually have two to three cubs at a time. The cubs weigh less than two pounds at birth, which is especially small compared to the size of the mother. Yet this is not surprising, because the mother is pregnant during the denning period, when she does not eat.

Brown bears are the most widely distributed species of bear. They are found in North America, Europe, and Asia, from the Canadian tundra to the Iraqi desert. However, brown bears have disappeared from about 50 percent of their original

Grizzly bears are characterized by a fatty lump over their shoulders, which helps sustain them through their winter hibernation. (PhotoDisc)

Grizzly Bear Facts

Classification:
Kingdom: Animalia
Subkingdom: Bilateria
Phylum: Chordata
Subphylum: Vertebrata
Class: Mammalia
Order: Carnivora
Family: Ursidae (bears)
Genus and species: Ursus arctos horribilus
Geographical location: Inland northwestern United States and Canada
Habitat: Varied, including temperate and Arctic grasslands, temperate forests, mountainous regions
Gestational period: A five-to-seven-month embryonic diapause, followed by a sixty-day gestation
Life span: Average twenty to twenty-five years in wild; record, thirty-five years
Special anatomy: A hump above the shoulders

range due to human activities. Accurate census data are difficult to obtain because of their wide distribution, large individual territories and solitary nature. In Japan, the Middle East, and Western Europe, several isolated populations have less than one hundred individuals. Larger populations, of a few thousand, are found in Turkey and Eastern Europe. Most of the world's brown bears live in Russia (around thirty thousand), Alaska (around thirty-five thousand) and Canada (around fifteen thousand). In the lower forty-eight states, brown bears have disappeared from 99 percent of their original habitat and are estimated to number around one thousand. Most of the United States and Canadian brown bears are grizzly bears.

—*Laura A. Clamon*

See also: Bears; Estivation; Fauna: Asia; Fauna: Europe; Fauna: North America; Hibernation; Polar bears; Pregnancy and prenatal development.

Bibliography

Brown, Gary. *The Great Bear Almanac.* New York: Lyons & Burford, 1993. This is a comprehensive, general source of bear facts, with an emphasis on brown bears.

Parker, Janice. *Grizzly Bears.* Austin, Tex.: Raintree/Steck-Vaughn, 2000. For younger readers, this book is well illustrated and packed with information.

Servheen, Chris. *The Status and Conservation of the Bears of the World.* Knoxville, Tenn.: International Association for Bear Research and Management, 1989. Also available at http://www.bearbiology.com/bearstat.html. This scholarly work is readable for nonscientists and includes the most comprehensive study of bear population status available.

Stirling, Ian. *Bears.* San Francisco: Sierra Club, 1995. Although written for a general audience, this book includes chapters by the world's leading bear researchers. It has fantastic photographs.

GROOMING

Type of animal science: Behavior
Fields of study: Ethology, zoology

Grooming comprises the various actions by which animals clean and maintain their body surface and is vital to the health of individuals. In many species, grooming also plays an important social role in the interactions of animals and in the maintenance of the social organization of the species.

Principal Terms

ALLOGROOMING: mutual grooming or grooming between two individuals

ALTRUISM: a behavior that increases the fitness of the recipient individual while decreasing the fitness of the performing individual

CONSORTSHIP: a pairing of a male and a female

PHEROMONE: a chemical produced by one individual that influences the behavior of another individual of the group

SYMBIOSIS: a relationship between two species of organisms which is not necessarily advantageous or disadvantageous to either organism

THERMOREGULATION: the process by which animals maintain body temperatures within a certain range

Grooming serves a number of purposes. It is health related in that the activity cares for the skin, feathers, or fur. Grooming also serves a social function. Between members of a community, grooming reduces stress, communicates and signals social status, spreads pheromones, achieves thermoregulation or pain relief, increases or decreases arousal, self-stimulates, and prevents sexually transmitted diseases.

As grooming is similar through the various levels of animal taxa, it has been conjectured that grooming behavior is evolutionarily ancient. Most animals (mammals, birds, and insects) groom by moving their limbs over their own bodies or mouthing or licking their bodies. In some birds, sandbathing is quite common. With fish, a species with no limbs, it is not uncommon to see them rubbing or simply moving against rocks, branches, or sand, generally accomplishing what a sandbath does for a bird.

A bird preening its feathers, a cat licking its paws, or a bear brushing his back against a tree truck are all self-grooming. This is where the animal, alone, takes care of the grooming behavior without help from another animal. However, mutual grooming is quite common. In mammalian species, this is a form a display behavior, which helps to cement the social bonds between members of the group. Yet another kind of cleaning behavior, called cleaning symbiosis, occurs between certain species of fish and shrimp. Here, one species will eat the parasites off another species. The cleaner gets food, the recipient remains debris- and parasite-free.

Grooming as Cleaning Behavior

Observations and experiments performed on laboratory rats showed that pregnant rats spent an increasing proportion of their time grooming their ventral surface, which includes the nipple lines and anogenital areas. When the rats were fitted with collars so they could not reach these regions, their mammary development was inhibited. The conclusion to be drawn from this is that self-stimulating grooming is a necessary part of the preparation process for nursing.

A similar conclusion was reached when looking at the function of grooming in male rats.

Prepubertal males engaged in self-grooming of the anogenital area more than their female counterparts. When collars restricted the rats from grooming, their sexual development was significantly hampered. Again, the conclusion is that male rats needed to perform self-stimulating autogrooming in order to prepare for reproduction.

Another very important fact about self-grooming behavior is the appearance of a grooming pattern. For example, depending on the species, self-grooming may start around the head area and progresses downward until the entire body has been groomed. In experiments, it has been noted that a particular animal will not vary its grooming pattern. Studies on hamsters show that specific grooming tendencies evolve as the animal matures. Certain types of grooming behavior are always the result of some external stimuli, or they are the consequence of an aspect of the animal's natural behavior. A male rat invariably performs genital grooming after copulation. A dog, when aroused by some external stimuli (fear or excitement) will often groom its genitalia for de-arousal. Many animals, when afraid or nervous, will gnaw on their paws. Pigs often chew on other pigs' tails, it is thought, when they are bored or have an excess of nervous energy. The expression "licking one's wounds" originates from the observation that most animals will tend to an injury by licking the area, spreading an antiseptic found in saliva to reduce pain and decrease chances of infection. Insectivores, like solendons and shrews, spend a good deal of time grooming, using only their hind legs, to ensure thermoregulation.

Regardless of the lack of similarity between grooming species, grooming behavior is remarkably similar; reiterating that there has not been a great deal of diversification in the evolution of grooming behavior. Sea lions, seals, and walruses, regardless of whether the pelage is sparse or

Social grooming, where one animal grooms another, promotes social bonds and maintains hierarchies. (PhotoDisc)

Preening—More than Good Looks

Preening, for a human, conjures up a vision of someone preparing and dressing themselves carefully in preparation for a social event. People observing birds preen themselves tend to carry over that interpretation for explaining the observed activity. One function of a bird's preening is to remove various ectoparasites, such as fleas, lice, and ticks, as well as other debris on the body. Nonetheless, not all preening rituals are for cleaning alone. When birds preen themselves, they are actually discharging the secretions of the preen gland, also known as the urophgial or oil gland, onto the surface of the skin through several nipplelike pores. The preen gland is the only prominent skin gland in birds and is most developed in aquatic birds, including ducks and geese. Most birds have the gland, but it is almost completely lacking in the emu, ostrich, and several species of parrots, among others. The most common preening action for a bird is to rub its bill and head over the preen gland pore and then rub the collected oil over the feathers of the wings and body and the skin of the legs and feet. In this way, the down or underplumage, which is buoyant and serves as insulation for the bird, remains water-resistant, thereby allowing for thermoregulation. Some birds may preen themselves more than a dozen times a day. It is thought that the oil from the preen gland contains a chemical that is the antecedent to vitamin D. In the presence of sunlight, this substance is converted to vitamin D and the skin absorbs it. This promotes the health of the bird by preserving the integrity of its feathers, bill, and scales.

dense, spend much time grooming. This is typical of marine mammals. They accomplish this by a doglike scratching motion with their hind flippers and using their fore flipper to rub their head and neck while balancing on the other front flipper. They also nibble on their fur, much like dogs. It is common for these animals to rub against rocks or each other, similar to the activity of bears when they rub up against trees. Fish will also use this rubbing behavior against rocks or coral to scour their bodies free from debris.

Grooming as Social Behavior

Mutual grooming, also called allogrooming or allopreening, is when an animal cares for the body surface of another. Mutual grooming in animals is more a form of communication, a social act, than it is a cleansing one. Researchers have come to this conclusion because the time allocated to mutual grooming exceeds what is necessary for simple cleansing and sanitation. This behavior, especially in primates and social birds, promotes social bonding and establishes and maintains the hierarchy among members of the community.

Mutual grooming also brings attention to whether there is such a thing as animal altruism, which is the selfless delivery of service from one animal to another. Whenever one observes a parent grooming its young, the conclusion is that the motive behind the act is to tend to the health of the infant or young, which promotes the proliferation of the species. Nearly all mammals and many species of birds display this behavior. With marsupial births, the young of the Virginia opossum are in a semiembryonic state. The mother licks the embryon at birth so that the membrane encasing it will break. She then licks a trail from the birth canal to the pouch so that the neonate, using its developed olfactory senses, can find its way to the pouch without further aid by the mother. In altricial and semialtricial placental mammals, the young are usually born naked or with little fur. Most of these animal types, such as cats, dogs, mice, shrew, rats, and hamsters, lick the newborn to remove the birth membrane, break the umbilical cord, and eat the placenta after birth. They lick the newborn to clean the perineal region and to remove urine and feces, also aiding in thermoregulation.

When one observes animal and birds of a similar peer group performing grooming rituals, what appears to be unilateral or altruistic grooming behavior may in fact be mutual aid, in that over the long term, repayment may be expected. It is the expectation of reciprocation that establishes and maintains social status. For example, studies on nonhuman primates (mostly Old World monkeys such as rhesus macaques, stump-tailed macaques, and baboons) have shown that a competition exists among members of a group for animals considered to be excellent groomers. Alliances are formed between groomer and groomed; the groomed may provide excellent protection or be a skilled food gatherer. There is evidence of ranking among members of primate society and mutual grooming appears to assist in maintaining the social order and structure.

In primates that form consortships (Cercepethecines), grooming is important in the copulatory sequence, which is not the case in primates that do not form consortships. In general, sexual grooming serves as a male strategy to increase receptivity of an estrus female. In olive baboons and hamadryas, there is a high correlation between the age of the male and the grooming of the females, implying that this may be an alternative method of securing female cooperation in mating when direct agonistic approaches fail or are no longer an option.

Equids are well-known for mutual grooming. Horses, zebras, and similar animals groom each other by pairing off and standing nose-to-nose or head-to-tail, scratching and nibbling each other's neck, back, and tail with their teeth. The nose-to-nose greeting and rubbing of noses is also typical of tapirs and rhinoceros.

Interspecies Grooming

Humans have successfully exploited the horses' need for bonding by touch and grooming. Trainers often start training horses in spring when the animals are shedding. By brushing and grooming, the trainer develops a friendship and level of trust with the animal. This is an example of interspecies bonding through grooming. This behavior is widely noticed between any combination of humans, dogs, and cats. Sometimes licking another animal or human is simply a show of affection.

Symbiotic grooming, briefly mentioned above, occurs among some species where they pair off to assist one another in the grooming process. Pilot fishes and remoras are commensal fishes (a type of symbiosis) that attach themselves to sharks and other fish. Apart from eating the remnants of the host's meal, they also feed on the external parasites that plague the host fish. Fleas and lice are symbiotic groomers as they eat harmlessly on dead skin or feathers of mammals and birds.

Grooming behavior is an integral part of existence for all higher-order animals. It is such an important part of human behavior that if proper grooming is not taught at an early age and not performed, there are negative repercussions from peers and society in general. Grooming is such an important part of human community, that in the United States alone, the annual sales of men's and women's toiletries is a multibillion dollar industry. Scientists have only recently begun to study grooming behavior closely.

—Donald J. Nash

See also: Altruism; Communication; Copulation; Courtship; Displays; Emotions; Ethology; Habituation and sensitization; Hormones and behavior; Imprinting; Insect societies; Learning; Mammalian social systems; Offspring care; Pair-bonding; Rhythms and behavior; Symbiosis; Territoriality and aggression.

Bibliography

Colbern, Deborah L., and Willem H. Gispen, eds. *Neural Mechanisms and Biological Significance of Grooming Behavior.* New York: New York Academy of Sciences, 1988. This collection of papers attempts to investigate grooming behavior in mammals through experimentation. It is very technical.

Hinde, Robert A. ed. *Primate Social Relationships: An Integrated Approach.* Boston: Blackwell Scientific Publications, 1983. This collection of papers discusses primate behavior, revealing new complexities of social behavior and structure. It is also very technical.

Peterson, Richard, and George A. Bartholomew. *The Natural History and Behavior of the California Sea Lion*. Stillwater, Okla.: American Society of Mammalogists, 1967. An entertaining study of sea lions with pictures. It is non-technical.

Poole, Trevor B. *Social Behavior in Mammals*. New York: Chapman and Hall, 1985. This book is written for advanced undergraduates and postgraduates in mammalogy, animal behavior, productive biology, ecology, sociobiology, and experimental psychology.

Rubenstein, Daniel I., and Richard W. Rangham, eds. *Ecological Aspects of Social Evolution: Birds and Mammals*. Princeton, N.J.: Princeton University Press, 1986. This collection of papers recognizes that animal societies can be better understood when different types of social relationships are clearly distinguished from one another.

GROUPS

Types of animal science: Behavior, ecology, classification
Fields of study: Ecology, ethology, zoology

Animals form groups, or aggregations, for a variety of purposes. These groups can be temporary in nature, occurring only during specific seasons, or can last throughout the lifetime of the members and lead to highly organized societies.

Principal Terms

COMPETITION: interactions among individuals that attempt to utilize the same limited resource

CONSPECIFICS: members of the same species

DILUTION EFFECTS: the reduction in per capita probability of death from a predator due to the presence of other group members

ENCOUNTER EFFECTS: the reduction in the probability of death from a predator due to a single group of *N* members being more difficult to locate than an equal number of solitary individuals

INTERFERENCE: the act of impeding others from using some limited resource

PREDATION: the act of killing and consuming another organism

RESOURCE DEFENSE: the control of a resource indirectly or directly

SOCIALITY: the tendency to form and maintain stable groups

Some animals spend most of their time alone because the presence of other conspecifics would interfere with the use of a particular resource or a suite of resources. These animals only come together with another solitary individual to pair for reproduction. Others form groups ranging from pairs of animals to large herds. Finally, some animals are brought together by phenomena over which they have no control (winds, currents, or tides) and are simply clumped in space. Groups formed for whatever reason can be either temporary or permanent, and are generally theorized as being beneficial for a variety of reasons. Some associations are simply the result of congregating around a common food resource. Other associations arise for specific functions, such as finding mates, caring for young, providing for a learning environment for developing young, providing protection from the elements, thermoregulation or huddling, locomotory efficiency (swimming or flight), locating and subduing food items, resource defense against other groups or competing species, division of labor, population regulation, predator vigilance, and reduced predation risk via dilution, confusion, encounter, or group defense effects. While potential benefits can be many, aggregation can also result in distinct disadvantages for individuals making up the group. Such disadvantages include increased competition for resources (such as mates, food, or shelter), increased risk of disease and spread of parasites, and interference in reproductive behaviors.

Reproduction and Rearing Young

Living in groups can increase an individual's chance of finding a mate, but it often results in increased aggression between males who must compete for females. Because all females may come into estrus within a short time period, grouping together at specific mating territories ensures that all females will be mated. In such species, courtship rituals are often common. These rituals serve a dual purpose: They provide the male with information about the sexual receptivity of the female and they allow the female to assess the quality of the male prior to pairing with him. In some cases, the rituals also serve to bind the pair together for a breeding season, or in some species for longer periods of time. In other cases, males simply congregate at display grounds, attract, and court females. Females leave the display grounds after mating to nest elsewhere, while the males remain to court other females. This latter grouping strategy is known as lekking.

In some species, which both aggregate during mating seasons and form pair bonds, both male and female can care for the young. In some cases, a mating pair may have helpers at the nest—other members of the species (usually offspring) which aid the parents in raising the young. Helpers at the nest greatly contribute to the breeding success of the parental birds and gain experience themselves in rearing young. They can then use their experience to be successful parents in the next breeding season. Other examples, such as lions and elephants, include kin groups (generally sisters or a mother and her daughters) that help to raise whatever offspring are present in the group.

Colonial nesting and thus synchronous egg-laying produces offspring in large numbers, who are vulnerable to predators for only a short period of time. In this way, each parent lessens the chance that their offspring will be the ones taken by any predator (dilution effect). However, colonial nesting also presents the possibility that offspring may grow more slowly or run the risk of parasitism from fleas and mites. Some evidence exists that offspring of bank swallows gained weight more slowly if they were from large colonies versus small colonies, suggesting that large colonies were depleting their resources more rapidly. Nests from large colonies were also more often infested with fleas than those in small colonies.

Rearing young in a group gives the young opportunities to learn from more than one adult. It can also provide them with practice in tasks that later prove important when the offspring are on their own. Cooperative hunting can provide young with the opportunity to learn hunting skills from their elders. Generally, this benefit results in longer-lived species that produce only a few young per year.

Survival

Cold temperatures can cause physiological problems, particularly for animals that are ectothermic (relying upon the environment for heat). During the day, such animals can bask in the sun and maintain high body temperatures. At night, however, they may have difficulty staying warm, so many huddle together in large groups. This type of behavior allows ectotherms to continue to digest their food (a process requiring heat).

Ectothermic species are not the only ones that huddle—many birds and mammals do so as well. Some birds flock together on trees on particularly cold nights to reduce the surface area of their body that is exposed to the elements. Voles, which are normally asocial, huddle in groups during the winter to keep warm. Marmots can huddle underground in groups of twenty for up to seven months to avoid freezing temperatures in the mountain regions they occupy.

Lowered individual predation risk is theorized as a primary benefit leading to the evolution of gregarious behavior, particularly in the absence of kinship between group members. Three likely outcomes of grouping that can reduce rates of attack are the dilution effect, the confusion effect, and the encounter effect. In the dilution effect, the probability of a particular individual being killed or injured is reduced by the presence of other group members that might be attacked first. This helps to explain why ostriches lay eggs in communal clutches which only the first laying female incubates. The first female is not necessarily acting in an altruistic manner, because she is diluting the chance that predators will find her eggs and eat them or, after the chicks hatch, that predators will capture her chicks instead of those of another female. Individuals in a group may also benefit by putting other animals between themselves and the predator (the selfish herd effect). Grouping provides the opportunity to decrease the area of danger around each individual. If individuals within the group are acting in a selfish herd manner, the groups formed tend to be tightly clumped, as all individuals attempt to put other individuals on all sides around them.

Confusion effects benefit group members because predators may have difficulty in fixing upon a particular individual for attack. The time it takes for the predator to discern a particular individual from a mass of surrounding individuals may be enough for the entire group to scatter and to further confuse the predator.

In contrast, the encounter effect results when it is more difficult for a predator to find a single group of prey than an equal number of scattered individuals due to the apparent rarity of the grouped individuals. However, the actual result of aggregation may be increased risk of detection if the group becomes a more conspicuous entity, which is detrimental to the individuals making up the group, rather than beneficial.

Groups can also benefit by collective defense, which is seen in animals possessing weaponry such as horns or other piercing appendages. Commonly these groups will form a line (phalanx) or a rosette (circular structure) with weapons pointed outward toward the attacker(s). Common examples of these defensive groups include musk oxen, elephants, and spiny lobsters. The use of weapons in these defensive lines may enhance the probability of survival above that of mere dilution, since each defending animal is capable of inflicting damage on a predator.

An individual has to partition its time between foraging and being vigilant for potential predators. However, if that individual is within a group, it can spend more timing on foraging and less time on vigilance because its scan frequency for predators can decrease proportionate to the number of members of the group. Furthermore, because of the many eyes within the group, detection of predators with subsequent alerting of the danger is enhanced. Some species go so far as to alert other group members to danger by alarm calling. This may have evolved for several reasons: The

Lions are one of the few feline species that live in groups. These prides are composed of related females who share responsibility for raising all the young in the group. (PhotoDisc)

caller benefits because it knows where the predator is and can position itself appropriately within the group to avoid being a target; the caller may enhance the probability of becoming the target, but at the same time, it reduces the probability that its kin will be targets (this works in kin groups); and/or the caller may attract attention to itself at the time of the call, but if it survives, it is entitled to a payback at a later date from some other individual in the group doing the same.

Communal Foraging and Hunting

Frequently, food being sought for exploitation is clumped in space. As a result, the animals that feed upon this food are also clumped, and this promotes aggregation into group structures, such as herds. Because animals can observe others of their species feeding, they can follow successful foragers to feeding groups and group for no other reason. This phenomenon may allow such foragers to exploit food resources in a more systematic way. Some evidence suggests that grasses clipped by herbivore herds actually grow faster and are more productive than grasses not so clipped, and by proceeding from one patch to another, the herds actually allow the grasses to grow back in a systematic fashion. This allows the food time to replenish itself before the herd passes by it again, and would be less likely to happen if individuals exploited the resource, isolated in time and space.

Other animals form groups to facilitate hunting and capture success. However, some of these groupings can be highly variable in time, based on the food supply and how a kill is partitioned between group participants. For example, sociality in lions is controlled by food supply. When wildebeests or zebra are especially numerous, lions concentrate their efforts upon capturing them, but solitary hunters are successful only 15 percent of the time, while groups of five are 40 percent successful. Furthermore, groups of five lions are better able to protect their kill from scavengers than are groups of two or three lions, even though a group of two or three lions will maximize their daily intake of food per successful kill, while a group of five will only secure the minimum daily requirement of food. Despite their propensity for sociality, lions will hunt alone or in pairs when the migratory wildebeests or zebras leave and only resident gazelles are left. Again, the success rate for a solitary lion hunting a gazelle is only 15 percent, while that for a pair is 30 percent; however, a gazelle can provide the minimum daily requirement of food for only two lions.

Similarly, groups of dogs (the African wild dog, Asian dhole, dingoes, or wolves) are able to kill prey larger than themselves by cooperatively hunting. These packs are composed of kin (parents and their offspring) and cover large distances in order to hunt their prey. The individual dogs will spread out around the prey in a phalanx and then approach until one member selects a victim and runs after it. Other members of the pack will run after the prey in relays until the prey is exhausted and can be subdued.

Communal spiders build webs larger than a single individual could spin and capture prey larger than any one could capture alone. They communally feed on the captured prey. Most spider species have a short interval after the spiderlings have hatched where they remained clumped in a group and live in a communal web. After a period of several days, the spiderlings disperse to take up a solitary life. In communal species, however, adults of the same species come together to form colonies of up to one thousand individuals.

Coordinated group hunting is also known in marine mammals, particularly killer whales (orcas) and humpback whales. Orcas live in matriarchal societies (pods) of two to twelve members and hunt other marine mammals. Single orcas will charge sea lions in the surf of a beach, while other pod members will wait offshore to ensnare any sea lions that respond to the charge by entering the water. Adults will also train juvenile orcas, in the process of play-stranding, to capture seals or sea lions by throwing a dead seal to a beached juvenile. Orcas will also attack baleen whales by surrounding them and biting and holding the whale underwater so that it will drown. Humpback whales are known for bubble feeding, where members of the pod surround a school of small

Migration and Crowd Control

The urge to migrate is under control of an internal clock that is species-specific and that regulates physiological and behavioral rhythms. While individuals making up a population in a specific locale may be solitary during most of the year, they begin to manifest migratory restlessness at the appropriate time to migrate. This results in all individuals expressing the migratory urge at the same time and causes even nonsocial species to group together in a migratory exodus. Some bird species, as well as monarch butterflies, migrate in large groups to warmer climates to avoid the cold; other birds migrate to colder climates to avoid the heat of summer; and monarch butterflies migrate up into the mountains to avoid the heat in the lowlands. Some fish come together to locate home streams (salmon) or other breeding groups; some pinnipeds form large herds on communal pupping grounds; herds of ungulates follow a specific resource such as water or food. For those groups that follow a food or water resource, there is frequent mixing of several species that all exploit the same resource.

Birds are thought to fly in V-shaped formations as a means of reducing the drag on the birds behind the lead flyers. This may be so, since vortex rings are shed off of the wings during the flight stroke as there exists a circulation pattern of air on the wing. Birds flying behind the lead birds in a flock would be buffeted by these vortex rings, and thus the drag on their bodies would increase. Moreover, the vortex rings may create turbulent patterns of airflow behind the lead birds that would detrimentally affect the laminar flow needed to create lift on the bodies and wings of the birds behind. To avoid this situation, birds will often fly in a V formation that allows birds to avoid the disruptions to the airflow made by their fellow members during migration, when energy expenditures are critical considering the long distances covered. Likewise, the individuals in a fish school are arranged so that one fish does not encounter the vortices shed from the lead fish's tail.

fish and release bubbles while spiraling upward. This bubble net concentrates the terrified fish into a tight ball, which the whales then eat as they approach the surface and open their mouths. Groups of dolphins herd fish into shallow waters and surround them so they can be easily picked off.

Even groups of mixed species are known to cooperatively hunt. This is common in shorebirds, where one species might herd the prey while another species either dives to prevent the prey's escape from below or stabs at the prey to prevent its escape along the banks of the water. Pelicans and cormorants, as well as grebes and egrets, are known to form these associations.

Colonial species, particularly those that are sessile and need space to spread out, need to be able to defend their resources. Among invertebrates, larger colonies may have the competitive advantage in excluding newcomers from unoccupied space, as well as pushing other colonial species out of the way, so that the space once occupied can be overrun. Corals are well known for their warlike actions as the space on a reef becomes scarce—they will use their stinging cells to attack adjacent coral species in an attempt to kill the polyps making up the other colony. Dense colonies of bryozoans are much more able to withstand overgrowth by other invertebrates than are bryozoan colonies that are more spread out. Mussel beds can overrun barnacle colonies in the intertidal zone if their main predator is not present in sufficient numbers to keep the mussel population low.

Division of Labor

In eusocial insects, colony members divide the labor amongst themselves. In worker bees, the labor done is dependent on the age of the bee. After emergence, a worker bee's first job is cleaning the hive. She then tends the brood, builds up the honeycomb, and guards the nest. Her final task is to forage for pollen and nectar. The change in her duties correlates to physiological changes in her nurse glands and wax glands.

Ant colonies comprise thousands of individuals, divided into workers, brood, and queen. In army ants, workers vary in size, and these size differences determine their role in the colony. Smaller workers spend most of their time tending to and feeding the larval broods; medium-sized workers make up the majority of the population and are responsible for making raids on other colonies and locating food. The largest workers are called soldiers because of their powerful jaws; they accompany the raiding parties but carry no food.

Naked mole rats are the only known vertebrate that lives in colonies like those of the social insects. Only one female breeds. Larger individuals remain in the colony and huddle to keep the entire colony warm. Smaller individuals are the worker caste and are responsible for nest building and foraging.

These examples serve to illustrate that grouping behavior of animals can serve a myriad of purposes, or can simply be the result of phenomena beyond the animal's control (patchy food resources, physical forces of nature). The functions that lead to cooperative activities are usually best explained by groups being composed of kin; however, cooperation can also evolve in the absence of kinship, provided the benefits to group members outweigh the costs.

—*Kari L. Lavalli*

See also: Communication; Communities; Competition; Courtship; Defense mechanisms; Ethology; Herds; Hierarchies; Insect societies; Mammalian social systems; Nesting; Offspring care; Pair-bonding; Population analysis; Predation; Symbiosis; Territoriality and aggression; Thermoregulation; Zoos.

Bibliography

Bertram, B. C. R. "Living in Groups: Predators and Prey." In *Behavioral Ecology: An Evolutionary Approach*, edited by J. R. Krebs and N. B. Davies. Sunderland, Mass.: Sinauer Associates, 1978. This article provides a good background of how living in groups can benefit both predators and their prey and how grouping behavior may have evolved for this purpose.

Halliday, Tim, ed. *Animal Behavior*. Norman: University of Oklahoma Press, 1994. It highlights the major activities performed by all animals and provides a series of examples to illustrate the variety of ways in which animals accomplish their goals.

Hamilton, W. D. "Geometry for the Selfish Herd." *Journal of Theoretical Biology* 31 (1971): 294-311. This landmark paper was the first to explore the idea that grouping may be the result of selfishly behaving individuals that are trying to avoid becoming the predator's meal.

Janson, C. H. "Testing the Predation Hypothesis for Vertebrate Sociality: Prospects and Pitfalls." *Behavior* 135 (1998): 389-410. This paper explores explanations of sociality among vertebrates, primarily primates, by attempting to invoke a predation hypothesis. It clearly articulates the problems involved in attempting to show that predation is driving the formation of groups.

Krebs, J. R., and N. B. Davies. *An Introduction to Behavioral Ecology*. 3d ed. Cambridge, Mass.: Blackwell Scientific, 1993. This classic book provides an excellent background tying together behavior and ecology in an attempt to explain why animals do what they do.

Pulliam, H. R., and T. Caraco. "Living in Groups: Is There an Optimal Group Size?" In *Behavioral Ecology: An Evolutionary Approach*, edited by J. R. Krebs and N. B. Davies. 2d ed. Sunderland, Mass.: Sinauer Associates, 1984. This seminal paper attempts to explain what might drive animals to form groups and what the benefits and disadvantages are to grouping.

GROWTH

Type of animal science: Development
Fields of study: Cell biology, zoology

Animal development begins with a single cell, the fertilized egg. Continued growth and formation through numerous cellular replications result in the formation of the complex animal body.

Principal Terms

DIFFERENTIATION: the process during development by which cells obtain their unique structure and function

FERTILIZATION: the union of two gametes (egg and sperm) to form a zygote

GAMETE: a functional reproductive cell (egg or sperm) produced by the adult male or female

GROWTH: the increased body mass of an organism that results primarily from an increase in the number of body cells and secondarily from the increase in the size of individual cells

MITOSIS: the process of cellular division in which the nuclear material, including the genes, is distributed equally to two identical daughter cells

ZYGOTE: a fertilized egg

Animal development has been a source of wonder for centuries. Development involves the slow, progressive changes that occur when a single cell—the zygote, or fertilized egg—undergoes mitosis. Mitosis is the process by which a cell divides into identical daughter cells. During development, mitosis occurs repeatedly, forming multiple generations of daughter cells. These cells increase in number and ultimately form all the cells in the body of a multicellular animal, such as a frog, mouse, or elephant. The simple experiment of opening fertile chicken eggs to observe the embryos on successive days of their three-week incubation period illustrates the process of embryonic development. A narrow band of cells can be seen increasing in number and complexity until the body of an entire, but immature, chick is seen.

Animal Growth and Development

An organism's growth occurs because of the increasing number of cells that form as well as because of the increasing size of individual cells. For example, a mouse increases from a single cell, the zygote, to about three billion cells during the period from fertilization to birth. Embryology is the study of the growth and development of an organism occurring before birth. Growth and development, however, continue after birth and throughout adulthood. Growth ceases only at death, when the life of the individual organism is ended. The bone marrow of human adults initiates the formation and development of millions of red blood cells every minute of life. About one gram of old skin cells is lost and replaced by new cells each day.

Development produces two major results: the formation of cellular diversity and the continuity of life. Cellular diversity, or differentiation, is the process that produces and organizes the numerous kinds of body cells. The first cell that determines an individual's unique identity, the zygote, ultimately gives rise to varying types of cells having diverse appearances and functions. Muscle cells, red blood cells, skin cells, neurons, osteocytes (bone cells), and liver cells are all examples of cells that have differentiated from a single zygote.

Reproduction

Morphogenesis is the process by which differentiated cells are organized into tissues and organs. The continued formation of new individual organisms is called reproduction. The major stages of animal development include fertilization, embryology, birth, youth, adulthood—when fertilization of the next generation occurs—and death. A new individual animal is begun by the process of fertilization, when the genetic material from the sperm, produced by the father, and the egg, produced by the mother, are merged into a single cell, the zygote. Fertilization may be external, occurring in freshwater or the sea, or internal, occurring within the female's reproductive tract. While fertilization marks the beginning of a new individual, it is not literally the beginning of life, since both the sperm and egg are already alive. Rather, fertilization ensures the continuation of life through the formation of new individuals. This guarantees that the species of the organism will continue to survive in the future.

Following fertilization, the newly formed zygote undergoes embryological development consisting of cleavage, gastrulation, and organogenesis. Cleavage is a period of rapid mitotic divisions with little individual cell growth. A ball of small cells, called the morula, forms. As mitosis continues, this ball of cells hollows in the middle, forming an internal cavity called the blastocoel. Gastrulation immediately follows cleavage. During gastrulation, individual cell growth as well as initial cell differentiation occur. During this time, three distinct types of cells are formed: an internal layer called the endoderm, a middle layer, the mesoderm, and an external layer, the ectoderm. These cell types, or germ-cell layers, are the parental cells of all future cells of the body.

Cells from the ectoderm form the cells of the nervous system and skin. The mesoderm forms the cells of muscle, bone, connective tissue, and blood. The endoderm forms cells that line the inside of the digestive tract as well as the liver, pancreas, lungs, and thyroid gland. The transformation of these single germ layers into functional organs is called organogenesis. Organogenesis is an extremely complex period of embryological development. During this time, specific cells interact and respond to one another to induce growth, movement, or further differentiation; this cell-to-cell interaction is called induction. Each induction event requires an inducing cell and a responding cell.

In the formation of the brain and spinal cord, selected cells from the ectoderm form a long, thickened plate at the midline of the developing embryo. Through changes in cell shape, the outer edges of this plate fold up and fuse with each other in the middle, forming a tubular structure (a neural tube). This tubelike structure then separates from the remaining ectoderm. At the head region of the embryo, the neural tube enlarges into pockets that ultimately form brain regions.

For differentiation and development to occur, cells must be responsive to regulatory signals. Some of these signals originate within the responding cell; these signals are based in the genetic code found in the cell's own nucleus. Other signals originate outside the cell; they may include physical contact with overlying or underlying cells, specific signal molecules, such as hormones, from distant cells, or specialized structural molecules secreted by neighboring cells that map out the pathway along which a responding cell will migrate.

Postnatal Development

Embryological development climaxes in the formation of functional organs and body systems. This period is concluded by birth (or hatching, in the case of some animals). Following birth, development normally continues. In some animals, such as frogs, newly hatched individuals undergo metamorphosis during which their body structures are dramatically altered. Newly hatched frogs (tadpoles), for example, are transformed from aquatic, legless, fishlike creatures into mature adults with legs that allow them to move freely on land.

In mammals, development and growth occur primarily after birth, as the individual progresses through the stages of infancy, childhood, adolescence, and adulthood. Mature adulthood is attained when the individual can produce his or her own gametes and participate in mating behavior.

Embryonic growth is especially impressive because the rate of cellular mitosis is so enormous. In the case of the mouse embryo, thirty-one cell generations occur during embryonic development. Thus, the zygote divides into two cells, then four, then eight, sixteen, thirty-two, and so on. This results in a newborn mouse consisting of billions of cells—produced in a period of only twenty-one days. When the newborn passes through its life stages to adulthood, its body cells may number more than sixty billion. One marine mammal, the blue whale, begins as a single zygote that is less than one millimeter in diameter and weighs only a small fraction of a gram. The resulting newborn whale (the calf) is about seven meters long and weighs two thousand kilograms: The embryonic growth represents a 200-millionfold increase in weight. Yet, for some animals, impressive growth periods also occur in the juvenile and adolescent stages of life.

In many cases, once an individual animal reaches its typical adult size, the rate of mitosis slows so that the number of new cells simply replaces the number of older, dying cells. At this maintenance stage, the individual no longer grows in overall size even though it continuously produces new cells. Since most of the cells in the mature adult have reached a final differentiated state, the function of mitosis is simply to replace the degenerating, aging cells. The slowing of the rate of cellular mitosis during this time may be attributable to the presence of specialized cell products called chalones. Chalones are thought to be local products of mature cells that inhibit further growth or mitosis.

Studying Growth

Historically, much study of animal development and growth was performed by simple observation. Aristotle, perhaps the first known embryologist, opened chick eggs during varying developmental periods. He observed and sketched what appeared to be the formation of the chick's body from a nondescript substance. With the invention of lenses and microscopes, growth and development could be studied on a cellular level. The concept of cellular differentiation arose, since investigators could see that embryonic muscle cells, for example, looked different from embryonic nerve cells. Again, much of the investigative information was descriptive in nature. Embryologists detailed the existence of the three germ-cell layers in gastrulation as well as the various tissues and primitive cells involved in organogenesis.

Experimentation as a method of investigating animal growth and development began during the nineteenth century. Lower animal species, such as the sea urchin and frog, were frequently investigated; their developmental patterns are simpler than those of mammals, their development occurs outside the maternal body, and they can be found in abundant numbers. Many of these experiments used separation or surgical techniques to isolate or regraft specific tissues or cells of interest. An attempt was made to determine how one tissue type would interact with and influence the development of another tissue type. Thus, the ideas of induction, in which some tissues affect other tissues, came into being. During this time, the descriptive and comparative observations resulting from these experimental manipulations were the major contributions of investigators.

The embryologists of the early twentieth century paid little attention to genetics. They believed that the major influences on development and growth were embryological mechanisms, although genes were thought to provide some nonessential peripheral functions. Chemical analyses of embryos attempted to establish the chemical basis for the cell-to-cell interactions that were seen during development and differentiation. During the middle portion of the twentieth century, geneticists began to investigate the role of the gene in cell function. The function of genes in the cellular manufacturing of specific proteins led to the hypothesis that each kind of cellular protein was the product of one gene. During this time, bacteria and fruit flies (*Drosophila*) were primary organisms of study because of their relatively simple genetic makeups.

In the latter part of the twentieth century, molecular biology techniques were applied to the study of development. Using techniques for transferring and replicating specific genes, researchers have greatly clarified the central importance of genes in development. Scientists came to believe that all the major developmental and differentiation influences that control cell growth are regulated through specific genes that are turned off or on.

Developmental Biology

The combination of molecular biology techniques with embryological investigations has led to a new field of study—developmental biology. New methods have been developed and used. Radioactive tracer technology has allowed the investigator to label particular genes or gene products and trace their movements and influences on cell growth through several generations. Recombinant deoxyribonucleic acid (DNA) technology has allowed the isolation and replication of significant genes that are important in development. Immunochemistry uses specific proteins (antibodies) to bind to differentiating cell products and quantify them. Cell-cell hybridization allows the introduction of specific genes into the nuclei of cells in alternate differentiation pathways.

Developmental biology, with its multidisciplinary approach, is solving many of the fundamental questions of development. As scientists become better able to understand the role of genetics and cell-to-cell interactions, they gain insight into the mechanisms that control cell growth and development. Consequently, the potential to control undesirable growth or to enhance underdeveloped growth is within reach.

The problem of cell aging is also under investigation. Questions about why mature cells stop dividing and growing and what the causes of aging are constitute important areas of developmental research. While various theories have been presented, the fundamental key to cellular aging remains to be discovered. One of the most challenging areas of continuing research is the determination of how developmental patterns guide evolutionary changes. Developmental principles may provide the answer to why evolution has given rise to animal diversity. In addition, developmental biology may give scientists the information needed to predict and determine future evolutionary trends. The individual animal is a growing organism that begins as a zygote and passes through the stages of embryonic development, birth, youth, adulthood, aging, and death. Preservation of the species depends on adult individuals' producing gametes that result in the formation of a future generation of zygotes and individuals. Remarkably, each zygote contains the necessary genetic instructions to regulate the orderly processes of growth and development. Thus, animal life continues from generation to generation.

—*Roman J. Miller*

See also: Cell types; Cleavage, gastrulation, and neurulation; Cloning of extinct or endangered species; Determination and differentiation; Development: Evolutionary perspective; Embryology; Fertilization; Gametogenesis; Morphogenesis; Multicellularity; Parthenogenesis; Pregnancy and prenatal development; Regeneration; Reproduction.

Bibliography

Alberts, Bruce, Dennis Bray, Julian Lewis, Martin Raff, Keith Roberts, and James D. Watson. *Molecular Biology of the Cell*. 3d ed. New York: Garland, 1994. This encyclopedic college text is greatly enhanced by its superior diagrams and illustrations. While all aspects of cell biology are covered in great detail, of special interest are the chapters on cellular mechanisms of development and differentiated cells and the maintenance of tissues. Highly recommended for the serious student who wishes to understand developmental processes on a cellular level.

Balinsky, B. I. *An Introduction to Embryology*. Philadelphia: Saunders College Publishing, 1981. While somewhat dated, this very well-written text covers the essentials of classical animal embryology while integrating with them the cellular and molecular mechanisms that regulate them. Of specific interest is the chapter on growth, which integrates this topic into the study of developmental biology.

Baserga, Renato. *The Biology of Cell Reproduction*. Cambridge, Mass.: Harvard University Press, 1985. This illustrated book can be read by the layperson who is interested in understanding more fully the role of cell division and reproduction in growth. The basics of the cell cycle are covered in detail, as are many of the influences that regulate it. In the latter portion of the book, the author describes the genetic mechanisms involved in growth and development.

Gilbert, Scott F. *Developmental Biology*. 6th ed. Sunderland, Mass.: Sinauer Associates, 2000. This is a college-level text written for the upper-level student who has a serious interest in development. The author's approach to development uses a historical experimental analysis of the progress of investigators in the research field. About half of the text deals with specific cell or genetic mechanisms that regulate growth and development. An excellent resource for advanced and accurate information regarding the field of development.

_______, ed. *A Conceptual History of Modern Embryology*. Baltimore: The Johns Hopkins University Press, 1994. A collection of essays concentrating on historical aspects of research on embryonic induction and the relationship between experimental embryology and genetics.

Gilbert, Scott F., and A. M. Raunio. *Embryology: Contructing the Organism*. Sunderland, Mass.: Sinauer Associates, 1997. A college-level textbook on comparative embryology. The first two chapters introduce general terms and concepts, while the following twenty chapters each concentrate on a different animal group.

Hartwell, Leland H., and Ted A. Weinert. "Checkpoints: Controls That Ensure the Order of Cell Cycle Events." *Science* 246 (November 3, 1989): 629-634. An excellent review article describing the biochemical controls of cell division and cell proliferation. Information is a bit technical and is written at the level of the more advanced reader who has some background in biology.

Wessels, Norman K., and Janet L. Hopson. *Biology*. New York: Random House, 1988. The chapter on animal development gives an overview of classic embryology, including the role of growth. The chapter on developmental mechanisms and differentiation takes the reader through the major cellular and genetic regulatory mechanisms involved in development. This extremely well written introductory college text is easily read; it is well illustrated with color photographs and self-explanatory diagrams.

HABITATS AND BIOMES

Type of animal science: Ecology
Fields of study: Anthropology, environmental science, invertebrate biology, zoology

The biosphere is the sum total of all habitats on earth that can be occupied by living organisms. Descriptive and experimental studies of habitat components allow scientists to predict how various organisms will respond to changes in their environment, whether caused by humans or nature.

Principal Terms

ABIOTIC: the physical part of an ecosystem or biome, consisting of climate, soil, water, oxygen and carbon dioxide availability, and other physical components

BIOME: one of the widespread types of ecosystem on the earth, such as the Arctic tundra or the desert

BIOSPHERE: the sum of all the occupiable habitats for life on earth

BIOTIC: the living part of an ecosystem or biome, consisting of all organisms

COMMUNITY: a population of plants and animals that live together and make up the biotic part of an ecosystem

ECOSYSTEM: a relatively self-sufficient group of communities and their abiotic environment

ENVIRONMENT: the habitat created by the interaction of the abiotic and biotic parts of an ecosystem

HABITAT: the specific part of the environment occupied by the individuals of a species

POPULATION: a group of all the individuals of one species

SPECIES: a group of similar organisms that are capable of interbreeding and producing fertile offspring

Life in the form of individual organisms composed of one or more living cells is found in a vast array of different places on earth, each with its own distinctive types of organisms. Life on earth has been classified by scientists into units called species, whose individuals appear similar, have the same role in the environment, and breed only among themselves. The space in which each species lives is called its habitat.

Habitats, Communities, Ecosystems, and Biomes

The term "habitat" can refer to specific places with varying degrees of accuracy. For example, rainbow trout can be found in North America from Canada to Mexico, but more specifically they are found in freshwater streams and lakes with an average temperature below 70 degrees Fahrenheit and a large oxygen supply. The former example describes the macrohabitat of the rainbow trout, which is a broad, easily recognized area. The latter example describes its microhabitat—the specific part of its macrohabitat in which it is found. Similarly, the macrohabitat of one species can refer to small or large areas of its habitat. The macrohabitat of rainbow trout may refer to the habitat of a local population, the entire range of the species, or (most often) an area intermediate to those extremes. While "habitat," therefore, refers to the place an organism lives, it is not a precise term unless a well-defined microhabitat is intended. The total population of each species has one or more local populations, which are all the individuals in a specific geographic area that share a common gene pool; that is, they commonly interbreed. For example, rainbow trout of two adjacent states will not normally interbreed unless they are part of local populations that are very close to one another. The entire geographic distribution of a species, its range, may be composed of many local populations.

On a larger organizational scale, there is more than only one local population of one species in any habitat. Indeed, it is natural and necessary for many species to live together in an area, each with its own micro- and macrohabitat. The habitat of each local population of each species overlaps the habitat of many others. This collective association of populations in one general area is termed a community, which may consist of thousands of species of animals, plants, fungi, bacteria, and other one-celled organisms.

Groups of communities that are relatively self-sufficient in terms of both recycling nutrients and the flow of energy among them are called ecosystems. An example of an ecosystem could be a broad region of forest community interspersed with meadows and stream communities that share a common geographic area. Some ecosystems are widely distributed across the surface of the earth and are easily recognizable as similar ecosystems known as biomes—deserts, for example. Biomes are usually named for the dominant plant types, which have very similar shapes and macrohabitats. Thus, similar types of organisms inhabit them, though not necessarily ones of the same species. These biomes are easily mapped on the continental scale and represent a broad approach to the distribution of organisms on the face of the earth. One of the more consistent biomes is the northern coniferous forest, which stretches across Canada and northern Eurasia in a latitudinal belt. Here are found needle-leafed evergreen conifer trees adapted to dry, cold, windy conditions in which the soil is frozen during the long winter.

Clear-cutting forests is one of the many causes of habitat loss worldwide. (PhotoDisc)

North American Biomes

The biomes of North America, from north to south, are the polar ice cap, the Arctic tundra, the northern coniferous forest; then, at similar middle latitudes, eastern deciduous forest, prairie grassland, or desert; and last, subtropical rain forest near the equator. Complicating factors that determine the actual distribution of the biomes are altitude, annual rainfall, topography, and major weather patterns. These latter factors, which influence the survival of the living, or biotic, parts of the biome, are called abiotic factors. These are the physical components of the environment for a community of organisms.

The polar ice cap is a hostile place with little evidence of life on the surface except for polar bears and sea mammals that depend on marine animals for food. A distinctive characteristic of the Arctic tundra, just south of the polar ice cap, is its flat topography and permafrost, or permanently frozen soil. Only the top meter or so thaws during the brief Arctic summer to support low-growing mosses, grasses, and the dominant lichens known as reindeer moss. Well-known animals found there are the caribou, musk-ox, lemming, snowy owl, and Arctic fox.

The northern coniferous forest is dominated by tall conifer trees. Familiar animals include the snowshoe hare, lynx, and porcupine. This biome stretches east to west across Canada and south into the Great Lakes region of the United States. It is also found at the higher elevations of the Rocky Mountains and the western coastal mountain ranges. Its upper elevation limit is the "treeline," above which only low-growing grasses and herbaceous plants grow in an alpine tundra community similar to the Arctic tundra. In mountain ranges, the change in biomes with altitude mimics the biome changes with increasing latitude, with tundra being the highest or northernmost.

Approximately the eastern half of the United States was once covered with the eastern deciduous forest biome, named for the dominant broad-leaved trees that shed their leaves in the fall. This biome receives more than seventy-five centimeters of rainfall each year and has a rich diversity of bird species, such as the familiar warblers, chickadees, nuthatches, and woodpeckers. Familiar mammals include the white-tailed deer, cottontail rabbit, and wild turkey. The Great Plains, between the Mississippi River and the Rocky Mountains, receives twenty-five to seventy-five centimeters of rain annually to support an open grassland biome often called the prairie. The many grass species that dominate this biome once supported vast herds of bison and, in the western parts, pronghorn antelope. Seasonal drought and periodic fires are common features of grasslands.

The land between the Rockies and the western coastal mountain ranges is a cold type of desert biome; three types of hotter deserts are found from western Texas west to California and south into Mexico. Deserts receive fewer than twenty-five centimeters of rainfall annually. The hot deserts are dominated by many cactus species and short, thorny shrubs and trees, whereas sagebrush, grass, and small conifer trees dominate the cold desert. These deserts have many lizard and snake species, including poisonous rattlesnakes and the Gila monster. The animals often have nocturnal habits to avoid the hot, dry daytime.

Southern Mexico and the Yucatán Peninsula are covered by evergreen, broad-leaved trees in the tropical rain-forest biome, which receives more than two hundred centimeters of rain per year. Many tree-dwelling animals, such as howler monkeys and tree frogs, spend most of their lives in the tree canopy, seldom reaching the ground.

Aquatic biomes can be broadly categorized into freshwater, marine, and estuarine biomes. Freshwater lakes, reservoirs, and other still-water environments are called lentic, in contrast to lotic, or running-water, environments. Lentic communities are often dominated by planktonic organisms, small, drifting (often transparent) microscopic algae, and the small animals which feed on them. These, in turn, support larger invertebrates and fish. Lotic environments depend more on algae that are attached to the bottoms of streams, but they support equally diverse animal communities. The marine biome is separated into coastal and pelagic, or open-water, environments, which have plant and animal communities somewhat similar to lotic and lentic freshwater environments, respectively. The estuarine biome is a mixing zone where rivers empty into the ocean. These areas have a diverse assemblage of freshwater and marine organisms.

The Biosphere

All the biomes together, both terrestrial and aquatic, constitute the biosphere, which by definition is all the places on earth where life is found. Organisms that live in a biome must interact with one another and must successfully overcome and exploit their abiotic environment. The severity and moderation of the abiotic environments determine whether life can exist in that microhabitat. Such things as minimum and maximum daily and annual temperature, humidity, solar radiation, rainfall, and wind speed directly affect which types of organisms are able to survive. Amazingly, few places on earth are so hostile that no life exists there. An example would be the boiling geyser pools at Yellowstone National Park, but even there, as the water temperature cools at the edges to about 75 degrees Celsius, bacterial colonies begin to appear. There is abundant life in the top meter or two of soil, with plant roots penetrating to twenty-two meters or more in extreme cases. Similarly, the mud and sand bottoms of lakes and oceans contain a rich diversity of life. Birds, bats, and insects exploit the airspace above land and sea up to a height of about 1,200 meters, with bacterial and fungal spores being found much higher. Thus, the biosphere generally extends about 10 to 15 meters below the surface of the earth and about 1,200 meters above it. Beyond that, conditions are too hostile. A common analogy is that if the earth were a basketball, the biosphere would constitute only the thin outer layer.

Studying Habitats and Biomes

Abiotic habitat requirements for a local population or even for an entire species can be determined in the laboratory by testing its range of tolerance for each factor. For example, temperature can be regulated in a laboratory experiment to determine the minimum and maximum survival temperatures as well as an optimal range. The same can be done with humidity, light, shelter, and substrate type: "Substrate preference" refers to the solid or liquid matter in which an organism grows and/or moves—for example, soil or rock. The combination of all ranges of tolerance for abiotic factors should describe a population's actual or potential microhabitat within a community. Furthermore, laboratory experiments can theoretically indicate how much environmental change each population can tolerate before it begins to migrate or die.

Methods to study the interaction of populations with one another or even the interaction of individuals within one local population are much more complicated and are difficult or impossible to bring into a laboratory setting. These studies most often require collecting field data on distribution, abundance, food habits or nutrient requirements, reproduction and death rates, and behavior in order to describe the relationships between individuals and populations within a community. Later stages of these field investigations could involve experimental manipulations in which scientists purposely change one factor, then observe the population or community response. Often, natural events such as a fire, drought, or flood can provide a disturbance in lieu of a manipulation caused by man.

There are obvious limits to how much scientists should tinker with the biosphere merely to see how it works. Populations and even communities in a local area can be manipulated and observed, but it is not practical or advisable to manipulate whole ecosystems or biomes. To a limited extent, scientists can document apparent changes caused by civilization, pollution, and long-term climatic changes. This information, along with population- and community-level data, can be used to construct a mathematical model of a population or community. The model can then be used to predict the changes that would happen if a certain event were to occur. These predictions merely represent the "best guesses" of scientists, based on the knowledge available. Population

Habitat Destruction

One of the leading causes of species extinction or threatened extinction is habitat loss and destruction. Although there are many natural disturbances, such as earthquakes, volcanoes, and predators, that degrade or destroy habitats, the problem is most closely correlated with the increase in human population. Between 1850 and 2000, the population of the world increased from about one billion people to over six billion people. Accompanying that increase was the largest destruction of animal habitats and the largest extinction of animal species in the history of the world.

The greatest number of animal species occur in the tropical countries of Africa, South America, Central America, and Asia. Unfortunately, it is in these countries that human population growth continues at an alarming rate, posing an increasing threat in the form of the loss and destruction of animal habitats. In many tropical islands, most of the original habitats are gone, while over 50 percent of the natural habitats in approximately two-thirds of the countries in Asia and Africa have been destroyed.

Particularly threatened are the habitats in rain forests, wetlands, mangrove forests, and grasslands. Although rain forests occupy a small portion of Earth's surface area, they contain over 50 percent of its plant and animal species. Every year, rain forest area exceeding the size of the state of Florida is being lost to commercial logging, conversion to agriculture, and fuel wood production. At the present rate of loss, it is estimated that there may be no more rain forests by the year 2040.

One of the most important habitats for a large variety of vertebrate species are the wetlands and aquatic areas. In the United States, wetland areas account for about 15 percent of the threatened and endangered species. Many of these areas are being drained worldwide for development projects, agriculture, and dam projects. Over 50 percent of the wetlands that existed in the United States in 1800 are now destroyed.

Many aquatic species rely on mangrove forest habitats as breeding areas. Over 50 percent of the commercial species of fish and shrimp in the tropics depend on these areas, as well as many different species of birds. In addition, these special forest habitats, which exist along shallow intertidal coasts, are very important because they control erosion produced by storms. In order to produce more rice, agricultural products, firewood, and construction materials, many of the mangrove forests are being destroyed. In some countries, over 30 percent of these habitats have been lost or destroyed.

Grassland habitats are extremely important for supporting the biodiversity of many vertebrate species. Because these habitats generally occur in gentle terrain with excellent soils, they are frequently exploited for crop production, grazing land, and livestock production. Consequently, as the human population has grown, these areas have drastically shrunk in size.

Another contributing factor to the destruction of a habitat is the problem of habitat fragmentation. This occurs when contiguous habitats are broken into fragments, producing an increasing number of edge boundaries relative to the original habitat. Since many predators, including wolves, foxes, raccoons, and humans, often concentrate on habitat edges for hunting, more and more species become endangered as habitats are fragmented and eventually destroyed.

—Alvin K. Benson

ecologists often construct reasonably accurate population models that can predict population fluctuations based on changes in food supply, abiotic factors, or habitat. As models begin to encompass communities, ecosystems, and biomes, however, their knowledge bases and predictive powers decline rapidly. Perhaps the most complicating factor in building and testing these large-scale models is that natural changes seldom occur one at a time. Thus, scientists must attempt to build cumulative-effect models that are capable of incorporating multiple changes into a predicted outcome.

The biosphere, then, can be studied at different levels of organization, from the individual level through populations, communities, ecosystems, and biomes to the all-encompassing biosphere. Each level has unique relationships that require different methods of inquiry; in fact, these levels describe many of the subdisciplines within the science of ecology.

Biosphere and Biodiversity

Understanding the organization of the natural world of which man is a part is essential to the continued success of humankind. By understanding the abiotic and biotic relationships within and between each ecological level, from microhabitat to biosphere, scientists can partially explain why so many species of organisms have evolved over the last 3.5 billion years. This study of biodiversity may eventually be a key to maintaining a stable biosphere, in which there would be no drastic changes in climate or community relationships. For example, compare the diversity of microhabitats and species in a natural grassland biome with the established monoculture practices of agriculture, with the latter's emphasis on one species. In a wheat field, there are fewer microhabitats available, but those few are available in abundance. This leads to an increase in the population size of "pest" species that compete with man for an abundant food resource, wheat. Understanding the microhabitat requirements of pest species can lead to the reduction of crop losses.

The goal of the study of habitats, biomes, and the biosphere is the construction of predictive models. Once scientists have a general understanding of natural ecosystem processes, mathematical models may be able to predict future changes in the environment caused by the activities of civilization or natural climatic changes. For example, they may indicate whether increased human population size or increased large-scale agriculture in or near desert biomes will lead to the spread of desertlike conditions. On a biosphere scale, they may show whether the increase in the carbon dioxide content of the atmosphere caused by man's activities will lead to global warming. Both these effects are being predicted by many scientists. Prior to the relatively recent growth in the science of ecology and the general interest in it, man had little concern about the effects of the exploitation of natural resources such as forests, of synthetic chemical pollutants such as dichloro-diphenyl-trichloroethane (DDT), or even of the rapid growth of the human population size. With an increasing knowledge and understanding of habitat requirements, natural community interrelationships, and cycling of nutrients and pollutants within the biosphere, however, scientists have greater predictive power concerning the effects of economic development and human population growth. Ecologists are studying the effects of changing rain forests into agricultural land, the extinction of species, the loss of the ozone layer, and many other phenomena that have the potential to change the abiotic environment and therefore affect the stability of biotic communities. These results will be incorporated into future predictive models, giving them increased accuracy.

The challenge—both to scientists and to human civilization as a whole—is to use an understanding of the biosphere to maintain a level of economic growth that is ecologically sustainable. The study of communities and ecosystems may discover ways that civilization can better adapt to its current environment rather than attempt to mold the environment to fit its own preconceived, established ideas.

—*Jim Fowler*

See also: Biodiversity; Communities; Deserts; Ecological niches; Ecology; Ecosystems; Fauna: Africa; Fauna: Antarctica; Fauna: Arctic; Fauna: Asia; Fauna: Australia; Fauna: Caribbean; Fauna: Central America; Fauna: Europe; Fauna: Galápagos Islands; Fauna: Madagascar; Fauna: North America; Fauna: Pacific Islands; Fauna: South America; Forests, coniferous; Forests, deciduous; Grasslands and prairies; Lakes and rivers; Mountains; Paleoecology; Population fluctuations; Population growth; Rain forests; Savannas; Swamps and marshes; Territoriality and aggression; Tidepools and beaches; Tundra; Urban and suburban wildlife; Wildlife management; Zoos.

Bibliography

Allaby, Michael. *Biomes of the World*. 9 vols. Danbury, Conn.: Grolier International, 1999. Each volume offers systematic coverage of a biome of group of related biomes, including physical description, characteristic animal life, and environmental issues and threats. Illustrations, maps, photos, further reading lists, lists of Web sites, glossary, and index.

Bradbury, Ian K. *The Biosphere*. New York: Belhaven Press, 1991. Emphasizes the living organisms within the global ecosystem, including basic cell chemistry, genetics, and evolution. Outlines the way in which ecosystems work. Does not require a background in biology to understand.

Cox, George W. *Conservation Biology: Concepts and Applications*. 2d ed. Dubuque, Iowa: Wm. C. Brown, 1997. An introductory college-level textbook. Discusses terrestrial and aquatic ecosystems, including each system's characteristics, plant and animal life, and environmental problems.

Hanks, Sharon La Bonde. *Ecology and the Biosphere: Principles and Problems*. Delray Beach, Fla.: St. Lucie Press, 1996. A college-level textbook aimed at non-science majors. Covers what science is and what students need to know about it; what the biosphere is, how it works, and what its current problems are; and what students can do about the problems.

Luoma, Jon R. *The Hidden Forest: The Biography of an Ecosystem*. New York: Henry Holt, 1999. Focuses on the Andrews Experimental Forest, a 16,000 acre primal forest in central Oregon, which has been the subject of an intensive thirty-year study. A case study in the minutia of ecosystem research.

Miller, G. Tyler, Jr. *Living in the Environment*. 5th ed. Belmont, Calif.: Wadsworth, 1988. A college textbook that serves as an introduction to environmental science. Good coverage of the biosphere, ecosystem, and biome levels of ecology as well as an extensive annotated bibliography for each chapter.

Scientific American 261 (September, 1989). A special issue on managing the earth, with articles on the changing atmosphere, changing climate, threats to biodiversity, the growing human population, strategies for agriculture, and an ecologically sustainable economy. Emphasis is on the application of ecological knowledge.

Sutton, Ann, and Myron Sutton. *Eastern Forests*. New York: Alfred A. Knopf, 1986. Describes all the forest community types from the eastern deciduous and northern coniferous forest biomes. Unique in that it contains natural history information on insects, spiders, mushrooms, trees, wildflowers, reptiles, amphibians, birds, and mammals. Prefaced by a general description of each community type, written for the layperson.

Wetzel, Robert G. *Limnology*. 2d ed. Philadelphia: W. B. Saunders, 1983. The standard college textbook for freshwater lake ecology; contains extensive coverage of the abiotic environment of lakes. There is also a chapter on the interaction of zooplankton, fish, and bottom-dwelling animal communities. The book is somewhat technical, and background study in biology is helpful.

HABITUATION AND SENSITIZATION

Type of animal science: Behavior
Field of study: Ethology

Habituation is learning to ignore irrelevant stimuli that previously produced a reaction. Results of habituation studies have been used to explain, predict, and control behavior of humans and other organisms.

Principal Terms

ACETYLCHOLINE: a neurotransmitter produced by a nerve cell that enables a nerve impulse to cross a synapse and reach another nerve or muscle cell

APLYSIA: a large sluglike mollusk that lives in salt water and has been used in habituation experiments; its outer covering is called the mantle

IMPULSE: a "message" traveling within a nerve cell to another nerve cell or to a muscle cell

MOTOR NEURON: a nerve cell that causes a muscle cell to respond

NEUROTRANSMITTER: a chemical substance which enables nerve impulses to cross a synapse and reach another nerve cell or muscle cell

ORIENTING REFLEX: an unspecific reflex reaction caused by a change in the quantity or quality of a stimulus; it will disappear or decrease after repeated presentations of the stimulus

SENSITIZATION: an arousal or an alerting reaction which increases the likelihood that an organism will react; also, a synonym for loss of habituation with increased intensity of response

SYNAPSE: the minute space or gap between the axon of one nerve cell and the dendron of the next; also, the gap between a nerve cell and a muscle cell

Habituation is a simple form of nonassociative learning that has been demonstrated in organisms as diverse as protozoans, insects, *Nereis* (clam worms), birds, and humans. The habituated organism learns to ignore irrelevant, repetitive stimuli which, prior to habituation, would have produced a response. With each presentation of the habituating stimulus, the responsiveness of the organism decreases toward the zero, nonresponse level. If habituation training continues after the zero-response level, the habituation period is prolonged. Habituation to a particular stimulus naturally and gradually disappears unless the training continues. If training is resumed after habituation has disappeared, habituation occurs more rapidly in the second training series than in the first. Habituation is important for survival of the individual. Many stimuli are continuously impinging upon it: Some are important, others are not. Important stimuli require an immediate response, but those which result in neither punishment nor reward may be safely ignored.

Stimulus and Response

When a new stimulus is presented (when a sudden change in the environment occurs), the organism—be it bird, beast, or human—exhibits the "startle" or "orientation" response. In essence, it stops, looks, and listens. If the stimulus is repeated and is followed by neither reward nor punishment, the organism will pay less and less attention to it. When this happens, habituation has occurred, and the organism can now respond to and deal with other stimuli. On the other hand, if, during habituation learning, a painful consequence follows a previously nonconsequential stimulus, the organism has been sensitized to that stimulus and will respond to it even more strongly than it did before the learning sessions, whether they are occurring in the laboratory or in the field.

Young birds must learn to tell the difference between and respond differently to a falling leaf and a descending predator. A young predatory bird must learn to ignore reactions of its prey which pose no danger, reactions that the predator initially feared.

A theory known as the dual-process habituation-sensitization theory was formulated in 1966 and revised in 1973. It establishes criteria for both habituation and sensitization. Criteria for habituation (similar to those proposed by E. N. Sokolov in 1960) are that habituation will develop rapidly; the frequency of stimulation determines the degree of habituation; if stimulation stops for a period of time, habituation will disappear; the stronger the stimulus, the slower the rate of habituation; the frequency of stimulation is more important than the strength of the stimulus; rest periods between habituation series increase the degree of habituation; and the organism will generalize and therefore exhibit habituation to an entire class of similar stimuli. Stimulus generalization can be measured: If a different stimulus is used in the second habituation series, habituation occurs more rapidly; it indicates generalization.

Sensitization

Sensitization, a very strong response to a very painful, injurious, or harmful stimulus, is not limited to stimulus-response circuits but involves the entire organism. After sensitization, the individual may respond more strongly to the habituating stimulus than it did prior to the start of habituation training.

There are eight assumptions about sensitization in the dual-process theory. Sensitization does not occur in stimulus-response circuits but involves the entire organism. Sensitization increases during the early stages of habituation training but later decreases. The stronger the sensitizing stimulus and the longer the exposure to it the greater the sensitization; weaker stimuli may fail to produce any sensitization. Even without any external intervention, sensitization will decrease and disappear. Increasing the frequency of sensitization stimulation causes a decrease in sensitization. Sensitization will extend to similar stimuli. Dishabituation, the loss of habituation, is an example of sensitization. Sensitization may be time-related, occurring only at certain times of the day or year.

According to the dual-process theory, the response of an organism to a stimulus will be determined by the relative strengths of habituation and sensitization. Charles Darwin, the father of evolution, observed and described habituation, although he did not use the term. He noted that the birds of the Galápagos Islands were not disturbed by the presence of the giant tortoises, *Amblyrhynchus*; they disregarded them just as the magpies in England, which Darwin called "shy" birds, disregarded cows and horses grazing nearby. Both the giant tortoises of the Galápagos Islands and the grazing horses and cows of England were stimuli which, though present, would not produce profit or loss for the birds; therefore, they could be ignored.

The Neurology of Stimulus Response

Within the bodies of vertebrates is a part of the nervous system called the reticular network or reticular activating system; it has been suggested that the reticular network is largely responsible for habituation. It extends from the medulla through the midbrain to the thalamus of the forebrain. (The thalamus functions as the relay and integration center for impulses to and from the cerebrum of the forebrain.) Because it is composed of a huge number of interconnecting neurons and links all parts of the body, the reticular network functions as an evaluating, coordinating, and alarm center. It monitors incoming message impulses. Important ones are permitted to continue to the cerebral cortex, the higher brain. Messages from the cerebral cortex are coordinated and dispatched to the appropriate areas.

During sleep, many neurons of the reticular network stop functioning. Those that remain operational may inhibit response to unimportant stimuli (habituation) or cause hyperresponsiveness (sensitization). The cat who is accustomed to the sound of kitchen cabinets opening will sleep through a human's dinner being prepared (habituation) but will charge into the kitchen when she hears the sound of the cat food container opening (sensitization).

Researcher E. N. Sokolov concluded that the "orientation response" (which can be equated with sensitization) and habituation are the result of the functioning of the reticular network. According to Sokolov, habituation results in the formation of models within the reticular activating system. Incoming messages that match the model are disregarded by the organism, but those that differ trigger alerting reactions throughout the body, thus justifying the term "alerting system" as a synonym for the reticular network. Habituation to a very strong stimulus would take a long time. Repetition of this strong stimulus would cause an even stronger defensive reflex and would require an even longer habituation period.

The Role of Neurotransmitters

Neurotransmitters are chemical messengers that enable nerve impulses to be carried across the synapse, the narrow gap between neurons. They transmit impulses from the presynaptic axon to the postsynaptic dendrite(s). E. R. Kandell, in experiments with *Aplysia* (the sea hare, a large mollusk), demonstrated that as a habituation training series continues, smaller amounts of the neurotransmitter acetylcholine are released from the axon of the presynaptic sensory neuron. On the other hand, after sensitization, this neuron released larger amounts of acetylcholine because of the presence of serotonin, a neurotransmitter secreted by a facilitory interneuron. When a sensitizing stimulus is very strong, it usually generates an impulse within the control center—a ganglion, a neuron, or the brain. The control center then transmits an impulse to a facilitory interneuron, causing the facilitory interneuron to secrete serotonin.

Increased levels of acetylcholine secretion by the sensory neuron result from two different stimuli: direct stimulation of the sensory neurons of the siphon or serotonin from the facilitory interneuron. Facilitory interneurons synapse with sensory neurons in the siphon. Serotonin discharged from facilitory interneurons causes the sensory neurons to produce and secrete more acetylcholine.

On the molecular level, the difference between habituation and adaption—the failure of the sensory neuron to respond—is very evident. The habituated sensory neuron has a neurotransmitter in its axon but is unable to secrete it and thereby enable the impulse to be transmitted across the synapse. The adapted sensory neuron, by contrast, has exhausted its current supply of neurotransmitter. Until new molecules of neurotransmitter are synthesized within the sensory neuron, none is available for release.

In 1988, Emilie A. Marcus, Thomas G. Nolen, Catherine H. Rankin, and Thomas J. Carew published the multiprocess theory to explain dishabituation and sensitization in the sea hare, *Aplysia*. On the basis of their experiments using habituated sea hares that were subjected to different stimuli, they concluded that dishabituation and sensitization do not always occur together; further, they decided, there are three factors to be considered: dishabituation, sensitization, and inhibition.

Habituation Studies

Habituation studies have utilized a wide variety of approaches, ranging from the observation of intact organisms carrying out their normal activities in their natural surroundings to the laboratory observation of individual nerve cells. With different types of studies, very different aspects of habituation and sensitization can be investigated. Surveying the animal kingdom in 1930, G. Humphrey concluded that habituation-like behavior exists at all levels of life, from the simple one-celled protozoans to the multicelled, complex mammals.

E. N. Sokolov, a compatriot of Ivan P. Pavlov, used human subjects in the laboratory. In 1960, he reported on the results of his studies, which involved sensory integration, the makeup of the orientation reflex (which he credited Pavlov with introducing in 1910), a neuronal model and its role in the orientation reflex, and the way that this neuronal model could be used to explain the conditioned reflex. Sokolov measured changes in the diameter of blood vessels in the head and finger, changes in electrical waves within the brain, and changes in electrical conductivity of the skin. By lowering the intensity of a tone to which human subjects had been habituated, Sokolov demonstrated that habituation was not the result of fatigue, because subjects responded to the lower-intensity tone with the startle or orientation reflex just as they would when a new stimulus was introduced. Sokolov concluded that the orientation response (which is related to sensitization) and habituation are the result of the functioning of the reticular network of the brain and central nervous system. Sokolov emphasized that the orientation response was produced after only the first few exposures to a particular stimulus, and it increased the discrimination ability of internal organizers. The orientation response was an alerting command. Heat, cold, electric shock, and sound were the major stimuli that he used in these studies.

Ivan Pavlov's groundbreaking work on habituation and sensitization, especially in dogs, earned him a Nobel Prize in Medicine in 1904. (Nobel Foundation)

E. R. Kandell used the sea hare, *Aplysia*, in his habituation-sensitization studies. *Aplysia* is a large sluglike mollusk, with a sheetlike, shell-producing body covering, the mantle. *Aplysia* has a relatively simple nervous system and an easily visible gill-withdrawal reflex. (The gill is withdrawn into the mantle shelf.) Early habituation-sensitization experiments dealt with withdrawal or absence of gill withdrawal. Later experiments measured electrical changes that occurred within the nerve cells that controlled gill movement. These were followed by studies which demonstrated that the gap (synapse) between the receptor nerve cell (sensory neuron) and the muscle-moving nerve cell (motor neuron) was the site where habituation and dishabituation occurred and that neurohormones such as acetylcholine and serotonin played essential roles in these processes. Kandell called the synapse the "seat of learning."

Charles Sherrington used spinal animals in which the connection between the brain and the spinal nerve cord had been severed. Sherrington demonstrated that habituation-sensitization could occur within the spinal nerve cord even without the participation of the brain. Pharmaceuticals have also been used in habituation-sensitization studies. Michael Davis and Sandra File used neurotransmitters such as serotonin and norepinephrine to study modification of the startle (orientation) response.

Habituation studies conducted in the laboratory enable researchers to control variables such as genetic makeup, previous experiences, diet, and the positioning of subject and stimulus; however, they lack many of the background stimuli present in the field. In her field studies of the chimpanzees of the Gombe, Jane Goodall used the principles of habituation to decrease the distance between herself and the wild champanzees until she was able to come close enough to touch and be accepted by them. The field-experimental approach capitalizes on the best of both laboratory and field techniques. In this approach, a representative group of organisms that are in their natural state and habitat are subjected to specific, known stimuli.

Learning to Survive

Habituation is necessary for survival. Many stimuli are constantly impinging upon all living things; since it is biologically impossible to respond simultaneously to all of them, those which are important must be dealt with immediately. It may be a matter of life or death. Those which are unimportant or irrelevant must be ignored.

Cell physiologists and neurobiologists have studied the chemical and electrical changes that occur between one nerve cell and another and between nerve and muscle cells. The results of those studies have been useful in understanding and controlling these interactions as well as in providing insights for therapies. Psychologists utilize the fruits of habituation studies to understand and predict, modify, and control the behavior of intact organisms. For example, knowing that bulls serving as sperm donors habituate to one cow or model and stop discharging sperm into it, the animal psychologist can advise the semen collector to use a different cow or model or simply to move it to another place—even as close as a few yards away.

Conservationists and wildlife protectionists can apply the principles of habituation to wild animals, which must live in increasingly closer contact with one another and with humans, so that both animal and human populations can survive and thrive. For example, black-backed gulls, when establishing their nesting sites, are very territorial. Males which enter the territory of another male gull are rapidly and viciously attacked. After territorial boundaries are established, however, the males in contiguous territories soon exhibit "friendly enemy" behavior: They are tolerant of the proximity of other males that remain within their territorial boundaries. This has been observed in other birds as well as in fighting fish.

—*Walter Lener*

See also: Communication; Defense mechanisms; Hearing; Instincts; Intelligence; Learning; Nervous systems of vertebrates; Reflexes; Rhythms and behavior; Sense organs; Smell; Vision.

Bibliography

Alcock, John. *Animal Behavior: An Evolutionary Approach*. 6th ed. Sunderland, Mass.: Sinauer Associates, 1998. A comprehensive study of the behavior of many different animals. Throughout, the authors emphasize how behaviors of different organisms have developed and changed and how these modifications have enabled different groups to survive.

Alkon, Daniel L. "Learning in a Marine Snail." *Scientific American* 249 (July, 1983): 70-84. A detailed description of experiments in which the saltwater snail, *Hermissenda crassicornis*, is conditioned to associate two different stimuli. Carefully and precisely written, though it abounds with technical terms, it is readily understandable because of the many line drawings, diagrams, and physiographic recordings.

Barash, David P. *Sociobiology and Behavior.* New York: Elsevier, 1982. An easily understandable introduction to the evolution of social behavior in different organisms, with many examples. Sociobiological hypotheses are tested and test results evaluated.

Drickamer, Lee C., Stephen H. Vessey, and Doug Meikle. *Animal Behavior: Mechanisms, Ecology, Evolution*. 4th ed. Dubuque, Iowa: Wm. C. Brown, 1996. A comprehensive upper-level college textbook in animal behavior.

Eckert, Roger, and David Randall. *Animal Physiology*. 4th ed. San Francisco: W. H. Freeman, 1997. Includes an excellent, lavishly illustrated discussion of how nervous systems function. Highly recommended for background reading as well as for detailed explanations of more technical terms and concepts.

Gould, James L. *Ethology: The Mechanisms and Evolution of Behavior*. New York: W. W. Norton, 1980. An excellent ethology textbook that provides an in-depth analysis of the most complex organisms. Describes the functioning of their life processes and body parts. The evolutionary approach provides insights into the behavior of organisms from the simplest to the most complex. Persuasive, well written, and readily understandable.

Gould, James L., and Peter Marler. "Learning by Instinct." *Scientific American* 256 (January, 1987): 74-85. The authors maintain that a sharp distinction cannot be made between instinct and learning. Gould and Marler argue that learning occurs in many animals, both invertebrates and vertebrates, and that human learning evolved from a few types which appear in other organisms. Specific examples are given of learning in bees, birds, and humans. This well-illustrated article is easy to read and understand.

Halliday, Tim, ed. *Animal Behavior*. Norman: University of Oklahoma Press, 1994. Describes how animals relate to their own and other species, covering mating rituals, heirarchy, foraging and food storage, and migration. Color photographs.

Klopfer, P. H., and J. P. Hailman. *An Introduction to Animal Behavior: Ethology's First Century*. Englewood Cliffs, N.J.: Prentice-Hall, 1967. An easily understandable historical perspective on the scientists and the studies that established ethology as a science. The major ideas are discussed in terms of their originators.

Slater, P. J. B., and T. R. Halliday, eds. *Behavior and Evolution*. New York: Cambridge University Press, 1994. A textbook for upper-level undergraduate or graduate students. The first four chapters provide a general overview of the relationship between evolution and animal behavior. The final five chapters cover social systems, sexual behavior, kinship, and the evolution of intelligence.

HADROSAURS

Types of animal science: Classification, evolution
Fields of study: Anatomy, evolutionary science, paleontology, systematics (taxonomy)

Hadrosaurs are a group of bipedal dinosaurs that were the most diverse and widespread large-bodied herbivores of the Late Cretaceous. The presence of eggs and nests provides evidence for hadrosaur social behavior.

Principal Terms

ALTRICIAL: animals that need parental care after hatching

CLADISTICS: a method of analyzing biological relationships in which advanced characters of organisms are used to indicate closeness of origin

CRETACEOUS: a period of time that lasted from about 146 to 65 million years ago, the end of which was marked by the extinction of the dinosaurs

LAMBEOSAURINES: hadrosaurs in which the skull bears a large tubular crest

ORNITHISCHIA: one of the two main dinosaur groups, characterized by a pelvis in which the pubis is swung backward

PUBIS: one of the three bones that make up the pelvis (the others are the ischium and ilium)

The hadrosaurids, or "duck-billed" dinosaurs, were large, bipedal dinosaurs, with body lengths up to fifteen meters, that lived during the Upper Cretaceous across North and Central America, South America, Europe, and Asia. It is thought that they originated in Asia, but once they arose, they spread and diversified worldwide, quickly becoming the primary constituent of herbivorous dinosaur faunas. The first hadrosaur remains to be found were represented by a few fragments from Montana and South Dakota, described by Joseph Leidy in 1856. Shortly afterward, a partial skeleton was found in New Jersey, the first nearly complete dinosaur found anywhere, and described by Leidy in 1858 as *Hadrosaurus foulkii*.

Hadrosaurs are ornithopod (bird-footed) dinosaurs, a term that refers to their three-toed feet. They are thus part of the Ornithischia, one of the two major dinosaur subdivisions, and characterized by a pubis (one of the three bones of the pelvis) that is inclined backward. In cladistic terms, they are considered to be a monophyletic group; that is, they are all derived from a common ancestor. They are particularly characterized by a toothless front to the mouth, which is flared outward to form a broad, flat "bill," prompting early re-

Hadrosaur Facts

Classification:
Kingdom: Animalia
Subkingdom: Bilateria
Phylum: Chordata
Subphylum: Vertebrata
Class: Reptilia
Order: Ornithischia
Suborder: Ornithopoda (bipedal herbivores)

Geographical location: Western and eastern North America, central and southern South America, Europe, Asia

Habitat: Terrestrial habitats

Gestational period: Unknown

Life span: Estimated at twenty-five years

Special anatomy: Large, bipedal dinosaurs in which the snout was developed into a beak and a crest was often developed on the head

searchers to dub them the duck-billed dinosaurs. This bill was covered by a thick, horny sheath, and the rest of the jaws bore closely packed batteries of grinding teeth, up to four hundred on each side of each jaw. The postcranial anatomy of hadrosaurs is generally very similar and the really obvious differences lie in the crests that many species bear on the top of their heads. These are formed of outgrowths of the nasal bones, are frequently hollow, and are found particularly in the hadrosaurs known as lambeosaurines.

There are several theories about the function of the hadrosaurs' crests. Some have suggested that it served as a kind of snorkel, allowing the dinosaur to breathe while foraging under water, while others suggest that it contained olfactory sensors, or served as a resonator for vocalizations. (©John Sibbick)

Hadrosaurid Lifestyle

More is known about the life history of hadrosaurids than about any other group of dinosaurs, due to information from skin impressions, trackways, and eggs and nesting sites. They were originally thought to be amphibious animals, inhabiting marshy areas and swimming and diving for soft aquatic plants. This was partly based on skin impressions from mummified specimens which appeared to show webbing between the digits. It was even suggested that the crested forms used the crests as snorkels and were able to breathe through them while underwater. More recent analyses of stomach contents and the function of the tooth batteries suggest, however, that they were adapted to efficiently process fibrous vegetation of low nutritional value, such as coniferous needles and twigs. The crests are now thought to have developed for a variety of purposes, including increased sensory area for an improved sense of smell, and to act as resonators for producing distinctive calls.

In 1979, juvenile hadrosaurid bones were discovered in the Two Medicine Formation in Montana by Jack Horner. Subsequently, he discovered nesting sites with eggs that provide evidence for reproductive strategies in the hadrosaur *Maiasaura*. These animals appear to have built large, circular nesting mounds with concave centers in common nesting areas. There is evidence of site fidelity, indicating that the nesting sites were used from season to season. Up to twenty eggs were laid in the nest in a circular pattern, then covered by vegetation that kept the eggs warm as it decomposed. Bones of young have been found in some of the nests and these indicate from tooth wear that the animals had already been feeding on vegetation. Also, the ends of the bones were still poorly formed, showing that they were altricial or nest-bound for some period after they had hatched. During this period they would have been fed by adults, which in turn suggests social behavior in which the eggs were protected by adults who then fed and raised the nest-bound hatchlings.

—*David K. Elliott*

See also: *Allosaurus*; *Apatosaurus*; *Archaeopteryx*; Dinosaurs; Evolution: Animal life; Extinction; Fossils; Ichthyosaurs; Nesting; Paleoecology; Paleontology; Prehistoric animals; Pterosaurs; Sauropods; Stegosaurs; *Triceratops*; *Tyrannosaurus*; Velociraptors.

Bibliography

Benton, Michael J. *Vertebrate Palaeontology*. 2d ed. London: Chapman and Hall, 1997. General vertebrate paleontology text that devotes one chapter to the hadrosaurids.

Carpenter, Kenneth, Karl F. Hirsch, and John R. Horner, eds. *Dinosaur Eggs and Babies*. New York: Cambridge University Press, 1994. A comprehensive coverage of knowledge on the eggs, nests and young of dinosaurs.

Currie, Philip J., and Kevin Padian. *Encyclopedia of Dinosaurs*. San Diego, Calif.: Academic Press, 1997. Excellent coverage of all aspects of dinosaur biology including that of hadrosaurids.

Horner, John R., and James Gorman. *Digging Dinosaurs*. Reprint. New York: Perennial Library, 1990. Well-illustrated and easy-to-read text that covers the work of Jack Horner on hadrosaurid eggs and nests.

Norman, David. *The Illustrated Encyclopedia of Dinosaurs*. New York: Crescent Books, 1985. Although old, this book has wonderful illustrations and an excellent text with extensive coverage of hadrosaurids.

THE HARDY-WEINBERG LAW OF GENETIC EQUILIBRIUM

Type of animal science: Evolution
Fields of study: Ecology, evolutionary science, genetics

The Hardy-Weinberg law of genetic equilibrium is one of the foundations of mathematical population genetics. A description of the genetic makeup of a population under ideal conditions, it acts as a benchmark against which the effects of natural selection or other evolutionary forces can be measured.

Principal Terms

ALLELE: one of several alternate forms of a gene; the deoxyribonucleic acid (DNA) of a gene may exist as two or more slightly different sequences, which may result in distinct characteristics

ALLELE FREQUENCY: the relative abundance of an allele in a population

DIPLOID: having two chromosomes of each type

GENE: a section of the DNA of a chromosome, which contains the instructions that control some characteristic of an organism

GENE POOL: the array of alleles for a gene available in a population; it is usually described in terms of allele or genotype frequencies

GENOTYPE: the set of alleles an individual has for a particular gene

GENOTYPE FREQUENCY: the relative abundance of a genotype in a population

HAPLOID: having one chromosome of each type

POPULATION: the individuals of a species that live in one place and are able to interbreed

RANDOM MATING: the assumption that any two individuals in a population are equally likely to mate, independent of the genotype of either individual

Genetics began with the study of inheritance in families: Gregor Mendel's laws describe how the alleles of a pair of individuals are distributed among their offspring. Population genetics is the branch of genetics that studies the behavior of genes in populations. The population is the only biological unit that can persist for a span of time greater than the life of an individual, and the population is the only biological unit that can evolve. The two main subfields of population genetics are theoretical (or mathematical) population genetics, which uses formal analysis of the properties of ideal populations, and experimental population genetics, which examines the behavior of real genes in natural or laboratory populations.

Population genetics began as an attempt to extend Mendel's laws of inheritance to populations. In 1908, Godfrey H. Hardy, an English mathematician, and Wilhelm Weinberg, a German physician, each independently derived a description of the behavior of allele and genotype frequencies in an ideal population of sexually reproducing diploid organisms. Their results, now termed the Hardy-Weinberg principle, or Hardy-Weinberg equilibrium, showed that the pattern of allele and genotype frequencies in such a population followed simple rules. They also showed that, in the absence of external pressures for change, the genetic makeup of a population will remain the same, at an equilibrium. Since evolution is change in a population over time, such a population is not evolving. Modern evolutionary theory is an outgrowth of the "New Synthesis" of R. A. Fisher, J. B. S. Haldane, and Sewall Wright, which was done in the 1930's. They examined the significance of various factors that cause evolution by examining the degree to which they cause deviations from the predictions of the Hardy-Weinberg equilibrium.

Assumptions and Predictions

The predictions of the Hardy-Weinberg equilibrium hold if the following assumptions are true: The population is infinitely large; there is no differential movement of alleles or genotypes into or out of the population; there is no mutation (no new alleles are added to the population); there is random mating (all genotypes have an equal chance of mating with all other genotypes); and all genotypes are equally fit (have an equal chance of surviving to reproduce). Under this very restricted set of assumptions, the following two predictions are true: Allele frequencies will not change from one generation to the next, and genotype frequencies can be determined by a simple equation and will not change from one generation to the next.

The predictions of the Hardy-Weinberg equilibrium represent the working through of a simple set of algebraic equations and can be easily extended to more than two alleles of a gene. In fact, the results were so self-evident to the mathematician Hardy that he at first did not think the work was worth publishing.

If there are two alleles (A, a) for a gene present in the gene pool, let p = the frequency of the A allele and q = the frequency of the a allele. As an example, if $p = 0.4$ (40 percent) and $q = 0.6$ (60 percent), then $p + q = 1$, since the two alleles are the only ones present and the sum of the frequencies (or proportions) of all the alleles in a gene pool must equal 1 (or 100 percent). The Hardy-Weinberg principle states that at equilibrium the frequency of AA individuals will be p^2 (equal to 0.16 in this example), the frequency of Aa individuals will be $2pq$, or 0.48, and the frequency of aa individuals will be q^2, or 0.36.

The basis of this equilibrium is that the individuals of one generation give rise to the next generation. Each diploid individual produces haploid gametes. An individual of genotype AA can make only a single type of gamete, carrying the A allele. Similarly, an individual of genotype aa can make only a gametes. An Aa individual, however, can make two types of gametes, A and a, with equal probability. Each individual makes an equal contribution of gametes, since all individuals are equally fit and there is random mating. Each AA individual will contribute twice as many A gametes as each Aa individual. Thus, to calculate the frequency of A gametes, add twice the number of AA individuals and the number of Aa individuals, then divide by twice the total number of individuals in the population (note that this is the same as the method to calculate allele frequencies). That means that the frequency of A gametes is equal to the frequency of A alleles in the gene pool of the parents.

The next generation is formed by gametes pairing at random (independent of the allele they carry). The likelihood of an egg joining with a sperm is the frequency of one multiplied by the frequency of the other. AA individuals are formed when an A sperm joins an A egg; the likelihood of this occurrence is $p \times p = p^2$ (that is, $0.4 \times 0.4 = 0.16$ in the first example). In the same fashion, the likelihood of forming an aa individual is $q^2 = 0.36$. The likelihood of an A egg joining an a sperm is pq, as is the likelihood of an a egg joining an A sperm; therefore, the total likelihood of forming an Aa individual is $2pq = 0.48$. If one now calculates the allele frequencies (and hence the frequencies of the gamete types) for this generation, they are the same as before: The frequency of the A allele is $p = (2p^2 + 2pq)/2$ (in the example, $(0.32 + 0.48)/2 = 0.4$), and the frequency of the a allele is $q = (1 - p) = 0.6$. The population remains at equilibrium, and neither allele nor genotype frequencies change from one generation to the next.

Ideal Versus Real Conditions

The Hardy-Weinberg equilibrium is a mathematical model of the behavior of ideal organisms in an ideal world. The real world, however, does not approximate these conditions very well. It is important to examine each of the five assumptions made in the model to understand their consequences and how closely they approximate the real world.

The first assumption is infinitely large population size, which can never be true in the real world, as all real populations are finite. In a small population, chance effects on mating success over many generations can alter allele frequencies. This effect is called genetic drift. If the number of breeding adults is small enough, some genotypes will not get a chance to mate with one another, even if mate choice does not depend on genotype. As a result, the genotype ratios of the offspring would be different from the parents'. In this case, however, the gene pool of the next generation is determined by those genotypes, and the change in allele frequencies is perpetuated. If it goes on long enough, it is likely that some alleles will be lost from the population, since a rare allele has a greater chance of not being included. Once an allele is lost, it cannot be regained. How long this process takes is a function of population size. In general, the number of generations it would take to lose an allele by drift is about equal to the number of individuals in the population. Many natural populations are quite large (thousands of individuals), so that the effects of drift are not significant. Some populations, however, especially of endangered species, are very small: The total population of California condors is less than twenty-five, all in captivity.

The second assumption is that there is no differential migration, or movement of genotypes into or out of the population. Individuals that leave a population do not contribute to the next generation. If one genotype leaves more frequently than another, the allele frequencies will not equal those of the previous generation. If incoming individuals come from a population with different allele frequencies, they also alter the allele frequencies of the gene pool.

The third assumption concerns mutations. A mutation is a change in the DNA sequence of a gene—that is, the creation of a new allele. This process occurs in all natural populations, but new mutations for a particular gene occur in about one of 10,000 to 100,000 individuals per generation. Therefore, mutations do not, in themselves, play much part in determining allele or genotype frequencies. Yet, mutation is the ultimate source of all alleles and provides the variability on which evolution depends.

The fourth assumption is that there is random mating among all genotypes. This condition may be true for some genes and not for others in the same population. Another common limitation on random mating is inbreeding, the tendency to mate with a relative. Many organisms, especially those with limited ability to move, mate with nearby individuals, which are often relatives. Such individuals tend to share alleles more often than the population at large.

The final assumption is that all genotypes are equally fit. Considerable debate has focused on the question of whether two alleles or genotypes are ever equally fit. Many alleles do confer differences in fitness; it is through these variations in fitness that natural selection operates. Yet, newer techniques of molecular biology have revealed many differences in DNA sequences that appear to have no discernible effects on fitness.

Theoretical and Experimental Genetic Studies

The field of population genetics uses the Hardy-Weinberg equations as a starting place, to investigate the genetic basis of evolutionary change. These studies have taken two major pathways: theoretical studies, using ever more sophisticated mathematical expressions of the behavior of model genes in model populations, and experimental investigations, in which the pattern of allele and genotype frequencies in real or laboratory populations is compared to the predictions of the mathematical models.

Theoretical population genetics studies have systematically explored the significance of each of the assumptions of the Hardy-Weinberg equilibrium. Mathematical models allow one to work out with precision the behavior of a simple, well-characterized system. In this way, it has been possible to estimate the effects of population size or genetic drift, various patterns of migration, differing mutation rates, inbreeding or other patterns of nonrandom mating, and many different patterns of natural selection on allele or genotype frequencies. As the models become more complex, and more closely approximate reality, the mathematics becomes more and more difficult. This field has been greatly influenced by ideas and tools originally devised for the study of theoretical physics, notably statistical mechanics. Some of the most influential workers in this field were trained as mathematicians and view the field as a branch of applied mathematics, rather than biology. As a consequence, many of the results are not easily understood by the average biologist.

Experimental population genetics tests predictions from theory and uses the results to explain patterns observed in nature. The major advances in this field have been determined, in part, by some critical advances in methodology. In order to study the behavior of genes in populations, one must be able to determine the genotype of each individual. The pattern of bands on the giant chromosomes found in the salivary glands of flies such as *Drosophila* form easily observed markers for groups of genes. Since these animals can be easily manipulated in the laboratory, as well as collected in the field, they have been the subjects of much experimental work. Using population cages, one can artificially control the population size, amount of migration, mating system, and even the selection of genotypes, and then observe how the population responds over many generations. More recently, the techniques of allozyme or isozyme electrophoresis and various methods of examining DNA sequences directly have made it possible to determine the genotype of nearly any organism for a wide variety of different genes. Armed with these tools, scientists can address directly many of the predictions from mathematical models. In any study of the genetics of a population, one of the first questions addressed is whether the population is at Hardy-Weinberg equilibrium. The nature and degree of deviation often offer a clue to the evolutionary forces that may be acting on it.

Understanding Genotypes

As the cornerstone of population genetics, the Hardy-Weinberg principle pervades evolutionary thinking. The advent of techniques to examine genetic variation in natural populations has been responsible for a great resurgence of interest in evolutionary questions. One can now test directly many of the central aspects of evolutionary theory. In some cases, notably the discovery of the large amount of genetic variation in most natural populations, evolutionary biologists have been forced to reassess the significance of natural selection compared with other forces for evolutionary change.

In addition to the great theoretical significance of this mathematical model and its extensions, there are several areas in which it has been of practical use. An area in which a knowledge of population genetics is important is agriculture, in which a relatively small number of individuals are used for breeding. In fact, much of the early interest in the study of population genetics came from the need to understand the effects of inbreeding on agricultural organisms. A related example, and one of increasing concern, is the genetic status of endangered species. Such species have small populations and often exhibit a significant loss of the genetic variation that they need to adapt to a changing environment. Efforts to rescue such species, especially by breeding programs in zoos, are often hampered by an incomplete consideration of the population genetics of small populations. A third example of a practical application of population genetics is in the management of natural resources such as fisheries. Decisions about fishing limits depend on a knowledge of the extent of local populations. Patterns of allele frequencies are often the best indicator of population structure. Population genetics, by combining Mendel's laws with the concepts of population biology, gives an appreciation of the various forces that shape the evolution of the earth's inhabitants.

—Richard Beckwitt

See also: Cloning of extinct or endangered species; Evolution: Animal life; Gene flow; Genetics;

Natural selection; Neutral mutations and evolutionary clocks; Nonrandom mating, genetic drift, and mutation; Population analysis; Population fluctuation; Population genetics; Punctuated equilibrium and continuous evolution; Reproduction; Wildlife management; Zoos.

Bibliography

Audesirk, Gerald, and Teresa Audesirk. *Biology: Life on Earth*. 5th ed. Upper Saddle River, N.J.: 1999. An introductory college textbook designed for nonscience majors. The chapter on the processes and results of evolution includes a complete explanation of basic population genetics, presented in a nontechnical way. The chapter is well illustrated and includes a glossary and suggestions for further reading.

Avers, Charlotte J. *Process and Pattern in Evolution*. New York: Oxford University Press, 1989. A text that introduces modern evolutionary theory to students who already have a background in genetics and organic chemistry. Covers basic population genetics and introduces most of the techniques used in the study of evolution. Includes references to original research, as well as other suggested readings.

Ayala, Francisco J., and John A. Kiger, Jr. *Modern Genetics*. 2d ed. Menlo Park, Calif.: Benjamin/Cummings, 1984. This genetics text assumes an audience that has had college-level biology and some chemistry. It provides a good description of classical as well as molecular genetics. Covers most of the methods and major results of population genetics. Chapters include a bibliography as well as problem sets, and there is a glossary.

Dobzhansky, Theodosius. *Genetics of the Evolutionary Process*. New York: Columbia University Press, 1970. An older book, this text is an introduction to experimental population genetics, by one of the architects of the field. There are numerous references to the original literature and many examples. The book is suitable for anyone with at least some introduction to biology and provides one of the clearest explanations of the major concepts of modern evolutionary thought.

Futuyma, Douglas J. *Evolutionary Biology*. 3d ed. Sunderland, Mass.: Sinauer Associates, 1998. An advanced text in evolution for students with previous exposure to calculus and a strong biology background, including genetics and various courses in physiology and ecology. The great strength of this book is in the presentation of areas of current research and argument in evolution, rather than a cut-and-dried array of "facts." There are numerous references to original research and a glossary.

Hartl, Daniel L. *A Primer of Population Genetics*. 3d ed. Sunderland, Mass.: Sinauer Associates, 2000. This text is intended for students with a college-level knowledge of biology but does not require prior exposure to genetics, statistics, or higher mathematics. There are examples of the significance of population genetics ideas in many areas of biology and medicine, and each chapter has problem sets with answers. There are numerous references to original research.

Nagylaki, Thomas. *Introduction to Theoretical Population Genetics*. New York: Springer-Verlag, 1992. A college-level textbook, using elementary mathematics to investigate theoretical population genetics. Uses calculus and linear algebra, but does not require a background in genetics.

Starr, Cecie, and Ralph Taggart. *Biology: The Unity and Diversity of Life*. 9th ed. Pacific Grove, Calif.: Brooks/Cole, 2001. A textbook for an introductory college biology course. The chapter on population genetics, natural selection, and speciation covers population genetics and mechanisms of evolution. The book is well provided with examples and many striking photographs.

Svirezhev, Yuri M., and Vladimir P. Passekov. *Fundamentals of Mathematical Evolutionary Genetics*. Translated by Alexey A. Voinov and Dmitrii O. Logofet. Boston: Kluwer Academic Publishers, 1990. Offers a clear exposition of the mathematical material in historical perspective. Part 1 covers deterministic models and part 2 uses stochastic models.

HAWKS

Type of animal science: Classification
Fields of study: Anatomy, ornithology

Hawks are a diverse group of birds adapted to exploit a wide variety of habitats, prey, and climatic conditions. All hawks are distinguished by sharp, strongly hooked bills, a fleshy cere, and strong legs with sharp talons. Most are active and efficient hunters that use their keen vision to target and track suitable prey.

Hawks are a diverse group of birds adapted to exploit a wide variety of habitats, prey, and climatic conditions. Although varying in size from the 75-gram male tiny hawk (*Accipiter superciliosus*) and pearl kite to the 6.5-kilogram female harpy eagle (*Harpia* spp.), all hawks are distinguished by sharp, strongly hooked bills, a fleshy cere, and strong legs with sharp talons or claws. Most are active and efficient hunters that use their keen vision to target and track suitable prey.

The 237 species of hawks are placed in the avian order Falconiformes. Families within this order include the booted eagles, harpy eagles, buteo hawks, subbuteos, chanting goshawks, accipiters, harriers, and kites. Only the fifty-four species of accipiters are strictly considered true hawks, but most raptor biologists and hawk enthusiasts also include the buteo hawks, called buzzards in Europe, within the category of hawks, while the most liberal definition incorporates the kites, harriers, and eagles as well.

Principal Terms

CERE: the covering of the nostrils, typical of all hawklike birds of prey.

CREPUSCULAR: active in the twilight periods of dawn and dusk

FLEDGING: time period when young hawks and owls develop the ability to fly

FOVEA: specific area with an exceptionally dense concentration of light-sensitive cells in eyes of animals

FRATRICIDE: deliberate killing of one sibling by another sibling

HABITAT SELECTION: process of choosing a home range, territory, nesting site, or feeding site on the basis of specific features of the habitat that the raptor is best adapted to exploit

HOME RANGE: geographic area used by an individual, pair, or group for their daily, seasonal, and sometimes their yearly activities; the defended portion of the home range is called a territory

TALONS: the long, curved, sharply pointed claws of a bird of prey, used for slashing and killing, holding, and carrying prey, and for defense

TERRITORIALITY: all behaviors involved with the establishment, proclamation, and defense of a specific area

WING LOADING: ratio of weight to lifting area of wing

Habitats and Lifestyles

Hawks are a widespread and successful group that occurs on all continents and the larger islands. They are absent only from Antarctica, the ice-covered mountaintops, and the most remote oceanic islands. Hawks are primarily birds of woodland and woodland-edge habitats, and reach their greatest diversity in the rain forests of the tropics. They are components of all temperate woodlands of varying density and diversity, but some species occur in a wide variety of more open habitats, including chaparral, grasslands, desert, and tundra. Many species take readily to conifer plantations and ornamental conifer stands, provided that sufficient food is nearby. Black kites (*Milvus migrans*), Cooper's (*Accipiter cooperi*), red-tailed (*Buteo jamaicensis*), and Swainson's hawks (*Buteo swainsoni*) exemplify species that tolerate human-modified habitats, especially farmlands, pastures, and orchards. Given their need for large territories and abundant prey, few species have either the tolerance or ability to exploit urban landscapes, although wooded suburbs and city open-space habitats do attract an occasional furtive nesting or roosting pair of red-tailed hawks or brahminy kites (*Hiliastur indus*).

The daily lives of hawks center around finding food and avoiding enemies. To ensure an adequate food base, most hawks maintain large home ranges for at least part of the year, and some remain on territory throughout the year. If prey populations suffice, hawks are sedentary, but if they decline, hawks must move elsewhere. Some species are nomadic, often wandering widely in search of food, but most northern species move southward in spectacular migrations along traditional flyways each fall. Red-tailed hawks migrate only a few hundred miles but others, such as the lesser spotted eagle (*Aquila pomarina*) and Eurasian buzzard (*Buteo buteo*), fly thousands of miles to their wintering grounds. Their return migration in spring is timed to ensure arrival on the breeding grounds when weather conditions and prey populations are optimal.

Most hawks are entirely carnivorous, or nearly so, and use their remarkable eyesight and intimate knowledge of their territory to locate, pursue, and capture food. Their large eyes are set forward in the head with overlapping fields of vision. Each eye has a long focal length from lens to retina, which produces a telescopic vision that enables hawks to spot prey at great distances. Hawk eyes also have two concentrations of visual cells, called fovea, instead of one, as in other birds and mammals. The central fovea and lateral fovea apparently permit the simultaneous estimate of both distance to and movement of prey, thereby increasing hunting efficiency when pursuing agile, fast-moving animals.

Most hawks are larger—some considerably larger—than the prey that they take. Many species exhibit pronounced sexual dimorphism, with the female being larger than the male. Bigger females can better incubate eggs and protect young, but the sexual size difference enables exploitation of a much wider prey range. Sexual dimorphism is best illustrated in the bird-eating accipiter hawks, in which the larger female takes grouse and duck-sized prey while the smaller male pursues the smaller and more agile birds. Conversely, insectivorous and scavenger hawks show little or no sexual dimorphism.

Almost all hawks that hunt live animals have comparatively light bodies on large wings for a low wing loading, which facilitates extended periods of soaring in search of prey and also for carrying prey. Many species show subtle differences in overall size and wing shape that reveal differences in hunting techniques or diet. For example, the long, narrow wings and light bodies of swallow-tailed (*Elanoides forficatus*) and plumbeous (*Ictinia plumbea*) kites enhance their aerial maneuverability for catching insects just above the rain forest canopy. In contrast, the long, broad wings of hawks and eagles maximize lift needed to soar for hours or to carry medium-sized mammals and birds, along with an occasional reptile or amphibian. The larger hawks and eagles take correspondingly bigger prey up to the size of rabbits, hares, hyraxes, sloths, small antelope, foxes, and young deer.

The long tails and rounded wings of bird hawks such as goshawks, sparrow hawks, hawk-eagles (*Spizaetus* spp.) and bat hawks (*Macheirhamphus alcinus*) provide just the right combination of speed and agility to pursue birds through the often cluttered microhabitat just beneath the woodland canopy.

The long, dihedral wings of harriers (*Circus* spp.) provide a stable flying platform for slow quartering flight over open grasslands and marshes. Harriers use both eyes and ears to hunt small mammals in the tall grasses and, like owls, have a distinctive facial ruff that helps gather sounds to pinpoint prey in the tall grasses.

Hunting and dietary specializations abound in this group. At least twelve species of kites are insectivorous or nearly so, but many other hawks and eagles feed opportunistically on abundant locusts, and grasshoppers, ants, termites, and locust swarms invariably attract a variety of hawks and other birds. Reptile specialists include the snake eagles (*Circateus* spp) and serpent eagles (*Spilornatus* spp.), which sit quietly for hours, patiently watching for the slow and stealthy movements of snakes. Possibly the most extreme dietary specialist is the rufous crab-hawk (*Buteogallus aequinoctialis*), which picks through aerial roots of mangroves in search of crabs, but snail kites (*Rostrhamus sociabilis*) and slender-billed kites (*Rostrhamus hamatus*) focus almost exclusively for pulmonate snails, which they extract with their long, hooked bills.

Hawk Facts

Classification:
Kingdom: Animalia
Phylum: Chordata
Subphylum: Vertebrata
Class: Aves
Order: Falconiformes
Families: Accipitridae
Subfamilies: Elaninae (white-tailed kites, five genera, eight species); Milvinae (kites, eight genera, fifteen species); Accipitrinae (hawks, five genera, twenty-two species); Circinae (harriers, three genera, eight species); Pandioninae (ospreys, one genus, seven species)
Geographical location: All continents except Antarctica
Habitat: Woods and woodland edges, rain forests; may also inhabit chaparral, grasslands, deserts, and tundra
Gestational period: One to two months, depending on species
Life span: Eighteen to fifty years, depending on species
Special anatomy: Talons; exceptionally keen eyesight; nostril coverings (cere); sharp, hooked bills

Courtship and Nesting

Most hawks breed each year, but some of the larger tropical eagles and snake eagles may breed once every several years. All phases of the breeding cycle of hawks are dependent on prey abundance: In low prey years pairs often do not nest and may not even establish a territory, while in high prey years nesting success, measured by the number of eggs produced, hatched, and young fledged, is highest. Tropical hawks may nest at any time of year, but usually do so during the dry season. Temperate species are more strictly tied to a spring and early summer nesting cycle when prey abundance is maximal and longer days permit extended hunting periods.

The breeding cycle of temperate species begins when males or mated pairs claim and advertise territories by often extravagant displays that may include sky dancing, dives and swoops, perches, calling, and posturing. Unmated males court females with aerial displays such as circling, following flights, sky dancing, and food transfers from males to females. As courtship proceeds, the pair bond is cemented by synchronized flights, following flights, mutual preening, food begging by the female, and courtship feeding.

Territory size depends on food needs and prey density. Smaller tropical hawks may limit their breeding activities to a hundred acres or so, while the larger eagles that inhabit open habitats, such as the Scottish golden eagles, sometimes establish home ranges of thousands of acres on the moors and grasslands. Most hawk territories are cone-shaped, with little or no overlap permitted near the base of the cone, which is typically centered near the nest site or feeding site.

Hawks build stick nests in trees (for example, Cooper's hawks, sharp-shinned hawks), on cliffs (ferruginous hawks, *Buteo regalis*), bluffs and outcrops (Swainson's hawks, *Buteo swainsoni*), on structures (ospreys, red-tailed hawks), or directly on the ground (harriers). Depending on habitat, some species are flexible in choice of substrate;

Hawks, such as this red-tailed hawk, spend most of their day flying over fields, using their keen eyesight to search for the small animals that are their prey. (PhotoDisc)

thus, red-tailed hawks may build their nests in trees, on ledges, less frequently on structures, or directly on the ground. Many hawks (ferruginous hawks, Swainson's hawks) readily accept artificial platforms such as bridge abutments, piers, telephone poles, and towers.

Both members of a pair bring sticks to the nest site, but the female does the actual work of nest building, arranging the materials into a compact nesting platform with a shallow bowl. Many species refurbish one or more nests year after year, sometimes resulting in massive nesting platforms of sticks and branches after many years of use. Nests are generally lined with dried grasses or mosses and many may decorate them with sprigs of greenery.

Depending on species, from one to ten eggs are laid at two- to four-day intervals, and incubation begins immediately. In most species, the female incubates the eggs while the male supplies her with food, which he may deliver to her or to a nearby site. In some species the male may replace her for brief periods of time to allow her to rest, roost, or forage.

The female is often highly secretive during this nesting phase as she hunkers down to incubate the eggs. The young hatch in about one month in the smaller accipiters and kites, and up to two months in the larger eagles. Eggs hatch asynchronously, producing a nest of uneven-aged young. The female keeps constant company with the newly hatched young, brooding them and feeding them with food supplied by the male. As the young grow, their food demands increase and both male and female spend most of their time hunting food to feed the young.

Adults become increasingly aggressive with young in the nest. Generally, both adults participate in nest defense, although the female is almost invariably the most aggressive. Warning cries (most hawks), power dives (many buteos), and determined slashing attacks (goshawks) can drive all but the most determined intruders away, but unfortunately have little effect on humans intent on destruction of the nest and young. Otherwise, the only natural predators of eggs and young are nest-robbing birds, chiefly corvus species, mammals, other larger diurnal raptors, and the larger owls, especially the *Bubo* species, which take young during the night.

Survival of the young depends entirely on the ability of the adults to find sufficient food. If prey is sufficient, all of the young successfully fledge, but if prey populations decline, one or all of the smallest and weakest young of a nest die of starvation. Nestling fratricide has been observed in many species and may help ensure survival of the strongest young at the expense of their younger or weaker nestlings.

After leaving the nest, the young usually stay with the adults for several weeks or longer—a year or more in some tropical eagles—learning how to perfect their hunting skills and avoid enemies. Many retain a streaked or spotted juvenile plumage for several years that may help conceal them from potential enemies. This plumage is replaced at sexual maturity by adult coloring that probably promotes sexual advertisement and territorial displays.

Conservation and Economic Importance of Hawks

Humans are the most consistent and potent threat to hawks. Historically, huge numbers of hawks were slaughtered by farmers, hunters, gamekeepers, and trappers for sport and as control measures designed to increase game and poultry populations. Many hawks still fall victim to illegal shooting by hunters and sportsman who prefer live target practice, while accipiter hawks are sometimes shot by pigeon fanciers or grouse managers to protect their interests. Generally, the larger buteo hawks and eagles are at greater risk because of their size and habit of selecting open, conspicuous perches for territorial posturing and while hunting. Globally, some hawks are still slaughtered for the taxidermy market, some are taken for falconry, and some are captured as part of the lucrative international trade in zoo specimens, despite the fact that they are all federally protected in most countries of the world.

Hawks also fall victim to pesticides and industrial wastes, which accumulate within their tissues and eventually reach harmful levels that result either in death of the bird or reduction of its productivity. Toxic levels of chemicals have been documented in almost every hawk species on every continent and have particularly affected populations and productivity of fish-eating hawks, such as the osprey and bald eagle.

While all of these forms of deliberate and inadvertent human persecutions take an annual toll, the most persistent threat to hawk populations results from habitat destruction or fragmentation and deforestation. Habitat losses are especially critical for hawks, which require large areas of undisturbed habitat for home range to provide a food base for themselves and their young.

Although no species have become extinct in the last several hundred years, twenty-nine species are on the current list of endangered or threatened species, and it took concerted efforts by state and federal programs on many levels to save several species, such as the bald eagle and osprey. Bans on dichloro-diphenyl-trichloroethane (DDT) and other environmental poisons, along with extensive programs of captive breeding, nesting platforms, and reintroduction programs have helped reestablish bald eagles (*Haliaetus leucocephalus*) in North America and red kites and white-tailed sea eagles (*Haliaetus albicilla*) in Scotland. Conservation programs continue to target the protection and recovery of many species throughout the world, including the harpy eagle in Latin America and the Philippine eagle (*Pithecophaga jefferyi*) of the Philippine Islands.

—*Dwight G. Smith*

See also: Beaks and bills; Birds; Eagles; Feathers; Flight; Molting and shedding; Nesting; Respiration in birds; Wildlife management; Wings.

Bibliography

Brown, Leslie, and Dean Amadon. *Eagles, Hawks, and Falcons of the World*. 2 vols. New York: McGraw Hill, 1968. Somewhat dated now, but still considered by many to be the defining work on diurnal birds of prey by two of the acknowledged world experts. Each species is exhaustively treated with range maps. Includes great paintings of each of the world's diurnal raptors.

Del Hoyo, Josep, Andrew Elliott, and Jordi Sargatal, eds. *New World Vultures to Guineafowl*. Vol. 2 in *Handbook of the Birds of the World*. Barcelona, Spain: Lynx Edicions, 1994. Twenty authors contributed to this massive, oversized volume that summarizes information about all of the diurnal birds of prey of the world. Each group—falcons, accipiters, eagles, and so forth—is preceded by a detailed summary that includes information on habits and habitat, relationships to humans, conservation status, and other timely information.

Johnsgard, Paul A. *Hawks, Eagles, and Falcons of North America*. Washington, D.C.: Smithsonian Institution Press, 1990. Interesting and entertaining descriptions of all of the diurnal birds of prey of North America. Drawing extensively from the scientific literature, Johnsgard summarizes all that is known about the distribution, subspecies, description, identification, habitats and ecology, diet, social behavior, breeding biology, evolutionary relationships, and status of each species. Also includes several introductory chapters tracing evolution and comparing foods and feeding, behavior and reproductive biology, and ecology. Appendices contribute a key, origins of names, glossary, twelve pages of drawings detailing anatomy and field identification, and an extensive list of references.

Newton, Ian. *Population Ecology of Raptors*. Vermillion, S.Dak.: Buteo, 1979. Good analysis of factors affecting raptor populations. Of the many chapters in this book, the sections on population biology and behavior are particularly good. Excellent presentation of appropriate figures, charts, and tables complement the major points.

Palmer, Ralph S., ed. *Handbook of North American Birds*. 5 vols. New Haven, Conn.: Yale University Press, 1962- . The ecology and life history of each North American species of diurnal raptor is succinctly summarized in volumes 4 and 5. Includes a very complete review of the literature.

HEARING

Types of animal science: Anatomy, physiology
Fields of study: Anatomy, biophysics, neurobiology, physiology

One of the five major senses, hearing is utilized by animals to detect sound in the environment through the transduction of sound waves to the universal language of the nervous system, the nerve impulse.

Principal Terms

AUDITORY NERVE: the cranial nerve that conducts sensory impulses from the inner ear to the brain

PITCH: the frequency of sound—the higher the frequency, the greater its pitch

SOUND FREQUENCY: the distances between crests of sound waves measured in hertz

SOUND INTENSITY: the loudness of a sound directly related to the amplitude of the sound waves measured in decibels

SYNAPSE: the functional connection between nerve cells or an effector cell, such as a sensory receptor and a nerve cell

TETRAPODS: vertebrates with four limbs

Sensory perception provides the only means of communication between the external world and the nervous system. The process of sensory perception begins in the sense organs, where specially designed receptor cells are stimulated by various types of energy. These receptor cells are highly selective for specific forms of stimulus energy. For example, the photoreceptors in the eye are specific for light energy and largely ignore other forms of stimuli. Each receptor cell transduces (changes) the stimulus into an electrical charge (nerve impulse) which travels through nerve fibers to the brain, where the electrical impulse is translated into a particular sensation. The major types of sensory receptor cells are the chemoreceptors (sense chemical energy), mechanoreceptors (sense mechanical energy), photoreceptors (sense light energy), thermoreceptors (sense thermal energy), and electroreceptors (sense electrical energy). The sensory receptors for the organs of hearing are called hair cells, which are extraordinarily sensitive mechanoreceptors. Tiny filaments (like hair follicles, only much smaller) called cilia project from the ends of the receptor cell. This filament bends in response to mechanical pressure, and the bending generates the nerve impulse.

Through the course of evolution, the sensory systems developed from single, independent receptor units into complex sense organs, such as the vertebrate ear, in which receptor cells are organized into a tissue associated with accessory structures. The organization of the receptor cells and the architecture of the accessory structures allow far more intricate and accurate sampling of the environment than is possible by independent, isolated receptor cells. While invertebrates possess receptor cells that sense vibrational (mechanical) energy, the true sense of hearing originated with the vertebrates.

The Vertebrate Ear

All vertebrates possess a pair of membranous labyrinths (cavities lined by a membrane) embedded in the cranium, lateral to the hindbrain. This region is often referred to as the inner ear. Each labyrinth consists of three semicircular canals, a utriculus, and a sacculus, which, during development, become filled with a fluid called endolymph. The semicircular canals and utriculus are primarily associated with equilibrium, but the sacculus has evolved into an organ of hearing. In fishes, the lagena, a depression in the floor of the sacculus, has its own maculae (patches of hair cells) which respond to vibratory stimuli of relatively high frequency (sound waves). In tetrapods, the lagena has evolved into an additional fluid-filled duct of the labyrinth called the spiral duct or cochlea. With the evolution of the cochlea, the sensory region enlarged into the organ of Corti. The hair cells are located in the organ of Corti. Most of the structures of the ear assist in the transformation of sound waves (airborne vibrations) into movements of the organ of Corti, which stimulate the hair cells. These hair cells then excite the sensory neurons (nerve cells) of the auditory nerve.

Mammals are the only vertebrates to possess a true cochlea, but birds and crocodilians have a nearly straight cochlear duct that contains some of the same features, including an organ of Corti. Detection of sound in the lower vertebrates that have no cochlear ducts is carried out by hair cells associated with the utriculus and lagena. The cochlea is coiled somewhat like the shell of a snail and is divided into three longitudinal compartments. The two outer compartments, the scala tympani and the scala vestibuli, are filled with a fluid called perilymph and are connected to one another by a structure called the helicotrema. The scala media, filled with endolymph, is located between the two outer compartments and is bound by the basilar membrane and Reissner's membrane. The organ of Corti lies within the scala media and sits upon the basilar membrane. Four rows of hair cells are present in adult mammals—one inner and three outer rows. The cilia of the inner row are thought to be sensitive primarily to the velocity (speed) at which they are displaced by sound waves. The cilia of the outer rows are more sensitive and can detect the degree of deflection as well as the speed.

Sound waves are transported to the inner ear via the outer ear and middle ear. The outer ear consists of the tympanic membrane (ear drum), which is situated on the surface of the head in frogs and toads. In reptiles, birds, and mammals, the tympanic membrane is located deeper in the head at the dead end of an air-filled passageway called the outer ear canal or external auditory

Comparison of Hearing Organs

Jawless fishes: These primitive vertebrates have only two semicircular canals within the inner ear. In jawless fishes, the structures of the inner ear are more associated with equilibrium than hearing.

Gnathostome fishes: All fishes possess membranous labyrinths embedded in the auditory region of the skull. The labyrinths consist of the utriculus and sacculus and three semicircular canals. The inner ear serves primarily as an organ of equilibrium in fishes, but a few species have developed an accessory function of detecting sound transmissions.

Amphibians: The hearing organ of amphibians is considerably advanced over that of the fishes. In frogs and toads, the outer ear consists of a tympanic membrane in the adults that lies flat on the head. The middle ear is an air-filled chamber with a small movable bone. The inner ear consists of a utriculus, a sacculus, a lagena, and three semicircular canals filled with endolymph and perilymph. The hair cells are located at the base of each canal.

Reptiles and birds: Hearing is well developed in reptiles and especially birds. There is a tympanic membrane, a well-developed middle ear, and a straight-tubed cochlea and spiral organ. Some birds can hear ultralow frequencies that may be important in navigating.

Mammals: The membranous labyrinth and the sense of hearing are highly developed in mammals. Mammal ears consist of all of the components discussed in the text. Mammals are keenly adapted to hearing. Humans can hear in a range of sixteen to twenty thousand hertz, but bats can detect frequencies up to one hundred thousand hertz.

meatus. In mammals, there is also an outer appendage, the pinna, which collects sound waves and directs them into the outer ear canal. The tympanic membrane makes contact with the bones (ossicles) of the air-filled inner ear. In amphibians, birds, and reptiles, there is a single bone called the columella (or stapes). In mammals, there is a series of three bones. The malleus (hammer) is in contact with the tympanic membrane at one end and articulates with a second bone, the incus (anvil). The incus then articulates with a third bone, the stapes (stirrup), which connects to a structure called the oval window of the cochlea.

Mammals have a mobile outer ear appendage, called a pinna, which catches and funnels sound waves into the interior ear to increase hearing acuity. (Digital Stock)

Detection of Sound

Sound waves striking the tympanic membrane cause it to vibrate. These vibrations are transmitted through the auditory ossicles of the middle ear and through the oval window to the perilymph. The bones of the middle ear amplify the pressure of the vibrations set up in the eardrum by airborne vibrations. Vibrations reaching the oval window pass through the cochlear fluids and the Reisner's and basilar membranes separating the cochlear compartments before dissipating their energy through the membrane-covered round window of the cochlea. The distribution of pertubations (disturbances) within the cochlea depends on the frequencies of the vibrations entering the oval window. Very long, low frequencies travel through the perilymph of the scala vestibuli, across the helicotrema to the scala tympani, and finally toward the round window. Short wave frequencies take a shortcut from the scala vestibuli through the Reissner's membrane and the basilar membrane to the perilymph of the scala tympani. Movement of perilymph from the scala vestibuli to the scala tympani produces displacement of both Reissner's membrane and the basilar membrane. Movement of Reissner's membrane does not directly contribute to hearing, but displacement of the basilar membrane is required for pitch discrimination. Displacement of the basilar membrane into the scala tympani produces vibrations of the basilar membrane. Each region of the basilar membrane vibrates with maximum amplitude to a different sound frequency. Sounds of higher frequency (pitch) cause maximum vibrations of the basilar membrane at the apical region (closest to the stapes), while sounds of low frequency produce maximum vibrations at the distal region of the basilar membrane.

The sensory hair cells are situated on the basilar membrane with the cilia projecting into the endolymph of the cochlear duct. The cilia of the outer hair cells are embedded with the tectorial membrane located above the hair cells within the cochlear duct. Displacement of the cochlear duct by pressure waves of perilymph produces a shearing force between the basilar membrane and the tectorial membrane. This causes the cilia to bend, and the bending of the cilia produces a nerve impulse in the sensory nerve endings that synapse with the hair cells. The higher the intensity of the sound, the greater the displacement of the basilar membrane, which results in greater bending of the cilia of the hair cells. Increased bending of the cilia produces a higher frequency of nerve impulses in the fibers of the cochlear nerve that synapse with hair cells. Since a specific region of the basilar membrane is maximally displaced by a sound of a particular frequency, those nerve cells that originate in this region will be stimulated more than nerve cells which originate in other regions of the basilar membrane. This mechanism results in a neural code for pitch discrimination. Within the brain, sensory neurons of the eighth cranial (auditory) nerve synapse with neurons in the medulla which project to the inferior colliculus of the brain. Neurons from this region of the brain project into the thalamus, which in turn sends nerve fibers to the auditory cortex of the temporal lobe of the brain. Through this pathway, neurons in different regions of the basilar membrane stimulate neurons in corresponding areas of the auditory cortex. Hence, each area of this cortex represents a different part of the basilar membrane and a different pitch.

—D. R. Gossett

See also: Antennae; Brain; Communication; Ears; Eyes; Nervous systems of vertebrates; Noses; Physiology; Sense organs; Smell; Vision; Vocalizations.

Bibliography

Campbell, Neil A., Lawrence G. Mitchell, and Jane B. Reece. *Biology: Concepts and Connections*. 3d ed. San Francisco: Benjamin/Cummings, 2000. An outstanding introductory biology text that gives a clear concise description of the sense of hearing.

Feldhamer, G. A., L. C. Drickamer, S. H. Vessey, and J. F. Merritt, eds. *Mammalogy: Adaptation, Diversity, and Ecology.* Boston: WCB/McGraw-Hill, 1999. This text gives a good discussion of hearing among the different groups of mammals.

Fox, Stuart Ira. *Human Physiology.* 6th ed. Boston: WCB/McGraw-Hill, 1999. An excellent treatment of the physiology of hearing in humans, but applicable to other mammals.

Linzey, Donald W. *Vertebrate Biology.* Boston: McGraw-Hill, 2001. A very good text for the comparative description of hearing in the different classes of vertebrates.

Randall, David J., Warren Burggren, Kathleen French, and Russell Fernald. *Eckert Animal Physiology: Mechanisms and Adaptions*. 4th ed. New York: W. H. Freeman, 1997. An advanced text that gives an excellent description of the hearing process in animals.

HEART

Types of animal science: Anatomy, physiology
Fields of study: Anatomy, biochemistry, cell biology, genetics, histology, pathology, physiology, zoology

The heart is a muscular pump that moves blood throughout the body to deliver oxygen and nutrients to the tissues, remove metabolic waste products, and warm the animal by distributing metabolic heat.

Principal Terms

ATRIUM (AURICLE): a thin-walled muscular heart chamber that receives blood returning to the heart

CLOSED CIRCULATION: a circulation system made of arteries, capillaries, and veins that returns blood flow back to the heart

DIASTOLE: period of heart muscle relaxation and declining pressure

HEART: general term for a muscular vessel segment or organ that contracts and provides for unidirectional movement of blood or hemolymph

OPEN CIRCULATION: an open-ended sinus or arterial vessel system in which the circulation system does not return blood or hemolymph directly back to the heart

SYSTOLE: period of heart muscle contraction and peak pressure

VENTRICLE: a thick-walled muscular heart chamber that receives blood from the atrium and pumps blood into the circulation

The heart beats with a lifelong, continual rhythm to supply a constant flow of oxygen and nutrients, as well as to remove metabolic waste products from every cell in the body. The interstitial fluid that surrounds each cell keeps a constant supply of these factors and prevents a buildup of waste products by communicating with the blood. However, the blood can only keep the interstitial fluid in balance by remaining in motion. If the blood flow stops, the oxygen and nutrients are quickly depleted and harmful waste products accumulate. In adult humans, the heart beats about seventy times per minute, or one hundred thousand times per day. Each heartbeat moves about seventy milliliters of blood, which equals over seven thousand liters of blood per day, or enough blood to fill 7,400 quart-sized milk cartons.

Heart Structure

Hearts vary in their complexity, from simple pulsing vessel segments in insects to the complex four-chambered heart in birds and mammals. Hearts typically have a rhythmic contraction rate determined by specialized muscle cells and can even beat outside the body. A strong, saclike structure called the pericardium encloses, supports, and aids the heart in refilling during diastole by creating negative pressures in the atria and ventricles. Fluid is secreted into the space between the outside wall of the heart and the pericardium. This fluid lubricates and reduces friction during contraction and relaxation. When the heart contracts and ventricular pressure develops, pressure valves made from strong, thin, fibrous tissue are arranged to open and close in such a manner that blood or hemolymph flows in one direction out of the heart (unidirectional flow). When the heart relaxes and ventricular pressure decreases, the valves close to prevent flow reversal from the arteries, and other valves open to allow flow into the heart from the tissues.

Heart chambers fill with blood during diastole and pump blood during systole. One complete heart cycle involves one phase of relaxation and one phase of contraction. The highest pressure developed in the ventricle during contraction is called the systolic pressure, and the lowest pressure in the arteries just before the next contraction is called the diastolic pressure.

The fluid pumped by the heart can be divided into two categories: blood and hemolymph. Blood has two major components: the liquid plasma and cells. The red cells contain hemoglobin for oxygen transport, while the white cells defend against invading germs and viruses. Hemolymph, on the other hand, lacks red blood cells and has hemoglobin freely dissolved in the circulating fluid. In addition, hemolymph flows directly around each cell, whereas blood exchanges nutrients and oxygen across the capillary wall to the interstitial fluid that surrounds cells.

Heart Rate

A heart will rhythmically contract even when removed from the animal. This property is called automaticity and results from a specialized collection of cells in the upper part of the right atrium called the sinoatrial node. The sinoatrial nodal cells spontaneously generate an electrical impulse or action potential that travels throughout the heart by specialized heart muscle cells called conduction fibers. The conduction fibers provide for a uniform and effective heart contraction. Intercalated disks connect each heart muscle cell together and conduct the electrical impulse. Intercalated disks assure all of the heart muscle cells will contract together and effectively pump the blood.

Heart rate varies by animal size. Elephants have a slow, thirty-five beats per minute rate, while the smallest mammal, the shrew, has a heart rate of over six hundred beats per minute. The tiny hummingbird's heart beats at over 1,200 beats per minute while in flight.

The rate at which the sinoatrial node generates impulses can be modified by the autonomic nervous system. The sympathetic branch of the autonomic nervous system increases the heart rate, for example, during exercise or when frightened, while the parasympathetic branch slows the heart down, for example, during sleep.

Different Beats for Different Beasts

One of the simplest hearts is a rhythmically contracting muscular section in the dorsal vessel of the annelid worm, grasshopper and other arthropods. This "heart" contracts and pumps hemolymph into open-ended sinuses that distribute the hemolymph throughout the body. The hemolymph fluid does not directly return to the heart, but is drawn in through ostia or pores in the heart during the relaxation phase (open circulation).

Mollusks and octopuses have a greater diversity of heart structure, with many having two chambers. The thin-walled chamber that collects blood from venous vessels is called an auricle or atrium. The thick-walled muscular chamber that receives blood from the atrium and propels blood to the body is called a ventricle. The open circulation system in these animals delivers blood flow through several main arteries to important regions such as the kidney, head, and digestive tract.

The two-chambered fish heart has one atrium and one ventricle and represents a slight increase in complexity compared to mollusks and crustaceans. The sinus venosus collects blood at the convergence point of great veins and empties into the atrium. The bulbus arteriosus in teleosts (bony fishes), or the conus arteriosus in elasmobranchs (cartilaginous fishes), stores pressure energy from the ventricular ejection and helps to maintain a steady blood flow through the circulation during diastole.

The fish atrium receives deoxygenated blood and the ventricle pumps blood to gills, where the blood releases carbon dioxide and picks up oxygen from the water. The oxygenated blood then leaves the gills and circulates through the rest of the body. A significant amount of pressure is lost as the blood travels through the small capillaries in the gills. So, movements of the fish's body are important to help move the blood through the body and back to the heart.

Fish have a single path of circulation, from the

ventricle through the gills and then to other body organs and tissues. In contrast, air-breathing animals have two separate circulations, one through the lungs (pulmonary circulation) and another to the rest of the body (systemic circulation). Lungfish (dipnoan lungfish) have the first indications of two separate circulations where oxygenated and deoxygenated blood are separated by a ridge of tissue in the ventricle. This diverts deoxygenated blood to the lungs and oxygenated blood to the rest of the body or systemic circulation.

Amphibians, Reptiles, Birds, and Mammals

Amphibians have an increased distinction between systemic and pulmonary circulations, with two separate atria. The right atrium receives deoxygenated blood from the systemic circulation, and the left atrium receives oxygenated blood from the lung or pulmonary circulation. The ventricle is partially divided into a right and left side by a muscular wall called a septum. The septum keeps the oxygenated blood coming in from the lungs separate from the deoxygenated blood returning from systemic circulation. Differential flow timings also help to separate deoxygenated and oxygenated blood. The ventricle delivers blood into a divided artery, with the pulmonary circuit leading to the lungs and skin (frogs receive some oxygen from the skin circulation on their backs), and the systemic circuit leading to all organs except the lungs. The amphibian heart is relatively weak, with spongy, poorly muscularized ventricles compared to mammals.

Most reptiles also have two separate atria, similar to amphibians; however, the sinus venosus is no longer a distinct structure and is incorporated into the right atrium. In contrast, adult crocodiles have a completely divided ventricle, thus having

Heart Valves

The heart valves provide for unidirectional blood flow and separate the atria from the ventricles and the heart from the arterial blood vessels. The valves are made of thin, strong, fibrous tissue, and each flap is called a leaflet or cusp. The valve cusps close by pressure differences across their opening. The valves that separate the atria from the ventricles are called the atrioventricular valves. In the four-chambered mammalian heart, the right atrioventricular valve is called the tricuspid valve, and the left atrioventricular valve is called the mitral or bicuspid valve. As indicated by their names, the tricuspid valve has three cusps, while the bicuspid or mitral valve has two. Another characteristic of the atrioventricular valves is their support structures, the chordae tendinae and the papillary muscles. The chordae tendinae attach to the valve cusps and, during the high pressures created by ventricular contraction, prevent the valve cusps from being forced open into the atrium. The papillary muscles are specialized heart muscles that arise from the inside of the heart and attach to the chordae tendinae. These muscles contract with each heartbeat and pull down the leaflets to keep the valve closed, and further prevent their opening into the atria.

The valves that separate the heart from the pulmonary artery and the aorta are called the semilunar valves. These valves prevent blood that has been ejected into the arteries from flowing backward into the ventricles during diastole, when the ventricular muscle is relaxed. The semilunar valves, in contrast to the atrioventricular valves, lack additional supportive structures such as the chordae tendinae and papillary muscles. Instead, their three cusps meet together and form a tight seal between the artery and ventricle.

The rhythmic "lub-dub" sounds associated with the heartbeat result from the vibration of the valve cusps and rapid blood flow through the valve just as it narrows and then closes. The first heart sound, or the "lub," is associated with the closing of the atrioventricular valves. The second heart sound, or the "dub," is associated with the closing of the semilunar valves. The term "heart murmur" refers to the hissing sound of blood squirting through a defective valve.

completely separate pulmonary and systemic circulations, similar to birds and mammals.

Birds and mammals are characterized by two completely separate atria and ventricles. The right atrium receives deoxygenated venous blood from the systemic veins, and the right ventricle pumps blood through the pulmonary circulation to pick up oxygen. The left atrium receives the oxygenated blood from the pulmonary veins, and the left ventricle pumps this blood throughout the systemic circulation (all organs except the lungs).

With the complete separation of the right and left ventricles, the pressures between the two can be different. The ventricles are highly muscularized, with the left ventricle being approximately six times thicker than the right ventricle, and this reflects the greater pressures needed to circulate blood through the systemic and pulmonary circulations. However, the volume of blood pumped by each ventricle needs to be the same, because each ventricle returns blood to the other. For example, the right ventricle pumps blood through the pulmonary circulation that returns to the left ventricle. If the right ventricle pumps a larger volume of blood than the left, blood will accumulate in the pulmonary circulation. In humans, a heart attack involving the left ventricle weakens the muscle, which decreases the volume of blood pumped. The right ventricle pumps normally and blood accumulates in the pulmonary circulation, causing breathing problems.

During flight, birds can have an increase in oxygen demand of ten- to fifteen-fold to power their flight muscles. Birds have very powerful hearts that can deliver five to seven times more blood to active muscles compared to similarly sized mammals.

—Robert C. Tyler

See also: Anatomy; Beaks and bills; Bone and cartilage; Brain; Circulatory systems of invertebrates; Circulatory systems of vertebrates; Claws, nails, and hooves; Digestive tract; Ears; Endoskeletons; Exoskeletons; Eyes; Fins and flippers; Immune system; Kidneys and other excretory structures; Lungs, gills, and tracheas; Muscles in invertebrates; Muscles in vertebrates; Nervous systems of vertebrates; Noses; Physiology; Reproductive system of female mammals; Reproductive system of male mammals; Respiratory system; Sense organs; Skin; Tails; Teeth, fangs, and tusks; Tentacles; Wings.

Bibliography

Pough, F., H. Heiser, J. B. Heiser, and W. N. McFarland. *Vertebrate Life*. 5th ed. Upper Saddle River, N.J.: Prentice Hall, 1999. Emphasis on the differences in animals based on the sequences of evolution.

Schmidt-Nielson, Knut. *Animal Physiology: Adaptation and Environment*. 5th ed. New York: Cambridge University Press, 1997. This classic textbook in animal physiology is a standard for many high school and college courses and focuses on physiologic function and adaptation to different environmental conditions.

Weibel, Ewald R. *The Pathway for Oxygen*. Cambridge, Mass.: Harvard University Press, 1984. Ewald Weibel's research presented in the book established many of the concepts and facts currently known about the heart and circulatory system structure and function.

Willmer, Pat, Graham Stone, and Ian Johnston. *Environmental Physiology of Animals*. Malden, Mass.: Blackwell Science, 2000. This textbook has simple-to-understand diagrams and explanations of animal organ systems and their adaptation to different environments.

Withers, P. C. *Comparative Animal Physiology*. Fort Worth, Tex.: W. B. Saunders, 1992. Gives a broad overview of animal physiology comparing the characteristics of each species.

HERBIVORES

Types of animal science: Anatomy, classification, ecology, reproduction
Fields of study: Anatomy, ecology, zoology

Herbivores, animals which eat only plants, include insects and other arthropods, fish, birds, and mammals. They keep plants from overgrowing and are food for carnivores or omnivores.

Principal Terms

CARNIVORE: any animal that eats only the flesh of other animals

GESTATION: the term of pregnancy

METAMORPHOSIS: insect development into adults, passing through two or more dissimilar growth forms

RUMINANT: a herbivore that chews and swallows plants, which enter its stomach for partial digestion, are regurgitated, chewed again, and reenter the stomach for more digestion

Herbivores are animals whose diets consist entirely of plants. They have two ecological functions. First, they eat plants and keep them from overgrowing. Second, they are food for carnivores, which subsist almost entirely upon their flesh, and omnivores, which eat both plants and animals. Herbivores live on land or in oceans, lakes, and rivers. They can be insects, other arthropods, fish, birds, or mammals.

Wild Herbivores

Insects are the largest animal class, with approximately one million species. Fossils show their emergence 400 million years ago. Insects occur worldwide, from pole to pole, on land and in fresh or salt water. They are the best developed invertebrates, except for some mollusks. They mature by metamorphosis, passing through at least two dissimilar stages before adulthood. Metamorphosis can take up to twenty years or may be complete a week after an egg is laid.

Many insects are herbivores. Some feed on many different plants; others depend on one plant variety or a specific plant portion, such as leaves or stems. Relationships between insects and the plants they eat are frequently necessary for plant growth and reproduction. Among the insect herbivores are grasshoppers and social insects such as bees.

Artiodactyls are hoofed mammals, including cattle, pigs, goats, giraffes, deer, antelope, and hippopotamuses. Most are native to Africa, but many also live in the Americas, Europe, and Asia. Artiodactyls walk on two toes. Their ancestors had five, but evolution removed the first toe and the second and fifth toes are vestigial. Each support toe—the third and fourth—ends in a hoof. The hippopotamus, unique among artiodactyls, stands on four toes of equal size and width.

Artiodactyls are herbivores, lacking upper incisor and canine teeth, but pads in upper jaws help the lower teeth grind food. Many are ruminants, such as antelope, cattle, deer, goats, and giraffes. They chew and swallow vegetation, which enters the stomach for partial digestion, is regurgitated, chewed again, and reenters the stomach for more digestion. This maximizes nutrient intake from food.

Deer are hoofed ruminants whose males have solid, bony, branching antlers that are shed and regrown yearly. The deer family, approximately 40 species, occurs in Asia, Europe, the Americas, and North Africa. Deer live in woods, prairies, swamps, mountains, and tundra. Their size ranges from the seven-foot-tall moose to the one-

Buffalo, like all herbivores, spend much of their time grazing. The low nutrient content of grass requires most herbivores to opt for quantity over quality in their digestive strategy. (PhotoDisc)

foot-tall pudu. Deer first appear in the fossil record ten million years ago. Deer eat the twigs, leaves, bark, and buds of bushes and saplings, and grasses. Females have one or two offspring after ten-month pregnancies. Common species are the white-tailed and mule deer in the United States; wapiti in the United States, Canada, Europe, and Asia; moose in North America and Europe; and reindeer in Russia, Finland, and Alaska.

Antelope, a group of approximately 150 ruminant species, have permanent, hollow horns in both sexes. Most are African, although some are European or Asian. They eat grass, twigs, buds, leaves, and bark. There are no true antelope in the United States, where their closest relatives are pronghorns and Rocky Mountain goats (goat-antelope with both goat and antelope anatomic features). The smallest antelope, the dik-dik, is rabbit-sized. Elands, the largest antelope, are ox-sized. Unlike deer horns, antelope horns are unbranched. Most antelope run rather than fighting, and are all swift. Antelope live on plains, marshes, deserts, and forests. Females birth one or two offspring per pregnancy. Impala and gazelles, such as the springbok, are found in Africa. In Asia, Siberian saigas and goat antelope (takin) inhabit mountain ranges. Chamois goat antelope live in Europe's Alps.

Giraffes and hippos are unusual artiodactyls. Giraffes inhabit dry, tree-scattered land south of the Sahara. Their unusual features are their very long legs and necks. Males are over sixteen feet tall, including the neck. Both sexes have short, skin-covered horns. Long necks, flexible tongues, and upper lips pull leaves—their main food—from trees. Giraffes have brown blotches on buff coats and blend with tree shadows. They live for up to twenty years. They have keen senses of smell, hearing, and sight, and can run thirty-five miles per hour. Due to their two-ton weights, they live on hard ground. Giraffes rarely graze, and go for months without drinking, getting most of their water from the leaves they eat, because it is difficult for them to reach the ground or the surface of a river with their mouths. Females have one offspring after a fifteen-month gestation.

The unusual feature of hippos is that they walk on all four toes of each foot. Perhaps this is because they weigh three to four tons. Hippos are short-legged, with large heads, small eyes, small ears, and nostrils that close underwater. Huge hippo mouths hold long, sharp incisors and canines in both jaws. Hippos once lived throughout Africa. Now they are rarer, due to poaching for ivory. A hippo can be fifteen feet long and five feet high at the shoulder. Semiaquatic, hippos spend most daylight hours nearly submerged, eating aquatic plants. At night they eat land plants. Females bear one offspring at a time.

Elephants

Herbivorous elephants, the largest land animals, eat three hundred pounds per day of grass, fruits, leaves, flowers, roots, twigs, branches, and bark. Elephants are African or Asian. African males reach twenty-five feet long, eleven feet tall, and weigh seven tons. Females are shorter and half the weight of males. Asian elephants exhibit anatomic differences, including smaller ears, tusks, and overall size.

Elephants lift trees with their trunks and have long been domesticated to carry things. Elephants also grab food from the ground and trees via the ends of their trunks, with nostrils and grasping lips. They also suck water into their trunks and squirt it into their mouths or on their bodies. The showers keep them cool. Their ivory tusks, several feet long, are used to dig roots, peel edible tree bark, and fight enemies and male competitors. The use of tusk ivory in jewelry leads poachers to kill males. Elephant ears have keen hearing and act as cooling fans.

Adult males live alone. Females and young form herds, led by a sagacious female. Herd members breed with visiting males, protect each other, and raise young. Gestation, two years, produces one offspring, which nurses for three years. Elephants can live for sixty years in the wild and eighty years in captivity, but only 10 percent of wild elephants reach age forty.

Aquatic Herbivores

Fish are aquatic vertebrates, having gills, scales and fins. They include rays, lampreys, sharks, lungfish, and bony fishes. The earliest vertebrates, 500 million years ago, were fishes. They comprise over 50 percent of all vertebrates and have several propulsive fins: dorsal fins along the central back; caudal fins at tail ends; and paired pectoral and pelvic fins on sides and belly. Fish inhabit lakes, oceans, and rivers, even in Arctic and Antarctic areas. Most marine fish are tropical. The greatest diversity of freshwater species is found in African and rain-forest streams.

Fish vary in length from half an inch to fifty feet, and some weigh seventy-five tons. Many, including giant whale sharks, are herbivores. Fish respiration uses gills, through which blood circulates. When water is taken in and expelled, oxygen enters the blood via the gills and carbon dioxide leaves. Fishes reproduce by laying eggs that are fertilized outside the body, or by internal fertilization and development with the birth of well-developed young.

Domesticated Artiodactyls

Bovids are domesticated artiodactyls. Most have horns and hooves. Bovid horns are spiraled, straight, tall, or L-shaped from the sides of the head. All have hooves to help them grip the ground. Most are ruminants. Their breeding habits are similar. Males fight over females and the strongest wins. Gestation, four to eleven months, yields two to three young. The young nurse for several months and then join the herd. Young males leave female herds to live with other bachelors.

Cattle are domesticated bovids, raised for meat, milk, and leather. Modern cattle come from European, African, and Asian imports. Breeding modern cattle began in Europe in the mid-1800's.

Today, three hundred breeds exist. Dairy cattle, such as Holsteins, make copious milk. Beef cattle, such as Angus, were bred to yield meat. As of 1990, about 1.3 billion cattle were found worldwide.

Sheep are also artiodactyls. Wild sheep occur in some places, such as the North American bighorn and Mediterranean mouflon. Sheep were domesticated eleven thousand years ago from mouflon. Today, domesticated sheep have a world population of approximately 1.3 billion and inhabit most countries, being more widely distributed than any other domesticated animal. These ruminants have paired, hollow, permanent horns. Male horns are massive spirals; those of females are smaller. Adult body length is five feet and weights are 250 to 450 pounds. Females birth two or three young after five-month gestations. Sheep can live for up to twenty years.

Sheep provide wool, meat, and milk. About eight hundred domesticated breeds exist, in environments from deserts to the tropics. Those bred for wool, half the world's sheep, live in semiarid areas, are medium sized, and produce fine wool. Most are in Australia, New Zealand, and South America. Mutton-type sheep, 15 percent of the world sheep population, produce meat. Fat-tailed sheep, 25 percent of the sheep population, produce milk. In 1990, the five leading sheep countries were Australia, China, New Zealand, India, and Turkey. The United States raised less than 1 percent of the world total.

Goats are ruminants, closely related to sheep, but have shorter tails, different horn shape, and bearded males. They eat grass, branches, and leaves and breed from October to December. A five-month gestation yields two offspring. Numerous goat breeds are domesticated worldwide for meat and milk, and as pets and burden carriers. Domesticated Angora goats yield silky mohair. Goat milk is as nutritious as cow milk and used in cheese-making.

It is clear that wild herbivores are ecologically important to food chains. This is because they eat plants, preventing their overgrowth, and they are eaten by carnivores and omnivores. Domesticated herbivores—cattle, sheep and goats, used for human sustenance—account for three to four billion living creatures. Future production of better strains of domesticated herbivores via recombinant deoxyribonucleic acid (DNA) research may cut the numbers of such animals killed to meet human needs. Appropriate species conservation should maintain the present balance of nature and sustain the number of wild herbivore species living on earth.

—Sanford S. Singer

Horses, Donkeys, and Zebras

The horse, donkey, and zebra (HDZ) family lives in habitats from grasslands to deserts. Wild specimens inhabit East Africa and the Near East to Mongolia. Domesticated horses and donkeys are used for food, meat, leather, drayage, and racing worldwide. Zebras are too savage to domesticate. All HDZ have long heads and necks, slender legs, and manes on their necks, as well as long tails to swat insects. They have good wide-angle day and night vision. Their sense of smell is also excellent. The smallest family members are African wild donkeys, 4.0 feet tall, 6.5 feet long, with 1.5-foot tails, weighing five hundred pounds. The largest are Grevy's zebras, 5 feet tall at the shoulder, 9 feet long, with 1.5-foot tails, and weights up to nine hundred pounds.

Zebras have black or brown and white vertically striped coats. All HDZ family members are herbivores. Donkeys eat bark, leaves, buds, fruits, and roots. Horses and zebras mostly eat grasses, and some bark, leaves, buds, and fruits. Wild HDZ spend their waking hours foraging for and eating food. When attacked, they bite, kick, or run off at speeds up to forty miles per hour.

Wild donkeys and horses live in herds with a male leader and female mates. Young stay in a herd until two years old for females or four years old for males. Males join bachelor groups until winning a herd. Females join another herd. Donkeys usually live alone. Most HDZ mate in spring and summer. A twelve-month gestation yields one offspring, which nurses for a year.

See also: American pronghorns; Antelope; *Apatosaurus*; Beavers; *Brachiosaurus*; Camels; Carnivores; Cattle, buffalo, and bison; Coevolution; Deer; Digestion; Digestive tract; Donkeys and mules; Ecosystems; Elephants; Elk; Food chains and food webs; Giraffes; Goats; Gophers; Hadrosaurs; Herds; Hippopotamuses; Horns and antlers; Horses and zebras; Hyraxes; Ingestion; Kangaroos; Koalas; Mammoths; Manatees; Metabolic rates; Mice and rats; Moles; Moose; Nutrient requirements; Omnivores; Pandas; Pigs and hogs; Plant and animal interactions; Platypuses; Porcupines; Predation; Rabbits, hares, and pikas; Reindeer; Rhinoceroses; Rodents; Ruminants; Sauropods; Sheep; Squirrels; Stegosaurs; Teeth, fangs, and tusks; *Triceratops*; Ungulates.

Bibliography

Gerlach, Duane, Sally Atwater, and Judith Schnell. *Deer.* Mechanicsburg, Pa.: Stackpole Books, 1994. This book discusses deer habits, habitats, species, and much more, and has a useful bibliography.

Gullan, P. J., and P. S. Cranston. *Insects: an Outline of Entomology.* 2d ed. Malden, Mass.: Blackwell Science, 2000. The book is a reference on insects, including herbivores.

Olsen, Sandra L., ed. *Horses Through Time.* Boulder: Roberts Rinehart, 1996. This interesting book covers horse history, evolution, and more.

Rath, Sara. *The Complete Cow.* Stillwater, Minn.: Voyageur Press, 1990. This book covers cattle, their history, and their breeds.

Shoshani, Jeheskel, and Frank Knight. *Elephants: Majestic Creatures of the Wild.* Rev. ed. New York: Checkmark Books, 2000. This book covers many topics on elephants, including natural history.

HERDS

Types of animal science: Behavior, ecology
Fields of study: Ecology, ethology, population biology, wildlife ecology, zoology

Herds are associations of animals, usually the same species, that keep, feed, or move together. Generally, the use of the word "herd" is reserved for large, terrestrial animals.

Principal Terms

CARNIVORES: animals that eat the flesh of other animals

DILUTION EFFECTS: the reduction in per capita probability of death from a predator due to the presence of other group members

HERBIVORES: animals that eat plants and show specializations of teeth and digestive tracts to do so

PHALANGES: the free toes of the foot; some can be modified to bear claws, hoofs, or nails

SUBUNGULATES: nonhoofed mammals that support their weight on more than the terminal phalanges; some, such as elephants and hyraxes, have pads under their metatarsals, and others, such as the sirenians, have forelimbs modified into flippers

UNGULATES: hoofed mammals that support their weight only on the hoof-clad terminal phalanges and have teeth specialized for clipping vegetation

Herding in animals serves several purposes, most commonly exploitation of food resources and reduction of the probability of predation via dilution effects. When food is spaced irregularly into patches and cannot be easily defended, individuals of a particular species will form a herd, simply as a result of coming together to feed on the same resource in the same location. Sizable grazing mammals are well known for forming some of the largest and most dense herds seen. The most familiar of these include African ungulates (wildebeests, zebra, gazelle, wild horses, rhinos, hippopotami) and subungulates (elephants, sirenians, the extinct mastadons); bison, buffalo, caribou, elk, moose, and deer in North America; kangaroo in Australia; and deer, elk, moose, antelope, wild horses, sheep, goats, pigs, boars, and peccaries in Eurasia. In addition to large mammalian herbivores, small grazers that exploit vegetation in herds include marine and freshwater snails, tortoises and turtles, geese, and hyrax.

Those grazers that feed indiscriminately tend to form herds that vary in density with season and density of foliage. Furthermore, these herds tend to vary their densities and vegetation utilization patterns with the type of vegetation exploited and its productivity potential. In seasons or areas where rainfall is common, herds tend to congregate together and exploit small patches of grass, which they clip into grazing lawns. Maintaining grazing lawns increases the productivity of the plants via increased nitrogen content and increased digestibility in freshwater, marine, and terrestrial environments. When rainfall is scarce, the herds disperse over vast areas, which allows some grasses to grow into tall meadows. The benefits to individuals in a herd from grazing a larger area and creating a grazed lawn are greater than those to a lone individual that would only be able to graze a small patch of grass and that would, presumably, be at greater risk of predation than one individual surrounded by many others. Furthermore, if the lone individual were subject to a greater risk of predation, there would be less likelihood of it being able to return to the same patch over and over again to keep it as a grazed lawn. Thus, by grouping, an individual benefits twice: once by the gain in nutrition from a grazed lawn and again from the dilution effect, where the risk of predation is diluted by the number of members forming the group.

Herding to Avoid Predation

The large herbivores that most commonly form herds are clearly visible to predators. Predation risk reduction via the dilution effect or the selfish herd effect is the other main benefit of herding. In the dilution effect, the probability of a particular individual being killed or injured is reduced by the presence of other group members that might be attacked first. In other words, there is safety in numbers. Individuals in a group may also benefit by putting other animals between themselves and the predator (the selfish herd effect). Grouping thus provides the opportunity to decrease the area of danger around each individual. If individuals within the group are acting in a selfish herd manner, the groups formed tend to be tightly clumped, as all individuals attempt to put other individuals on all sides around them and move into a central location. Finally, formation of a herd can make it more difficult for a predator to select one specific individual from the crowd for attack, or once an attack is initiated and herd members scatter during flight, can make it difficult for the predator to decide who it should chase.

Herd members also benefit by a reduction in the time necessary to scan for predators while foraging. An individual foraging has to split its time between consuming as much food as possible and avoiding becoming another animal's meal. Thus, a solitary individual has to be much more vigilant than does a member of a group who can rely on many other eyes and thus reduce its own scan

Musk oxen form a tightly packed clump characteristic of the selfish herd effect, where animals try to put as many others as possible between themselves and potential predators. (Corbis)

Dinosaur Herds

Dinosaur fossils have been known for over a century and yet little is known about their daily life when they were alive. Paleontologists, however, argue that discoveries made since the late 1970's indicate that some, if not many, dinosaurs might have been social organisms. Fossil beds of duck-billed dinosaurs not only revealed scooped-out, mud-filled nests with as many as fifteen offspring, but also showed that the nests were closely spaced, strongly suggesting communal nesting grounds, as are seen in birds. Within the nests were offspring of varying ages, suggestive of parental care.

Dinosaur footprints are also used to infer that many species grouped together in herds, some of which may have migrated long distances in what are known as trackways. Trackways preserve the footprints of a number of individuals in mud, which was later fossilized. These trackways show dozens of animals traveling in the same direction at roughly the same speed. Such trackways have been found for *Eubontes* (a Jurassic dinosaur), *Triceratops*, *Iguanodon*, and others.

The final piece of evidence that paleontologists use to infer herding in dinosaurs is the presence of horns and superfluous appendages. Observations of extant herding species indicate that males most often display horns, tusks, or other features that are used in male dominance fights over sexually receptive females within the herd. Paleontologists use these modern-day rituals to hypothesize that dinosaurs may have used their horns in a similar fashion.

rate. Because of this advantage and the advantage in the reduction of the rate of per capita predation, even herds that simply congregate around a resource benefit greatly.

Migratory and Social Herds

Many large herbivores migrate in response to food availability in different seasons. Red deer, caribou, wildebeest, mountain goats, northern fur seals, and humpback whales are examples of animals that all migrate in response to seasonal changes in rainfall or food abundance. Some of these migrations are over incredible distances: Wildebeests travel about six hundred miles; northern fur seals and humpbacks can travel three to four thousand miles. During these migrations, small herds unite with bigger herds to form even larger herds. While terrestrial herbivores migrate to follow food, some marine mammals, such as gray whales and humpbacks, migrate south to calf and breed (but do not feed, living off food reserves instead) and migrate north to abundant feeding grounds.

Some herding species display social organization beyond that expected by mere association. This social organization tends to break a herd into smaller units (matrilineal groups, harems, or small territories controlled by one or several males) that are clustered within the entire herd. These species include the horses, zebras, pronghorn sheep, walruses, sea lions, seals, and elephants. Horse and zebra herds are composed of a number of small groups of females and their foals. These individual groups are overseen by a single stallion; young males leave these groups to form bachelor herds. Group members distinguish each other via a "corporate smell." Stallions generally control a group of females and will fight with challenging stallions in elaborate rearing displays. If a stallion is challenged and loses after inseminating the females in his group, the challenger will mount the females and rape them to induce them to abort. They will then come into estrus quickly and he will remate them. Stallions will groom females to cement their relationship with them; likewise, mares will groom their foals for the same purpose. Male impalas and gazelle maintain harems during the breeding season; male wildebeest do the same, but if the herd is large, the defense of the harem may be accomplished by several males, rather than one. Male elk are divided into four main categories: Primary bulls are the first to establish harems, but as they become exhausted from challenges and herding and mating females, all the while not eating, the harems are taken over by secondary bulls. Once the secondary bulls become exhausted, tertiary bulls take over. The fourth category, opportunistic bulls, only mate with females by chance.

Pinniped herds (sea lions, seals, and walruses) form breeding herds, where males establish territories. In some groups, females are herded together in harems to remain in the male's territory; in other groups, females are free to move from territory to territory. Males vigorously defend their territories against intruding males.

Sirenians, the dugong and manatee, have proved difficult to study, but dugongs often form large herds for unknown reasons. During mating, a series of males follows a receptive female to form a cluster of up to twenty animals. Males then initiate fighting to determine who will mount the female. Following mating, the main social unit is the female and her calf. A similar situation exists in the manatee, with females and their calves being the main social unit and females in estrus becoming the focus of a mating herd of males.

Herding serves important purposes for animals: utilization of a common resource (usually food) and predator reduction via a safety-in-numbers principle. Herds can also be formed during migrations or for breeding purposes. While many herds are simply loose associations of animals, some are more highly organized into a number of social units such as harems or matrilineal groups.

—*Kari L. Lavalli*

See also: Communities; Competition; Defense mechanisms; Ecological niches; Groups; Hierarchies; Mammalian social systems; Migration; Offspring care; Population analysis; Predation; Reproductive strategies.

Bibliography

Drickamer, L. C., S. H. Vessey, and D. Meikle. *Animal Behavior: Mechanisms, Ecology, and Evolution*. 4th ed. Dubuque, Iowa: Wm. C. Brown, 1996. Describes the mechanisms, ecology, and evolution that support a comprehensive study of behavior.

Fryxell, J. M. "Forage Quality and Aggregation by Large Herbivores." *American Naturalist* 138, no. 2 (August, 1991): 478-498. This article examines three major hypotheses for why herbivores form herds and demonstrates that the mechanism operating depends largely on the type of herbivore and the circumstances it faces within its specific environment.

Hamilton, W. D. "Geometry for the Selfish Herd." *Journal of Theoretical Biology* 31 (1971): 294-311. This landmark paper was the first to explore the idea that grouping may be the result of selfishly behaving individuals that are trying to avoid becoming the predator's meal.

McNaughton, S. J. "Grazing Lawns: Animals in Herds, Plant Form, and Coevolution." *American Naturalist* 124, no. 6 (December, 1984): 863-886. This seminal article examines the effect that grazers have upon a relatively productive grassland in the Serengeti plain and explains why it may be beneficial to feed in a herd configuration.

Reynolds, J. E., and D. K. Odell. *Manatees and Dugongs*. New York: Facts on File, 1991. This book is one of the few that concentrates on sirenians and provides an overview of their distribution and biology.

Wallace, J. *The Rise and Fall of the Dinosaur*. New York: Gallery Books, 1987. This book provides an overview of the hypotheses surrounding the evolution of dinosaurs, their extinction, and their biology, as drawn from fossil evidence.

HERMAPHRODITES

Types of animal science: Anatomy, development, physiology, reproduction
Fields of study: Developmental biology, ethology, invertebrate biology, marine biology, neurobiology, reproduction science, zoology

Hermaphrodites are individuals having both male and female reproductive organs. Depending on the species, individuals may be simultaneously both male and female, or may change from one sex to the other.

Principal Terms

GONAD: the organ that produces reproductive cells (sperm or eggs)

PROTANDRY: the condition of starting out male with the potential to become female

PROTOGYNY: the condition of starting out female with the potential to become male

SEQUENTIAL HERMAPHRODITE: species or individual with the potential to change from one sex to the other

SEX-LIMITED TRAITS: features that are only expressed in one sex

SEXUAL DIMORPHISM: the existence of anatomical, physiological, and behavioral differences between the two sexes of a species

SIMULTANEOUS HERMAPHRODITISM: the condition of being simultaneously male and female

In most species, reproduction involves sex—that is, the joining of genetic material from two individuals to create new, genetically unique offspring. In sexually reproducing species, females are those individuals which produce relatively large sex cells that are full of nutrients (eggs), and males are those individuals which produce relatively small sex cells that have little or no nutrients, but which can be produced in much greater numbers (sperm). Generally there are other differences between the sexes as well—differences in hormones, anatomy, body shape, size, color, and behavior. Such distinguishing features are referred to as sex-limited traits because they appear in only one sex; collectively, they result in sexual dimorphism—literally the two sexual forms of a single species.

Whether an individual animal develops into a male or a female depends on which genes get turned on early in development. In mammals, genes for maleness get turned on in individuals with an X and a Y chromosome, while genes for femaleness get turned on in individuals with two X chromosomes. In other animals, sex determination may depend on other factors. For example, in many reptiles, development into a male or a female depends upon the temperature of the embryo as it develops in its egg.

Hermaphrodites are individuals that have both male and female reproductive organs—that is, they have both ovaries to produce eggs and testes to produce sperm. In most animals, hermaphroditism is a result of abnormal development and is extremely rare. In some species, however, hermaphroditism is normal.

Simultaneous Versus Sequential Hermaphroditism

Depending on the species, hermaphroditism can be found in either simultaneous or sequential form. In species with simultaneous hermaphroditism, individuals are simultaneously both male and female; each adult has the ability to produce both sperm and eggs. Depending on the species, a single reproductive encounter between simultaneous hermaphrodites may involve both partners exchanging sperm and eggs (for example, earthworms) or may involve the partners taking turns as male and female (for example, some coral reef fishes).

On the other hand, individuals of species with sequential hermaphroditism start life as either male or female, but have the ability to change sex at some later point. In protandrous species, individuals start out as male and have the potential to later change to female. These are typically species which require large body bulk before they can produce eggs, so individuals start out as male and change to female only if they get old enough and large enough to make eggs. In protogynous species, individuals start out as female and have the potential to later change to male. These are typically species in which males must defend a harem or a territory in order to mate, so individuals start out as female and change to male only if they get large enough to fight and win.

Moray eels are sequential hermaphrodites, beginning their adult lives male and later becoming female. (Rob and Ann Simpson/Photo Agora)

Triggers for Sex Change

In both protandrous and protogynous species, the trigger for sex change may relate not only to body size but also to the social structure of the individual's community. For example, a large female of a protogynous species may not change to male if there is an even bigger male present who would clearly win every fight. On the other hand, a relatively small female might change to male if she is the biggest female around and the larger local males suddenly died or disappeared. Thus, body size is a relative, not an absolute trigger for sex change, and in a very few species individuals can revert back to their original sex if social circumstances change again.

When a sequential hermaphrodite changes sex, it changes not only its reproductive organ or gonad, but also all the other sexually dimorphic aspects of its anatomy, physiology, and behavior. Hormones and behavior are the first things to change; then, over a period of time, the hormonal changes induce changes in the gonad and other tissues, including the brain. As a hermaphroditic fish, for example, switches from one sex to another, it may change the size and shape of its fins, its color, its aggression level, and its sexual preferences and rituals. These miraculous changes provide visible proof of the (generally silent) presence in both sexes of sex-limited genes for both male and female attributes.

—Linda Mealey

See also: Asexual reproduction; Breeding programs; Cleavage, gastrulation, and neurulation; Cloning of extinct or endangered species; Copulation; Courtship; Determination and differentiation; Development: Evolutionary perspective; Estrus; Fertilization; Gametogenesis; Hydrostatic skeletons; Mating; Parthenogenesis; Pregnancy and prenatal development; Reproduction; Reproductive strategies; Reproductive system of female mammals; Reproductive system of male mammals; Sexual development.

Bibliography

Forsyth, Adrian. *A Natural History of Sex*. New York: Charles Scribner's Sons, 1986. Superbly written popular science book on sex in the wild. One chapter covers hermaphroditism.

Mealey, Linda. *Sex Differences: Developmental and Evolutionary Strategies*. San Diego, Calif.: Academic Press, 2000. A user-friendly but information-packed textbook on sex differences and sexual development in humans and other animals.

Milius, Susan. "Hermaphrodites Duel for Manhood." *Science News* 153 (February 14, 1998): 101. Very brief nontechnical reports characterize this weekly magazine. This article summarizes some of the more interesting behaviors of hermaphrodites (in this case, worms) which reproduce by forcefully injecting sperm through their partner's skin.

Warner, R. R. "Mating Behavior and Hermaphroditism in Coral Reef Fishes." *American Scientist* 72 (1984): 128-135. Articles in this magazine are written by the experts, but are intended for the general public. This article covers an example of hermaphroditism in fishes.

HETEROCHRONY

Type of animal science: Evolution
Fields of study: Development, evolutionary science, physiology

Neoteny is a heterochronic effect in which descendants grow more slowly than their ancestors and thus appear to be juveniles relative to their ancestors. It is one of several different heterochronic phenomena that can involve differential rates and duration of growth for ancestral and descendant species.

Principal Terms

ACCELERATION: a faster rate of growth during ontogeny that causes a particular characteristic to appear earlier in a descendant ontogeny than it did in the ancestral ontogeny

HETEROCHRONY: any phenomenon in which there is a difference between the ancestral and descendant rate or timing of development

HYPERMORPHOSIS: a phenomenon in which the rate and initiation of growth in the descendant are the same as in the ancestor but the cessation of development takes place later

ONTOGENY: the life history of an individual, including both its embryonic and postnatal development

PHYLOGENY: the history of a lineage of organisms, often illustrated by analogy to the branches of a tree

POSTDISPLACEMENT: a form of paedomorphosis in which the initiation of growth in a descendant occurs later than in the ancestor, ensues at the ancestral rate, and ceases at the ancestral point

PREDISPLACEMENT: a form of peramorphosis in which the initiation of growth in a descendant occurs earlier than in the ancestor, ensues at the ancestral rate, and ceases at the ancestral point

Heterochronic phenomena are processes by which changes in the timing, rate, and duration of an ancestral pattern of growth and development result in changes in descendants. One of the six types of heterochrony is neoteny, in which a descendant's slower rate of development causes it to be an immature or truncated expression of its ancestor.

Paedomorphosis and Peramorphosis

The six types of heterochrony fall into two general patterns: paedomorphosis and peramorphosis. In these patterns, ancestral and descendant ontogenetic trajectories are compared. If the descendant morphology exceeds or surpasses the ancestral morphology, this is called peramorphosis. There are three types of peramorphosis: acceleration, hypermorphosis, and predisplacement. In acceleration, the beginning and end of development occurs at the same time in both the ancestral and descendant ontogenetic trajectories. The rate of development is faster in the descendant, however, so its morphology transcends that of the ancestor. In hypermorphosis, development begins at the same time and ensues at the same rate in both the ancestor and descendant; however, the descendant continues development for a longer interval—that is, it stops growing later. In predisplacement, the descendant begins development earlier than the ancestor, and the rate of development and the time at which growth ceases remain the same. The result of both predisplacement and hypermorphosis is a longer interval of growth. In acceleration, however, the interval of growth is the same in both the ancestor and its descendant; only the rate of growth has changed.

If a descendant morphology is a truncated or an abbreviated version of ancestral morphology, this is called paedomorphosis. There are also three types of paedomorphosis: neoteny, progenesis, and postdisplacement. In progenesis, descendant development begins and proceeds at the ancestral rate but stops sooner. In postdisplacement, the ancestral rate of development and the time of cessation are the same in the descendant; however, development begins later. Thus, for postdisplacement and progenesis, the time interval over which the descendant morphology develops is truncated when compared to the ancestral ontogenetic trajectory. In contrast, neoteny is characterized by a slower descendant rate of development but an unchanged time interval for growth and development.

The Contributions of Von Baer and Haeckel

Two scientists of the nineteenth century were especially important contributors to the early ideas concerning development and its impact on evolutionary change. Karl Ernst von Baer suggested that the features that appear early in ontogeny are those that are shared by the most organisms, whereas the features that appear later in ontogeny are those that are shared by successively smaller groups of organisms. This maxim has been called von Baer's law. Ernst Heinrich Haeckel held a different, more restricted, concept of ontogeny. He popularized the phrase "ontogeny recapitulates phylogeny." By this he meant that the ontogenetic or developmental phases through which an organism passes can be interpreted as being equal to the sequence of events that occurred during the evolutionary history of that particular organism. Haeckel's narrower concept of the relationship between ontogeny and phylogeny has been rejected by modern biologists in favor of von Baer's law.

Von Baer and Haeckel laid the foundation for ideas of heterochrony, but it was not until the later twentieth century that the subject of heterochrony was once again an area of active research. An important book by Stephen Jay Gould entitled *Ontogeny and Phylogeny* (1977) was instrumental in reopening the discussion initiated a century earlier. One of the most important things accomplished in Gould's book was the distinction that was made between different types of paedomorphosis. Frequently, in earlier works dealing with species with a juvenile appearance, no distinction was made among neoteny, postdisplacement, and progenesis: All three were discussed as neoteny. This confusion prevented generalization.

Organisms for which neoteny is used as an explanation for their origins include some salamanders of the family Ambystomatidae and the primates of the family Hominidae (the family that includes humans). In the latter case, *Homo sapiens* is considered a juvenilized anthropoid; the reduction in the amount of hair and the longer period of infancy can be used as evidence for this supposition.

Studying Heterochronic Change

Two kinds of information are necessary for documenting heterochronic processes of change. Because frequent reference must be made to ancestral and descendant ontogenetic trajectories, it is of crucial importance to have available, for the organisms under study, an estimate of their phylogenetic, or evolutionary, history. This is not an easily obtained body of evidence, and much scientific debate has occurred over the procedures that should be followed in seeking to unravel phylogenetic history. Some researchers even doubted that it would ever be possible to obtain evidence sufficient for the task. Nevertheless, some small consensus has emerged. This is especially gratifying for students of heterochrony, who depend so much upon the phylogenies that anchor their conclusions.

It is also necessary to document ontogeny, development, and/or growth in either a quantitative or qualitative sense. The quantitative measurement of growth is relatively straightforward. Measurements are taken from specimens of known ages. It is most desirable that measurements be taken from the same individual specimens throughout their ontogenies. This is not always possible, nor is it always possible to obtain accurate ages for the available specimens. Another problem is that the ages of the specimens of different species might not be directly comparable. These are among the factors that complicate the otherwise simple process of measuring the sizes and shapes of specimens over time—that is, measuring in various places along an ontogenetic trajectory. Assuming that appropriate measurements have been gathered, a variety of statistical procedures can then be applied to the accumulated data. These procedures include some rather sophisticated multivariate statistics that have only become feasible with the advent of computers.

The qualitative documentation of an ontogenetic trajectory suffers from many of the same sources of complication. In this approach, ontogenies are conceptualized as a series of discrete stages, phases, events, or appearances. These sequences are determined for the organisms under examination, and the stages that bear close resemblance are sometimes considered identical. The problem with this view is that it conceives of ontogeny as being composed of static sequences; this is a largely Haeckelian view. In fact, ontogenies are best viewed as dynamic, which makes it conceptually difficult to compare isolated parts of an indivisible ontogeny.

An unanswered question of evolutionary biology concerns the processes by which new morphologies and new organisms originate. Although many promising inroads have been made, none has been more hopeful than the idea of heterochrony. The conjecture is that small shifts in the timing, rate, and duration of ontogeny contribute to the appearance of new structures or perhaps even new organisms. This approach has so far proved both effective and promising, but it has not yet been applied to a sufficient range of organisms. Thus, it is not yet possible to tell whether heterochrony will prove to be a universally applicable approach to the study of the origin of new structures and species.

—Charles R. Crumly

See also: Convergent and divergent evolution; Development: Evolutionary perspective; Evolution: Animal life; Gene flow; Growth; Metamorphosis; Morphogenesis; Phylogeny; Population genetics.

Bibliography

Alberch, Pere, Stephen Jay Gould, George F. Oster, and David B. Wake. "Size and Shape in Ontogeny and Phylogeny." *Paleobiology* 5 (1979): 296-317. This paper expands upon and formalizes a model based upon the earlier work of Gould (1977). The authors try to present a unified concept of morphological evolution by fusing ideas of developmental biology and evolutionary ecology. Their model is technical and includes mathematical equations as its foundation. Examples used to illustrate their approach include limb development patterns in salamanders, reptiles, and birds, skull ontogeny in salamanders, and gonopodial development in a fish.

Gould, Stephen Jay. *Ontogeny and Phylogeny*. Cambridge, Mass.: Belknap Press, 1977. This volume was responsible for the reawakening of interest in heterochrony. The text is divided into two sections: The six chapters in the first section are a history of ideas pertaining to ontogeny, and the last five chapters recast these older ideas in a modern form. Although extremely specialized material is covered, the writing can be easily understood by the layperson. More than seventy illustrations, and many references to earlier work. Bibliography, glossary, and index.

Humphries, C. J., ed. *Ontogeny and Systematics*. New York: Columbia University Press, 1988. Seven separately authored articles deal with the ways in which ontogenetic information might be treated by a systematist. Issues discussed include homology and epigenetics. Each contribution is illustrated and includes a separate section of literature cited. Index.

McKinney, Michael L., ed. *Heterochrony in Evolution: A Multidisciplinary Approach.* New York: Plenum Press, 1988. This book includes sixteen specialized chapters by recognized authorities in the study of heterochrony. Chapters are grouped into three sections. The first section includes six chapters and deals mostly with the means by which one measures ontogenetic trajectories. The second section (seven chapters) is devoted to specific studies of heterochrony in selected groups, including some plants, ammonites, gastropods, rodents, and primates. The final section includes three chapters that summarize understanding of heterochrony. One chapter discusses how variation in heterochronic parameters (initiation and cessation of growth, rate of growth) are genetically controlled; another surveys the fossil record for instances of heterochrony.

McKinney, Michael L., and Kenneth J. McNamara. *Heterochrony: The Evolution of Ontogeny.* New York: Plenum Press, 1991. Defines categories of changes in timing of development, with examples for each category, and attempts to correlate types of developmental timing changes with ecological changes. Includes an extensive literature review. Illustrated.

McNamara, Kenneth J. *The Shapes of Time.* Baltimore: The Johns Hopkins University Press, 1997. Focuses on heterochrony as the "missing link" between changes in gene frequency and natural selection in the evolution of organisms.

_______, ed. *Evolutionary Change and Heterochrony.* New York: John Wiley & Sons, 1995. Overviews of current research in heterochrony in the fields of biology, paleontology, and anthropology, written by experts in their fields. Covers molecular and cell biology, macroevolution, speciation, ecology, sexual selection, and behavior.

HIBERNATION

Type of animal science: Behavior
Fields of study: Biochemistry, genetics, zoology

Hibernation is a dormancy into which some animals enter as winter approaches, remaining in a state of suspended animation until the weather moderates.

Principal Terms

ANESTHESIOLOGIST: a physician who administers anesthetics during surgical procedures

CELSIUS: a scale for measuring temperature in which freezing is zero degrees and boiling is one hundred degrees, abbreviated C

FAHRENHEIT: a scale for measuring temperature in which freezing is 32 degrees and boiling is 212 degrees, abbreviated F

HIBERNACULA: the winter habitats of brown bats

SYNCHRONIZATION: causing events to occur simultaneously

VERTEBRATES: animals with brains and spinal cords

As winter approaches, some animals enter a barely living state. Whereas the body temperature of warm-blooded animals is generally about 37 degrees Celsius (98.6 degrees Fahrenheit), some of the larger vertebrates, such as bears, enter a restful state for several of the colder months, during which body temperatures sink to about 30 degrees Celsius (88 degrees Fahrenheit).

In another class of smaller animals, notably the brown bat, the body temperature hovers just above freezing, usually between 1 and 3 degrees Celsius (34 to 38 degrees Fahrenheit). At these levels, the animal is barely alive. It does not bleed when cut. It breathes infrequently. Its heartbeat drops dramatically.

Groundhogs and chipmunks are found in areas where there may be sporadic periods of warm weather during the winter, at which times these animals awaken from their torpor temporarily. Their body temperatures, which will have sunk to about 10 degrees Celsius (50 degrees Fahrenheit), increase as the temperature outside their underground lairs rises. When the temperature sinks again, these animals resume their sleep.

Animals such as bears, brown bats, some rodents, hummingbirds, whippoorwills, chipmunks, ground squirrels, skunks, and marmots have built-in mechanisms that prevent their temperatures from sinking below the levels their systems can withstand. When their temperatures approach life-threatening levels, they begin to shiver, thereby maintaining or raising their temperatures without wakening them from their slumbers.

The Why and How of Hibernation

During hibernation the body shuts down and requires little energy. All of its systems, including endocrine, circulation, respiration, and elimination, are reduced to the barest essentials necessary to maintain life. Animals living in this state require almost no nutrition. They draw what nourishment they need from stores of fat that they accumulate by eating a great deal immediately before hibernating. They conserve their body heat by rolling themselves into balls.

The mechanisms that trigger hibernation mystify scientists. Some sort of internal clock, probably responding to light and temperature, clicks in at given times, determining the beginning of hibernation and its extent. This mechanism is far from precise. Some animals that have been maintained from year to year under consistent conditions may lose their synchronization, entering hibernation possibly in spring or summer.

The Light Sleepers

On hearing the word "hibernation," people think immediately of bears, who enter periods of dormancy as winter approaches. By the time they enter their lair for their winter's sleep, they have gained considerable weight and have added to their bodies layers of fat to provide them with the nutritional reserves they will require to survive the next three or four months.

During this period of dormancy, the bear's temperature drops less than 10 degrees Fahrenheit. Some scientists resist designating as hibernation the period during which such animals as bears, chipmunks, raccoons, and skunks sleep beneath ground. This period is sometimes called "winter lethargy."

The Hibernation Inducement Trigger

Scientists studying hibernation note that the hibernation inducement trigger (HIT) springs into action in three specific circumstances, occurring singly or in combination: when the days grow shorter in autumn, restricting the amount of light to which organisms are subjected; when temperatures drop, which often happens when the days grow shorter; and when food becomes scarce.

Questions about how HIT works have increased rather than decreased in recent years. Blood extracted from squirrels during hibernation that is preserved and injected into other squirrels in the spring, when hibernation would normally be ending, triggers an offseason onset of hibernation in the second group of squirrels. This experiment points to the existence of some sort of biochemical basis for hibernation, although its precise workings are little understood. Although the results are known, the causes remain shrouded in mystery.

Animals whose body temperatures do not drop dramatically during their dormant period may waken from their sleep several times during the winter. When the weather moderates, they often leave their lairs and scurry about seeking food. When cold weather resumes, they return to their lairs to continue their slumbers.

The eastern chipmunk is among the light sleepers. Unlike bears, it does not accumulate excess body fat to see it through its three or four dormant months. Rather, it stores food in its burrow as winter approaches, building its nest on top of the food it has gathered. It wakens frequently in winter to eat and to defecate in a section of the burrow away from the food supply.

The male chipmunk usually ends its sleep late in February. Leaving the burrow, it first seeks food and water, then looks for a mate, who produces from two to five babies one month after mating.

Skunks are also among the light sleepers. When the outside temperature approaches 10 degrees Celsius (50 degrees Fahrenheit), skunks retreat to their dens for the winter. They may take over an abandoned woodchuck's nest, but often they build their own dens below the frost line, at depths of between six and twelve feet. They line the den with dried leaves and grasses, creating a cozy nest. Although skunks are solitary in summer, they often live in groups during the winter, huddling together to keep warm.

Raccoons enter a dormant state in cold climates but are active throughout the year in milder ones, undergoing fewer body changes in winter than other true hibernators. Their body temperature drops minimally. Their heartbeat, while decreasing slightly, remains close to normal. Raccoons sleep as long as cold weather persists. They stir during warm spells but sleep again when the temperature drops.

In winter, raccoons, usually solitary dwellers, change that pattern and share their dens with other raccoons for their body warmth. Regardless of the weather, male raccoons become active late in January, which is mating season. After mating, they return to their dens for more sleep before spring.

During hibernation, the body temperature of ground squirrels drops approximately 50 degrees Fahrenheit, to a point only a few degrees above freezing. (Corbis)

The Heavy Sleepers

Some species undergo significant changes during dormancy that qualify them as true hibernators. In these animals, heartbeat, temperature, and respiration drop so dramatically during winter that life is barely sustained. These heavy sleepers do not respond to temporary increases in outside temperature, rather sleeping soundly through the months of their hibernation.

The most renowned of the heavy sleepers is the woodchuck, also known as the groundhog. People watch groundhogs' burrows every February 2 to determine whether the groundhog will see his shadow. If he does, legend has it, he will return to his den and there will be six more months of winter. The groundhog actually leaves its burrow in February seeking a mate.

Woodchucks accumulate as much body fat as possible before hibernation, sometimes achieving a weight of ten pounds. During their deep sleep, they take nourishment from their layers of fat, although they may waken at times to nibble seeds stored in their dens. By the time they emerge from hibernation, most have lost between 35 to 50 percent of their weight.

The den into which woodchucks retreat when winter arrives may have a forty-foot tunnel, camouflaged at its entry, that leads into the den. The winter den is below the frost line, usually about five feet deep. The winter rooms are high in the tunnel so that they cannot flood.

When the woodchuck enters its den for hibernation, it seals off the tunnel leading into it. The outer tunnel may become the winter dwelling of skunks or rabbits. During hibernation, the woodchuck, which normally breathes about thirty-five times a minute, breathes about once in five minutes. Its body temperature drops from about 36.7 degrees Celsius (98 degrees Fahrenheit) to 3.3 degrees Celsius (38 degrees Fahrenheit), only slightly above freezing. Its heart, beating in warm weather at about eighty times a minute, now beats four times a minute.

Woodchucks have a layer of brown fat that builds up around their vital organs. When it is time for hibernation to end, the animal receives an instant jolt of energy from this brown fat. It begins to shiver, which gradually warms it up. Its supply of oxygen increases steadily, resulting in increased blood flow, which warms the woodchuck. It takes a few hours for it to move from its dormant state to its spring time state.

In mild climates, ground squirrels are active all year. The bodies of those that do hibernate change slowly from their summer to their winter phase. Their body temperature, around 32.2 degrees Celsius (90 degrees Fahrenheit) in summer, drops a couple of degrees every day until it approaches 4.4 degrees Celsius (40 degrees Fahrenheit). An internal mechanism keeps the squirrel's body temperature from dropping below that number. Brown bats enter hibernation when the air temperature stabilizes around 10 degrees Celsius (50 degrees Fahrenheit), entering their winter quarters, or hibernacula, in swarms. In these caves, the temperature remains constant and is above freezing, which is crucial to the bats' survival, because bats subjected to lower temperatures develop fatal ice crystals in their blood.

Brown bats hang upside down in their caves in winter. Their bodies become stiff and appear to be dead. Their heartbeats drop from over five hundred beats a minute to between seven and ten beats a minute. They hibernate for three to four months, waking occasionally for water and any insects they can find to devour. In dormancy, brown bats' bodies assume the temperature of the atmosphere surrounding them.

The jumping mouse is another true hibernator. Late in October, it seals itself into its den, having gorged on all the food it can find in the weeks prior to hibernation. It curls up into a ball, placing its head between its hind legs. During hibernation, it breathes just once every fifteen minutes.

—R. Baird Shuman

See also: Estivation; Metabolic rates; Metamorphosis; Rhythms and behavior; Sleep; Thermoregulation; Warm-blooded animals.

Humans and Hibernation

Considerable research related to hibernation and its implications for humans is ongoing. Noting that when animals' bodies approach the freezing point, they feel little pain and bleed little, if at all, when they are cut, anesthesiologists and other physicians have begun to simulate some conditions of hibernation in the operating room.

Only in the last third of the twentieth century was it possible to perform some kinds of delicate, lifesaving surgery by packing the body in ice and chilling it to the point that most vital signs are significantly reduced. Open-heart and transplant surgery have advanced considerably by using such techniques. The loss of blood accompanying such surgery in earlier eras has been substantially reduced by operating on chilled bodies, then gradually warming them to normal temperatures. The need for anesthetics also has been greatly reduced, thereby eliminating much of the risk that anesthetization frequently carries.

Hibernation among humans would be useful in space travel, where reducing their vital functions for extended periods would curtail the need for oxygen and food among space venturers.

Bibliography

Brimner, Larry Dane. *Animals That Hibernate*. New York: Franklin Watts, 1991. A thorough presentation of the major hibernators. Aimed at a juvenile audience. Precise and well illustrated.

Busch, Phyllis S. *The Seven Sleepers: The Story of Hibernation*. New York: Macmillan, 1985. Written for juvenile readers, this book examines the hibernation patterns of vertebrates and insects.

Carey, Cynthia, et al., eds. *Life in the Cold: Ecological, Physiological, and Molecular Mechanisms*. Boulder, Colo.: Westview Press, 1993. A well-balanced collection of essays focusing on hibernation. More for specialists than beginners.

Lyman, Charles P. *Hibernation and Torpor in Mammals and Birds*. New York: Academic Press, 1982. Perhaps the most authoritative source on hibernation in vertebrates.

Lyman, Charles P., and Albert R. Dawe, eds. *Mammalian Hibernation: Proceedings*. Cambridge, Mass.: Museum of Comparative Zoology, 1960. Somewhat dated but still a reliable resource for basic information about hibernation.

HIERARCHIES

Type of animal science: Behavior
Fields of study: Anthropology, evolutionary science, population biology

Hierarchies, systems of establishing dominance and subordination, are important in maintaining social order in many species of animals.

Principal Terms

ADAPTION: in evolutionary biology, any structure, physiological process, or behavior that gives an organism an advantage in survival or reproduction in comparison with other members of the same species

AGGRESSION: a physical act or threat of action by one individual that reduces the freedom or genetic fitness of another

COMPETITION: the active demand by two or more organisms for a common resource

DOMINANCE: the physical control of some members of a group by other members, initiated and sustained by hostile behavior of a direct, subtle, or indirect nature

FITNESS: in the genetic sense, the contribution to the next generation of one genotype in a population relative to the contributions of other genotypes

SOCIOBIOLOGY: the study of the biological basis of the social behavior of animals

All animal species strive for their share of fitness. In this struggle for reproductive success, there is often competition among the individuals that make up the population. This competition is generally for some essential resource such as food, mates, or nesting sites. In many species, the competition over resources may lead to actual fighting among the individuals. Fighting, however, can be costly to the individuals involved. The loser may suffer real injury or even death, and the winner has to expend energy and still may suffer an injury. In order to prevent constant fighting over resources, many animal species have adapted a system of dominance, or what students of sociobiology call a dominance hierarchy or social hierarchy. The dominance hierarchy is a set of aggression-submission relationships among the animals of a population. With an established system of dominance, the subordinate individuals will acquiesce rather than compete with the dominant individuals for resources.

General Characteristics of Hierarchies

The simplest possible type of hierarchy is a despotism where one individual rules over all other members of the group and no rank distinctions are made among the subordinates. Hierarchies more frequently contain multiple ranks in a more or less linear fashion. An alpha individual dominates all others, a beta individual is subordinate to the alpha but dominates all others, and so on down to the omega individual at the bottom, who is dominated by all of the others. Sometimes, the network is complicated by triangular or other circular relationships where two or three individuals might be at the same dominance level. Such relationships appear to be less stable than despotisms or linear orders.

Hierarchies are formed during the initial encounters between animals through repeated threats and fighting, but once the issue of dominance has been determined, each individual gives way to its superiors with little or no hostile exchange. Life in the group may eventually become so peaceful that the existence of ranking is hidden from the observer until some crisis occurs to force a confrontation. For example, a troop of baboons can go for hours without engaging in sufficient hostile exchanges to reveal their ranking, but in a moment of crisis such as a quarrel over food the hierarchy will suddenly be evident. Some species are organized in absolute dominance hierarchies in which the rank orders remain constant regardless of the circumstances. Status within an absolute dominance hierarchy changes only when individuals move up or down in rank through additional interaction with their rivals. Other animal societies are arranged in relative dominance hierarchies. In these arrangements, such as with crowded domestic house cats, even the highest-ranking individuals acquiesce to subordinates when the latter approach a point that would normally be too close to their personal sleeping space.

The stable, peaceful hierarchy is often supported by status signs. In other words, the mere actions of the dominant individual advertise his dominance to the other individuals. The leading male in a wolf pack can control his subordinates without a display of excessive hostility in the great majority of cases. He advertises his dominance by the way he holds his head, ears, and tail, and the confident face-forward manner in which he approaches other members of his pack. In a similar manner, the dominant rhesus monkey advertises his status by an elaborate posture which includes elevated head and tail, lowered testicles, and slow, deliberate body movements accompanied by an unhesitating but measured scrutiny of other monkeys he encounters. Animals not only utilize visual signals to advertise dominance, but they also use acoustic and chemical signals. For example, dominant European rabbits use a mandibular secretion to mark their territory.

Special Properties of Dominance Hierarchies

A stable dominance hierarchy presents a potentially effective united front against strangers. Since a stranger represents a threat to the status of each individual in the group, he is treated as an outsider. When expelling an intruder, cooperation among individuals within the group reaches a maximum. Chicken producers have long been aware of this phenomenon. If a new bird is introduced to the flock, it will be subjected to attacks for many days and be forced down to the lowest status unless it is exceptionally vigorous. Most often, it will simply die with very little show of fighting back. An intruder among a flock of Can-

Pecking Orders in Domestic Fowl

Among common barnyard chickens, social behavior is a relatively simple system based on the dominance hierarchy. A power struggle begins soon after a new flock is established. The chickens quickly form a hierarchy that is quite literally a pecking order. The chickens establish their status by pecking or by threatening actions toward an opponent with the obvious intention of attacking in this manner. Superior genetic fitness is attained by the high-ranking birds. They enjoy more freedom of movement, and they have priority of access to food, better nesting sites, and favored roosting places. While the dominant males mate far more frequently than the subordinates, the dominant females actually mate less, because the subordinate females more readily display submissive and receptive postures to the males. The fitness of the dominant females is still enhanced due to the advantages gained in access to food and nesting sites. Males establish a separate hierarchy above that of the females. The adaptive advantage to this behavior lies in the fact that males who are subordinate to females will not be able to mate. It has been well documented that fighting ability among fowl is an inherited trait, and significant genetic variation in this trait can be observed both between and within the various species of fowl. It is advantageous for chickens to live in a stable hierarchy. If the hierarchy of the flock is intentionally disrupted, the chickens will eat less food, lose more weight, and lay fewer eggs.

ada geese will be met with the full range of threat displays and repeated mass approaches and retreats.

In some primate societies, the dominant animals use their status to stop fighting among subordinates. This behavior has been observed in rhesus and pig-tailed macaques and in spider monkeys. This behavior has been observed even in animal societies, such as squirrel monkeys, that do not exhibit dominance behavior. Because of the power of the dominant individual, relative peace is observed in animal societies organized by despotisms, such as hornets, paper wasps, bumblebees, and crowded territorial fish and lizards. Fighting increases significantly among the equally ranked subordinates as they vie for the dominant position when the dominant animal is removed.

Young males are routinely excluded from the group in a wide range of aggressively organized mammalian societies such as baboons, langur monkeys, macaques, elephant seals, and harem-keeping ungulates. At best, these young males are tolerated around the fringes of the group, but many are forced out of the group and either join bachelor herds or wander as solitary nomads. As would be expected, these young males are the most aggressive and troublesome members of the society. They compete with one another for dominance within their group and often unite into separate bands that work together to reduce the power of the dominant males. Males in the two groups show different behaviors. Among the Japanese macaques, the dominant males stay calm and aloof when introduced to a new object so as to not risk loss of their status, but the females and young males will explore new areas and examine new objects.

Nested hierarchies are often observed in some animal species. Societies that are divided into groups can display dominance both within and between the various components. For example, white-fronted geese establish a rank order of several subgroups including parents, mated pairs without young, and free juveniles. These hierarchies are superimposed over the hierarchy within each of the subgroups. In wild turkeys, brothers establish a rank order among their brotherhood, but each brotherhood competes for dominance with other brotherhoods on the display grounds prior to mating.

Dominants and Subordinates

To be dominant is to have the priority of access to the essential resources of life and reproduction. In almost all cases, the superior dominant animals will displace the subordinates from food, mates, and nest sites. In the matter of obtaining food, for example, wood pigeons are flock feeders. The dominant pigeons are always found near the center of the flock when feeding and feed more quickly than the subordinate birds at the edge of the flock. The birds at the edge of the flock accumulate less food and often obtain just enough to sustain them through the night. Among sheep and reindeer, the lowest-ranking females are also the worst-fed animals and among the poorest of mothers. Baby pigs compete for teat position on the mother and once established will maintain that position until weaning. Those piglets that gain access to the most anterior teats will weigh more at weaning than those who have to settle for posterior teat positions. In gaining access to mates, one study with laboratory mice has shown that while the dominant males constituted only one third of the male population, they sired 92 percent of the offspring.

Life is still not all that hopeless for the subordinates. Oftentimes the loser in the battle for dominance is given a second chance, and in some of the more social species, the subordinate only has to await its turn to rise in the hierarchy. In some species, cooperation among subordinate groups, especially kin groups, can lead to the formation of a new colony and a new opportunity to establish dominance. In other species, it may well be advantageous for the subordinate to stay with the group. For example, individual baboons and macaques will not survive very long if they are away from the group's sleeping area, and they will have no opportunity to reproduce. It has been shown that even a low-ranking male eats well if he is part of a troop, and he may occasionally have the opportunity to mate. In addition, the dominant male will eventually lose prowess, and the subordinate will have a chance to move up in the dominance hierarchy.

—D. R. Gossett

See also: Communication; Communities; Competition; Courtship; Defense mechanisms; Ethology; Groups; Herds; Insect societies; Mammalian social systems; Nesting; Offspring care; Pair-bonding; Population analysis; Predation; Symbiosis; Territoriality and aggression; Zoos.

Bibliography

Barash, David P. *Sociobiology and Behavior.* 2d ed. New York: Elsevier, 1982. An older book but contains an excellent discussion of biological basis for dominance hieracrchies.

Campbell, Neil A., Lawrence G. Mitchell, and Jane B. Reece. *Biology: Concepts and Connections.* 3d ed. San Francisco: Benjamin/Cummings, 2000. An outstanding introductory biology text that includes a clear concise chapter on animal behavior and hierarchies.

Feldhamer, G. A., L. C. Drickamer, S. H. Vessey, and J. F. Merritt. *Mammalogy.* Boston: WCB/McGraw-Hill, 1999. This text gives a good discussion of hierarchies among the different groups of mammals.

Ridley, Mark. *Animal Behavior: An Introduction to Behavioral Mechanisms, Development, and Ecology.* 2d ed. Boston: Blackwell Scientific Publications, 1995. An authoritative look at the principles of animal behavior with excellent examples and clear illustrations.

Wilson, E. O. *Sociobiology: The Abridged Edition.* Cambridge, Mass.: Belknap Press of Harvard University Press, 1980. One of the best presentations on the biological significance of dominance hierarchies.

Wittenberger, James F. *Animal Social Behavior.* Boston: Duxbury Press, 1981. A very good text on the social behavior in animals with a very good chapter on hierarchies.

HIPPOPOTAMUSES

Type of animal science: Classification
Fields of study: Anatomy, wildlife ecology, zoology

The Hippopotamidae family has two genera, each with a single species. The two genera differ greatly in size, but both are characterized by a round tubular body, short stocky legs, broad snout, and a very large mouth.

Principal Terms

BLOOD SWEAT: oily secretion from mucous glands serving a protective function
CALF CRÈCHE: group of newborn calves in a place of protection
DUNG SHOWERING: behavior by bulls to show dominance over other males
NONRUMINATING: digesting grasses without chewing cud

The name hippopotamus means "river horse," and these mammals spend most of their time in the water. They have inflated-looking bodies that resemble barrels, supported on short, pillarlike legs with four toes ending in hooflike nails. The tail is short and bristled, with flattened sides. The belly is carried only a few inches above the ground. The eyes are raised on top of a flat head, the ears are small, and the nostrils are slits high up on the muzzle and can be closed when the animal is submerged. The two species, *H. amphibius* and *C. liberiensis* (pygmy hippopotamus) differ greatly in size, with the former up to 5.5 feet high and weighing seven thousand to ten thousand pounds. The pygmy hippopotamus is much smaller, with an average height of 2.5 to 3 feet, and weighing 350 to 600 pounds.

Hippopotamus Life

Hippopotamuses must submerge frequently because their naked skin is vulnerable to overheating and dehydration. They can stay submerged up to thirty minutes. Their skin has a brown to gray-purple coloration with pinkish creases. The pygmy hippopotamus is black-brown to purple in color with the cheeks often tinted pink. Unlike the pygmy hippopotamus, which is a solitary mammal dwelling in rivers and forests, *H. amphibius* is a huge animal that can be found in herds of up to eighty members.

Seeking food, *H. amphibius* travels at night from the rivers for grazing, but will return before dawn to spend the day digesting and socializing in the riverbeds. During the forays from the water,

Hippopotamuses spend most of their day in the water because their bare skin makes them prone to dehydration in the hot African sun. (Corbis)

Hippopotamus Facts

Classification:
Kingdom: Animalia
Phylum: Chordata
Subphylum: Vertabrata
Class: Mammalia
Order: Ariodactyla
Suborder: Suiformes
Family: Hippopotamidae
Genus and species: Hippopotamus amphibius (hippopotamus), *Choeropsis liberiensis* (pygmy hippopotamus)
Geographical location: Although once numerous in the rivers throughout Africa, *H. amphibius* can now be found only south of Khartoum and north of the Zambezi River; the pygmy hippopotamus is found in West African lowland rain forests
Habitat: Hippopotamuses live in short grasslands, rivers, and lakes; pygmy hippopotamuses live in lowland forests and swamps
Gestational period: Eight months for hippopotamuses, seven months for pygmy hippopotamuses
Life span: Up to forty-two years in the wild, past fifty in captivity
Special anatomy: The genera differ greatly in size, but both have a broad snout, a very large mouth, a short round body, and short stocky legs; the smooth, hairless skin is covered with special pores that secrete a pinkish substance known as blood sweat, which is protective when in the water or dry land; large canine teeth enlarge into tusks that grow continually; the stomach is three-chambered but is nonruminating; hippopotamuses appear to have good eyesight, hearing, and smell

the animals typically travel two to three miles. *H. amphibius* eats up to ninety pounds of grass on a nightly basis, often mowing twenty-inch-wide swaths with its muscular lips and mouth. The pygmy hippopotamus prefers to seek food on high, dry ground and is most active between 6 P.M. and midnight. They have home ranges that may cover between one hundred and four hundred acres. Most movements are along established paths in their home range, and rarely do they cross paths with others of their own species. During these forays, they seek water plants, grasses, fallen fruits and leaves.

For *H. amphibius*, their watery homelands are partitioned into individual mating territories by mature bulls that defend defined sections. These territories can remain fixed for years. Dung showering is used to mark territories and express dominance. Other behaviors that signal threats can include water scooping, head shaking, grunting, roaring, explosive exhaling, and charging. Submission is signaled by turning tail, approaching in a crouched position, lying prone on the land, or diving and swimming away from the dominant male.

Reproduction

Herds usually breed between ten to fifteen hippopotamuses. Nonbreeding males are tolerated in the territories if they do not bother the cows. Cows and calves associate in nursery herds and establish calf crèches, which serve as protection against predation from crocodiles, lions, and hyenas. Mating takes place during the dry season while the animals are in the water. After the birth of a calf, either on land or in the water, the cow remains with the calf for about a month before returning to the herd. Baby hippopotamuses are able to nurse underwater until they are weaned, at around eight months of age.

During mating season, the solitary pygmy hippopotamus seeks out a receptive female who tolerates the male's presence when in heat. One to four copulations may take place over a period of two days on both land and in the water. The young are born on land or in the water and remain concealed for three to four weeks. The main predator for the pygmy hippopotamus is the leopard. The pygmy hippopotamuses are considered to be in jeopardy for survival because of hunting and destruction of habitat by logging.

—Frank J. Prerost

See also: Fauna: Africa; Herds; Lakes and rivers.

Bibliography

Alden, P. *National Audubon Society Field Guide to African Wildlife*. New York: Alfred A. Knopf, 1995. Written in an authoritative fashion, this book provides important facts about the hippopotamuses.

Brust, B. W. *Zoobooks: Hippos*. San Diego, Calif.: Wildlife Education, Ltd., 1989. Offers useful information about the behavior and characteristics of hippos for school reports.

Burton, M., and R. Burton. *The Marshall Cavendish International Wildlife Encyclopedia*. Rev. ed. New York: Marshall Cavendish, 1990. Aimed at elementary school students, the book includes some interesting close-up pictures of hippos in the wild.

Eltringham, S. Keith. *The Hippos: Natural History and Conservation*. San Diego, Calif.: Academic Press, 1999. Comprehensive coverage of hippo natural history and biology, with an emphasis on conservation.

Happold, D. C. *The Mammals of Nigeria*. New York: Oxford University Press, 1987. Provides basic information about hippopotamuses and their habitat.

Kingdon, J. *The Kingdon Field Guide to African Mammals*. San Diego, Calif.: Academic Press, 1997. Good illustrations and facts about hippos directed to the general audience.

HOME BUILDING

Type of animal science: Behavior
Fields of study: Entomology, marine biology, ornithology, zoology

A home is the place or physical structure in which an animal lives, eats, finds a safe haven, and raises its family. The nature of animal homes varies greatly.

Principal terms

ARTIODACTYL: a hoofed mammal with an even number of toes
CARNIVORE: an animal that eats only animal flesh
HERBIVORE: an animal that eats only plants
MOLE HILL: an earth mound a mole dug up in search of food.
OMNIVORE: an animal that eats plant and animal matter
SOLITARY: living alone

The term "home" designates the area, place, or physical structure in which an animal lives, finds a safe haven, and raises its family. The nature of animal homes varies greatly. The very simplest example might be a spot on an ocean bottom where a marine sponge attaches to the sand and grows. Other sea animals, such as gastropods (snails) live in their shells and take their homes wherever they go. Crustaceans, such as lobsters, or land arthropods, such as scorpions, live in simple burrows in the oceans or in the earth. However, a large number of lake, ocean, and river dwelling species, such as fish and whales, have no fixed homes.

Land animals often live in more complex homes. For example, social insects such as bees, termites, and wasps inhabit nests or hives. Birds also build nests. They nest in trees, on the ground, or on rocky mountain terrain. In contrast, grazing animals such as deer and antelope live on the ground, wherever their search for food takes them on a given day. In the case of small mammals, individuals or groups often live in complex underground burrows such as mole holes. Larger carnivores and omnivores often inhabit dens that are underground burrows, as do members of the weasel family, or in caves and other natural formations, as do bears and big cats.

Insect and Bird Homes

Social insects such as bees, ants, and some wasps live in nests of differing sizes which are made of wax, paper, or dried mud. Social wasps, for example, live in spherical paper nests that are a foot in diameter. These nests are seen in trees or under the porch roofs of human habitations. A wasp colony lasts only one year because wasps do not store food as do ants and bees. Only a few fertilized females, queens-to-be, survive the winter to begin new colonies in spring. Many solitary wasps live alone, except for breeding. Then, females build small brood nests of materials other than paper. For example, potter wasps use mud and saliva, and stone-working wasps mix small pebbles, mud, and saliva.

Termites, also social insects, are known for damaging wood homes. Most species are tropical, but some inhabit the Americas and Europe. They live in huge, long-lasting colonies that may hold millions of inhabitants. These colonies (called nests or termitaries) vary greatly. Tropical species build huge mounds with walls of soil particles and dried saliva. Inside the mounds are many chambers, passages, and good ventilation and drainage systems. Termites are often subterranean, burrowing up into logs and wood structures.

Wasps construct nests in tree branches or under roof eaves. (Kevin and Bethany Shank/Photo Agora)

Many birds, such as the commonly seen crows, robins, sparrows, doves, and other small to medium birds, nest in trees or warm places around human homes. For instance, doves may nest atop porch lights. Nests of such birds are most often made of intertwined pieces of grass, twigs, and human trash. The nests are often abandoned yearly and in some cases are reused by other species. Also of interest are the nests of woodpeckers, which are located in tree holes.

Flamingos nest along shores of shallow, saltwater lagoons and lakes. The nest is a foot-tall mound of mud with a depression at its top. Flamingos are monogamous, and couples use the same nest over and over. In contrast, vultures usually live on bare ground under mountain overhangs or in caves, building no nests, and laying eggs on bare rock of these spartan home sites. Vultures are also monogamous and will live in a home site for up to forty years. An exception to the "bare rock" rule occurs with lammergeiers of Europe, Asia, and Africa. These bearded vultures build several nests per pair. They are conical in shape, located on rock ledges or in caves, and are used many times, in cycles, as home sites and to raise families.

Mammal Homes

Mammals have a wide variety of home sites. In many cases, such sites are temporary and are simply the last place the mammal found itself each day. Creatures living in this way are usually herbivores, ranging from hippopotamuses to deer and other artiodactyls. This is because every day, these animals range over areas of several square miles or more seeking food. The other end of the home site range is seen with many omnivores and carnivores. These creatures have specific home sites or dens, which may be burrows in the ground, caves, logs, and natural crevices. Small mammals that live like this include gophers and moles. They dig burrows or tunnels with the sharp claws of their front feet. Gophers store food in chambers in the burrows. They are solitary and territorial, coming together only to breed. Females use their burrows to live in and raise young.

The homes of moles are more complex. Moles are voracious and solitary, continually burrowing in the ground for food, which includes insects, worms, slugs, snails, and spiders. They defend their homes when other moles—even of their own species—intrude. Moles only socialize when entering tunnels of females to mate.

Mole burrows or holes are close to the ground surface and may be recognized by large, central earth mounds. These "mole hills" are the earth that has been dug up in search of food. The burrows are very elaborate, holding warmly lined central nest chambers, connected galleries, bolt holes that allow escape from enemies, and many passageways.

Wolves and spotted hyenas are somewhat similar, related species. These carnivores tend to live in dens and claim large hunting territories. Wolf and hyena homes and living habits are different. Wolves live in dens or lairs that may be caves, hollow tree trunks, crevices under large fallen logs, or holes they dug in the ground. Few improvements are made in the natural dens wolves inhabit. They are shelters used for safety, protection from the elements, and for raising offspring.

Spotted (laughing) hyenas, in contrast, are much more communal. They live in clans of up to one hundred individuals and inhabit shared communities comprising many dens. In the dens they sleep, mate, and socialize. Dens may be caves on rocky ground or holes dug by individuals, as a clan grows. The individual cave or tunnel dens are most often inhabited by one individual or a female with cubs. This is because extended pairing is unusual in spotted hyenas.

Lions and tigers, the largest predatory land carnivores, often roam through large territories in search of game and inhabit dens of varying permanence. The dens are dense thickets, groups of rocks surrounded by thick underbrush, or caves whose entrances are screened by thorn bushes or dense underbrush. Very often, dens are used to birth offspring and protect the big cats from the elements.

Primate Homes

Monkeys, which live in Africa, Asia, and South and Central America, live in bands which most often inhabit trees, sheltering in the forks between branches. African baboons are large, more highly organized, ground-living primate species. They live in groups called troops, which are often found living in rocky terrain or on cliffs. In many cases, group members inhabit convenient caves.

Gorillas, the largest, strongest, rarest apes, look almost human. They inhabit West African forests from lowlands to altitudes of ten thousand feet. Gorillas live in bands of up to twenty individuals. Each band claims a territory, which may be viewed as the band's neighborhood. A band forages over several square miles each day and lacks permanent dwellings. Instead, its members build temporary shelters each night after a day of foraging for the honey, eggs, plants, berries, bark, and leaves that are their diet. When terrain and time permit, females and young sleep on temporary tree platforms made of branches and leaves. Mature males nest at the bases of these trees, to protect them.

Beavers: Engineers of the Animal World

An example of animal engineering at its best is the beaver home. These herbivores live in streams, ponds, or lakes near woods. Beaver home building starts with making dams where water flows steadily but slowly. Dams create deep water pools used for food storage. Beavers work in groups—usually extended families—to build their dams. Their sharp incisor teeth are used to cut down nearby softwoods such as aspens and willows. The trees are interwoven with mud, stones, and sticks to form and seal the dam.

Inside or near the dam, the beavers build lodge chambers and tunnels to live in. Nearby they store piles of tree branches underwater to use as winter food. A lodge may be on an island, a pond bank, or the shore of a lake. An island lodge holds a central chamber, with its floor a little above water level. A pond lodge, built partly on the bank, has a front wall rising up from the pond bottom. A lake lodge is built on the lake shore. Beaver lodges are made of sticks, grass, and moss, interwoven and plastered with mud. Their interiors are often nine feet square and over three feet tall.

Homes, Lifestyles, and Forms

The homes of animals depend upon their lifestyles, habitats, and forms. Many herbivores and omnivores range widely to find sustenance. Hence, they often lay themselves down to sleep wherever the search for food takes them. Animals that live in hot, relatively dry climates often sleep out of doors. However, similar species inhabiting cool to cold climates very often build or find burrows or dens to live in. Some animals have forms and eating habits that cause them to live outdoors regardless of world location. This is typical of animals, such as zebras and reindeer, that graze daily over large areas and cannot restrict themselves to homes where they return each evening. In contrast, animals that hunt often prefer to have a safe haven where they can bring their catch home to devour in peace, while animals that live in cold climates use their homes to store food for the long winter.

—*Sanford S. Singer*

See also: Biogeography; Communities; Domestication; Ecological niches; Ethology; Groups; Habitats and biomes; Herds; Insect societies; Mammalian social systems; Mark, release, and recapture methods; Nesting; Offspring care; Packs; Population analysis; Reefs; Territoriality and aggression; Urban and suburban wildlife; Zoos

Bibliography

Goodman, Billie. *Animal Homes and Societies*. Boston: Little, Brown, 1991. An introduction to animal societies and the ways in which animals interact, including home building.

Robinson, W. Wright. *Animal Architects*. Woodbridge, Conn.: Blackbirch Marketing, 1999. A series of children's books, with separate volumes looking at the homes of insects, spiders, mammals, birds, and animals that live in shells. Well illustrated and wide-ranging.

Shipman, Wanda. *Animal Architects: How Animals Weave, Tunnel, and Build Their Remarkable Homes*. Mechanicsburg, Pa.: Stackpole Books, 1994. A look at how animals manipulate their habitats in order to survive.

Whitfield, Philip, ed. *The Simon and Schuster Encyclopedia of Animals: A Who's Who of the World's Creatures*. New York: Simon & Schuster, 1998. A compendium of useful information on a great many animal species.

HOMEOSIS

Type of animal science: Development
Fields of study: Developmental biology, embryology, genetics, physiology, reproduction science

Homeotic genes determine the identity of different body segments early in development. Research on these genes provides insight into how complex body patterns are established during development and how these patterns have evolved.

Principal Terms

EMBRYO POLARITY GENES: genes whose expression in maternal cells results in products being stored in the egg that establish polarity, such as the anterior-posterior axis, after fertilization

GAP RULE GENES: expressed in the zygote, these genes divide the anterior-posterior axis of fruit flies into several regions

HOMEOBOX: one of 180 nucleotide pairs that code for a protein called the homeodomain, found in such diverse organisms as insects, frogs, and humans; they are known to influence body plan formation in fruit flies

HOMEOSIS: a process that results in the formation of structures in the wrong place in an organism, such as a leg developing in place of a fly's antenna

HOMEOTIC SELECTOR GENES: genes that determine the identity and developmental fate of segments established in fruit flies by a hierarchy of genes

IMAGINAL DISK: a small group of cells that differentiate adult fruit fly structures after the last larval molt

PAIR-RULE GENES: segmentation genes of fruit flies that divide the anterior-posterior axis into two-segment units

SEGMENTATION GENES: these include the gap rule, pair-rule, and segment polarity genes

The body plans of advanced animals and plants can be viewed as a series of segments with unique identities. This is especially obvious in the annelids (segmented worms), but even in vertebrates the muscular regions and the backbone are segmented. Occasionally, a segment takes on the identity of another segment of the organism. This is called homeosis, and it was first described in 1984. Numerous examples of homeosis have been cited, including the Antennapedia mutant of the fruit fly, which has a leg developing in the antennal socket of the head. Much has been learned about developmental patterns in organisms, and about the evolution of these patterns, from the study of homeotic mutants.

The Genetic Control of Body Plans

The role of homeosis in elucidating genetic control of overall body plans is best illustrated in the development of the fruit fly. Early in its embryogenesis, the basic plan for the adult fly form is established, and this information is stored in imaginal disks (an adult fly is called an imago) through three larval stages and associated molts. Imaginal disks are small groups of cells that differentiate adult structures after the last larval molt. The early determination of these imaginal disks for specific developmental fates is controlled by a hierarchy of genes. This hierarchy of gene regulation has been carefully documented for the establishment of the anterior-posterior axis (a line running from the head to the abdomen) of larval and adult fruit flies. Three levels of genetic control—egg polarity genes, segmentation genes, and homeotic selector genes—result in an adult fly with anterior head segments, three thoracic segments, and eight abdominal segments.

Egg polarity genes are responsible for establishing the anterior-posterior axis. Mutations of these genes result in bizarre flies that lack head and thoracic structures or lack abdominal structures. Maternal egg polarity genes are transcribed, and the resulting ribonucleic acid (RNA) is translocated into the egg and localized at one end. This RNA is not translated into protein until after fertilization. Following translation, the proteins are dispersed unequally in the embryo, forming an anterior-posterior gradient that regulates the expression of the segmentation genes.

Segmentation genes represent the second tier of the genes that establish the body plan of the fly. Within the segmentation genes, there are three levels of control—gap genes, pair-rule genes, and segment polarity genes—resulting in progressively finer subdivisions of the anterior-posterior axis. While there is hierarchical control within the segmentation genes, genes in a given level also interact with one another. Gap genes of the embryo respond to the positional information of the gradient established by the maternal egg polarity genes. Gap genes form boundaries that specify regional domains, and several gap genes with distinct regions of influence have been identified.

Pair-rule genes follow next in the sequence; they appear to function at the level of two-segment units. Mutant pair-rule genes are responsible for flies that have half the normal number of segments. Ultimately, the larval fly body is divided into visible segments, but while the patterns are forming, genes appear to exert their influence on parasegments. A parasegment is half a segment that is "out of phase" with the visible adult segments. Parasegments include the posterior of one segment and the anterior of the adjacent segment. Developmental programming of parasegments ultimately gives rise to visibly distinct segments. The final level of control of the segmentation genes focuses on individual segments and is controlled by the segment polarity genes. In response to the pair-rule genes, the segment polarity genes subdivide each segment into anterior and posterior compartments. Thus, the segmentation genes create a series of finely tuned boundaries.

Homeotic Selector Genes

It is the third tier of genes, however, the homeotic selector genes, that actually specify segment identity. Segmentation gene mutations result in missing body parts, whereas mutations of the homeotic selector genes result in a normal number of segments, but segments with abnormal identities. Homeotic selector genes are found in two gene clusters, the *Antennapedia* complex and the bithorax complex, both of which were identified based on mutant phenotypes. Genes associated with the *Antennapedia* complex appear to determine the fate of segments associated with the anterior body segments, whereas the bithorax complex is responsible for the more posterior segments, such as the abdominal segments. The pattern that arises is modulated both by interactions among the homeotic selector genes and by interactions with the segment polarity genes.

The genes found within the *Antennapedia* and bithorax complexes have been identified based on mutations that have arisen. These genes function within smaller regions of the anterior or posterior axis, much as pair-rule genes subdivide regions established by the gap genes. It is intriguing that the genes within the two complexes appear to be lined up in the same order that they function spatially. That is, the position of an *Antennapedia* complex gene on the chromosome relative to other *Antennapedia* complex genes correlates with the actual position of the segments controlled by the gene. Recent analyses of homeotic mutants in beetles indicate that the homeotic genes are also physically organized in a left-to-right sequence corresponding to the location of the segments they control on the anterior-posterior axis. Unlike the fruit fly, however, the beetle has a single homeotic gene complex controlling the entire anterior-posterior axis.

The interactions between homeotic selector genes and other genes in the hierarchy may, in part, be controlled by the homeodomain protein coded for by the "homeobox" associated with many of these genes. The homeobox is a 180-nucleotide-pair sequence that is included in many of the homeotic selector genes as well as in some segmentation genes (including the pair-rule gene). The homeodomain protein has a unique structure that may bind the deoxyribonucleic acid (DNA) and affect its transcription. It is possible that this allows genes to regulate expression of themselves and other related genes. For example, a pair-rule gene known as *fushi tarazu* (meaning "not enough segments" in Japanese) is found in the *Antennapedia* complex and has a homeobox, although *fushi tarazu* is not a homeotic selector gene. Its homeodomain can bind to the *Antennapedia* gene and thus can regulate when this gene is turned on and off. *Antennapedia*, a homeotic selector gene, was first identified when a mutation of it resulted in the replacement of an antenna with a leg. Thus, the ability of the homeodomain to bind to DNA provides a way for the hierarchical control of homeotic selector genes by segmentation genes to occur.

The homeotic selector genes are ultimately linked with gene expression that leads to the development of specific structures associated with different segments. It is not known exactly how homeotic selector genes regulate segment differentiation. Although the homeotic selector genes are active early in development, they appear to be involved in programming cells for fates that are not expressed until much later.

Studying Homeosis Through Mutations

Most of what is known about homeosis has been learned from studying mutations. Sometimes these mutations have arisen spontaneously; sometimes they have been induced by exposing organisms to mutagenic substances, such as chemicals or X ray or ultraviolet radiation. The large number of mutations identified in fruit flies accounts for the wealth of information on homeosis in this organism, in sharp contrast to the limited information on humans, for whom ethical considerations prohibit mutagenesis. Mutations affecting segmentation and determination of segment identity represent defective developmental switches and provide insight into the normal developmental sequence.

Classical genetic approaches have been used with homeotic mutants to map genes to chromosomes and to identify interactions between genes. For example, it can be determined whether two genes are on the same chromosome by making a series of specific matings between flies with mutations in these genes and wild-type ("normal") flies. If the genes are not on the same chromosome, offspring with one mutation will not necessarily have the other mutation. If the mutations are both on the same chromosome, they will be inherited together, except in rare situations where there is recombination between chromosomes. Geneticists use the frequency of recombination to assess how close together two genes are on a chromosome. This approach helped geneticists determine the order of genes within the *Antennapedia* and bithorax complexes. Matings between different mutants have also established the hierarchy of genetic control among egg polarity, segmentation, and homeotic selector genes. For example, a pair-rule mutant will have no effect on gap genes, but a gap gene mutant will affect pair-rule genes.

To visualize the results of these crosses of hierarchical mutants, researchers employed a second technique: in situ hybridization. To investigate the effect of gap genes on pair-rule genes, a wild-type fly embryo and one with a gap mutation affecting the middle section were exposed to radioactively labeled DNA that was a copy of the pair-rule gene *fushi tarazu*. The DNA hybridized (bound) to *fushi tarazu* RNA, and it thus labeled tissues where the *fushi tarazu* gene was turned on and was making RNA copies of itself. Excess radioactive DNA was washed away, and a photographic emulsion that was then placed over the tissue was exposed by the bound radioactive DNA. This permitted researchers to see that the *fushi tarazu* gene was being expressed in the middle of the wild-type embryo, but not in the gap mutant embryo. When this experiment was repeated using radioactive gap gene DNA in a pair-rule mutant, no effect on gap gene expression was observed.

In situ hybridization has also provided information on how egg polarity genes provide segmentation genes with positional information. The egg polarity gene bicoid was identified by mutations resulting in a fly with abdominal structures but no head or thoracic structures. When radioactive DNA copies of the bicoid gene were hybridized to eggs, it was found that all the RNA was located at the anterior tip after being transferred from the mother. When this RNA was translated into protein, the protein was tagged with an antibody specific for the bicoid protein. Tagging the protein with an antibody is similar to tagging RNA with a radioactive DNA segment; both techniques allow researchers to see how the RNA or protein is distributed in tissues. In this case, the bicoid protein formed an anterior to posterior gradient after the egg was fertilized, with more protein being found at the anterior end.

Homeotic genes have been isolated and used for in situ hybridization studies. In addition, the sequence of nucleotides in the DNA in these genes has been established. A variety of techniques is available for DNA sequencing. Generally the DNA is broken into smaller segments that can be more readily identified. Cutting the DNA yields overlapping segments, and the overall sequence can be established by piecing these overlapping fragments back together. The presence of the homeobox was established by comparing DNA sequences from different genes in flies and other organisms. It was evident, based on these types of data, that the homeobox sequence differs by only a small number of nucleotides even between very distantly related organisms.

The role of the homeobox was investigated by inserting DNA containing a homeobox from the *fushi tarazu* gene into a bacteria in such a way that the bacteria then produced large amounts of the homeodomain protein. This protein was then tested for its ability to bind the DNA and was found to bind to specific fragments of DNA that were involved in homeosis, such as the *Antennapedia* gene. Homeodomain protein from a mutant *fushi tarazu* gene was defective in its DNA binding ability. This provided evidence that the homeobox may regulate gene expression via the direct binding of the homeodomain to DNA. This is only one of numerous examples illustrating how researchers have utilized genetic mutants and molecular biology to investigate the role of homeotic genes in the development of segmented organisms.

Implications of Homeotic Research

Research on homeosis and homeoboxes has made significant contributions to the fields of developmental biology and evolution. Since all higher animals and plants exhibit some form of segmented development, the common link of the homeobox has intrigued scientists interested in how body plans are established. Questions concerning the evolution of segmentation patterns in animals have also arisen as more is understood about the genes affecting segmentation and how they are regulated.

Significant similarities both among the homeoboxes identified in fruit flies and among the homeoboxes of distantly related species have been found. In fruit flies, homeoboxes have been identified in homeotic selector genes, segmentation genes, and egg polarity genes. Whether homeoboxes affect gene regulation and segmentation in other organisms is an exciting, and open, question. Segmented worms, leeches, earthworms, sea urchins, frogs, mice, and humans all have homeoboxes. Different mouse homeobox genes have been shown to be expressed in specific regions during embryogenesis. Relatively little is known about the process of segmentation in vertebrates, so it is more difficult to interpret the results of such experiments. Also, it appears that segmentation in vertebrates and in fruit flies (and other invertebrates) evolved separately. Thus, the presence of a homeobox may or may not indicate a common developmental process for segmentation among animals.

There is an intriguing possibility that addition and modification of homeotic selector genes were responsible for the evolution of insects from segmented worms. The presence of two homeotic gene complexes in fruit flies, contrasted with one in beetles, suggests that duplication of the gene complex, followed by subsequent specialization, may have allowed for greater fine-tuning of segmental identity. Evolutionary alterations of the initial homeotic gene complex may have been responsible for the addition of legs to a wormlike creature composed of similar segments, giving rise to a millipede-like creature. The reduction of all legs except for the walking legs in the thoracic region and, ultimately, the addition of wings to the thoracic region could also reflect changes in homeotic selector genes. Parts of this evolutionary journey can be reconstructed with homeotic fruit fly mutants that bear similarities to their ancestors. This is exemplified by the deletion of the *Antennapedia* gene, which results in a wingless fly; winged insects presumably arose from nonwinged insects. It is impossible, however, to create a millipede from a fly via mutations of the homeotic selector genes, so caution must be taken in speculating on the role of homeotic selector genes in insect evolution.

—*Susan R. Singer*

See also: Anatomy; Cell types; Determination and differentiation; Development: Evolutionary perspective; Embryology; Genetics; Metamorphosis; Morphogenesis; Mutations; Physiology; Pregnancy and prenatal development; Regeneration; Reproduction.

Bibliography

Alberts, Bruce, Dennis Bray, Julian Lewis, Martin Raff, Keith Roberts, and James D. Watson. *Molecular Biology of the Cell*. 3d ed. New York: Garland, 1994. Provides details on the molecular analysis of segment formation and specification. Illustrations of in situ hybridizations are especially helpful. Numerous examples using specific mutants are informative for college students.

Duboule, Denis, ed. *Guidebook to the Homeobox Genes*. New York: Oxford University Press, 1994. Provides a comprehensive overview of research on the homeobox genes, summarizing dispersed findings and listing key references. Gives an historical account of the discovery of the homeobox, an introduction to its role in development, and the classification and structural organization of the genes, plus an alphabetical description of individual genes.

French, Vernon, Phil Ingham, Jonathan Cooke, and Jim Smith, eds. *Mechanisms of Segmentation*. Madison, Conn.: Research Books, 1988. Provides a good explanation of homeodomain binding studies and details on the spatial expression of mouse homeodomain genes. Clear illustrations and accompanying reference lists.

Gilbert, Scott F. *Developmental Biology*. 6th ed. Sunderland, Mass.: Sinauer Associates, 2000. Chapter 18 provides extensive coverage of segmentation in fruit flies with a somewhat general approach. The section on the evolution of homeotic genes would be good preparation for reading the text by Raff and Kaufman.

Raff, Rudolf A., and Thomas C. Kaufman. *Embryos, Genes, and Evolution*. New York: Macmillan, 1983. Chapter 8 is an exciting chapter focusing on the role of homeotic selector genes in the evolution of insects from segmented worms. Excellent illustrations provide the reader with a clear understanding of ancestral morphologies. Suitable for college students.

Russo, V. E. A., et al. *Development: The Molecular Genetic Approach*. New York: Springer-Verlag, 1992. A collection of essays covering molecular events responsible for specific processes in the major model systems, control of fundamental processes, the regulation of genes in time and space, the roles of homeobox and oncogenes, and the potential practical benefits of developmental biology.

Villee, Claude A., Eldra Pearl Solomon, Charles E. Martin, Diana W. Martin, Linda R. Berg, and P. William Davis. *Biology*. 2d ed. Philadelphia: W. B. Saunders, 1989. An introductory college text that is also suitable for high school students. Chapter 16 provides details on the development of the fruit fly from egg to adult and summarizes the hierarchy of gene control involved in segmentation. There is an excellent photograph of the *Antennapedia* mutant.

Watson, James D., Nancy H. Hopings, Jeffrey W. Roberts, Joan Argetsinger Steitz, and Alan M. Weiner. *Molecular Biology of the Gene*. Menlo Park, Calif.: Benjamin/Cummings, 1987. Provides a well-illustrated analysis of the genes associated with segmentation and the determination of segment identity. A clear explanation of maternal effects on embryo development is presented that provides additional insight into egg polarity genes. Suitable for an introductory-level college student.

HOMINIDS

Type of animal science: Anthropology
Fields of study: Anatomy, anthropology, evolutionary science, human origins, systematics (taxonomy)

Though understanding of human ancestry is rudimentary at best, an astonishing series of discoveries since the nineteenth century has created a lively field of knowledge where there was none before.

Principal Terms

APES: large, tailless, semierect anthropoid primates, including chimpanzees, gorillas, gibbons, and orangutans, and their direct ancestors—but excluding man and his ancestors

AUSTRALOPITHECINES: nonhuman hominids, commonly regarded as ancestral to present-day humans

DRYOPITHECINES: extinct Miocene-Pliocene apes (sometimes including *Proconsul*, from Africa) found in Europe and Asia; their evolutionary significance is unclear

HUMANS: hominids of the genus *Homo*, whether *Homo sapiens sapiens* (to which all varieties of modern man belong), earlier forms of *Homo sapiens*, or such presumably related types as *Homo erectus* and the still earlier (and more problematic) *Homo habilis*

PRIMATES: placental mammals, primarily arboreal, whether anthropoid (humans, apes, and monkeys) or prosimian (lemurs, lorises, and tarsiers)

The idea that humankind might be significantly older than the six thousand years previously allotted by biblical scholars, who tried to calculate the generations of man since Adam, was not widely maintained until 1859, when human stone tools and the bones of extinct animals were found lying close to each other in France. Charles Darwin's *On the Origin of Species* appeared on November 1 of the same year, but it suggested only that "light will be thrown on the origin of man and his history" by the theory of evolution he had just proposed.

A series of important and widely noticed books then followed, including J. Boucher de Perthes' *De l'homme antédiluvien et de ses œuvres* (1860; of antediluvian man and his works); Thomas Henry Huxley's *Evidence as to Man's Place in Nature* (1863); and Charles Lyell's *Antiquity of Man* (1863), with others later by John Lubbock, James Geikie, and W. Boyd Dawkins. In *The Descent of Man* (1871), Darwin sagely hypothesized that the human line evolved in Africa (not Asia, as had previously been assumed) from a long-tailed, probably arboreal, ancestor. Yet, in Darwin's time only two fossil apes were known at all, together with some controversial bones of a creature known as Neanderthal man. Extinct species such as the australopithecines and *Homo erectus* had not been discovered.

Humans and Other Primates

The scientific name for mankind is *Homo sapiens sapiens* (wise man); the taxonomic family is the Hominidae, which also includes chimpanzees and gorillas. This classification is reasonable because, bone for bone, their skeletons are almost identical to human skeletons. On other evidence as well, man and his cousins appear to be remarkably alike. The protein sequences in chimpanzee and human hemoglobin, for example, are identical; there are only two chemical differences between gorilla and human hemoglobin. Between humans and all other animals, there are more than two. Strands of deoxyribonucleic acid (DNA) from chimpanzees and humans, moreover, are 99 percent identical. Chimpanzees, finally, are second only to humans in intelligence; their brains closely resemble those of humans. These remarkable similarities attest a common ancestry for all the Hominidae, as the classification itself would imply, and a fairly recent differentiation among its members. Some biochemists have argued that man, the chimpanzee, and the gorilla shared a common ancestor no more than six or eight million years ago.

Whatever the timing may have been, it is almost universally accepted that the link between humans and their protosimian ancestors was a now extinct genus of ape-men, the australopithecines (southern apes). The first of these, called, originally, the Taung child, was discovered in South Africa by Robert Dart in 1924. Its identity and significance remained controversial until 1936, when further discoveries by Robert Bloom convinced skeptical professionals and the public. Today, *Australopithecus* and *Homo* (man) are often grouped together as Hominini, as opposed to the Pongidae (apes), to which the chimpanzee and gorilla belong.

The Australophithecines

The australopithecines arose at least 4 million years ago, probably from dryopithecine ancestors. They lasted until two million years ago, evolving into a series of species. Of these, *Australopithecus afarensis* (found in the Afar region of Ethiopia) was the oldest and smallest. Males stood no taller than four feet, and the females were smaller. Most significantly, however, *afarensis* was fully bipedal; unlike the apes, it walked upright, with increasingly specialized hands, and with legs a bit longer than arms. Chimpanzee-like hips, together with curved toe and finger bones, suggest that it was essentially a tree-dweller living on fruits and seeds. A remarkably complete skeleton of *afarensis*, familiarly called Lucy, was found in 1974; it is about three million years old and is the oldest hominine skeleton yet found. *Afarensis* died out around 2.5 million years ago; it is thought to be ancestral to the later australopithecines and to modern humans.

Australopithecus africanus, deriving from Africa some three to one million years ago, probably evolved from *afarensis*. Most specimens come from Sterkfontein in South Africa, though others have been found in Ethiopia, Kenya, and Tanzania. This species was about the same size as *afarensis*, but it had a less apelike face. The arms were proportionately longer than a modern human's, yet shorter than those of *afarensis*; hands and teeth show similar "modernization." Dart's Taung child was the first (and is still the most famous) example of this species.

Australopithecus robustus (robust southern ape), found only in South African caves thus far, was once thought identical to *Australopithecus boisei*; in older literature, it was also known as *Paranthropus* (past man). Larger and more strongly built than *africanus*, *robustus* was more than a foot taller and had a larger brain. His teeth indicate that *robustus* was a plant eater. There is also a remarkable specimen (discovered by C. K. Brain at Swartkrans) of a child's skullcap in which the imprint of a leopard's lower canines can be seen. Since an exact-fit leopard's jaw was found nearby, it is assumed that the leopard killed the child and was then itself killed by an adult *robustus* armed with some kind of weapon. (A diorama at the Transvaal Museum, Pretoria, reconstructs this hypothetical incident.) *Australopithecus boisei*—a famous discovery by Mary Leakey, named for the Leakeys' sponsor, Charles Boise—called *Zinjanthropus*, was even bigger than *robustus* and lived at the same time, though in East Africa. A *boisei* skull discovered in 1985 in Kenya proved not only to be particularly massive but also considerably older (2.5 to 2.6 million years) than any known *robustus* specimen. *Australopithecus boisei* must therefore have been a separate and earlier species probably not descended from *africanus*. If so, then there was more than one australopithecine lineage, and the previously held idea that the australopithecines became increasingly robust through time must be reversed. As a result of this one find, there no longer are widely accepted ideas as to who gave rise to whom.

Throughout their history, the australopithecines manifest a regular progression from apelike characteristics to human ones. All the australopithecines walked upright, a fact that evidently encouraged increasing height, hand specialization, and brain development. Gradual changes in australopithecine dentition, moreover, suggest not only changing diet but revised habits as well. With advanced hands and evolving arms and shoulders, *Australopithecus* probably carried loads and used weapons, regardless of whether he was capable of making them. Though australopithecines may have done some hunting, they probably depended primarily upon foraging and scavenging—filching from leopard kills, for example.

The Evolution into *Homo*

At what point *Australopithecus* evolved into *Homo* is unclear, in part because the distinction between them is rather arbitrary. Though still regarded by some researchers as an advanced form of *Australopithecus* (that is, as more than one species) *Homo habilis* (handy man), another Leakey family discovery, is otherwise usually accepted as the earliest member of a distinctly human line. He was still only about five feet high (perhaps an optimum height for the environmental conditions), but had a larger brain, a rounder head, a less projecting face, advanced dentition, reduced jaws, and essentially modern feet. Stone artifacts have been found in close association with his remains; *habilis*, who lived between 2 and 1.5 million years ago, almost certainly made tools, hunted, built shelters, gathered plants, and scavenged. He is assumed to be ancestral to *Homo erectus*, and may have exterminated *Australopithecus*.

In 1859, when the prehistory of humankind was first broadly acknowledged, the only remains then known belonged to Neanderthal man (now called *Homo sapiens neanderthalensis*). Until 1924, when *Australopithecus* was discovered, all the intervening finds (excluding "Piltdown man," a deliberately planted fake) have since been classified as varieties of *Homo erectus* (erect man—a designation assuming that its predecessors stooped). Of these, the two best-known are Java man, discovered in 1891 by Eugène Dubois in Java, and Peking man, found in China by Davison Black in 1926. Only after a number of specimens had accumulated was it realized that Java man and Peking man were examples of the same species. An exceptionally complete *Homo erectus* skeleton was discovered in Kenya in 1984 and dated at 1.6 million years old. Overall, *Homo erectus* lived from some time before 1.6 million years ago to as recently as two hundred thousand years ago. He probably evolved in Africa but migrated from there (as no previous hominid had) to Europe and the Pacific shores of Asia. This species was as tall as modern humans, but more robust overall, with a noticeably thicker, somewhat "old-fashioned" skull that still included prominent brow ridges and a sloping forehead. He had large, projecting jaws, no chin, teeth that were larger than ours; he also had a bigger brain. *Homo erectus* was not only widespread in distribution, but also showed considerable regional variation. His success as a colonizer was attributable in large part to his intelligence, which was manifested in standardized but increasingly sophisticated toolkits, big-game hunting (almost certainly cooperative), the use of fire, and advanced housing. He lived during the Pleistocene, or glacial, epoch, and was probably stimulated to use his creative abilities by the deteriorating environments he sometimes encountered. Very late examples of *Homo erectus* are sometimes alternatively classified as *Homo sapiens*. Despite some continuing opposition, it is now usually accepted that *Homo erectus* gave rise to *Homo sapiens*.

Searching for the Fossil Remains

Considering the efforts that have been made to find them, hominid fossils are remarkably few. This is the case for three main reasons: First, the early hominids (unlike modern humans) did not exist in huge numbers; second, the majority of their bones were not preserved as fossils; and third, it is certain that only a small proportion of the hominid fossils that do exist have been found.

The hominid line evolved in Africa, and its earlier members (including the australopithecines and *Homo habilis*) have been found only there. *Homo erectus* was both more widespread and more numerous, but none of these types practiced ritual burials (Neanderthals were the first to do that), so the most usual agent of presentation was some sort of nonhuman carnivore. Predators, such as the large cats, might actually have hunted the early hominids; in any case, they certainly scavenged hominid carcasses. Hyenas and other cleanup animals then grabbed what they could, taking the leftover pieces to their dens in limestone caves. (There must be some truth to this scenario, because leopards and hyenas have left their toothmarks on australopithecine bones.) The gnawed bones, now thoroughly disarticulated, were scattered about the cave—and eventually solidified by limy deposits into a bone breccia.

When twentieth century investigators find such embedded hominid bones, or suspect their presence, they collect chunks of the breccia and dissolve the limy matrix with acetic acid, a procedure that does no harm to fossil bones. It is unlikely that any of the latter will be whole. Once the fragmented bones have been freed from the matrix, they are cleaned, preserved, sorted by type, and tallied. Whether big or small, routine or not, each must be identified. By far the great majority of the bones will belong to antelope of various kinds; less than one in a thousand, normally, proves to be hominid.

Such procedures are standard when dealing with cave deposits at the famous South African australopithecine sites of Sterkfontein, Kromdraai, and Swartkrans, all of which are adjacent to each other and to Pretoria. At Olduvai Gorge in Tanzania, where the Leakeys and others have found both australopithecine and *Homo habilis* remains, the geology is entirely different. Here, the

Australopithecus africanus *lived in Africa between 3 and 1 million years ago.* (©John Sibbick)

erosive power of a now-defunct river has exposed primarily volcanic sediments that once bordered a shifting saline lake. The disadvantage of this site is that it lacked any obvious place for the location of bones; years of determined effort were required to locate productive sites. The advantages of the site were that the presence of early humans was virtually assured because stone tools were scattered about plentifully (whole campsites were eventually found); the bones involved had not been dragged about and disarticulated by animals; and the involved stratigraphy made fairly precise dating at least theoretically possible. In general, a specimen from Olduvai Gorge brings with it more useful information than does one found in the South African caves. Still other sites have provided additional unique information.

The collection, preservation, and interpretation of hominid fossil bones is very much a multidisciplinary effort. In particular, detailed geological understanding of the site is essential. Only through stratigraphical analysis, usually, can the age and situation of the discovered fossil be understood. Stratigraphy aside, certain rocks can also be dated according to the radioactive elements they contain. More often, bones can be dated approximately because they occur in association with a particular assemblage of animal bones, the animal species themselves being of reliable short-term ages. In some cases, pollen samples have been of use. All this additional information, together with comparative anatomical analysis, helps to give hominid fossils a defensible identity.

Homo sapiens Facts

Classification:
Kingdom: Animalia
Subkingdom: Metazoa
Phylum: Chordata
Subphylum: Vertebrata
Superclass: Tetrapoda
Class: Mammalia
Subclass: Theria
Infraclass: Eutheria
Order: Primates
Suborder: Anthropoidea
Superfamily: Hominoidea
Family: Hominidae
Genus and species: Homo sapiens

Geographical location: Originally Africa, now spread to all continents

Habitat: Originally savannas, now spread to all habitats

Gestational period: Nine months

Life span: Originally a maximum of twenty-five years; now over one hundred years, with averages from forty to eighty years, depending on habitat

Special anatomy: Large, well-developed brain; upright, bipedal posture; opposable thumbs; larynx, vocal chords, and tongue adapted to produce a wide variety of sounds for language

The Search for Ancestors and Origins

Thinking human beings have always been fascinated with the concept of origins, and of all origins, none has been of more interest than that of humans. Human groups have asserted deeply meaningful identities by attributing their present being to a particular origin. Most of these psychologically necessary genealogies relied upon some divine agency to explain human existence. From a surprisingly early time, however, civilized humans (such as the Greeks) recognized that there had been a time when humankind did not know the use of metals. Subsequently, many thinkers took the concept of cultural evolution for granted.

By the seventeenth century C.E., anatomy had become a popular field of study. Comparisons soon established how like human anatomy that of the higher primates was. By the mid-eighteenth century, Carolus Linnaeus, the originator of modern biological classification, even ventured to place man and the apes within the same family. Yet this classification did not imply any necessary common ancestry. Linneaus and others of his time created the notion that individual species arose through a special, divine plan. The idea of special creation lost credibility when the fact of extinction became established at the end of the eighteenth century and as the diversity of species and varieties came increasingly to be appreciated. Nature, moreover, was no longer seen to be a benign reflection of its creator. As opinions of a distinctly human nature likewise declined, the realization that humans are animals encountered lessening resistance.

Before human evolution was generally accepted during the twentieth century, the evolution of cultures, language, law, institutions, and at least some animals had already been established. Though the idea was there, reliable evidence for the biological evolution of humans remained elusive. Before 1891, the only known prehistoric human bones belonged to Neanderthals; they were quickly and effectively dismissed as pathological freaks. Since Raymond Dart's Taung child of 1924 was likewise dismissed with ridicule, only a few specimens of what would later be recognized as *Homo erectus* (Java man and Peking man) survived to satisfy the now fashionable quest for a "missing link." (There are still innumerable missing links, but the essential connection between ancestral apes and man was confirmed by the discovery of australopithecines.) It was Robert Broom who, during the 1930's, established the reality of the australopithecines and, by implication, of human evolution.

Though the evolution of the hominid line is certainly a worthwhile scientific topic, it has always been regarded as much more than that, because the claims to ancestry define humanity. Yet the formal constraints of science are limited. In a remarkable series of thirty-nine papers (published between 1949 and 1965), for example, Raymond Dart promulgated an interpretation of the australopithecines as aggressive, predatory, and cannibalistic hunters. Because of the recurrent wars in which civilization had engaged during this period and shortly before, this image of human (or almost human) nature appealed to the popular imagination—so much so that less technical restatements of the same views by Robert Ardrey were not only commercially successful but also politically influential. Interpretations of australopithecine and early hominid behavior have subsequently changed, however—many now see these species as abject scavengers disputing the possession of already picked-over animal corpses with hyenas. It is arguable whether such changing interpretations are attributable to scientific advances or are the result of changing philosophical views of humanity.

—Dennis R. Dean

See also: Apes to hominids; Convergent and divergent evolution; Evolution: Animal life; Evolution: Historical perspective; Extinction; Fossils; Genetics; Homo sapiens and human diversification; Human evolution analysis; Neanderthals; Primates.

Bibliography

Brain, C. K. *The Hunters or the Hunted? An Introduction to African Cave Taphonomy.* Chicago: University of Chicago Press, 1981. In a sophisticated but readable style, Brain reports the evidence about the australopithecines—what they hunted and what hunted them. (Taphonomy is the study of entire bone assemblages.) His study is intended primarily for graduate students and professionals in the field.

Campbell, Bernard. *Humankind Emerging*. 8th ed. Boston: Allyn & Bacon, 2000. A readable college-level text that is useful for its treatments of historical, biological, and anthropological topics. No other book presents so much information so well. Graphics and other illustrations are excellent. Campbell also includes topical bibliographies and a glossary. Chapters 7 through 10 are especially relevant.

Day, Michael H. *Guide to Fossil Man*. 4th ed. Chicago: University of Chicago Press, 1986. Despite some abstruse language (a glossary is included), this authoritative compilation can be of significant help. Organized geographically, it lists and often illustrates all known hominid fossils. Conflicting interpretations are reported but seldom reconciled. This is not, therefore, a book for those seeking the fast answer. Bibliographies accompany each section of the text. For college-level students and professionals, this is a standard work.

Johanson, Donald, and Maitland A. Edey. *Lucy: The Beginnings of Humankind*. Reprint. New York: Simon & Schuster, 1990. Intended for a broad audience, this popular account re-creates the excitement and explains the significance of the finding of an extraordinary specimen, known technically as *Australopithecus afarensis*. It is the oldest hominid whole skeleton extant.

Johanson, Donald, Lenora Johanson, and Blake Edgar. *Ancestors: In Search of Human Origins*. New York: Villard Books, 1994. A companion to the three-part *Nova* television series. Chronological coverage of *Australopithecus*, *Homo habilis*, *Homo erectus*, archaic *Homo sapiens*, the Neanderthals, and *Homo sapiens*. Focuses on the australopithecines and early *Homo*. Lavishly illustrated.

Jordan, Paul. *Neanderthal: Neanderthal Man and the Story of Human Origins*. Stroud, U.K.: Sutton, 1999. A clear exposition of a complex subject, highlighting the ideas involved in the study of human evolution and the relationship of Neanderthal Man to *Homo sapiens*.

Kurten, Bjorn. *Our Earliest Ancestors*. Translated by Erik J. Friss. New York: Columbia University Press, 1993. A succinct overview of hominid evolution. Focuses on the major changes in anatomy and culture that have marked the development of hominids from *Homo habilis* through *Homo sapiens*, including the emergence of bipedalism, stone toolmaking, and articulate speech. Numerous maps, charts, and drawings.

Leakey, Maeve, et al. "New Hominin Genus from Eastern Africa Shows Diverse Middle Pliocene Lineages." *Nature* 410 (March 22, 2001): 433-440. A new skeleton, dated to 3.5 million years ago and differing from the roughly contemporary *Autralopithecus afarensis*, throws the lineage of modern man into even further confusion.

Lewin, Roger. *Bones of Contention: Controversies in the Search for Human Origins*. New York: Simon & Schuster, 1987. This lucid but sophisticated treatment reviews scholarly debates regarding the Taung child (*Australopithecus africanus*), *Ramapithecus* (a fossil ape) and "Lucy" (*Australopithecus afarensis*). Too specialized for most high school students, it will appeal primarily to college graduates and professionals. Lewin has also written *Human Evolution: An Illustrated Introduction* (1984) and coauthored three popular books with Richard Leakey: *Origins* (1977), *People of the Lake* (1979), and *The Making of Mankind* (1981).

Mellars, Paul, ed. *The Emergence of Modern Humans: An Archaeological Perspective*. Ithaca, N.Y.: Cornell University Press, 1990. A volume of proceedings from a 1987 conference on the origins of modern humans and the adaptations of hominids. Concentrates on the archaeology of hominids in western and central Europe.

Reader, John. *Missing Links: The Hunt for Earliest Man*. Boston: Little, Brown, 1981. Organized chronologically, chapters cover the discoveries of the major hominids. Outstanding original photographs by the author are featured. Though other treatments have been more detailed (and more technical), Reader's agreeable combination of history and science makes his book irresistible. For readers at all levels.

Szalay, Frederick S., and Eric Delson. *Evolutionary History of the Primates*. New York: Academic Press, 1979. Though not intended for beginners, and now somewhat dated in places, this remains a standard reference.

Tattersall, Ian. *The Fossil Trail: How We Know What We Think We Know About Human Evolution*. New York: Oxford University Press, 1995. The reconstruction of human evolution on the basis of interpreting fossil remains, including the Laetoli footprints, Lucy, Turkana Boy, and others. Good coverage of the development of dating techniques such as fluorine analysis and mitochondrial DNA.

Theunissen, Berg. *Eugène Dubois and the Ape-Man from Java: The History of the First "Missing Link" and Its Discoverer.* Boston: Kluwer Academic, 1989. Java man, discovered by Dubois in 1891 (with one important discovery the previous year), was the first example of *Homo erectus* to be recovered. Theunissen's account of this historic addition to our knowledge of early man is detailed but nontechnical, suitable for readers at the high school level and beyond.

HOMO SAPIENS AND HUMAN DIVERSIFICATION

Type of animal science: Anthropology

Fields of study: Anthropology, evolutionary science, ecology, genetics, human origins, systematics (taxonomy)

By studying fossil, cultural, and genetic evidence, scholars have attempted to trace the evolutionary development of the human species. It is believed that the earliest form of Homo sapiens *appeared about 350,000 years ago and that the first modern humans (*Homo sapiens sapiens*) appeared somewhat before 100,000 years ago, with racial diversification following thereafter.*

Principal Terms

GENE POOL: the total collection of genes available to a species

GENERALIZED: not specifically adapted to any given environment; used to describe one group of Neanderthal humans

HOMINID: any living or fossil member of the taxonomic family Hominidae ("of man") possessing a human form

HOMINOID: referring to members of the family Hominidae and Pongidae (apes) and to the taxonomic superfamily of Hominoidae

MORPHOLOGY: the scientific study of body shape, form, and composition

NATURAL SELECTION: any environmental force that promotes reproduction of particular members of the population that carry certain genes at the expense of other members

PLEISTOCENE EPOCH: the sixth of the geologic epochs of the Cenozoic era; it began about three million years ago and ended about ten thousand years ago

WÜRM GLACIATION: the fourth and last European glacial period, extending from about seventy-five thousand years ago to twenty-five thousand years ago

All human beings on the earth today are highly adaptive animals of the genus and species *Homo sapiens sapiens* (Latin for "wise, wise human"). In terms of physical structure and physiological function, *Homo sapiens sapiens*—modern humans—are classified taxonomically as members of the order Primates, which is part of the class Mammalia. Since humans and other members of Primates (monkeys and apes) are biologically related, scientists presume both groups to be the products of an evolutionary process similar to that which affected other divergent categories of animals. The evolutionary process that produced *Homo sapiens sapiens* from previously existing species is also believed to account for diversifications within the modern human population such as racial differentiation.

Modern humans and modern apes (the two most closely related of modern primate species) are believed to possess a common biological ancestry, or line, that diverged perhaps five or six million years ago. The scanty fossil record of this early period, in conjunction with modern genetic studies, seems to indicate that the branch of hominoid evolution which eventually led to *Homo sapiens sapiens* first gave rise to the earliest hominid type, called *Ramapithecus*.

Next to appear, several million years ago, during the late Pliocene epoch, were the early forms of *Australopithecus*, which existed in Africa. They share certain characteristics with both humans and apes. Their brains are larger than those of apes but smaller than those of humans. There have been four species of *Australopithecus* identified.

The Emergence of the Genus *Homo*

Examples of the first undisputed members of the genus *Homo*—true human (though not *sapiens*)—appear in the fossil record about 1.5 million years ago. Samples of *Homo erectus* ("upright human") have been found in China, Africa, Java, and Europe. This creature habitually walked upright, made shelters, and used sophisticated tools. *Homo erectus* is also very important, since it is the first hominid to have used fire purposefully. It was suggested by John E. Pfeiffer, in a 1971 article entitled "When *Homo erectus* Tamed Fire, He Tamed Himself," that this first domestication of a natural force was a tremendous evolutionary step, changing the fundamental rhythms of life and human adaptability to environments. Most scholars accept the premise that *Homo erectus* was the hominid grade intermediate between the australopithecines and *Homo sapiens*.

Exactly when, where, and how advanced members of the species *Homo erectus* evolved into *Homo sapiens* are key questions in the study of human evolution, and they are questions that resist resolution. It might be thought that the closer one comes, in terms of time, to modern man, the easier it would be to find the answers. In actuality, such is not the case. The ancestral line or lines leading to modern man become hazy beginning approximately 500,000 years ago. Direct fossil evidence of the earliest members of the species *Homo sapiens* is scarce; moreover, finds of modern human fossils in the Middle East have intensified the debate about the immediate ancestry of *Homo sapiens sapiens*. All the evidence indicates, however, that the middle to upper Pleistocene epoch (beginning about 350,000 years ago), known as the Paleolithic or old stone age in archaeological terms, witnessed the emergence of early *Homo sapiens*.

The Earliest *Homo Sapiens*

In 1965, hominid fossil remains were found at a site named Vértesszöllös, near Budapest. They consisted of some teeth and an occipital bone (a bone at the back of the skull). The site also yielded stone tools and signs of the use of fire. Several features of the find recall *Homo erectus*, but the estimated cranial capacity of 1,400 cubic centimeters is well into the normal range for *Homo sapiens*. The age of the site was established at 350,000 years B.P. (before the present). These remains have been attributed to a *sapiens-erectus* intermediate type on the grounds that the remains, and the site, show a mixture of elements reflective of the transitional hominid evolutionary process. Such an assessment places Vértesszöllös man at the root of the *Homo sapiens* evolutionary line, some 100,000 years earlier than other specimens.

A better-known example of early *Homo sapiens* comes from a gravel deposit at Swanscombe near London, England. In 1935, 1936, and 1955, three related skull pieces were unearthed that fit together perfectly to form the back of a cranial vault with an advanced (over *Homo erectus*) cranial capacity of about 1,300 cubic centimeters. This has been dated to around 275,000 to 250,000 years B.P. A more complete skull of approximately the same age (dated to the Mindel-Riss interglacial period about 250,000 years B.P.) was found at Steinheim, in southern Germany, in 1933. Swanscombe's and Steinheim's advanced morphological characteristics, in combination with relatively primitive ones, such as low braincase heights, suggest that they are primitive members of the species *sapiens* and are representatives of a population intermediate between *Homo erectus* and *Homo sapiens*.

The finds at Swanscombe and Steinheim have been augmented by others from France and Italy, and especially from the Omo River region in southern Ethiopia. One Omo skull displays more mixed features (between *erectus* and *sapiens*) including flattened frontal and occipital areas, a thick but rounded vault, large mastoid processes (pointed bony processes, or projections, at the base of the skull behind the ears), and a high cranial capacity. Another skull is more fully *sapiens*, or modern in appearance. Some paleoanthropologists assert that the Omo group of fossils also helps bridge the gap between advanced *Homo erectus* and *Homo sapiens*.

Neanderthal Man

The best-known examples of early *Homo sapiens* come from a group of fossils known collectively as Neanderthal man. Their name derives from the place where the first fossil type was discovered in 1865, the Neander Valley near Düsseldorf, Germany. Similar Neanderthal fossil types have been found at more than forty sites in France, Italy, Belgium, Greece, the Czech Republic Slovakia, the former Soviet Union, North Africa, and the Middle East.

Neanderthal fossils tend to show an aggregate of distinctive characteristics that at one time led to their being regarded as a separate human species, *Homo neanderthalensis*. They are generally regarded as a subspecies of humans, with the designation *Homo sapiens neanderthalensis*. The characteristic features of their morphology include large heads with prominent supraorbital tori (thick brow ridges), receding jaws, stout and often curved bones, and large joints.

Most important Neanderthal fossils disclose large brain capacities (1,500 cubic centimeters) and are found in sites revealing complex and sophisticated cultures. These two facts clearly separate Neanderthal humans from more primitive "presapiens" species that exhibit some of the same morphological features. Neanderthalers generally stood fully erect between 1.5 and 1.6 meters in height; they were not the stoop-shouldered brutes of early characterizations. They lived during the last glaciation (the Würm glacial stage) in Eurasia. The sites from which most examples of the Neanderthalers have been recovered have commonly yielded tools of the Mousterian complex, a stone-tool industry named for the kind found at Le Moustier, France, and dating from about 90,000 to about 40,000 years B.P.

In fact two groups of Neanderthal humans seem to have existed. The first are referred to as classic Neanderthalers from such sites as Germany, France, Italy, Iraq (Shanidar man), and the former Soviet Union. The second group, known as either generalized or progressive Neanderthalers, lived contemporaneously with, as well as later than, classic Neanderthal humans. They display a combination of modern *sapiens* features and typical Neanderthal characteristics (especially the prominent supraorbital torus, the forehead ridge). Included in this category for the sake of simplification are those specimens termed neanderthaloid. Examples include Rhodesian man (from Zambia, formerly Northern Rhodesia) and Solo man (from Java), both unearthed at upper Pleistocene deposits (100,000 to 10,000 B.P.). Neanderthalers were cave dwellers and were well adapted to cold conditions (especially the classic Neanderthal variety). They used fire, manufactured stone flake tools, and buried their dead with care. They also seem to have practiced fairly complex religious rituals.

Cro-Magnon Man

Neanderthals were a successful group for many thousands of years, flourishing from about 127,000 B.P. to 37,000 B.P., with a wide distribution geographically. Neanderthal traces suddenly and mysteriously disappear from the fossil record, however, and they seem to have been superseded around 37,000 B.P. by other *Homo sapiens* with a more advanced culture and different morphology. In Europe, these are known as the Cro-Magnon peoples, so named for the Cro-Magnon cave near Les Eyzies in southwestern France, where the first skeletons were found in 1868 and where more than one hundred skeletons have since been discovered. Indeed, Cro-Magnon skeletal anatomy is virtually the same as that of modern European and North African populations. The skull is relatively elongated, with a large cranial capacity of about 1,600 cubic centimeters; the brow ridges are only slightly projecting. The average height of Cro-Magnon man was between 1.75 and 1.8 meters.

Cro-Magnon humans produced a culture that, in variety and elegance, far exceeded anything created by their predecessors. They made weapons and tools of bone and stone, stitched hides for clothing, and lived in freestanding shelters as well as caves. Some Cro-Magnon people produced beautiful cave paintings (they have been found in southwestern France and northern Spain) and bone carvings, and they modeled in clay. Though Cro-Magnon samples are the best-known examples of early *Homo sapiens sapiens*, mounting fossil evidence from sites outside Europe as well as genetic research performed in the 1980's suggests a much older date of origin for the emergence of modern man.

At Qafzeh, a cave near Nazareth, Israel, anatomically modern fossils classified as *Homo sapiens sapiens* were discovered in 1988 and reliably dated to 92,000 years B.P. In addition, newer fossil finds of progressive Neanderthalers from Kebara Cave in Israel, taken together with earlier Neanderthal finds from the caves of et-Tabun and es-Skhul, also in Israel, make it certain that progressive Neanderthalers and modern humans coexisted for many thousands of years.

Anthropologists have puzzled over the disappearance of the Neanderthals and, more important, over where they fit in the human family tree. It appears unlikely that classic Neanderthal humans were in the direct ancestral line of modern *Homo sapiens sapiens*. Reasons for their sudden disappearance are believed to include a combination of factors: extinction because of disease, lack of adaptation to the warmer climate following glaciation, and annihilation by the more advanced sapient groups.

Many scholars have considered the classic Neanderthals to be a cold-adapted, specialized side branch from the modern human line that became extinct as the climate became warmer. The generalized or progressive Neanderthals are considered by some to have avoided this specialization, perhaps continuing to exist through adaptation and ultimately being absorbed by flourishing modern human populations during the late Pleistocene epoch.

The Emergence of Modern Humans

Although the exact time place, and mode of the origin of the modern human species cannot yet be determined, genetic studies point to a date before 100,000 years B.P. Examination of mitochondrial DNA (mtDNA) from a sampling of present-day humans representing five broad geographic regions has allowed researchers to propose a genetic family tree and calculate roughly (assuming a fairly constant mutation rate) a temporal origin for the modern human population. Further studies seem to indicate that the modern human ancestral line emerged between 280,000 years and 140,000 years B.P. Genetic evidence, in concert with fossil finds, makes it plausible that a common ancestral population for *Homo sapiens sapiens* appeared in sub-Saharan Africa or the Levant (in the eastern Mediterranean region). Regional differentiation occurred, followed by radiation outward to other areas. The range of genetic and anatomical variability exhibited by fossil remains of modern humans is no greater than that known for the extant races of modern times.

During the late Pleistocene epoch (approximately 40,000 to 11,000 years B.P.) five different racial groups seem to have developed on the Eurasian and African landmasses. The last glaciation, approximately 30,000 to 10,000 years B.P., absorbed enough water to lower the oceans ninety meters below present levels. Emerging land bridges allowed people to move from Asia into North America, Australia, and elsewhere. In time the major racial groups became subdivided into smaller ones that resulted in the major races seen in modern times.

This view of racial diversification emphasizes the effectiveness both of geographic barriers in reducing free gene flow among varied groups of *Homo sapiens* and of environmental pressure in selecting different adaptive responses from the gene pool. These are also key factors in the entire evolutionary process by which modern humans developed over epochs into their present taxonomic position in the animal kingdom.

The Study of Paleoanthroplogy and Physical Anthropology

The study of human evolution is primarily the concern of the physical anthropologist and the paleoanthropologist. Evolution may be defined as change in the genetic composition of a population through time. Because evolution is thought to operate according to several principles and factors, modern human evolutionary theory is studied in the light of ideas and practices taken from different disciplines, including archaeology, biochemistry, biology, cultural anthropology, ecology, genetics, paleontology, and physics.

Early investigations into human evolution sought to establish the sequence of the human ancestral line through chronological and morphological analyses of hominid fossil remains (bones and teeth), thus placing them in their proper phylogenetic context (their natural evolutionary ordering). This remains the principal method of study, but it has been augmented by sophisticated techniques in fossil dating and new avenues of exploration into the evolutionary process, such as genetic research.

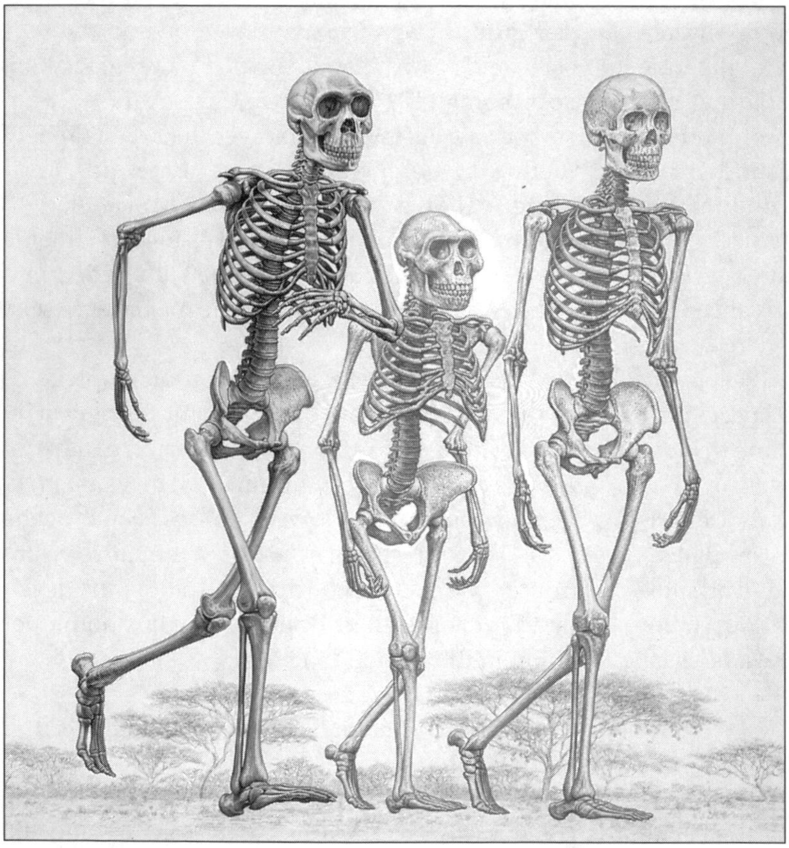

Determination of the accurate age of a fossil is most important, since it sets the fossil in a correct stratigraphic context that allows comparison with remains from the same geologic layer or level a great distance away. Accurate dating also has helped determine the order of succession for fossils that could not be established on morphological grounds alone.

The most valuable absolute dating methods are the radioactive carbon technique, which can effectively date specimens between 60,000 years B.P. and the present; the potassium-argon technique, which most easily dates material older than 350,000 years B.P.; and the fission-track method, which helps bridge the gaps between other methods. These methods are based on the constant or absolute rates at which radioactive isotopes of carbon, potassium, and argon decay. When absolute dating is impossible, investigators have ascribed a relative age to fossil remains by noting the contents of the layer of rock or the deposit in which the remains were found. A layer containing remains of extinct animals is likely to be older than one containing remains of present forms.

The anatomical differences between (left to right) Homo erectus, Australopithecus afarensis, *and* Homo sapiens *are small but significant: Increases in brain size and spinal cord size created the capacity for speech and increased mental properties.* (©John Sibbick)

In conjunction with dating, anatomical studies of fossil remains and comparisons with the morphological features of known hominid types, as well as comparisons to primate skeletal structures, have been primary approaches to the study of the evolutionary path of *Homo sapiens*. The species *Homo sapiens* (of which the modern human races compose a number of geographical varieties) may be defined in terms of the anatomical characteristics shared by its members. In general, these include a mean cranial capacity of about 1,400 cubic centimeters, an approximately vertical forehead, a rounded occipital (back) part of the skull, jaws and teeth of reduced size, and limb bones adapted to fully erect posture and bipedalism. Scientists assume that any skeletal remains which conform to this pattern and cannot be classified in other groups of higher primates must belong to *Homo sapiens*.

It is striking that the anatomical differences observed between *Homo erectus* and *Homo sapiens* have been confined to the skull and teeth. The limb bones thus far discovered for both are similar (though *erectus* appears more robust). Cranial capacity and morphology continue to be the dominant determining boundary separating sapient and presapient human species.

The Contribution of Other Sciences and Social Sciences

Human adaptability studies, using techniques from physiology, demographics, and population genetics, investigate all the biological characteristics of a population that are caused by such environmental stresses as altitude, temperature, and nutrition. It is believed that these normal stresses acted as genetic selectors in prehistoric times and continue to do so. Such racial variants as skin color and body hair are observable products of these stresses. The investigation of climatic changes during prehistoric epochs as revealed in the geologic record is important for understanding those pressures affecting the evolutionary history of man.

Genetic studies have become indispensable to the study of human evolution. Four forces have been identified as fundamental in the evolutionary process: mutation, natural selection, gene flow, and genetic drift. Since mtDNA is inherited through the female, it is possible to calculate how much time has elapsed since the mutations that gave rise to present variations originated in prehistoric populations.

Also important to the study of the evolution of *Homo sapiens* is the examination and classification of cultural remains preserved at hominid fossil sites. Not only can the relative date of a fossil be supported, but sometimes it is also possible to reconstruct the environmental situation that may have influenced the evolutionary process operating in a population. Cultural response is an integral part of hominid adaptation, and it in turn influences natural selection. Technology changes the physical and economic environment, and economic changes alter the demographic situation. Humans continue to promote or influence their own evolution by willingly or unwillingly altering the environment to which they must adapt.

The modern methods useful for investigating the evolutionary history of *Homo sapiens* are multidisciplinary. While each of them reveals an aspect of the emergence of modern man and complements the other methods of study, emphasis is placed on careful fieldwork, accurate dating, and comparative morphological analyses of hominid fossil remains. Increasing in importance, however, is the accumulating wealth of genetic data on human population relationships.

Human Evolution in the Context of Animal Evolution

Increasing attention is being given to the biological and behavioral changes that led to the emergence of *Homo sapiens sapiens*—the last major event in human evolution. Mounting evidence continues to push backward in time the point at which modern *Homo sapiens* made his appearance in the evolutionary scheme. The finds at Qafzeh, for example, indicated that modern man arose fifty thousand years earlier than had previously been thought. A clearer understanding of the evolutionary history of modern *Homo sapiens* has not only helped to define the place of the modern human species more accurately in relation to the rest of the animal kingdom but also helped to illuminate the pressures, adaptations, and changes that have made humans what they are.

The accumulating data on the evolutionary appearance of *Homo sapiens* have allowed biologists and anthropologists to see the rise of modern man as part of the evolutionary development of the animal kingdom in general and of primates in particular. The primates that exist today make up a remarkable gradational series that links *Homo sapiens sapiens* with small mammals of very primitive types.

Through the pressures and process of evolution (including adaptation and natural selection), *Homo sapiens* has become one of the most successful and adaptive animals that ever lived, because he came to possess an elaborate culture (culture is based on learned behavior). The key is *Homo sapiens'* superior mental capacity. Only human beings can assign arbitrary descriptions to objects, concepts, and feelings and can then communicate them unambiguously to others. In the late Middle Pleistocene, the hominid branch that gave rise to early *Homo sapiens* witnessed an increase in brain size, complex social organizations, continual use of fire, and perhaps even language. As to what initiated these changes, many have suggested tool use and, in turn, a hunting economy.

In a classic article published in 1960, entitled "Tools and Human Evolution," Sherwood Washburn argued that the anatomical structure of modern man is the result of the change, in terms of natural selection, that came with the tool-using way of life. He stated that tools, hunting, fire, and an increasing brain evolved together. Washburn also argued that effective tool use led to effective bipedalism—another significant characteristic of *Homo sapiens:* Man is different from all other animals because he became a user of increasingly complex tools.

The other behavioral pattern that is seen to have been of utmost importance to sapient evolution is big-game hunting. Early *Homo sapiens* was undoubtedly a big-game hunter, as were all of his successors until approximately 8,000 years ago. It has been argued that human intellect, interests, emotions, and basic social life are the evolutionary products of the success of hunting adaptations. Success in hunting adaptation dominated the course of human evolution for hundreds of thousands of years. The agricultural revolution and the industrial and scientific revolutions are only now releasing human beings from conditions characteristic of 99 percent of their evolutionary history.

Scholars have suggested that research into human origins and development is much more relevant than is often realized. It has been argued that, although man no longer lives as a hunter, he is still physically a hunter-gatherer. Some investigators in the field of stress biology, the study of how the human body reacts to stressful situations, feel that man is biologically equipped for one mode of life (hunting) but lives another. Thus, there would be some link between an emotional reaction, such as explosive aggression, and human evolutionary history. Tools and more efficient hunting helped produce great change in hominid evolution and made man what he is. Humans continue to be users of increasingly complex tools, such as computers, and perhaps this continued development of technology may determine the future evolutionary path of *Homo sapiens*.

—Andrew C. Skinner

See also: Apes to hominids; Convergent and divergent evolution; Evolution: Animal life; Evolution: Historical perspective; Extinction; Fossils; Gene flow; Genetics; Hominids; Neanderthals; Natural selection; Nonrandom mating, genetic drift, and mutation; Primates; Systematics.

Bibliography

Brauer, Gunter, and Fred H. Smith, eds. *Continuity or Replacement: Controversies in "Homo Sapiens" Evolution*. Brookfield, Vt.: A. A. Balkema, 1992. A collection of technical papers from a 1988 international symposium on the biological and cultural evolution of hominids.

Cann, R. L., M. Stoneking, and C. A. Wilson. "Mitochondrial DNA and Human Evolution." *Nature* 325 (January 1, 1987): 31-36. Though written for the knowledgeable student of human evolution, this important article presents the results of a major study using DNA to trace genetic differences in widely varied samples of the present population and, by constructing an evolutionary family tree, to propose a time and place for the emergence of modern man. It shows the increasing importance of genetics for the study of the modern human evolution.

Eldredge, Niles, and Ian Tattersall. *The Myths of Human Evolution*. New York: Columbia University Press, 1982. In fewer than two hundred pages the authors give the nonspecialist a brief, comprehensive look at human evolutionary history, while attempting to show that the once-standard expectations of evolution—slow, steady, gradual progress—are not supported by the evidence. According to the authors, fossil evidence shows human evolution to be the result of long periods of stability interrupted by abrupt change, occurring in smaller populations. The thesis is an important consideration presented in well-written form.

Johanson, Donald, Lenora Johanson, and Blake Edgar. *Ancestors: In Search of Human Origins*. New York: Villard Books, 1994. A companion to the three-part *Nova* television series. Chronological coverage of *Australopithecus*, *Homo habilis*, *Homo erectus*, archaic *Homo sapiens*, the Neanderthals, and *Homo sapiens*. Focuses on the australopithecines and early *Homo*. Lavishly illustrated.

Leakey, Richard E. *The Making of Mankind*. New York: E. P. Dutton, 1981. This book, by the son of Louis and Mary Leakey, is written from the perspective of one who grew up with famous physical anthropologists and scientists who were tracing human origins. The text is complemented with many color photographs. Chapters discussing the development of language and early *Homo sapiens* culture are important. It also distills and synthesizes important ideas of others.

Lewin, Roger. *The Origin of Modern Humans*. New York: Scientific American Library, 1998. Overview of the key concepts in the origin of modern humans, including the molecular biology and archaeology supporting the out-of-Africa hypothesis versus the multiregional hypothesis of human evolution.

Mellars, Paul, ed. *The Emergence of Modern Humans: An Archaeological Perspective*. Ithaca, N.Y.: Cornell University Press, 1990. A volume of proceedings from a 1987 conference on the origins of modern humans and the adaptations of hominids. Concentrates on the archaeology of hominids in western and central Europe.

Phenice, Terrell W. *Hominid Fossils*. Dubuque, Iowa: Wm. C. Brown, 1973. One of the best introductory illustrated keys of hominid fossil remains of the Pleistocene epoch. The drawings are all oriented the same way, and each fossil illustration is listed with its measurements and with information concerning tools found at the respective sites. This book (and similar types of atlases) is very helpful to the nonspecialist.

Tattersall, Ian. *The Fossil Trail: How We Know What We Think We Know About Human Evolution*. New York: Oxford University Press, 1995. The reconstruction of human evolution on the basis of interpreting fossil remains, including the Laetoli footprints, Lucy, Turkana Boy, and others. Good coverage of the development of dating techniques such as fluorine analysis and mitochondrial DNA.

Waechter, John. *Man Before History*. Oxford, England: Elsevier-Phaidon, 1976. This is a concise, readily accessible volume on the evolutionary history of humans. It contains some of the finest visual and graphic representations (photographs and illustrations) on the subject found in any single work on the evolutionary journey of the human species. It also contains an extensive glossary of terms with numerous illustrations. Particularly strong in its discussion and representation of late prehistoric culture (especially art), it is an excellent reference work geared toward the nonspecialist.

HORMONES AND BEHAVIOR

Type of animal science: Physiology
Fields of study: Ethology, neurobiology

Hormones exert a tremendous influence on behavior by heightening perceptions, altering motivational levels, and controlling drives, including hunger, sex, and the urge to migrate.

Principal Terms

ABLATION: the technique of removing a gland to determine its function and observe what effects its removal will precipitate

ANDROGENS: masculinizing hormones, such as testosterone, responsible for male secondary (anatomical) sex characteristics and masculine behavior

BEHAVIOR: an animal's movements, choices, and interactions with other animals and its environment

BIOLOGICAL CLOCK: a timekeeping mechanism that is "endogenous" (a part of the animal) and capable of running independently of "exogenous" timers such as day-night cycles or seasons, although the clock is normally set by them

ENDOCRINE SYSTEM: a collection of glands that secrete their products into the bloodstream

ESTROGEN: a feminizing hormone responsible for female secondary (anatomical) sex characteristics and sex-related behaviors

PROLACTIN: a hormone responsible for secretions of milk from the mammary glands of mammals and from the crops of birds

RECEPTOR MOLECULE: a molecule on the cell membranes of target tissues that binds to the hormone molecule and initiates the action of the hormone

Behavioral differences are usually attributed to two causes: differences in experience (learning) and differences in heredity (genes). Hormonal interaction involving an organism, its experience, and the highly specific behaviors that result are well illustrated by Daniel Lehrman's 1964 study of the ring dove, a small relative of the domestic pigeon.

A male ring dove begins courtship by bowing and cooing to the female when they are placed together in a cage. Toward the end of the first day of courtship, the birds choose a location and start building a nest. The doves court, nest-build, and mate, and the female lays an egg about five o'clock in the afternoon on the seventh to eleventh day. A second egg is laid about nine o'clock the next morning. The male sets for about six hours at midday, and the female occupies the nest the rest of the time. After fourteen days of incubation, the eggs hatch, and the parents feed the squabs crop milk, secreted from the lining of the adult dove's crop (enlarged gullet). At about two weeks of age the squabs (fledgling birds) begin pecking at grain, and the adults feed them less and less. With feeding chores diminished, the male begins bowing and courting, the pair start nest-building, and the cycle repeats itself.

The Hormonal Triggers of Behavior

The simplicity of this description belies the hormonal ferment going on beneath the placid exterior. The courtship ritual causes the production and release of estrogen and progesterone, hormones responsible for setting behavior and for the development of the oviduct. The oviduct develops from eight hundred milligrams, when the doves are placed together, to four thousand milligrams, when the first egg is laid. Birds presented with nests already containing eggs when they are first paired will build nests on top of the eggs. Even if the eggs are returned to the top of the nest by the investigator, the doves will not set until they have engaged in nest-building for five to seven days. On the other hand, if the doves are first injected with progesterone, 90 percent will set within three hours after pairing. Courtship and nest-building play a vital role in creating the hormonal conditions necessary for setting behavior. Development of the crop is, in turn, initiated when setting begins. Setting behavior causes the release of prolactin, which stimulates crop development and feeding behaviors. Unpaired doves injected with prolactin will develop crop milk and will feed squabs when exposed to them even if they have not mated. While they will feed squabs, however, they will not sit on eggs, because they have not courted, engaged in nest-building, or developed the hormonal balance necessary to support setting behaviors.

This ring dove study indicates that the behavior of one individual can alter the behavior and the hormonal balance of another individual; that an individual's behavior can alter its own hormonal balance and, hence, its own behavior; and that inanimate aspects of the environment (nest and eggs, for example) can alter an individual's hormonal balance and, hence, its behavior. The interactions among these three factors are complex. For example, the male's behavior (courting) changes the female's behavior (nest-building), which changes her hormonal balance (estrogen and progesterone), which causes her to alter the external environment (lay eggs in the nest), which affects the pair's behavior (setting), which changes their hormonal balance (prolactin), which stimulates them to feed the squabs.

Fetal Hormones Affecting Adult Behavior

Hormones produced by developing organisms have marked effects on adult behaviors. In mammals, the fetal testis becomes active and produces testosterone, then becomes inactive before birth and remains so until puberty. The changes resulting from this brief surge of testosterone are remarkable. They include the anatomical features that distinguish male and female genitalia, changes in neural anatomy, and the sensitivity of nervous (and other) tissues to adult hormones.

Rat and mouse fetuses have several littermates, which develop side by side in a common uterus, like peas in a pod. Male mice that develop between two male embryos are more aggressive as adults than males developing between females. Similarly, females that develop between two male embryos are more aggressive as adults than females that develop between other females. In another study, females from litters that were predominantly male showed more masculine behavior (such as the mounting of other females) than females from predominantly female litters. The explanation is that testosterone produced by the male embryos' testes is absorbed into the bloodstream of sibling embryos, altering their nervous systems and hence their behaviors. In cattle, testosterone produced by a bull calf twin affects the development of his heifer twin to the extent that she is usually sterile.

Scientists have shown that pregnant rhesus monkeys treated with testosterone produced offspring that showed rougher play and more threat behavior than usual. Male rhesus monkeys experience a decrease in blood testosterone levels within six hours after losing a fight to another male and are more submissive. These studies indicate that hormones play an important role in determining male-female behavioral differences.

Many biological phenomena are repeated or change intensity over and over again throughout the life of an individual. Examples include sleep-wake cycles, menstrual cycles, and the migration cycles of birds; these are repeated approximately once a day, once a month, and once a year, respectively. The regularity of these cycles led biologists to propose a "biological clock." The golden-mantled ground squirrel avoids freezing temperatures by going into hibernation once a year. Even if these squirrels are kept in constant conditions of light and temperature to deprive them of seasonal cues, they will enter hibernation once a year. These and other data lead researchers to believe that the clock resides within the animal. Although it can be reset by environmental cues, it can also run independently of them.

Hormones, Seasons, and Mating Behavior

A white-crowned sparrow, nesting in central Alaska, experiences dramatic seasonal changes and migrates to the southern United States or Mexico (more than three thousand kilometers) to avoid freezing. Central Alaska's short summer demands that the sparrow fly north as early in the spring as is safe and that it be prepared for mating and rearing chicks when it arrives. During the winter, the gonads atrophy to 1 percent or less of their breeding season weight. The bird's ability to sense the approach of spring depends on its sensing the increase in daylight. During the short winter days, the sparrow is content to stay in Arizona or Mexico, but as day length increases to fourteen or fifteen hours, the bird's hypothalamus releases hormones that stimulate the pituitary to release prolactin and gonadotropic hormones. The gonads respond by increasing in size and producing additional hormones, which stimulate the bird to begin its long migration.

When the male white-crown arrives at his breeding grounds in central Alaska, he chooses a nesting territory, attacks any male territorial intruders, and attempts to attract a mate with his constant singing. Each female chooses a mate and helps him defend the nesting site. In the next few days she feeds to gain nutrients for egg production, and her estrogen levels rise rapidly, stimulating her to solicit mating. Once the eggs are laid, the gonads of both birds begin to atrophy, estrogen and testosterone levels decline, and prolactin levels increase and stimulate feeding of the young. As the gonads atrophy, the birds become less aggressive and the male stops singing.

As the young become independent, both parents enter a "sexual refractory period," during which the gonads will not respond to artificially increased day length as they would in the spring. The birds feed voraciously, increasing body fat, which serves as fuel for the long trip south. In the next year, by early spring, the birds will have passed through the refractory period and be primed to respond to the increasing day length with a fresh hormonal flurry which will set them off on the long journey north.

Recognizing the existence of a refractory period is important. It underscores the ideas that while birds do respond to environmental conditions (day length), there is a given set of events through which the physiological machinery passes and that specific time parameters are dictated by the biological clock. White-crowned sparrows can be expected to show hormonal changes and migratory restlessness during springtime even if they had been caged and maintained in constant conditions. It is to the bird's advantage, however, to experience and recognize the seasonal changes in day length, because biological clocks tend to run a bit fast or slow. The actual measuring of day lengths allows the bird to reset that clock and arrive in Alaska at the most advantageous time for rearing a family of sparrows.

Studies of a closely related bird, the white-throated sparrow, indicate that the changes of behavior and physiology are primarily the result of two hormones: corticosterone from the adrenal cortex and prolactin from the anterior pituitary. Both hormones have daily peaks of secretion, but the timing of these daily peaks (relative to each other) changes with the seasons. If injections of these hormones are given with timing differences characteristic of specific seasons, the physiological and behavioral changes seen in the birds are characteristic of the seasons that the injections mimic.

Experimental Endocrinology

The earliest report of experimental endocrinology, in the mid-nineteenth century, demonstrated that replacements of testicular tissue would maintain comb growth and sexual behavior in castrated roosters. Techniques for determining endocrine function used today include ablation and replacement.

Hormones trigger many kinds of animal behavior, including bird migration. (Digital Stock)

Ablation (removal) of endocrine tissue results in deficiency symptoms. The effects of ablation are not always unambiguous. If the testes and accessory tissues are not completely removed when a horse is castrated, for example, tissue capable of producing testosterone remains, and the consequence is an infertile gelding that behaves like a stallion.

Hormones produced by different glands can have similar physiological effects. Both the adrenal glands and the testes produce androgens (masculinizing hormones). Sexually experienced male cats do not lose their sex drive if castrated, and researchers do not have a satisfactory answer as to why this occurs. Perhaps the adrenal hormones are sufficient to maintain established feline male sexual behavior but not sufficient to initiate it in inexperienced cats. The ablation of the adrenal glands, however, has severe consequences in terms of electrolyte and blood glucose imbalances that are life threatening. Replacement of ablated endocrine tissue can reinstate normal function. If a male cat is castrated as a kitten, it will not develop normal male sexual behaviors. If, however, a normal testis is later transplanted to the abdominal cavity (or elsewhere), normal behavior will develop.

In the early 1960's, Janet Harker became convinced that the "biological clock" controlling the daily activity cycle of the cockroach was contained in the subesophageal ganglion, a patch of nervous tissue the size of a pin head resting just below the esophagus. When she ablated this ganglion, the cockroach became arrhythmic. Harker removed the legs from a normal roach and glued the roach on top of the arrhythmic roach, surgically uniting their body cavities so that the same body fluids circulated through both roaches. The arrhythmic roach ran about the cage with an activity rhythm dictated by the rhythm of hormones released into the body fluids by the legless roach on its back.

Most hormone and behavior studies involve nonhuman species and most involve sexual behaviors. Most behaviors are oriented toward perpetuating one's species. Only those individuals with behaviors conducive to rearing offspring will provide the genetic basis for behaviors represented in the next generation. It may seem that wild, unrestrained mating would be selected for, but that makes no real sense. Mate selection, shelter-seeking, feeding, maintenance of social position, and a host of other behaviors are critical to the success of one's progeny. Individuals that produce more offspring than they can feed or protect usually rear fewer than those who produce fewer to begin with. Many predator species (owls and wolves, for example) tend not to mate in years when prey is scarce. This ultimately maximizes reproduction by reserving energies that can best be spent later. This restraint is mediated by adjusting hormonal levels. Hormones have been called the ultimate arbiters of sexual behavior.

The Endocine System, the Nervous System, and Behavior

The nervous system is usually thought of as the mediator of behavior, but the endocrine system is also a major player. Arguing that one system is more important would be like deciding whether height, width, or depth is more important in describing a box. It is useful, however, to discuss their differences. The nervous system has a shorter response time. Nerve impulses travel at speeds of up to 120 meters per second. Hormonal effects are much slower but are less transitory. If frightened by a false alarm, an animal may jump and run as a direct consequence of nervous system activity, but even after it recognizes that there is no real threat, it will be "keyed-up." This is a consequence of hormonal activity: Fright triggered the release of epinephrine (adrenaline) and norepinephrine from the adrenal glands. These hormones cause increased cardiac output; increased blood supply to the brain, heart, and muscles; decreased blood flow to the digestive tract; dilation of airways to breath more efficiently; and a significant increase in metabolic rate.

This is called the "fight or flight reaction," and it will affect behavior for several minutes and possibly for hours. These hormones enhance perceptions and elevate the responsiveness of the nervous system. In the final analysis, the understanding of hormones, what determines their ebb and flow, how they are affected by the environment, how hormones interact with one another, and how their levels are controlled by genetic programs of the individual and of the species is essential for the understanding of behavior.

—Dale L. Clayton

See also: Brain; Communication; Courtship; Endocrine systems in invertebrates; Endocrine systems in vertebrates; Hormones in mammals; Instincts; Mating; Nervous systems of vertebrates; Nesting; Pheromones; Reproduction; Reproductive systems of female mammals; Reproductive systems of male mammals; Rhythms and behavior; Sex differences: Evolutionary origins; Smell; Territoriality and aggression.

Bibliography

Adler, Norman T., ed. *Neuroendocrinology of Reproduction*. New York: Plenum, 1981. A chapter by E. Adkins-Regan, entitled "Early Organizational Effects of Hormones: An Evolutionary Perspective," presents a very comprehensive review of the developmental effects of hormones on behavior. It has broad taxonomic coverage, ranging from coelenterates to mammals.

Alcock, John. *Animal Behavior: An Evolutionary Approach*. 6th ed. Sunderland, Mass.: Sinauer Associates, 1998. This is a popular, well-written textbook covering the general area of animal behavior. Chapters 4 and 6 contain especially interesting and valuable material on hormones and sexual behavior.

Becker, Jill B., S. Marc Breedlove, and David Crews. *Behavioral Endocrinology*. Cambridge, Mass.: MIT Press, 1992. Introductory textbook for undergraduates, drawing on both vertebrate and invertebrate systems. Covers sexual, courtship, and parental behavior, aggression, stress response, ingestive behavior, and biological rhythms.

Crews, David, ed. *Psychobiology of Reproductive Behavior: An Evolutionary Perspective*. Englewood Cliffs, N.J.: Prentice-Hall, 1987. This book deals with the many facets of reproductive behavior—their proximate and ultimate causes. It has good material in several chapters about the effects of hormones on behavior.

Drickamer, L. C., and S. H. Vessey. *Animal Behavior: Concepts, Processes, and Methods*.4th ed. Boston: McGraw-Hill, 1996. Chapters on the nervous system and behavior and hormones and behavior will be of interest to anyone interested in the topic of hormones and behavior.

Nelson, Randy Joe. *An Introduction to Behavioral Endocrinology*. 2d ed. Baltimore: The Johns Hopkins University Press, 2000. A textbook for upper-level undergraduates and graduate students. Explores the evolution of the anatomy, physiology, and biochemistry of human and animal endocrine systems. Discusses sex differences, reproductive and parental behavior, biological rhythms, and the role of hormones in learning and memory.

Schulkin, Jay. *Neuroendocrine Regulation of Behavior*. New York: Cambridge University Press, 1999. Discusses the roles of steroids and neuropeptides in regulating behavior. Uses model animal systems to study the influence of hormones on the brain, and generalizes to other animals and humans. Aimed at upper-level undergraduate and graduate students, professors, and medical healthcare professionals.

Svare, Bruce B., ed. *Hormones and Aggressive Behavior*. New York: Plenum, 1983. A series of review articles pulling together much important work on the effects of hormones on aggressive behaviors to appear in the 1970's and 1980's.

HORMONES IN MAMMALS

Type of animal science: Physiology
Fields of study: Developmental biology, zoology

Mammals use chemical messengers called hormones for information transfer and control between different body regions. These molecules, produced and secreted from endocrine glands into the bloodstream for transport, play important roles in mammalian development from conception to death.

Principal Terms

ENDOCRINE SYSTEM: an array of ductless glands scattered throughout the mammalian body that produce and secrete hormones directly into the bloodstream

HOMEOBOX: a set of genes that encode proteins involved in development of a wide range of animal species, from nematodes to insects to mammals

HOMEOSTASIS: the maintenance of constant conditions within the internal environment of an organism, a process controlled by antagonistic hormone pairs

HYPOPHYSIS: the pituitary gland, or "master" gland, which produces and secretes at least eight protein hormones influencing growth, metabolism, and sexual development

HYPOTHALAMUS: a brain region just below the cerebrum that interconnects the nervous and endocrine systems of mammals, thereby controlling most hormone production and many body functions

NEUROTRANSMITTER: a signaling molecule that provides neuron-to-neuron communication in animal nervous systems; some double as hormones

PROTEIN HORMONE: a hormone type composed of protein, a long chain of amino acids encoded by a gene

STEROID HORMONE: a hormone type derived from cholesterol, a type of fat molecule

Mammals are vertebrate animals that have fur (hair) and that nourish their young with milk by means of mammary glands, which are modified sweat glands. Mammals include, among other animals, primates (such as humans and chimpanzees), cetaceans (such as whales and dolphins), and marsupials (such as kangaroos, koalas, and opossums). Mammals are sexually reproducing and diploid (having two copies of every chromosome). During fertilization, a sperm from the male parent unites with an egg from the female parent to produce a diploid single-celled zygote. The zygote contains all the genetic information for all the cells of the future individual. The zygote divides first into two cells, then four, eight, sixteen, thirty-two, and so on.

As the cells divide, they begin to specialize, so that different groups of cells assume unique functions. Some cells become nerve cells, others skin cells, and others blood cells. A number of factors, including hormones, contribute to differentiation.

Throughout the development of the organism, profound changes such as birth, puberty, menopause, aging, and death occur. Progressive changes in cell functions contribute to these sequential processes. Many of these developmental changes are controlled by hormones, chemical messengers that provide communication between cells located in different portions of the body via the bloodstream.

The Endocrine System

Hormones fall into the two principal categories of protein hormones and steroid hormones. Protein hormones are composed of protein—long chains of amino acids encoded by genes. Steroid hormones are derivatives of cholesterol. Both hormone types function in the same fashion: They control genes. Hormones are produced and secreted from a source endocrine gland and are then transported through the bloodstream to a target tissue, where they penetrate cells and concentrate on the control regions of genes located on chromosomes. Once at a control region, a given hormone either activates or inactivates the gene. If a gene is activated, messenger RNA will be produced, leading to protein production. If a hormone inactivates a gene, protein production will cease. A given hormone may activate certain genes and inactivate others.

Endocrine glands are ductless glands (glands that lack channels for secreting their products) that produce and secrete hormones into the bloodstream. Major mammalian endocrine glands include the hypothalamus, hypophysis, thyroid, parathyroids, thymus, pancreas, adrenals, and gonads. The hormones secreted from these glands influence many cells and each other during mammalian development.

Homeostasis, the maintenance of a constant internal environment, is a major objective of endocrine hormones. They work antagonistically (against each other) to maintain various body conditions (such as blood sugar and calcium levels) in equilibrium. Endocrine hormones work principally, but not exclusively, by negative feedback. For example, a region of the brain called the hypothalamus reacts to various body conditions by releasing hormones that stimulate the nearby hypophysis (the pituitary gland) to release certain of its hormones. The hypophyseal hormones direct other glands and tissues to respond in a particular fashion. Once bodily conditions are back to normal, the hypothalamus terminates its initial stimulatory hormones, thereby stopping the entire sequence of events.

The hypothalamus controls a number of critical body functions, including the activities of endocrine glands, body temperature, wake and sleep cycles, and appetite. Ultimately, all these functions involve some type of hormones. When various conditions occur in the body (for example, hyperthermia, which is increased body heat), genes in certain hypothalamic cells synthesize special proteins called releasing factors that are sent into the bloodstream to activate target cells in glands (in the example cited, sweat glands) located elsewhere in the body. Often, the target of hypothalamic releasing factors is a nearby endocrine gland called the hypophysis, also known as the pituitary, or "master gland."

Regulating the Reproductive Cycle

The eight hormones known to be released from the hypophysis gland are vasopressin (the "antidiuretic hormone"), oxytocin, prolactin, growth hormone, thyrotropin, adrenocorticotropic hormone (ACTH), follicle-stimulating hormone (FSH), and luteinizing hormone (LH). Vasopressin is released in response to low water levels in the blood; it stimulates the kidneys to retain water, reduce urine output, and increase blood pressure until blood water levels return to normal, upon which it will no longer be produced. Oxytocin causes muscular contractions in the uterus during childbirth and in the breast for the secretion of milk for an infant. Prolactin is present in males and females, but it is functional only in females. It stimulates milk production from fat deposits in the breast. Growth hormone causes growth in children; it is present in adults, but contributes only to the control of metabolic rate. Thyrotropin stimulates the thyroid gland to produce and secrete various hormones that control metabolism (examples: thyroxine and triiodothyronine). Adrenocorticotropic hormone stimulates the adrenal cortex, located above each kidney, to release its metabolism-controlling steroid hormones. At puberty, follicle-stimulating hormone stimulates the female ovarian follicle to mature and begin producing the steroid hormone estrogen; it directs the male testes to begin producing sperm. Also at puberty, luteinizing hormone stimulates the ovary to begin producing eggs and the steroid hormone progesterone; it directs the testes to begin producing the steroid hormone testosterone.

The hypophyseal hormones, all proteins encoded by genes, have a major impact upon metabolism and development in mammals. This is especially true for the sexual-cycle hypophyseal hormones FSH and LH. In females, puberty begins with the first menstrual cycle. Each menstrual cycle is the female body's way of preparing for a possible pregnancy. At the beginning of the cycle, the hypophysis produces high concentrations of FSH, which stimulates the ovarian follicle to develop and produce estrogen, a steroid hormone that increases body fat in regions such as the buttocks and breasts. Simultaneously, increased LH production matures the egg in the ovarian follicle and stimulates progesterone production. Progesterone causes the endometrium (the lining of the uterus) to increase its blood vessel content and thickness for receiving and maintaining a fertilized egg and for the subsequent long gestation period (nine months in humans). If the egg is fertilized by sperm, it will adhere to the endometrium, and progesterone will continue to be secreted to maintain the endometrium and the pregnancy. If the egg is not fertilized, progesterone levels will drop, estrogen levels will rise, the endometrium will be sloughed away (menstrual bleeding will occur), and the cycle will start all over again.

The primate female menstrual cycle is only one very complex example of how hormones are intricately involved in mammalian developmental processes. There are many subtler aspects of the menstrual cycle that still are not well understood, such as the identity of the hormonal signal from the fertilized egg that stimulates the female ovary to continue progesterone production for continuation of the pregnancy. All mammalian hormones are interconnected by cause-and-effect relationships. Tremendous research remains before a clear and complete picture of hormonally controlled mammalian development will emerge.

Regulating Other Functions

The thyroid gland, located in the throat region, produces several hormones (examples: thyroxine and triiodothyronine) that elevate the body's metabolic rate. The thyroid also secretes a hormone called calcitonin, which works antagonistically with the hormone parathormone produced by the adjacent parathyroid glands. When blood calcium is high (as in a condition called hypercalcemia), the calcitonin gene in thyroid cells begins producing the protein hormone calcitonin, which stimulates bone cells called osteoblasts to build more bone, thereby removing calcium from the bloodstream. Once blood calcium levels are back to normal, calcitonin production halts. In hypercalcemia, the parathormone gene in parathyroid cells begins producing parathormone (also a protein hormone) that stimulates bone cells called osteoclasts to break down bone, thereby restoring blood calcium levels but possibly contributing to osteoporosis (bones that are brittle because of calcium deficiency) and other bone-related disorders. Parathormone production is stopped once blood calcium levels are back to normal.

The islets of Langerhans in the pancreas secrete two antagonistic protein hormones, insulin and glucagon. In response to high glucose levels in the blood (as in hyperglycemia), genes in beta cells produce and secrete insulin, which directs body cells, especially liver cells, to absorb glucose and store it as a polysaccharide called glycogen. Insulin production will stop once blood glucose levels are reduced to normal. An insulin deficiency leads to prolonged hyperglycemia, a serious and often fatal disorder called diabetes mellitus. When blood glucose levels are too low (as in hypoglycemia), genes in the alpha cells of the islets of Langerhans produce and secrete glucagon, which directs body cells to break down their glycogen reserves and begin releasing glucose back into the bloodstream until normal blood glucose levels are reached, upon which glucagon production ceases.

Further endocrine glands include the adrenal cortex, located on top of each kidney, which secretes three major classes of steroid hormones: the glucocorticoids such as cortisol, which controls fat and protein metabolism; the mineralocorticoids such as aldosterone, which controls blood sodium levels; and the androgens (male sex steroids). The adrenal medulla, located internally to the adrenal cortex, is derived from nervous tissue and secretes two hormones, epinephrine and norepinephrine, that double as excitatory neurotransmitters at nerve axon endings; chemical energy transmission between nerve cells occurs at synapses, or gaps, between adjacent neurons. Neurotransmitters are protein hormones that relay electrical impulses from one neuron (nerve cell) to another throughout the trillion-cell nervous systems of mammals. Other neurotransmitters include the excitatory acetylcholine and inhibitors glycine, enkephalin, and gamma-aminobutyric acid.

The kidney secretes the protein hormone erythropoietin when the blood has a low red blood cell level; erythropoietin stimulates the undifferentiated stem cells called hemocytoblasts in the red bone marrow of flat bones (ribs, sternum) to differentiate and develop into mature red blood cells. Platelet-derived growth factor (PDGF) is released from damaged blood vessels to activate platelet cells to begin blood clotting. Macrophage colony stimulatory factor and eosinophil chemotactic factor are two hormones that both activate and attract certain respective immune system cells to the site of an infection or allergic reaction. Histamine is released from damaged tissue and causes blood vessel dilation, so that the vessels are more leaky, thus allowing hormones and other molecules to reach the injury site, eventually leading to the inflammation and itching associated with wound healing.

The hormone prostaglandin helps inflammation and contracts some smooth muscles located throughout the body; nerve growth factor stimulates the growth of sensory nerves throughout the body; and epidermal growth factor stimulates the growth of the epidermis, the outermost skin layer that is constantly being shed and replaced. Sunlight exposure to skin produces cholecalciferol, or vitamin D, which helps to stimulate bone growth and maintenance.

The hormone prolactin is responsible for stimulating milk production in mammals. (PhotoDisc)

Developing the Full Hormonal Picture

The tally of mammalian hormones extends well beyond the molecules just discussed. What is most puzzling is how the many hormones are interconnected during the control of development. Hormones control gene activities of target cells, and some simple hormone systems act antagonistically (calcitonin-parathormone, insulin-glucagon). Yet there has not emerged a clear and complete picture of the overall interactions. Hormones control an incredibly complex array of cellular activities from conception to death.

Mammalian developmental hormones have been studied using a variety of biochemical and physiological experiments: isolation and purification experiments, injection into experimental animals, studies of metabolic disorders in animals, and molecular genetics experiments. Protein structures based upon genetic and biochemical studies are well understood. Steroid hormone structures have been unraveled from studies of cholesterol biochemistry.

Dissections of experimental animals yield intact endocrine glands (such as the thyroid and pancreas) that can be used to show function. Chemical secretions from these glands can be extracted and separated into the various hormone components by several biochemical techniques such as electrophoresis, chromatography, and centrifugation. The isolated hormones can be further purified by rerunning them through these separatory techniques.

Electrophoresis involves the separation of molecules in an electric field based upon their sizes and charges. Large molecules move slowly, whereas small, compact molecules move more quickly. Protein hormones move from the negative pole to the positive pole in electrophoretic gels. Affinity chromatography involves placing membrane hormone receptor proteins on a vertical column containing a porous resin. The specific hormone type that binds to this particular target receptor protein will stick to the resin. Nonbinding hormones will wash through the column. Finally, ultracentrifugation separates molecules based upon size in incredibly high-spinning gravity fields measuring about 100,000 times the earth's gravity. These three techniques, plus a few others, are very effective in isolating and purifying hormones as well as other important molecules.

Isolated hormones have been injected into experimental organisms and organ extracts, followed by observation and recording of the animal's physiological responses. For example, injection of vasopressin reduces an animal's urine output while simultaneously producing a slight blood pressure rise. Injection of insulin lowers blood sugar levels, which is why diabetics are prescribed insulin. Such experiments require the use of experimental animals, and extracts from these animals, which has sparked considerable controversy and debate concerning animal rights. These studies are important in understanding the physiology of the human body.

Genetic studies such as cloning and DNA sequencing have identified genes that may encode other developmental hormones. Discovery of the homeobox within the genes of all mammals indicates that there are some proteins (hormones) that control basic pattern development in mammals during early embryonic development. Some researchers believe that there may be certain hormones that accelerate aging and cause death in later life.

Hormones as a Key to Understanding Genes

All mammals start as a single-celled zygote—an egg that has been fertilized by a sperm—that undergoes a rapid sequence of mitotic divisions until it reaches the stage of a hollow, microscopic ball of identical cells called the blastula. Signaling molecules called hormones stimulate the blastula to fold in upon itself and form layers of tissue that gradually become differentiated into organs because of the presence of other hormones, which affect the tissue in sequence. Still other hormones later influence the interactions of organ systems for the smooth function of the organism. Knowledge of the chemical mechanisms and sequences through which hormones exert their effects will provide the key to understanding the action of genes, which in a more fundamental way are responsible for the various stages of life. Hormones direct cellular differentiation and development in the organism for the rest of its life. Hormones will be crucially involved in fetal development, birth, early growth and development, puberty, reproductive cycles, aging, and eventually death. Hormones control virtually all aspects of an organism's life.

If certain relevant mammalian developmental hormones can be identified, then target cells—cells that respond to the hormones in discrete but interrelated ways—may also be determined. A "developmental profile" for any organism would then be a real possibility, and the control of any organism's development, even behavior, could result, although this may pose ethical problems.

Currently, there is some detailed knowledge of the functions of many developmental hormones. Many hormones remain to be identified, however, and the overall scheme of hormonal control of development is still sketchy. Extensive research will be needed in the future.

—David Wason Hollar, Jr.

See also: Aging; Communication; Development: Evolutionary perspective; Endocrine system of vertebrates; Estrus; Gametogenesis; Genetics; Growth; Hibernation; Lactation; Mammals; Mating; Morphogenesis; Regeneration; Reproductive system of female mammals; Reproductive system of male mammals; Rhythms and behavior.

Bibliography

Alberts, Bruce, Dennis Bray, Julian Lewis, Martin Raff, Keith Roberts, and James D. Watson. *Molecular Biology of the Cell*. 3d ed. New York: Garland, 1994. As an intermediate-level undergraduate textbook, this outstanding work is a comprehensive, clear discussion of molecular and developmental biology. It contains excellent photographs and diagrams. Chapter 13, "Chemical Signaling Between Cells," provides detailed information concerning hormones and neurotransmitters.

Gehring, Walter J. "The Molecular Basis of Development." *Scientific American* 253 (October, 1985): 152B-162. As part of a special *Scientific American* issue called "The Molecules of Life," this article is a discussion of research into the homeobox, a region of genetic molecules that control early embryonic development in a variety of organisms, from nematodes to humans. Gehring describes several important experiments and major theories concerning early development.

Malven, Paul V. *Mammalian Neuroendocrinology*. Boca Raton, Fla.: CRC Press, 1993. A concise presentation of the way that nervous and endocrine systems interact to regulate physiological processes, using a comparative approach with rodents, domestic ungulates, and primates. Summarizes the neural influences and endocrine feedback mechanisms of each hormone.

Sang, James H. *Genetics and Development*. London: Longman, 1984. This text is suitable both as a graduate level textbook and as a heavily cited reference source for developmental geneticists. It clearly and precisely presents current theories of determination, differentiation, and development. Chapter 8, "Tissue Origins," and chapter 11, "Sex Differences," are excellent discussions of development.

Snyder, Solomon H. "The Molecular Basis of Communication Between Cells." *Scientific American* 253 (October, 1985): 132-141. As part of a special *Scientific American* issue called "The Molecules of Life," this article is a clear, comprehensive survey of mammalian endocrine systems and hormones, including both protein and steroid hormones. Mechanisms of hormone action, hormone transport, and experimental studies are presented in this beautifully illustrated, well-diagrammed article.

Stryer, Lubert. *Biochemistry.* 4th ed. New York: W. H. Freeman, 1995. Stryer's classic work is an outstanding textbook for advanced undergraduate students in biology, biochemistry, and physiology. The book is very clearly written and well illustrated. It has tremendous value as a reference tool. Chapter 35, "Hormone Action," is a thorough survey of hormone types and their mechanisms of action.

Wallace, Robert A., Jack L. King, and Gerald P. Sanders. *Biosphere: The Realm of Life.* 2d ed. Glenview, Ill.: Scott, Foresman, 1988. This is an excellent introductory biology textbook for both science and nonscience majors. It is clearly written and comprehensive in its coverage, although it is not too detailed to be understood by the general reader. The chapter on chemical messengers in animals is a concise but thorough survey of mammalian hormones.

Zubay, Geoffrey. *Biochemistry.* 4th ed. Dubuque, Iowa: Wm. C. Brown, 1998. As a graduate biochemistry textbook, this work is very detailed, although the illustrations and tables are excellent. The chapter on hormone action provides considerable information for protein and steroid hormone structures and actions, including a good discussion of negative feedback homeostasis.

HORNS AND ANTLERS

Type of animal science: Anatomy
Fields of study: Anatomy, physiology, zoology

Horns are hard, pointed outgrowths from the frontal bones of the heads of male ruminant herbivores, used for protection and to attract mates. They are permanent (true horns) or temporary (antlers) appendages. Antlers are shed yearly. In a few cases, horns are made of fused hair.

Principal Terms

ANDROGEN: a male hormone (especially testosterone) made by the testes

FRONTAL BONE: the bone which, vertically, makes up the forehead and is important, horizontally, to formation of the top (roof) of the orbital and nasal cavities

KERATIN: a tough, fibrous protein which is a major component of hair, nails, hooves, and the outer covering of true horns

PEDICLE: a small bone spur from which an antler grows (for example, in deer)

RUMINANT: a hoofed animal, with a stomach divided into four parts, that chews a cud of regurgitated, partly digested food

TRUE HORN: the permanent horns found in animals such as cattle, sheep, and goats

Horns are found in many ruminant herbivores. They are hard, pointed outgrowths that arise at the front of the heads of many herbivores. Horned animals include cattle, deer, giraffes, antelope, and rhinoceroses. In some cases, horns are permanent appendages that enlarge the bony front of the skull, as in cattle. In others, they are temporary bones arising on—but separate from—the skull, falling off every year, as in deer. In yet other cases, such as rhinoceroses, they are permanent groups of fused hairs, which become very hard.

Regardless of species, horns serve very important protective functions, both for the individuals having them and for the family groups or herds to which the individuals belong. Often, males use their horns to fight other males in the group for ascendency. In many horned species, an ascendant male does not fight much but faces down rivals with gestures, including displaying and threatening to use his horns.

The protective function of horns may explain why, most often, it is the males within a species who have horns, while the females are hornless. The males protect the herds or other family groups from predators. Although horns are most often found in males, females may have horns as well, for example, giraffes and oxen. When females of a species have horns, they are almost always much smaller than those found in males of the same species.

The nature of horns also differs from species to species. For example, in deer, their temporary horns, or antlers, are made of naked bone. In contrast, the permanent horns of cattle are made of bone covered with hard tissue. In the giraffe, the bony horns are covered with hairy skin. Also, in some cases a horn is not made of bone—the rhinoceros horn is made of hair.

The True Horns

True horns are pointed, permanent, bony structures most often seen on the heads of male ruminant mammals. The horns of females, where present, are smaller. Most horned animals—cattle, sheep, goats, and antelope—have paired horns. True horns have cores of bone which are extensions of the skull's frontal bone. Atop the bone core is a layer of skin, rich in a very tough, fibrous protein, keratin. The high keratin content in this skin makes it an extraordinarily tough and durable covering for the underlying bone.

Males and females of many animal species grow horns. The horns range from straight spikes to elaborately curved varieties. However, except in pronghorn antelope, the horns do not form branches, as in deer and other animals which shed their antlers every year. Animals having horns keep them for life. In pronghorn antelope the horn coverings are shed and grow again every year. This allows horn enlargement that would otherwise not occur.

Antlers

Antlers are horns which are shed each year and then grow again. In most antlered species only males have antlers. Like true horns, antlers grow out of the bones of the skull. They arise from permanent frontal bone structures called pedicles. At first the antlers have soft, velvetlike coverings of skin over their bone. Instead of hardening, as in true horns, this covering dies off and is rubbed away by the animal. The rubbing is believed to be due to itching caused by the dead velvet. Many zoologists propose that an advantage associated with rubbing off dead velvet is the development of good spatial perception of a stag's antler size and shape that helps to keep him from entanglement in underbrush and forests. Antlers have the same functions as true horns: protection, and use or display to develop ascendancy in a herd.

Antlers begin their growth in early summer, on the skull's frontal bone. They begin as small nubs that increase in length over time. The growing antler, sensitive to the touch, is covered with velvet, soft, smooth, and full of blood vessels. This gives it an abundant blood supply, carrying to growing antlers all the nutrients required for the two to three months of the growing period. At first, antlers are made of connective tissue. As time goes on, this tissue calcifies and becomes solid bone.

Antler growth is greatest in the late summer. At that time, hard tissue grows around the base of each antler, cutting off its blood supply. This kills the velvet, which loosens and is rubbed off by the stag. Antlers reach their full glory in the mating season, when they can be used to impress potential mates and to vanquish other males. Zoologists observe that the blood running through antler velvet cools down and this cool blood makes stags and bulls more comfortable on hot days. Antlers are shed between January and February, just following the mating season, when they are no longer useful.

Bone

Horns and antlers are made mostly of bone. About 70 percent of the matter in bone is an inorganic mineral material, hydroxyapatite, composed of calcium phosphate and calcium carbonate. Much of the remainder is a fibrous protein, collagen. The mineral and the protein together are called bone matrix. Within bone matrix, three types of specialized cells ensure its formation, its remodeling as needed, and its continuity throughout the life of any organism having a bony skeleton.

The first cell type, the osteoblast, produces bone matrix and surrounds itself with matrix, making collagen and stimulating mineral deposition. The second type, the osteocyte, is branched and becomes embedded in bone matrix. Osteocytes are interconnected and act to control the mineral balance of the body. Finally, osteoclasts destroy bone matrix whenever it is remodeled (rebuilt) during skeleton growth or the repair of bone breaks and fractures.

The conversion of connective tissue to bone begins when connective tissue cells arrange themselves in rows, enlarge, and change. This is followed by the synthesis of collagen and by mineral deposition around them. Osteoblasts needed for bone formation develop below the inner surfaces of a membrane that surrounds bone-to-be. Simultaneously, osteoclasts set the stage for formation of additional bone.

Who Has Horns and Antlers?

Bighorn sheep have true horns in individuals of both genders. Rams' huge, back-curved horns show growth rings that tell their age. Ewes have smaller, lighter horns that they use to protect their offspring. Rams use their horns to attract females, assert ascendancy, and battle predators.

Caribou of both genders have antlers, though those of females are small. Males shed their antlers in the winter, while females do not do so until spring. Females may use this advantage to oust males from winter feeding areas, in order to assure the continuation of the species by their offspring.

Elk antlers occur only in males. Bull elk use them to battle any predators who seek to harm the herds they lead, display them to attract females, and use them to assert ascendancy.

Moose antlers occur only in males. The huge, scoop-shaped antlers are mostly for show and to attract mates. Moose do not travel in herds and have relatively weak familial instincts.

American pronghorns of both genders have permanent horns that share attributes of both true horns and antlers. Doe antlers are short spikes. In males, the pronghorns—true horns—can become quite large. The coverings, but not the horns, are shed yearly. This leaves bone cores attached to the head. Then the bare core enlarges and its covering grows again. The horns of males have the usual protective, display, and assertive functions.

The first growth of antlers occurs when a stag or bull elk is one to two years old. At first, the antlers are small, straight, and spikelike. At that time young males are often called spikehorns by hunters. As an antlered male grows older its yearly antler crop branches out, having more prongs (points). In old, antlered males these horns may spread to widths of six feet and be several feet tall.

Antler growth is regulated by hormone secretions from the pituitary gland and the testes. The growth begins outside of the breeding season, when the testes are inactive. When the testes begin to make androgens, in preparation for breeding, both antler calcification and velvet tissue death occur. Decreased androgen production in winter leads to shedding of antlers.

Antlers, Horns, and Species Life Cycles

In the fall (September to October), American elk—called wapiti by Native Americans—mate, after crashing battles between bull elk, who joust with many-pointed antlers. This identifies winners, who thoroughly impress cow elk. Then each bull mates with his harem of cows and they wander off to form part of a huge winter herd of elk of both genders and all ages.

By early winter, bull elk shed their heavy, four- to six-foot-wide antlers, unneeded burdens that would diminish their chances of surviving the cold winter. Winter passes and by the late spring (May and June) the cows give birth to young, usually one per mother. At this time, new antlers begin to sprout from the pedicles on foreheads of bulls over a year old.

The antlers of elk follow the same growth pattern as in deer, leaving the bull elk with a strong bony rack of antlers in the fall, just right for jousting at the mating season. With some variation, in other antlered species—pronghorns who never lose bone cores of horns, and horned sheep, goats and bovines—the overall story is about the same. Offspring are conceived in the fall after jousts using antlers, pronghorns or horns, and life goes on. In some species females have horns or antlers, and in others small family groups are formed. Sometimes males hardly interact with females after copulation. However, a common thread to all their lives is the possession of true horns, pronghorns, or antlers.

—Sanford S. Singer

See also: Anatomy; Beaks and bills; Bone and cartilage; Brain; Circulatory systems of invertebrates; Circulatory systems of vertebrates; Claws, nails,

Horns are used for display, defense, and to establish an individual's place in the social hierarchy. (Corbis)

and hooves; Digestive tract; Ears; Endoskeletons; Exoskeletons; Eyes; Fins and flippers; Immune system; Kidneys and other excretory structures; Lungs, gills, and tracheas; Muscles in invertebrates; Muscles in vertebrates; Nervous systems of vertebrates; Noses; Physiology; Reproductive system of female mammals; Reproductive system of male mammals; Respiratory system; Sense organs; Skin; Tails; Teeth, fangs, and tusks; Tentacles; Wings.

Bibliography

Bubenik, George A., and Anthony B. Bubenik, eds. *Horns, Pronghorns, and Antlers: Evolution, Morphology, Physiology, and Social Significance*. New York: Springer-Verlag, 1990. This authoritative text provides a wealth of information on the three main kinds of ruminant horns, including illustrations and a good bibliography.

Goss, Richard J. *Deer Antlers: Regeneration, Function, and Evolution*. New York: Academic Press, 1983. Quite extensive and thorough, and contains useful illustrations and bibliography.

Petersen, David. *Racks: The Natural History of Antlers and the Animals That Wear Them*. Santa Barbara, Calif.: Capra Press, 1990. The material in this book is mostly anecdotal. However, it covers some topical issues in an interesting, pithy fashion.

Russell, H. John. *The World of the Caribou*. San Francisco: Sierra Club Books, 1998. This nice book describes caribou and closely related reindeer. A solid bibliography is included.

Van Wormer, Joe. *The World of the American Elk*. Philadelphia: J. B. Lippincott, 1969. This book contains excellent information and pictures of American elk, as well as a useful bibliography.

HORSES AND ZEBRAS

Type of animal science: Classification
Fields of study: Anatomy, genetics, zoology

Horses and zebras are herbivorous, hoofed mammals that represent five species of the Equidae family.

Principal Terms

CECUM: pouch in intestinal tract
CROWN: external tooth surface above gums
GELDING: castrated male
HAND: ten-centimeter (four-inch) measuring unit
UNGULATE: hoofed animal

Horses and zebras, members of the Equidae family, share common anatomical traits such as hard hooves, strong spinal columns, muscular arching necks, sloping shoulders, wide hindquarters, high-crowned teeth, slender legs, and elongated heads. Size and coloration varies according to specific horse breeds developed through selective breeding. Similar differences in zebra species have evolved due to natural selection and environment. Horses and zebras are related to wild asses and donkeys and represent five species of the genus *Equus*.

Anatomy Development

Eohippus was the ancestor of both domestic and wild horse species. A leaf-eating mammal as small as a fox, *Eohippus* lived during the Eocene epoch and had several toes on its feet. Descendants gradually grew larger, lost toes, and became grass grazers. Przewalski's horse, indigenous to Mongolia but not seen in the wild since 1968, links ancient and modern horses. Because it has sixty-six chromosomes and domestic horses have sixty-four, Przewalski's horse is a separate wild horse species and not an ancestor of modern horses. A hybrid produced by a horse and a Przewalski has sixty-five chromosomes and is fertile.

Modern horses and zebras are distinguished by a single hoof on each foot. Vestigial remains of prehistoric toes are located above hooves. Horses and zebras have long skulls and jaws that hold approximately forty to forty-four permanent teeth, including incisors to bite grass and molars to chew roughage. The teeth have long crowns that slope with age and can be examined to determine how old an animal is. Horses use their teeth to eat, groom, and fight.

The horse's digestive system is bigger and more efficient than that of carnivores, and has approximately 40 meters (131 feet) of intestines with a meter-long cecum attached to the colon. Horses and zebras live on grasslands ranging between sea level and mountains and eat mostly fibrous food such as hay, grain, and oats. Food ferments in the cecum, which can hold as much as thirty-eight liters (ten gallons). Horses eat an average of sixteen hours per day to sustain their systems.

The average horse's heart weighs about 3 kilograms (6.6 pounds), and the surface area of a horse's lungs measures approximately 2,500 square meters (8,200 square feet). Horses' weight varies from minimums of nearly 500 kilograms (1,000 pounds) to maximums of more than 908 kilograms (1 ton). Fifty-one vertebrae are in horses' spines from the top of the skull to the base of the tail.

Horses' eyes are placed on the sides of their heads, enabling them to see the horizon without moving their head. Their eyesight is better than that of dogs. Horses depend on hearing more than sight, but rely most on the sense of smell. The flehmen response is when the horse raises its upper lip to indicate it has smelled something interesting.

Social Behavior

Depending on the age, female horses are called fillies and mares, and males are colts, stallions, and geldings. Horses attain sexual maturity at two to three years old and usually produce a single foal after eleven months of gestation. Multiple births are rare, with twins occurring on average once per 1,500 births. Foals are able to stand within an hour of birth. Stallions fight while competing for mares, and one stallion may mate with several mares during a season. Mares usually come into heat one week after giving birth. Foals are weaned before they are one year old.

Horses are herd animals and rely on this social relationship, as well as on speed and endurance, as protection from predators. Domesticated horses enjoy the companionship of horses and other animals. Wild horses form a family group of one stallion, several mares, and their offspring. Young stallions sometimes form a bachelor group led by an older stallion. These groups forage in an area where they can find food, water, and shelter, and a pecking order maintains a hierarchy of rank within the group. Several groups often share the same range. Horses communicate by whinnying and nickering.

Vulnerable to wild animals, horses can run an average speed of forty-five to sixty kilometers (twenty-eight to thirty-seven miles) per hour. Their legs are slender and angled forward so that their weight is carried forward to enhance quick motion. Horses' shoulders absorb shock, while their powerful hindquarters provide impulsion. Horses naturally move at the gaits of walk (four beat), trot and pace (two beat), canter (three beat), and gallop (an extended canter), with several other natural gaits including pacing and the running walk.

Horse and Zebra Facts

Classification:
Kingdom: Animalia
Subkingdom: Metazoa
Phylum: Chordata
Subphylum: Vertebrata
Class: Mammalia
Subclass: Eutheria
Order: Perissodactyla
Suborder: Hippomorpha
Family: Equidae
Genus: Equus
Subgenera: Equus (horses), *Asinus* (asses), *Hippotigris* (zebras), *Dolichohippus* (zebras)
Species: Equus przewalskii (Przewalksi's horse), *Equus caballus* (domestic horse), *Equus africanus* (African ass), *Equus hemionus* (Asiatic ass), *Equus burchelli* (plains zebra), *Equus zebra* (mountain zebra), *Equus grevyi* (Grevy's zebra)
Geographical location: Horses can be found on all continents except Antarctica; zebras live in southeastern Africa
Habitat: Grasslands in tropical, temperate, and subarctic regions
Gestational period: Eleven months for horses; thirteen months for some zebras
Life span: Ten to twenty-five years, up to thirty-five in captivity
Special anatomy: One-toed feet, high-crowned teeth

Appearance

Domesticated horses are described by three types: heavy or draft, light, and pony. Humans genetically developed specific breeds for different tasks and degrees of hardiness, stamina, and versatility. Most horse breeds are derivatives of Arabians and thoroughbreds. Draft horses (often referred to as cold-blooded, which is not physiologically accurate), usually stand over sixteen hands high at the withers (the ridge between the shoulder blades) and have sturdy, thick dimensions, useful for pulling loads. Light horses, sometimes called hot bloods, are at least 14.2 hands high and tend to be streamlined. Ponies are shorter than 14.2 hands and vary in conformation according to breed. Welsh ponies tend to be more delicate, while

Each zebra has a distinctive pattern of stripes as individual as human fingerprints. (Corbis)

Shetland ponies are more rotund. Miniature horses are extremely small ponies that are often less than one meter tall.

Coat colors range from shades of brown and red to solid black and white, with some horses, such as Appaloosas and pintos, having spots and others being gray or roan. Spotted patterns are linked to genes that also often produce mottled noses, eyelids, and genitalia, blue eyes, and striped hooves. Lipizzans are born black and turn completely white by age two years. Some horses have dorsal stripes and zebra stripes on their legs. Various white facial and leg markings also distinguish horses.

African Cousins

Zebras belong to the horse family and are represented by three species, the plains, Grevy's, and mountain zebras. Several subspecies exist according to geographic range, herd behavioral differences, and physical variations such as dewlaps. Zebras are smaller than horses and have erect manes, longer ears, and thinner tails. The average zebra stands 140 centimeters (55 inches) high and weighs 300 kilograms (660 pounds). These striped animals exist in herds that graze in southeastern Africa's grasslands and also live in nearby deserts.

Each zebra species is determined by a specific stripe pattern of black, brown, and white markings. Individual zebras have unique striping somewhat like the uniqueness of human fingerprints. Scientists have determined that zebras, from the moment of birth, are drawn to striped objects, a type of imprinting which has led to speculation that stripes might be a factor in herd cohesiveness and sociability. Previous hypotheses that stripes are for camouflage and to confuse predators have been discredited. The exact purpose of stripes remains a scientific mystery. Researchers are aware that abnormally striped zebras are often shunned by herds, which threatens their survival.

Herds can include several zebras, often a male and females with their offspring or a group of

North American Wild Horses

Considered the symbol of the freedom of the American West, mustangs roam prairies, in addition to isolated areas along the East Coast and in the Midwest. In the early twentieth century, the federal government ordered the extermination of mustangs. Velma "Wild Horse Annie" Johnston lobbied Congress in the 1950's to stop wild horse slaughters. Congress passed the Wild Free-Roaming Horse and Burro Act in 1971 to protect wild horses under the jurisdiction of the Bureau of Land Management (BLM). The Wild Horse Act of 1999 assured federal protection for a feral horse herd in Missouri.

Although wild horses are romanticized in books and movies, realistically they pose problems such as uncontrolled population increase, encroachment on urban areas, and destruction of rural landscapes. They also are at risk from animal predators, microscopic diseases, and human poachers. Droughts and wildfires in the late twentieth century depleted food sources, resulting in many mustangs starving. Roundups by helicopters drive wild horses toward government corrals, where they are fed and offered for adoption to the public. Approximately forty-six thousand wild horses roamed the American West in 1999, and seven thousand were adopted.

Such dispersal, however, is controversial. Some of the mustangs are actually sold for slaughter because horse meat is a popular food in Asia and Europe, due to concerns about bovine spongiform encephalopathy (mad cow disease). Critics of the BLM suggest that agency will inadvertently cause the extinction of America's wild horses.

young males. Herds can expand to hundreds of zebras. Typically passive animals, male zebras occasionally fight for females during breeding season by kicking, biting, and shoving each other. Female zebras attain sexual maturity at age three and usually reproduce annually throughout their lives, while males reach breeding age at five years. Gestation lasts approximately eleven to thirteen months, and twins are rarely foaled. The life span of wild zebras can extend to twenty-two years, and zebras kept in captivity can live longer.

Zebra foals weigh approximately thirty-two to thirty-six kilograms (seventy to eighty pounds) at birth. Able to stand soon after being born, zebra foals eat grass which adds about 0.45 kilograms (1 pound) of body weight per day until they reach physical maturity. Zebra foals develop more quickly than horse foals and become independent sooner.

Defense Mechanisms

The herd protects zebras from predators, primarily large cats, such as lions and cheetahs, in addition to hyenas. Humans also hunt zebras for their hides. While other zebras sleep, a guard zebra watches for potential hazards and is aided by its keen night vision, comparable to an owl's, and tall ears which can rotate to pick up sounds. Zebras' primary defense mechanism is to run away from danger, and they can reach speeds as high as sixty-five kilometers (forty miles) per hour, which is much slower than most of their enemies.

Zebras also are at risk from reduced water resources and grasslands due to ranching and farming and compete with livestock for basic nutritional needs. While the common zebra remains abundant, the Grevy's zebra and the mountain zebra are considered endangered species and another type of zebra, the quagga, became extinct in 1883. A century later, the Quagga Breeding Project attempted to revive the quagga because genetic researchers determined that quagga and plains zebra deoxyribonucleic acid (DNA) was similar, and hypothesized that the quagga was not a separate species but rather a variation of the plains zebra. Zebroids are horse-zebra hybrids that are sterile.

—Elizabeth D. Schafer

See also: Camouflage; Claws, nails, and hooves; Domestication; Endangered species; Fauna: Africa; Fauna: Asia; Herds; Imprinting; Locomotion; Ungulates.

Bibliography

Back, Willem, and Hilary M. Clayton, eds. *Equine Locomotion*. Philadelphia: W. B. Saunders, 2000. Provides scientific explanations of horses' movement written by experts in equine biomechanics.

Dossenbach, Monique, and Hans D. Dossenbach. *The Noble Horse*. Translated by Margaret Whale Sutton. New York: Barnes & Noble, 1997. Translation of original German edition which provides text and illustrations explaining the anatomy, behavior, and history of horses, and their use by humans.

Groves, Colin P. *Horses, Asses and Zebras in the Wild*. Hollywood, Fla.: R. Curtis, 1974. Excellent scientific source about various members of genus *Equus*.

Hendricks, Bonnie L., ed. *International Encyclopedia of Horse Breeds*. Norman: University of Oklahoma Press, 1995. Describes both familiar and unique types of horses found globally.

MacClintock, Dorcas. *A Natural History of Zebras*. New York: Charles Scribner's Sons, 1976. Written for younger readers; has basic information about zebra species and is illustrated with photographs taken by Ugo Mochi.

Rose, Reuben J., and David R. Hodgson, eds. *Manual of Equine Practice*. 2d ed. Philadelphia: W. B. Saunders, 2000. A comprehensive illustrated collection of articles about anatomical and veterinary topics concerning horses.

HORSESHOE CRABS

Type of animal science: Classification
Fields of study: Anatomy, invertebrate biology, marine biology, systematics (taxonomy)

Horseshoe crabs are not crabs at all, being more closely related to spiders and scorpions. They have been around for at least 400 million years—before dinosaurs walked the earth—and have changed relatively little in that time.

Principal Terms

CEPHALOTHORAX: the forward segment of the horseshoe crab's body
CHELATE: pincerlike
CHELICERAE: appendages with pincers
DIOECIOUS: having separate sexes
EXOSKELETON: the external shell of horseshoe crabs and their relatives
GNATHOBASE: the part of the leg closest to the horseshoe crab's body

There are four species of horseshoe crabs, and all are members of the class Merostomata, aquatic animals with two body segments and a spikelike telson at the tail end. Perhaps the best-known representative is *Limulus polyphemus*, the common horseshoe crab native to the northwest Atlantic coast and the Gulf of Mexico. These animals live in shallow water to depths of one hundred feet and prefer soft sand or mud bottoms, through which they slowly plow as they scavenge for food.

Horseshoe crabs, unlike true crabs, do not have antennae. However, like crabs, they do have jointed appendages and a hard shell, or exoskeleton, made of chitin, which must be periodically shed to accommodate the growing body of the individual.

Horseshoe Crab Anatomy

The body of horseshoe crabs is divided into two segments: a large, helmet-shaped, forward section called the cephalothorax or prosoma, and a rear abdomen or opisthosoma, to which is attached the lancelike telson. Despite its threatening appearance, the telson is not used for defense, but rather for pushing and righting the body if the animal is overturned.

There are two lateral and two median eyes on the upper surface of the prosoma. Although horseshoe crabs may be able to detect movement, there is little evidence that they can form images. The unique and relatively simple anatomy of horseshoe crab eyes make them favorite subjects for nervous system research.

Under the cephalothorax there is a pair of small, pincerlike chelicerae, followed by five pairs of walking legs. The first four pairs are chelate and the fifth pair is for pushing away mud and silt during burrowing. The first four pairs also have spines along the joints closest to the body. These gnathobases are used to shred and macerate food and move it toward the mouth.

The abdomen has six pairs of appendages, five of which are modified as thin, flaplike gills. In addition to providing oxygen to the animals, the gills function as paddles during upside-down swimming in small individuals.

Reproduction in Horseshoe Crabs

Horseshoe crabs are dioecious, meaning there are separate sexes. During warm months, females migrate into the intertidal zone to rendezvous with the smaller males. The males crawl onto the shell of the females and cling to them while the females scoop out depressions in the sand and deposit two hundred to three hundred small green eggs, which the male then fertilizes. The location of this

Horseshoe crabs must lay their eggs in the intertidal zone, where they will be kept moist but still receive enough oxygen to survive. (Rob and Ann Simpson/Photo Agora)

egg burying is of critical importance: too high up the beach and the eggs will dry out; too low and they will die in the oxygen-poor sand.

After a lunar month a small (about one centimeter), swimming larva hatches which little resembles the adult. After successive molts the adult body form is eventually achieved, with sexual maturity being reached after three years.

Economic and Scientific Importance of Horseshoe Crabs

The economic value of horseshoe crabs has been recognized since at least the nineteenth century, when millions were harvested annually from Delaware Bay to be ground up as fertilizer. By the 1950's, the population of horseshoe crabs had decreased to the tens of thousands. Since then, controls have been put in place to protect this animal. In Japan, it has been declared a national monument to shield it from extinction.

Today the horseshoe crab is used for bait in the fishing industry. It is also valuable in biomedical research because of its blue, copper-based blood. An extract of this blood, called limulus amebocyte lysate (LAL), is used in detecting bacterial contamination of drugs and medical devices. There are many other chemicals derived from horseshoe crabs that may prove useful against human diseases.

—*Robert T. Klose*

See also: Arthropods; Marine animals; Marine biology; Molting and shedding; Shells.

Bibliography

Cohen, Nancy Eve. "Relief for Horseshoes." *Audubon* 102, no. 4 (July/August, 2000): 113. A report on the threats to horseshoe crab populations along the eastern seaboard of the United States. For a general audience.

Ruppert, Edward E., and Robert D. Barnes. *Invertebrate Zoology*. 6th ed. Fort Worth, Tex.: Saunders College Publishing, 1994. The bible of invertebrate zoology. Clearly written, abundantly illustrated, logically organized. For the serious student.

Horseshoe Crab Facts

Classification:
Kingdom: Animalia
Subkingdom: Bilateria
Phylum: Arthropoda
Subphylum: Chelicerata
Class: Merostomata
Subclass: Xiphosura
Order: Xiphosurida
Suborder: Limulina
Family: Limulidae
Genus and species: Limulus polyphemus

Geographical location: *L. polyphemus*, the most common species, is distributed along the northwestern Atlantic coast and the Gulf of Mexico; other species are native to Asian coasts, from Japan and Korea south through the East Indies and the Philippines

Habitat: Shallow water with soft bottoms

Gestational period: One lunar month for the eggs to hatch

Life span: Approximately thirty-five years

Special anatomy: Two distinct body segments, a rigid telson at the tail end, four eyes, a median frontal organ

Sargent, William. *The Year of the Crab*. New York: W. W. Norton, 1987. A charming, reflective collection of essays dedicated to the horseshoe crab. Lovingly written, this is a wonderful introduction to these unique animals.

Sherman, Irwin W., and Vilia G. Sherman. *The Invertebrates: Function and Form*. 2d ed. New York: Macmillan, 1976. Arguably the finest invertebrate zoology laboratory manual ever published in English. Chapter seven contains instructions for studying both living and preserved horseshoe crabs.

Zorpette, Glenn. "Mesozoic Mystery Tour." *Scientific American* 278, no. 6 (June, 1998): 14-15. An illustrated first-hand account of horseshoe crab mating and spawning in Brooklyn, New York. Also discusses some of their unique characteristics. For a general audience.

HUMAN EVOLUTION ANALYSIS

Type of animal science: Evolution
Fields of study: Anthropology, evolutionary science, genetics, human origins, systematics (taxonomy)

The study of human evolution traces the descent of the hominids from their primate ancestors and focuses particularly on the most recent stages that led to modern Homo sapiens.

Principal Terms

DEOXYRIBONUCLEIC ACID (DNA): the large molecular chains of nucleic acid that make up genetic material

GRACILE: slender and light-framed, as opposed to robust

HOMINID: belonging to the taxonomic family Hominidae, which includes all humans and their evolutionary ancestors as far back as the split from the great apes

MITOCHONDRIA: self-replicating units in a cell that are responsible for the metabolic generation of energy for cell processes

About fifteen million years ago, there was a great flowering of diversity among the primates. One branch of this evolutionary spurt gave rise to the family Pongidae (gorillas, orangutans, and chimpanzees). A second branch led to the family Hominidae. Fossil evidence suggests that the hominids diverged from the common ancestral primate at about the same time as the pongids, but according to molecular analysis of the genes of modern humans and gorillas, the divergence may have been much later—perhaps as recently as two million or three million years ago. The earliest known examples of the hominid family have been named *Ramapithecus*. Fossils of this creature, more ape than human, have been found in India, Greece, Africa, Turkey, and Pakistan. It is likely that this genus evolved into another hominid group, *Australopithecus*. Many australopithicine fossils have been found, mostly in Africa, and the genus is divided into two types, the large, heavy-boned robust species and the smaller, gracile types exemplified by "Lucy," the nearly complete skeleton discovered by Donald Johanson in 1974. The current view is that the robust australopithicines became extinct but that the smaller forms gave rise to the genus *Homo*, leading to modern humans.

The earliest discovery of a fossil of the *Homo* genus was given the species name *erectus* by its discoverer, Eugène Dubois, in 1891. It was found not in Africa but in Java. Dubois believed that Java man, as he called his find, was the first hominid to walk erect. Later discoveries showed that the more primitive *Australopithecus* had already achieved the erect stance of modern humans, but the name *Homo erectus* persists for the link between *Australopithecus* and more modern hominids.

Tool Use and Culture

The genus *Homo* was the first to leave clear evidence of toolmaking. Pebble tools of *Homo erectus* were described by Louis Leakey, and he named the collection of artifacts "Oldowan" culture, from the Olduvai Gorge where some of his most famous excavations were done. On the basis of his collection of African fossils dated to between one million and two million years ago, Leakey proposed a new species, *Homo habilis*, which would be intermediate between *Homo erectus* and *Homo sapiens*, but there were too few fossils from this critical period to make the sequence clear. In any case, the simple pebble tools of the earliest members of the human genus were gradually supplanted by somewhat more sophisticated implements made by chipping both sides, along with a growing preference for flint over softer stones. There are many more examples of flint tools than there are of actual bones of human ancestors from this period. The techniques used for making these tools were evidently handed down from generation to generation, and the patterns were quite conservative, so tools from a given culture can be identified wherever they are found. The tools and artifacts of the later members of *Homo erectus* are known as the Acheulian culture. The most characteristic tool of this culture is the hand axe, used for chopping, cutting, scraping, and possibly even as a weapon.

The species to which modern humans belong first appeared between 200,000 and 300,000 years ago. In 1856, when the first example of *Homo sapiens* was discovered, it was named Neanderthal man because the skeleton was found in a cave by the Neander Valley near Düsseldorf, Germany. Neanderthals were a robust species with a somewhat larger brain than modern humans. They wandered widely, leaving their remains over much of Europe, Asia, and Africa. The name Mousterian culture was coined to describe their artifacts, which were much more sophisticated than Acheulian tools.

Finally, perhaps as recently as fifty thousand years ago, Neanderthals were replaced by fully modern humans. The two groups can be separated into the subspecies *Homo sapiens neanderthalensis* and *Homo sapiens sapiens*. At about the same time the Neanderthals disappeared, the Mousterian artifacts were replaced by much more complex and widely varied implements of the Aurignacian culture. The explosion of cultural innovation—including cave paintings, carvings, and other artistic and technological inventions—had a remarkable effect on life.

The first fossils of the *sapiens* subspecies were found in a limestone cliff at Cro-Magnon in southern France in 1868. They and numerous later finds from the same period were given the name Cro-Magnon man, and from the very beginning, they were recognized as being fully modern in form.

Evolutionary Controversies

Controversy over the relationship between the modern human subspecies and the Neanderthals centers around the fact that the two groups overlapped in time. Evidently some subpopulation of Neanderthal evolved into Cro-Magnon while the more primitive type was still flourishing. For some unknown period, both types existed, but eventually—quite abruptly on a geological time scale—Neanderthals vanished from the face of the earth, and *Homo sapiens sapiens* reigned alone. Some anthropologists put the divergence between Neanderthals and the *sapiens* subspecies quite early and suggest that most of the known fossils of Neanderthals represent a dead end of evolution and not human ancestors. A few fossils that appear to be of the modern human type are tentatively dated much earlier than fifty thousand years ago, which is the time most of the fossil evidence would indicate that modern humans first appeared. Whether humans evolved from early or late Neanderthals, there is certainly no other candidate for an immediate ancestor nor any evidence of an alternative link between *Homo erectus* and modern humans.

The Neanderthal question is one of three major controversies about the details of human evolution. Another is over the site of major stages in human evolution. The "out-of-Africa" view is that Africa, with its rich store of fossil evidence, had to be the place where, successively, *Australopithecus*, *Homo erectus*, and modern humans emerged. Countering this is the fact that fossils and artifacts of each stage of human evolution are found in many regions of the globe. Some authorities feel that it is more probable that this mobile, adaptive, wide-ranging species was already dispersed during the final million years or so of human evolution rather than exclusively in Africa.

The third major disagreement among anthropologists is about how to interpret evidence from molecular biology. The random mutation of deoxyribonucleic acid (DNA) chains provides a molecular clock because the mutations cause a genetic drift. In an interbreeding population, the genetic changes are shared by all members, but if a population divides, the genetic changes vary between the two groups, resulting in more and more differences as the time since the separation increases. Studies of DNA differences among humans and between humans and the great apes suggest that the genetic divergence between humans and the gorillas has taken only about three million years. This is a much shorter period than fossil evidence would indicate. However, the time scale for DNA changes is not well enough established to persuade many of the geologists and fossil hunters to give up their chronology. In 1987, the even more disturbing claim was made that all living humans had descended from one woman and that this "Eve" had lived in Africa as recently as 200,000 years ago. The work that led to this announcement was based on the study of mitochondrial DNA, which is much less variable than the DNA in the nucleus of the cell.

Each tiny mitochondrion has its own set of genes, and they divide independently of the cell nucleus. Because the sperm does not contribute any mitochondria to the fertilized egg, mitochondrial DNA is inherited only through the female line of descent. In one study, mitochondrial DNA from 147 individuals, representing populations in Africa, Europe, Asia, Australia, and New Guinea, was compared. The samples seemed to fall into two major groups when analyzed by a computer program: a group containing the African samples and one holding all the rest. The group containing the African individuals appeared to be the more primitive. From the rate of mutations in mitochondrial DNA, it was calculated that the observed variations in the samples would have occurred over a time span of about 200,000 years.

The convergence of molecular evidence and fossil interpretation has given a boost to the out-of-Africa hypothesis for human origins. Africa is certainly a rich source of fossils of the earliest progenitors of the human species. It is also the home of most of the great apes. The controversy is mainly about when the migration of humans or proto-humans into the rest of the world occurred. Fossils of every stage of human evolution have been found in many parts of Europe and Asia and even Australia. It is possible, of course, that sample populations of each stage of human evolution emigrated from Africa, leaving their bones and their tools scattered all over Eurasia, only to be replaced by successive waves of emigrants from later stages. There is no hypothesis to suggest why the further evolution of these intermediate species should not have taken place outside Africa.

Fossils and Paleoanthropology

The major technique for the study of human evolution is the analysis of fossils and the artifacts associated with human activities. Fossil hunting is more an art than a science, and many of the major finds have been accidental. However, knowing where to look and the use of systematic excavation have been very fruitful. Dating the fossils and artifacts is the most critical part. When fossils are excavated from undisturbed layers of sediment, the dating can be determined by knowledge of the geological strata. Layers of lava from volcanic eruptions, sediment from floods, the presence of other fossils of plants or animals of known periods all contribute to the determination of the age of the deposit. Radioactive decay provides another method. Radioactive carbon-14 is continuously produced in the upper atmosphere and is absorbed by plants. The carbonate in fresh bone contains about one atom of radioactive carbon for every one hundred forty atoms of the inactive isotope. The radioactive isotope decays at a constant rate and is reduced to half of its original concentration in about five thousand years. The radioactivity of bones (or charcoal or other organic debris) can be measured, and the older the sample, the smaller the proportion of radioactive to nonradioactive carbon atoms will be. Unfortunately, the sensitivity of the method limits its usefulness to materials less than about thirty thousand years old. Also, it is often hard to rule out the possibility of contamination by groundwater or organic sources of "fresh" carbon. Older materials can sometimes be dated by other radioactive isotopes such as argon. Another dating method is to use the accumulation of atomic dislocations caused by cosmic ray bombardment in hard materials, especially stone and pottery. Such dislocations "heal" when the material is heated, so measuring the amount of cosmic ray damage in a material can reveal its age since the last heating. This method is good for much longer time periods than radiocarbon dating.

Molecular biology is relatively new, and its use as a method for studying human evolution is still experimental. The basic premise on which it operates is that mutations of DNA molecules are random events, relatively independent of environmental factors, and therefore relatively constant in time. Although exposure to natural or artificial radiation or certain chemical mutagens can increase the rate of DNA change, it is assumed that such factors would not be likely to affect a whole population; therefore, the assumption of a uniform slow rate of genetic drift is probably justified statistically.

DNA molecules are very large, and with modern techniques, it is possible to detect small changes in the molecule that may not produce any observable mutation in the organism. This means that chemical comparison of DNA samples from two individuals is a much more sensitive measure of their degree of relatedness than simple comparison of visible features. Analysis of DNA differences (and similarities) correlates well with older techniques of classification. Human DNA, for example, is much more similar to sheep DNA than to earthworm DNA. Monkey DNA is much closer to human DNA than is sheep DNA. By assuming that DNA changes at a uniform rate, these differences can be interpreted in terms of evolutionary time since humans and monkeys or humans and sheep shared a common ancestor.

The Implications of Human Origins

The study of human origins has social, intellectual, religious, and philosophical implications. The fact that humans are all one subspecies, with a remarkably homogeneous genetic makeup, should minimize the importance of race, culture, or language differences. Anthropology is the discipline that studies the origins of humans. Anthropologists also study living human cultures, and there is a two-way transfer of information between cultural anthropologists and physical anthropologists. Knowledge of human descent deepens understanding of the diversity of cultures and capabilities among living humans. Studies of contemporary and recent cultures give great insight into how Stone Age ancestors might have lived and worked. Even the study of developmental psychology provides insights into the ways the modern mind might have evolved.

Naturalist Charles Darwin's theory of human evolution met with powerful resistance not only

Homo habilis *was probably the first hominid not only to use tools but also to make them.* (©John Sibbick)

because it was contrary to a literal interpretation of the Christian scriptural story of human creation but also because it seemed to make humans helpless products of random processes. Humans may be powerless to control their biological evolution, but they are fully in control of the spectacular cultural evolution that provided people with language, knowledge, and civilization.

Although many people still see a conflict between religious beliefs about human origins and the scientific view, others have been able to reconcile the two doctrines and are comfortable with both their religious faith and their scientific knowledge.

Philosophers have always discussed what it means to be human and the origins of humankind. Scientific knowledge of human origins has had immeasurable impact on philosophy. Modern philosophy emphasizes the importance of language to humans, and anthropological studies of human evolution also point to the critical importance of language development in the final stages of human evolution. There is a growing conviction that the acquisition of language was the single most important step in the final evolutionary jump from Neanderthal to modern human. Language made it possible for humans to communicate with others and to pass along the accumulated wisdom and experience of one generation to the next. Language helped the pace of cultural development accelerate to a point where further biological evolution became irrelevant. Clothing and fire substituted for fur in a cold climate. Weapons were better than sharp claws and long teeth. Human evolution thus became cultural evolution.

—Curtis G. Smith

See also: Adaptations and their mechanisms; Apes to hominids; Convergent and divergent evolution; Development: Evolutionary perspective; Evolution: Historical perspective; Gene flow; Hominids; *Homo sapiens* and human diversification; Language; Neanderthals; Nonrandom mating, genetic drift, and mutation; Population genetics; Primates.

Bibliography

Cann, R. L., M. Stoneking, and A. C. Wilson. "Mitochondrial DNA and Human Evolution." *Nature* 325 (1987): 31-36. This brief article presents a computer analysis of the mitochondrial DNA studies and makes the conclusion that all living humans descended from one African woman who lived about two hundred thousand years ago.

Johanson, D., and M. Edey. *Lucy: The Beginnings of Humankind*. New York: Simon & Schuster, 1981. This is the exciting story, told by a world-famous anthropologist, of his discovery of "Lucy," the diminutive three-and-a-half-million-year-old skeleton of one of humankind's australopithecine ancestors. Also gives a readable account of the state of modern research into human evolution, with careful explanations, diagrams of dating methods, and chronologies. Contains many pictures and diagrams, an appendix, and a good bibliography of books and scientific papers.

Jolly, Alison. *Lucy's Legacy: Sex and Intelligence in Human Evolution*. Cambridge, Mass.: Harvard University Press, 1999. Traces four major transitions in human evolution, all based in cooperation, and posits that we are in the process of undergoing a fifth, incorporating specieswide, global communication.

Leakey, Mary. *Disclosing the Past*. New York: Doubleday, 1984. This is Mary Leakey's autobiography, but it serves as an unsurpassed history of field research into human origins over the past fifty years, told by an individual who was at the forefront of that story. The book contains photographs, maps, diagrams, and a completely captivating narrative of human evolution. Contains a list of twelve books for further reading.

McKee, Jeffrey K. *The Riddled Chain: Chance, Coincidence, and Chaos in Human Evolution*. New Brunswick, N.J.: Rutgers University Press, 2000. In contrast to the traditional top-down approach, McKee takes a bottom-up view of human evolution, in which each genetic variation must be tested through successive levels within a species. He argues that the large size of the current human population poises us for a rapid rate of evolution.

McKie, Robin. *Dawn of Man: The Story of Human Evolution*. New York: Dorling Kindersley, 2000. Published in conjunction with the Learning Channel series. Clearly written for a general audience, with many photographs and illustrations.

Shreeve, J. *The Neanderthal Enigma*. New York: William Morrow, 1995. The author wrote this book after spending many months interviewing leading anthropologists. The book presents the controversies over Neanderthal interpretation in an exciting and readable fashion. One may disagree with the author's hypothesis, but he tells a truly fascinating story. Contains an extensive bibliography of both research articles and books.

Smith, C. G. *Ancestral Voices: Language and the Evolution of Human Consciousness*. Englewood Cliffs, N.J.: Prentice-Hall, 1985. This book is written for the general reader and presents a fusion of anthropology, linguistics, and neuroscience to describe the final emergence of the human species from its animal background. It emphasizes the importance of cultural evolution in shaping modern humans. Contains many illustrations, a list of references, and a list of suggested readings.

Solecki, R. S. *Shanidar, the First Flower People*. New York: Alfred A. Knopf, 1971. Solecki was the anthropologist who excavated a Neanderthal skeleton that pollen analysis revealed to have been laid on a bed of flowers. This is a poetic book, with interesting speculation about the "humanness" of the Neanderthals, and a very good story about the process of excavation. Solecki has been an authority in the field of late human evolution and gives a good account of the state of knowledge in the 1960's. Bibliography, pictures, and diagrams.

HUMMINGBIRDS

Types of animal science: Anatomy, behavior, classification, physiology, reproduction
Fields of study: Ornithology, physiology, population biology, wildlife ecology

Hummingbirds are found only in North and South America. Classified in the general order Apodiformes, they have unique coloration and many of their feathers are iridescent. The rapid beating of their wings, which enables them to hover, produces a distinctive, recognizable hum.

Principal Terms

BANDING: technique for studying the movement, survival, and behavior of birds
COVERTS: small feathers at the base of the wings and tail
GORGET: patch of feathers between the bird's throat and breast
PRIMARY FEATHERS: flight feathers on the outer joint of the wing
RECTRICES: tail feathers
SECONDARY FEATHERS: flight feathers on the inner wing

Over four hundred species of hummingbirds have been identified in North and South America, forming the Western Hemisphere's second largest family of birds. These exceptionally small birds have the greatest comparative energy output of any warm-blooded animal. In one day, they often consume more than half their total weight in food and twice their weight in water. The smaller species have the fastest wing beat of all birds. Unlike other birds, their wing upstroke is as powerful as their downstroke. They can fly forward, backward, and briefly, upside down.

The range of the hummingbird stretches from Alaska to Tierra del Fuego, Chile. Because of a constant supply of nectar and insects, tropical hummingbirds rarely migrate. Those hummingbirds that do migrate sometimes travel enormous distances. The rufous hummingbird flies over two thousand miles from its winter home in Mexico to the Pacific Northwest and Alaska.

Physical Characteristics of Hummingbirds

Hummingbirds are extremely small, weighing from two to twenty grams. The bee hummingbird is the world's smallest bird. Their long, slender bills, which are often slightly decurved, and their long bitubular tongues give them easy access to flower nectar, their main source of food.

Although their feather structure is among the most specialized of birds, hummingbirds have the fewest feathers. The primary flight feathers de-

The hummingbird's long, pointed beak and bitubular tongue allow it to suck nectar from deep within trumpet-shaped flowers. (Corbis)

Hummingbird Facts

Classification:
Kingdom: Animalia
Phylum: Chordata
Class: Aves
Order: Trochiliformes
Family: Trochilidae (hummingbirds, sixty-two genera, eighty-eight species)
Geographical location: North and South America, with most living near the equator
Habitat: Wooded areas, mountain slopes, plateaus, canyons, often near water
Gestational period: Usually one breeding cycle in a year
Life span: Average is five years, though observations of banded and captive birds show they can live ten or more years
Special anatomy: Ten primary flight feathers; six to ten secondary wing feathers; ten rectrices; extremely large sternums; long bills, often decurved

crease in size from the outer feathers inward toward the secondary feathers. Hummingbirds are able to rotate each of their wings in a circle. To hover, they move forward and backward in a repeated figure eight. They can move in any direction instantaneously. Their feet, which are more suited for perching than walking, have three toes directed forward and one pointed back.

Hummingbird feathers are iridescent. Depending on the viewing angle, the colors will change from red to gold or from green to turquoise. Adult males have the most intensely colored feathers and in full sunlight seem to glow. Some feathers are also modified to produce sound, so that as the birds fly, a soft humming sound is heard.

Reproductive Biology and Behavior

Hormonal changes prompt the female hummingbirds to begin nest building as ova ripen in their ovaries. In the male, hormonal changes enlarge their testes to many times their normal weight. Following mating, the female takes from one day to two weeks to complete building her nest, and soon after, she lays her eggs. Most hummingbirds lay two eggs, two days apart. The tiny, elliptical white eggs weigh less than 0.02 ounce. Depending on the species, incubation lasts fifteen to twenty days. Newly hatched hummers are featherless and do not open their eyes for two weeks. By two and a half weeks they are covered with feathers and can groom themselves. After testing their wings, at three weeks they are able to leave the nest, although the mother continues feeding them for two to four more weeks.

Hummingbirds are highly territorial. Particularly when they are migrating, they aggressively protect their sources of nectar. A nesting female will attack any interloper approaching her nest.

Hummingbirds have adapted to living in diverse areas. Like most birds, hummingbirds eat vast quantities of food. They have taste receptors on their tongues and salivary glands, and prefer flower nectars with a high sugar content. Throughout the day, hummingbirds eat frequently. At night, in order to conserve energy, they enter a state of torpor, which is a short-term form of hibernation. In this state they are unable to flee from predators. To resume a normal state requires enough energy to warm their organs and tissues. The birds need to monitor their energy reserves so they can recover from their torpid state.

—Susan E. Hamilton

See also: Beaks and bills; Birds; Domestication; Feathers; Flight; Molting and shedding; Nesting; Respiration in birds; Wildlife management; Wings.

Bibliography

Johnsgard, Paul A. *The Hummingbirds of North America*. 2d ed. Washington, D.C.: Smithsonian Institution Press, 1997. An excellent resource, the second edition includes a description of species that breed north of the Isthmus of Tehuantepec. Colored plates and black and white illustration augment the comprehensive text.

Long, Kim. *Hummingbirds: A Wildlife Handbook*. Boulder: Johnson Books, 1997. A practical, user-friendly general resource that includes myths and legends as well as behavior, biology, and individual characteristics.

Stokes, Donald, and Lillian Stokes. *The Hummingbird Book*. Boston: Little, Brown, 1989. A simple guide on how to attract and identify hummingbirds.

Tilford, Tony. *The World of Hummingbirds*. New York: Gramercy Books, 2000. A beautifully illustrated guide that includes a list of hummingbird hotspots, useful web sites, a glossary and bibliography, as well as a section on artificial feeding and gardens that attract hummingbirds.

Toops, Connie. *Hummingbirds: Jewels in Flight*. Stillwater, Minn.: Voyageur Press, 1992. Written and with photographs by a photojournalist, this scientifically researched volume describes cultural history and legends of hummingbirds in addition to discussing courtship, nesting, territorial and migratory behaviors.

HYDROSTATIC SKELETONS

Types of animal science: Anatomy, evolution, physiology
Fields of study: Anatomy, invertebrate biology, physiology, zoology

Hydrostatic skeletons use the incompressibility of water to create a transient skeleton. This is accomplished in animals by contracting muscles around a fluid-filled space.

Principal Terms

CIRCULAR MUSCLE: muscle fibers that run in a circular pattern around the body perpendicular to the long axis of the body

COELENTERON: the fluid-filled gastrovascular cavity of Cnidarians

COELOM: the body cavity of higher invertebrate and vertebrates, where mesodermal tissues enclose a fluid-filled space

LONGITUDINAL MUSCLE: muscle fibers that run along the longitudinal or anterior-posterior axis of the body

PSEUDOCOEL: a fluid-filled body cavity that is bounded by mesodermal muscle on the outside and endodermal epithelium on the internal boundary

The most primitive form of skeletal support evolved is mostly hydrostatic skeletons because no mineralization process is necessary for their formation. Hydrostatic skeletons have been evolved in a wide range of organisms including plants, protists, and animals. The basis of all hydrostatic skeletons has to do with the material properties of water. Water cannot be compressed under biological conditions and thus can act as a support and locomotory transient skeleton. Essentially, all hydrostatic skeletons act in a similar fashion, that is, water is contained in a compartment and is subjected to pressure. In this way, the compartment can become stiff and act as a skeleton. The method used to create the pressure differs among groups.

Among the Protozoa, hydrostatic skeletons are found in the members of the phylum Sarcomastigophora. Within this group are the amoeboid types that form pseudopodia, false feet that are transient structures formed by hydrostatic pressure within the cell. It is hypothesized that contractile proteins within the cell direct fluid toward the cell periphery in channels, and in so doing cause the cell membrane to bulge outward as a pseudopod. Thick pseudopodia, termed lobopodia, are found in the shell-less amoebas. In this case, there appears to be a chemical difference in the fluid used in pseudopodia formation. Plasmasol has less viscosity and is pushed into the pseudopodia; when it is distributed laterally it turns to plasmagel. In thin pseudopodia termed actinopodia, cytoplasm is forced along microtubules termed axonemes. These types of pseudopodia are normally found among foraminiferan and radiolarian amoebas.

Worm Hydrostatic Skeletons

In the animal phyla, hydrostatic skeletons are developed in the Cnidaria and function in relationship to the gastrovascular cavity. In the polyp forms of colonial forms and especially within the Anthozoa, such as sea anemones, the myoepithelial cells surrounding the coelenteron make up a circular muscle band and thus are able to put pressure on the water in the coelenteron to extend the body or maintain the form of the polyp. This coelenteron extends into the tentacles of these forms and thus, via contraction of the myoepithelium, can elongate the tentacles to capture food.

Various types of flatworms comprise the phylum Platyhelminthes. Despite the lack of a coelom, these worms are still considered to have a hydrostatic skeleton. Since the longitudinal and circular muscles lie external to fluid-filled parenchyma tissue, the pressure exerted by this muscle extends the body by elongating this tissue. In this way, the worms can undulate in swimming or have peristaltic motion while moving along a substrate.

The nematode worms as well as other members of the Pseudocoelomate group have developed a fluid-filled body cavity. This cavity in most members of this group has a space with an outer boundary of mesodermal muscle and an inner boundary composed of endodermal cells. In nematodes, the muscle layer is arranged in a longitudinal pattern beneath a cuticle that is composed of layers, including fibrous collagen to maintain the shape of the worm under muscular pressure. Although less flexible than in true worms, contraction of the muscle layer can act against the pseudocoel fluid, thereby causing the body to undulate. Most nematodes need to act against a surface to have effective locomotion.

A more effective method of locomotion using a hydrostatic skeleton has been developed by annelid worms. These worms are segmented and have developed a true coelom. This means that the body cavity or coelom is bounded on all sides by mesodermally derived tissue: circular and longitudinally arranged muscles on the external boundary and membranes wrapping the gut tube. In addition, the coelom is divided in each segment into right and left halves and each segment has a membrane that separates the coelom in one segment from that of another segment. Thus, the circular musculature can constrict one side of the body while the other side is relaxed and stretched. As a result, undulation is more effective in this group than in the pseudocoelomates. When such bending is coupled with setae or segmentally arranged parapodial extensions on the body, the undulations could be used for crawling and swimming, although the latter locomotory ability is poor in some groups.

Using the hydrostatic skeleton in burrowing necessitates another evolutionary strategy involving the coelom. Here the intersegmental partitions or septa are lost or are perforated. This accomplishes the movement of coelomic fluid between segments during muscle contraction. Thus, circular muscles in posterior segments can drive fluid anteriorly, swelling and elongating the anterior portion of the animal. The posterior segments left behind can catch up with the anterior segments when the longitudinally arranged fibers contract. In this way burrowing is effected. Similar contractile wave patterns are used by terrestrial oligochaetes such as earthworms to create peristaltic-type contractions that drive the animal forward. The contraction of circular and longitudinal muscles alternate to create thick and thin areas of the body. This corresponds with elongation and subsequent contraction of the body segments. In earthworms, the segments retain their intersegmental septa. The Hirudinida or leeches have done away with their intersegmental septa and thus the coelomic space is continuous. Constriction of the circular muscles extends the body forward and the subsequent contraction of the longitudinal muscles will bring the rest of the body to meet it. The movement is accomplished by first attaching the posterior sucker, then elongating, attaching the anterior sucker and then pulling the rest of the body forward in the direction of the anterior sucker.

The Hydrostatic Skeletons of Arthropods, Bivalves, and Echinoderms

The development of an exoskeleton requires a change from a coelom-driven locomotion to one that uses muscles. The reduction of the coelom is a characteristic of the diverse arthropod taxon, but hydrostatic skeletons are still used in certain body areas. Flying insects, when attaining their adult state, emerge from their cocoons with folded wings. The insect must pump hemolymph into veins within the wing in order to expand them before they harden. These veins remain in the wings, and their walls and hydrostatic pressure may act to maintain the shape of the wing during flight. The ability of spiders to run fast even though they do have eight legs may lie in the hydraulic systems in their legs. Spiders have replaced extensor muscles with hydraulic spaces that, under pressure, automatically extend the legs. Thus, muscles are normally only used for flexion of the leg segments, and the legs can be moved faster than if muscles were used in both extension and flexion of the leg segments.

Although mollusks have a reduced coelom, hydrostatic skeletons are developed in certain members of this phylum. Bivalves normally burrow into the substrate and send up siphons through which water is brought in and out of the clam. In some species, water pressure is used to open and extend these siphons. In addition, the foot of the clam contains a blood sinus that, under pressure, fills with blood and expands in two ways. The foot can be extended into the substrate with subsequent swelling of the distal end. This action anchors the foot so that the clam can pull itself into the substrate. Among other mollusks, it is thought that the extension of tentacles to capture prey by squid and cuttlefishes is based upon hydraulic action of muscles on these structures.

Most echinoderms have a well-developed water vascular system derived from the coelom. This system is composed of sieve plate opening to the water. This plate is then connected via a stony tube to a ring canal. In asteroids or sea stars, this ring canal extends into the arms via radial canals. From the radial canal located in each, there extend bilaterally arranged lateral canals that enter a tube foot structure. The tube foot has two functional parts, an upper, bulblike ampulla and a lower tube foot. Contraction of the ampulla drives water into the podium, extending it and causing its tip to form a suctionlike disc that attaches to the substrate or prey. Relaxation of the ampulla withdraws water back into the ampulla, retracting the podial portion of the tube foot. In this way many echinoderms move along the ocean bottom and manipulate prey.

How Hydrostatic Skeletons Work

Although hydrostatic skeletons operate on the basis of fluid pressure, there are certain constraints as to how this pressure can be generated and on the makeup of the body wall resisting the pressure.

First, sets of muscles that will exert the pressure on the body must be located external to the fluid compartment or tissue that will function as the hydrostatic skeleton. Second, there have to be reinforcement fibers arranged in a helical pattern around the hydrostatic tissue or space. If fibers are just arranged in a circular or longitudinal pattern, bending of the structure may cause the creation of kinks or possibly bulges in the wall, due to pressure exerted on the body wall. In both cases, deleterious effects can occur on body organs or an aneurysm may form in the wall that may lead to compromising the structure through its rupture. To prevent such adverse conditions, fibers are arranged in a helix around the body wall. This same type of reinforcement is seen in the construction of the walls of high-pressure hoses. Since the fibers are wound at an angle around the structure, bending does not cause an aneurysm to form and there is no chinking of the wall. This type of fiber arrangement is found in the walls of annelids, nemertines, and flatworms, among others. The fibers are often composed of collagen, a helical protein.

Vertebrate Hydrostatic Skeletons

Like mollusks and arthropods, vertebrates have largely abandoned the coelom. Their endoskeletons have taken the place of hydrostatic skeletons. However, hydrostatic skeletons still occur, particularly in the reproductive system. The penis or hemipenes of mammals and reptiles contain spongy tissue that can engorge with blood. Venous return of the blood is largely prevented, extending the length and stiffness of the intromittent organ so that it may be inserted into the female's reproductive tract.

—*Samuel F. Tarsitano*

See also: Endoskeletons; Exoskeletons; Locomotion.

Bibliography

Barrington, E. J. W. *Invertebrate Structure and Function*. New York: John Wiley and Sons, 1979. An invertebrate textbook that covers hydrostatic skeletons, among other anatomical features.

Brusca, R. C., and G. J. Brusca. *Invertebrates*. Sunderland, Mass.: Sinauer Associates, 1990. This text discusses the anatomy and physiology of Protists and invertebrates including the reproductive system, locomotory and feeding structures, and sense organs. The structure and mechanisms of hydrostatic skeletons are found throughout the text.

Kristan, W. B., Jr., R. Skalak, R. J. A. Wilson, B. A. Skierczynski, J. A. Murray, F. J. Eisenhart, and T. W. Cacciatore. "Biomechanics of Hydroskeletons: Lessons Learned from Studies of Crawling in the Medicinal Leech." In *Biomechanics and Neural Control of Posture and Movement*, edited by Jack M. Winters and Patrick Crago. New York: Springer-Verlag, 2000. This chapter gives an account of the neural control of the body elongation of the leeches. The hydroskeleton of these worms is discussed with relationship to the distribution of the coelom, its extension throughout the body of the leech, and how it functions in locomotion.

Ruppert, E. E., and R. D. Barnes. *Invertebrate Zoology*. 6th ed. New York: Saunders College Publishing, 1994. While this book is a standard textbook on invertebrate morphology and life habit strategies, Ruppert and Barnes take a functional approach to the underlying reasons why structures have evolved and how they work. This text has a section devoted to how hydroskeletons work in terms of the fiber arrangements that help contain the pressurized fluid.

HYENAS

Types of animal science: Classification, ecology, reproduction
Fields of study: Anatomy, ecology, zoology

Hyenas are wolflike carnivore-scavengers whose useful ecological functions derive from their diet of carrion and live animals from termites to antelope.

Principal Terms

CARRION: dead animals
GESTATION: time in which mammalian offspring develop in the uterus
MUSK: bad-smelling liquids made to mark territory or for self-defense
PERINEAL: located between scrotum and anus in males or equivalent region in females

Hyenas comprise four carnivore species of the family Hyaenidae, with body shapes similar to wolves; awkward-looking hind legs shorter than the front legs; good hearing; good vision; and a good sense of smell. Sizes range from small aardwolves to large spotted hyenas. All inhabit grasslands and shrubby areas in Africa, the Middle East, and Arabia.

Some species form packs; others live alone. Most scavenge anything they find, including carrion. Aardwolves eat carrion but prefer termites. Hyenas mate year round and have two- to four-month gestation periods, depending on species.

Physical Characteristics of Hyenas

The physical characteristics of three of the four hyena species are exemplified by the spotted hyena, the largest, strongest species. Their maximum length is six feet, height three feet, and weight 175 pounds. Spotted hyena hind legs are shorter than their front legs, making them look awkward. They also have four-toed paws and manes of coarse hair on the neck, shoulders, and back. The adults are a brown-gray with brown spots, and have large heads, bone-crushing jaws, and an eerie, "laughing" cry, like hysterical human laughter.

Laughing (spotted) hyenas were long thought to be carrion-eaters only. It is now clear that they are major predators of live herbivores such as zebras. They attack in packs at night, bite their victims, and hold on until the prey stumbles. They

Hyena Facts

Classification:
Kingdom: Animalia
Subkingdom: Bilateria
Phylum: Chordata
Subphylum: Vertebrata
Class: Mammalia
Order: Carnivora
Family: Hyaenidae (hyenas and aardwolf)
Genus and species: Crocuta crocuta (spotted hyena); *Hyaena brunnea* (brown hyena), *H. hyaena* (striped hyena); *Proteles cristatus* (aardwolf)
Geographical location: Asia, Africa, and parts of the former Soviet Union
Habitat: Deserts, grasslands, shrubby areas, forests, and mountains
Gestational period: Depending on species, two to four months
Life span: Fourteen to twenty-five years in the wild, twenty-five to forty-five years in captivity
Special anatomy: Manes; four- or five-toed paws; hind legs shorter than forelegs; most have teeth able to crush bones, but the aardwolf eats only insects

kill by tearing open the belly of the prey. Spotted hyenas kill other carnivores as well, such as striped and brown hyenas.

Striped and brown hyenas have manes, short hind legs, and bone-crushing teeth. They are smaller and less aggressive than laughing hyenas, inhabiting grassy and shrubby areas of Africa, India, and the former U.S.S.R. Their gray-brown, black-striped fur is fine camouflage. Like other hyaenids, striped hyenas eat carrion. They also eat fruit, small mammals, birds, and sometimes large herbivores, such as antelope. They grow to maximum lengths of 5.5 feet, heights of 2.5 feet, and weights of 125 pounds. Spotted hyenas hunt at night in small packs.

Brown hyenas are dark brown with gray heads and striped legs. Their maximum length is 4.5 feet, and they reach 120 pounds. They inhabit the dry, rocky Southern African deserts, usually traveling alone. As scavengers they eat anything available, including carrion and bones picked clean by vultures, using strong teeth to crack bones for marrow.

Aardwolves, honorary hyenas, inhabit much of Africa. They are hyaenid by appearance, as their backs slope down from shoulder to tail due to short hind legs. They have reddish, black-striped fur and manes on their necks and shoulders. When attacked, aardwolves erect the mane to look fiercer and spray evil-smelling musk from perineal glands.

Aardwolves differ from hyenas in having five-toed front feet. Aardwolf teeth are small and suitable only for eating their main food, termites and other insects. Their maximum length is 2.5 feet, their height is 1.5 feet, and their weight is twenty-five pounds. The termites that aardwolves eat are active at night, so aardwolves are nocturnal and eat termites with their long, sticky tongues. Aardwolves live alone and mark territories with musk, denning in empty burrows of other animals.

The spotted hyena's maniacal, eerie cry has led to it being called the "laughing" hyena. (Digital Stock)

The Life Cycle of Hyenas

There are similarities and differences in hyena species lives. Spotted hyenas form groups of up to one hundred: a few males, many females, and numerous young. Females conceive year round, birthing two or three pups that can see and run immediately, after a four-month gestation. Females, larger than males, select short-term mates. Striped hyenas fight within groups, sometimes killing group members. Maximum life spans are twenty-five years in the wild and forty years in captivity.

Like spotted hyenas, striped hyenas mate year round. They live in small groups or alone. A three-month gestation yields two to five young born with the eyes not yet open. Mothers nurse offspring until they can feed themselves. Life spans are up to twenty-five years in captivity. Brown hyena life cycles and reproductive habits are nearly the same as in striped hyenas. However, they live alone except when mating or nursing young.

Aardwolves live in groups. They have a mating season when males fight for females and winners mate. Gestation is two months and yields two to five young, nursed for two months. Aardwolves live for fifteen years in captivity.

Spotted, striped, and brown hyenas eat carrion, preventing its decay and its endangerment of humans and other animals. Aardwolves eat termites, preventing damage to the wilderness and human habitations. These activities are their main ecological function. As spotted and striped hyenas eat live food, they also kill injured or weak members of other species, helping the species eaten to enhance their long-term survival.

—*Sanford S. Singer*

See also: Carnivores; Dogs, wolves, and coyotes; Fauna: Africa; Packs; Predation; Scavengers.

Bibliography

Ammann, Karl, and Katherine Amman. *The Hunters and the Hunted*. London: Bodley Head, 1989. This illustrated book includes an article on hyenas.

Bougrain-Dubourg, Allain. *The Tender Killers*. Translated by Yann Arthus-Bertran. New York: Vendome, 1985. Useful hyena facts and photographs.

Caputo, Robert, and Miriam Hsia. *Hyena Day*. New York: Coward, McCann & Geoghegan, 1978. A juvenile book on a day in the life of laughing hyenas.

Mills, Gus, and Heribert Hofer. *Hyaenas: Status Survey and Conservation Action Plan*. Gland, Switzerland: IUCN, 1998. Has illustrations, information on hyenas, and a bibliography.

Mills, M. G. L. *Kalahari Hyenas*. Norwell, Mass.: Kluwer Academic, 1994. Focuses on feeding ecology and social behavior of hyenas.

Werdelin, Lars, and Nikos Solounias. *The Hyaenidae: Taxonomy, Systematics, and Evolution*. Oslo, Norway: Universitetsforlaget, 1991. This illustrated book covers the title issues well.

HYRAXES

Types of animal science: Anatomy, classification, reproduction
Fields of study: Anatomy, zoology

Hyraxes, rabbit-sized, furry mammals, look like rodents but are more closely related to elephants. Their foot pads optimize traction on rocks and in trees.

Principal Terms

DIURNAL: active during the day
GESTATION: duration of pregnancy
NOCTURNAL: active at night
PELAGE: animal fur
SOLITARY: lives alone

Hyraxes are rabbit-sized mammals that look like rodents. Strangely, they are closely related to elephants. There are three kinds of hyraxes, all found in Africa: rock hyraxes, bush hyraxes, and tree hyraxes. These vegetarians eat different kinds of food and differ in social interactions. Some live in groups and others are often solitary.

Tree hyraxes, unlike rock or bush hyraxes, are usually solitary. Although the hyrax is rodentlike in form, its closest animal relative is the elephant. (Lorenzo Lorente/Photo Agora)

Physical Characteristics of Hyraxes

Hyraxes are one to two feet long and weigh three to fourteen pounds. They have stumpy tails, brown to gray fur on their backs, and lighter pelage on their sides. Hyrax fur is short on individuals living in warm, dry regions, and thick and soft on those living in colder areas.

The three types of hyraxes walk on all fours and are excellent climbers, because their feet end in rubbery pads having sweat glands. When hyraxes run, their feet sweat, and the resultant lubrication improves traction on rocks or trees. However, hyraxes also have small hooves on the first and third toes of their hind feet. All hyraxes have a gland in the middle of the back, surrounded by a ring of erectile, dark brown to yellow fur. When a hyrax becomes excited, this hair stands on end.

Life Cycles of Hyraxes

Bush and rock hyraxes are social animals which live in family groups of up to three dozen members. Each group is led by a dominant male. Its other members are adult females and young of various ages. Group members care for each other. The dominant male marks off their territory and defends it, using scent markers to warn off other hyraxes.

Single groups of bush and rock hyraxes may share a territory peacefully, even using the same burrows. They cluster together for warmth. The young of both species play together. Part of the basis for this coexistence is that they do not compete for food. Although hyraxes can eat grasses or other soft plants, bush hyraxes eat soft plants but not grasses, and rock hyraxes eat grasses. However, while the two species live together, they do not interbreed, since both their mating behaviors and the anatomy of their sex organs differ.

Tree hyraxes differ from the other types in being usually solitary. However, they may live with one or two others. They never live in large groups or with other kinds of hyraxes. This is partly due to their habitats in trees. Rock and bush hyraxes are diurnal, while the tree hyraxes are nocturnal.

In all types of hyrax, mating season depends on species and habitat. Gestation is seven to eight months. All females in a family group of rock or bush hyraxes give birth within a few weeks of each other, each having one to four young which are nursed for five months. Tree hyraxes litter one or two young. The offspring of all species can mate at approximately sixteen months old. Females join rock or bush hyrax groups, and males leave by age 2.5 years. Tree hyrax offspring are solitary after weaning. Hyraxes can live for nine to twelve years.

The Three Types of Hyrax

Rock hyraxes (dassies) live in diurnal family groups among rocks and boulders, from dry lowlands to mountains 14,000 feet high. They inhabit and hide in rocks or their crevices. Dassies are one to two feet long and weigh four to fourteen pounds. Their back fur is light to dark brown and the erectile fur around their midback glands is dark brown or yellow-orange. Rock hyraxes eat grasses. Their eyes can look right into the sun, enabling them to escape avian predators. They can live for nine to twelve years.

Bush hyraxes are diurnal and live amid rocks and boulders or in hollow trees from south to northeast Africa. Up to two feet long, they weigh three to twelve pounds, have light gray back fur, and yellow fur around the midback glands. These hyraxes eat soft plants and live in groups of up to thirty-four. Often they share territory with rock hyraxes. They can live for ten to twelve years.

Tree hyraxes (dendrohyraxes) live in Africa. Eastern tree hyraxes inhabit Kenya's coast and Zanzibar. Southern tree hyraxes inhabit southeastern and East Africa. Western tree hyraxes live in West and Central Africa. All make tree nests in savannas, rain forests, and evergreen forests at altitudes up to 12,000 feet. They are up to two feet long, and weigh three to nine pounds, less than rock or bush hyraxes. Their long, dark brown back fur and dark yellow erectile fur around midback glands blend with shadows in trees. They eat soft plants, are nocturnal and hide during daylight, are most often solitary, but may live with one or two others.

Hyrax Facts

Classification:
Kingdom: Animalia
Subkingdom: Bilateria
Phylum: Chordata
Subphylum: Vertebrata
Class: Mammalia
Order: Hyracoidea (hyraxes)
Family: Procaviidae
Genera: Procavia (rock hyrax, five species); *Heterohyrax* (bush hyrax, three species); *Dendrohyrax* (tree hyrax, three species)
Geographical location: Africa and the Middle East up to Syria
Habitat: Among rocks and boulders, from dry lowlands to mountains fourteen thousand feet high, in trees near coasts, in rain forests, and evergreen forests twelve thousand feet high
Gestational period: Seven to eight months
Life span: Nine to twelve years
Special anatomy: Feet with rubbery pads containing sweat glands; rudimentary hooves on the first and third toes of the rear feet; a midback gland, surrounded by erectile fur

Hyraxes and Their Predators

Hyraxes are of relatively little interest to humans, though some Africans eat them. All hyraxes are preyed on by eagles, lions, leopards, jackals, hyenas, and snakes. Other predators are more selective in hyrax predation, due to their different habitats. For example, special predators of tree hyraxes include civet cats, servals, and caracals.

—*Sanford S. Singer*

See also: Elephants; Fauna: Africa; Groups; Nocturnal animals; Rodents.

Bibliography

Crispens, Charles G. *The Vertebrates: Their Forms and Functions*. Springfield, Ill.: Charles C Thomas, 1978. A book full of information on vertebrates, including anatomy, development, and natural history.

Parker, Sybil. *Grzimek's Encyclopedia of Mammals*. New York: McGraw-Hill, 1990. This standard encyclopedia of animal life has a quite extensive article on hyraxes.

Prothero, Donald R., and Robert M. Schoch. *The Evolution of Perissodactyls*. New York: Oxford University Press, 1989. A fine book describing, among other things, the relationship between hyraxes and elephants.

Ricciuti, Edward R. *What on Earth Is a Hyrax?* Woodbridge, Conn.: Blackbirch Press, 1997. An interesting brief book on hyraxes, their behavior, and their natural history.

ICHTHYOSAURS

Types of animal science: Anatomy, classification, evolution
Fields of study: Anatomy, paleontology

Ichthyosaurs are extinct marine reptiles of the Mesozoic era. They were highly adapted to the marine habitat, having evolved a dorsal fin, flippers, and a tail fin. These fleshy features are often preserved in spectacular fossils from the Holzmaden and Solnhofen deposits in Germany.

Principal Terms

CARANGIFORM SWIMMING: a method of swimming where the tail is moved while the body is held rigid

DIAPSID: having two bone openings in the temporal region used for jaw adductor attachment; a supratemporal opening and an infratemporal opening

EURYAPSID: having only a single temporal bone opening above the squamosal and postorbital bone; the supratemporal opening of a diapsid condition

HYPERDACTYLY: a condition whereby the number of digits is increased above the normal five to create a wing

HYPERPHALANGY: a condition whereby the number of phalanges is increased in each digit

INIA TYPE: ichthyosaurs with a tunalike tail fin and possible carangiform swimming or forelimb flyer

NEOCERATODUS TYPE: ichthyosaurs with undulating tails and forelimb wings

Ichthyosaurs were the most marine-adapted type of all the reptiles. When they first make their appearance in the fossil record in the Triassic, they are already recognizable as ichthyosaurs adapted to an aquatic existence. For this reason, their ancestry has remained a mystery since their discovery in the eighteenth century. Ichthyosaurs ranged in size from less than a meter to over ten meters in length. Early ichthyosaurs had already modified their forelimbs and hindlimbs into winglike structures. The limbs in ichthyosaurs had hyperphalangeal conditions, whereby there was an increase in the usual number of phalanges in each finger. Some ichthyosaurs expanded the width of the wing by adding extra rows of fingers, a condition known as hyperdactyly. The tails had reevolved a fin that either extended from the middorsal surface to the midventral surface or formed a semilunate tail. In these ichthyosaurs, the body plan already was porpoiselike, with a barrel-shaped thorax and tapering tail. Along these lines, icthyosaurs reevolved a dorsal fin. The skull, with its beaklike rostrum, had a narrow snout with large orbits, with sclerotic plates for maintaining the shape of the eyes. The nostrils had migrated backward to lie in front of the orbits and the temporal region was reduced. The skull behind the large eyes was much reduced. The ichthyosaurs retained one large temporal fenestra (bone opening) that housed the pseudotemporalis muscle. The other main jaw adductor, the pterygoideus, originated on the palate. Based upon the temporal fenestral pattern, these forms were classified as either their own class of reptiles or were considered to be a member of the Euryapsida. Evidence from the skull and reproductive strategy point to another ancestry. The ichthyosaur skull shows the loss of the lower temporal opening of the diapsid skull condition. In addition, ichthyosaurs were live bearers of their young, with a placenta-like structure. The only reptilian group with the loss of the lower temporal fenestra and loss of the eggshell is the diapsid Lepidosauria. All other reptilian groups use the eggshell for the main source of cal-

Ichthyosaur Facts

Classification:
Kingdom: Animalia
Phylum: Chordata
Subphylum: Craniata
Class: Reptilia
Subclass: Lepidosauria
Order: Ichthyopterygia
Family: Ophthalmosauria
Genus and species: Ophthalmosaurus icenicus
Geographical location: Oceans of the Mesozoic Era
Habitat: Marine
Gestational period: Unknown
Life span: Unknown
Special anatomy: Limbs transformed to flippers (wings) that employ hyperphalangy to elongate flipper; reevolve dorsal fin and tail fluke in some; smooth skin without scales; rostrum elongate with conical labyrinthodont teeth; live-bearing

cium for the skeletal development of the embryo, and thus the eggshell cannot be eliminated in their reproductive cycle.

Types of Ichthyosaurs

Ichthyosaurs are typed according to their mode of locomotion or by the morphology and shape of the pectoral flippers. The *Neoceratodus* type, or lungfish type, is so called because their straight tails resembled the tails of lungfish. These types had equally sized pectoral and pelvic flippers. It is considered that these forms had a flexible body with some degree of tail undulation possible for locomotion. The winglike forelimbs, with their hyperphalangy condition, gave them a hydrodynamic shape, and may have been their main propulsion system, acting as wings. Jurassic ichthyosaurs saw the development of a tunalike semilunate tail fin. These types, known as *Inia* type, named after the Amazon dolphin, were considered to be faster and more maneuverable swimmers than the *Neoceratodus* type. The hind limbs are reduced in these forms, and it is considered by some that the forelimbs were used as wings and the tail was used as a steering device. Others consider that the semilunate tail was used in a form of swimming used by tunas, known as carangiform swimming, using rapid movements of the tail. The vertebral column here extended into the lower lobe of the tail. A cartilage ray extended into the leading edge of the upper lobe of the tail, as seen in beautifully preserved tail found in the upper Jurassic deposits around Solnhofen, Germany. These forms also had a well-developed dorsal fin to aid in the prevention of rolling during swimming. It is thought that these ichthyosaurs had a more rigid body and thus had less drag on the body during swimming. Another ichthyosaur form, the *Leptopterygius* type, had well-developed hind limb flippers and pelvic girdle, and slightly reduced pectoral flippers. Undulation of the tail was con-

Ichthyosaurs were highly adapted to marine life, and in an example of convergent evolution, many of these dinosaurs exhibit body shapes similar to today's marine mammals, such as porpoises. (©John Sibbick)

sidered the main propulsion system of these ichthyosaurs. Finally, in the *Mixosaurus* type, undulation of the body seems to have been the mode of locomotion. In this body form, a series of elongated neural spines supported the tail fin. There seem to be two opposing views of ichthyosaur swimming speed and prey capture. Some authors believe that ichthyosaurs were high-speed swimmers, while others feel that ichthyosaurs were slower moving but highly maneuverable, and perhaps capable of short, high-speed swimming bursts.

Much is known about the diet of ichthyosaurs through the preservation of stomach contents. Many ichthyosaur species relied heavily on squid and, to a lesser degree, fish. There are reports of pterosaur remains preserved in the guts of ichthyosaurs as well.

Surprisingly, much is also known concerning the reproductive habits of ichthyosaurs. Because of their wing limbs and their round-girthed bodies, it is unlikely that ichthyosaurs came out of the water. How then did they give birth? The answer to this question lay in some of the most spectacular fossils of ichthyosaurs housed in the Stuttgart Museum in Germany. Female ichthyosaurs were preserved in the act of giving birth, whether from problems arising in the birthing process or possibly from poisonous dinoflagellate blooms that killed the ichthyosaurs during the birthing process. In some of these fossils the preserved offspring can be seen lying in a placenta-like structure expelled from the female's body. Other baby ichthyosaurs are often seen in the abdominal area of the ichthyosaur, and were previously thought to be the result of cannibalism. However, it is more likely that these were developing fetuses.

—Samuel F. Tarsitano

See also: *Allosaurus*; *Apatosaurus*; *Archaeopteryx*; Dinosaurs; Evolution: Animal life; Extinction; Fossils; Hadrosaurs; Paleoecology; Paleontology; Prehistoric animals; Pterosaurs; Sauropods; Stegosaurs; *Triceratops*; *Tyrannosaurus*; Velociraptors.

Bibliography

Degus, R. "Rulers of the Jurassic Seas." *Scientific American* (December, 2000): 52-59. Discusses the origin and evolution of ichthyosaurs, their eyesight, types, and diet.

Dixon, C., B. Cox, R. J. G. Savage, B. Gardiner, and M. C. McKenna. *The Macmillan Illustrated Encyclopedia of Dinosaurs and Prehistoric Animals*. New York: Macmillan, 1988. Basic information and photos of ichthyosaurs, with classification and sizes of the different types of ichthyosaurs.

McGowen, C. *Dinosaurs, Spitfires, and Seadragons*. Cambridge, Mass.: Harvard University Press, 1991. Reviews the origin and evolution of ichthyosaurs. Included are the modifications of the skeleton for swimming and classification based upon flipper types. Ichthyosaur diet and reproductive biology are discussed.

Massare, J. A., and J. M. Calloway. "The Affinities and Ecology of Triassic Ichthyosaurs." *Bulletin of the Geological Society of America* 102 (1990): 409-416. This paper discusses the phylogenetic relationships of the ichthyosaurs as well as their locomotion and habits.

Motani, R. "Phylogeny of the Ichthyopterygia." *Journal of Vertebrate Paleontology* 19 (1999): 472-495. This work is an extensive cladistic analysis of the ichthyosaurs. Phylogenetic trees are used to represent the relationships within this group with all genera listed.

Romer, A. S. *Osteology of the Reptiles*. Reprint. Chicago: University of Chicago Press, 1968. This is a classic book on the morphology of reptiles, including ichthyosaurs. Contained is a listing and classification of reptile genera.

IMMUNE SYSTEM

Type of animal science: Physiology
Fields of study: Cell biology, immunology, pathology

The immune system distinguishes "self" from "nonself" in the body, fighting off foreign invaders such as bacteria, viruses, and parasites. It works through the production of proteins called antibodies, and of cells that recognize and kill foreign pathogens.

Principal Terms

ANTIBODY: protein produced by lymphocytes, with specificity for a particular antigen

ANTIGEN: chemical that stimulates the immune system to respond in a very specific manner

CELL-MEDIATED IMMUNITY: production of lymphocytes that specifically kill cells with foreign antigens on their surfaces

HUMORAL IMMUNITY: production of antibodies specifically reactive against foreign antigens carried in body fluids (humors)

LYMPHOCYTE: white blood cell that produces either cell-mediated or humoral immunity in response to foreign antigens

MACROPHAGE: mature phagocytic cell that works with lymphocytes in destroying foreign antigens

An animal must keep itself distinct from its environment, recognizing its own tissues and keeping them from being invaded or mixed with tissues of other organisms. There are two types of protection used by animals in keeping out invaders and resisting foreign substances, nonspecific and specific defenses. In both types, the body distinguishes between cells that belong to the animal, which are "self," and anything that does not belong, or "nonself."

Nonspecific Defenses

Even animals as primitive as sponges have the ability to recognize and maintain self-integrity. Scientists have broken apart two different sponges of the same species in a blender, intermixing the separated cells in a dish. Cells crawled away from nonself cells and toward self cells, reaggregating into clusters of organized tissues containing cells of only one particular individual. Phagocytic cells that engulf and destroy foreign invaders were first identified by a scientist who had impaled a starfish larva on a thorn. He observed that, over time, large cells moved to surround the thorn, apparently trying to engulf and destroy it, recognizing it as nonself. Even earthworms have the ability to recognize and reject skin grafts from other individuals. If the graft comes from another worm of the same population, the skin is rejected in about eight months, but rejection of skin from a worm of a different population occurs in two weeks. Phagocytic cells in earthworms have immunological memory, enabling a worm to reject a second transplant from the same foreign source in only a few days.

Barriers, chemicals, and phagocytic cells are nonspecific protective mechanisms, which do not distinguish among different kinds of invaders. Tough outer coverings such as skin, hide, scales, feathers, or fur provide surface barriers. Nonspecific defenses also include secretions of mucus, sweat, tears, saliva, stomach acid, and urine, as well as body-fluid molecules, such as complement and interferon. Damaged tissues or bacterial invaders signal other cells to produce inflammation, a nonspecific response characterized by heat, redness, swelling, and pain. Cellular defenses associated with inflammation include phagocytes such as neutrophils and macrophages, which engulf and digest bacteria and debris, and natural killer cells, which destroy cancer cells or virally infected cells by poking holes in them.

Immunity

The only specific defense in vertebrates is provided by the immune system, in which the component parts react against particular antigens on invaders, such as individual strains of bacteria or types of viruses. This more sophisticated protection is produced by lymphocytes that provide either cell-mediated or humoral immunity against particular antigens. Cell-mediated immunity depends on T lymphocytes (T cells) that become mature as they pass through the thymus, from which they get the "T" of their name. Humoral immunity is the function of antibodies, proteins released by B lymphocytes (B cells) that have matured and developed into plasma cells. The B lymphocytes reach maturity in the bursa of Fabricius in birds, where they were first recognized and from which they were named. Other vertebrates lack the bursa of Fabricius, and B cells mature in the bone marrow instead, so the name "B lymphocyte" still applies.

Antigen molecules are usually proteins or glycoproteins (proteins with sugars attached) that generate either an antibody response or a cellular immune response when they are foreign to the responding animal. So-called self antigens are molecules on cell membranes that identify the cells as belonging to the animal itself. An animal would not normally produce an immune response against its own antigens, but the same antigens would generate an immune response if placed in another animal to whom they were foreign. These antigens are the means by which self and nonself distinctions are made by the immune system, so the system can determine whether to ignore cells or attack them. Occasionally, self antigens, for some reason, are no longer recognized by the animal's immune system, and are attacked as if they were foreign. This causes an autoimmune disease, where the immune system destroys the body's own tissues.

Scientific understanding of how the immune system functions is largely dependent on work done using laboratory animals, including rabbits,

Five Classes of Antibodies

IgM is produced first in a response to foreign antigen, but its concentration declines rapidly. With five Y-shaped monomer subunits forming a pentamer structure, IgM is very effective in binding many copies of the same antigen and agglutinating them, but is too big to cross the placenta.

IgG is the most abundant class of antibodies in circulating blood, a monomer capable of passing through vessel walls to protect cells and tissues. In some species, including humans, it crosses the placenta to pass on the mother's immune protection to the fetus. Produced after IgM in an immune response, it is much more effective against bacteria, viruses, and toxins.

IgA is secreted as a dimer (two subunits) into milk, sweat, saliva, and tears. It is especially important in colostrum, the secretion before milk production begins that is the only way some newborn animals receive their mother's antibodies. IgA prevents bacteria and viruses from binding to epithelial cell surfaces, especially in the digestive tract.

IgE antibodies bind to the surfaces of mast cells and basophils with the arms of the Y-shaped monomer extended. Foreign antigens bind to the ends of the Y arms and trigger these cells to release histamine and other chemicals that cause the inflammation of allergy. IgE is also the antibody that attacks parasites inside the body, such as worms.

IgD molecules are monomers located mainly on the surfaces of B cells, apparently acting as receptors for the antigen that is recognized by each B cell and triggers its activation.

mice, and hamsters. Laboratory mice have been highly inbred into strains where all the animals are genetically identical and their genes and antigens are well known. Studies on these mice have been essential in determining how the immune system normally works, and how it fails to work in autoimmune diseases and the inability to prevent cancer cells from proliferating.

Antigen Presentation and Receptors

Central to the functioning of the immune system in mammals is a system of genes called the major histocompatibility complex (MHC). These genes encode a collection of cell-surface glycoproteins that are the self antigens by which the immune system recognizes its own body cells. Class I MHC molecules are expressed on the surfaces of all nucleated cells, while Class II MHC markers are produced only by specialized cells, including cells of the thymus, B lymphocytes, macrophages, and activated T lymphocytes. Both Class I and Class II MHC molecules identify the cells bearing them as self, and these also serve as the context in which the immune system recognizes foreign antigens that are presented on the cell surface. Cells with self antigens are tolerated by the immune response of that individual animal, while cells that show foreign antigens are attacked and destroyed. Rejection of a graft or transplanted organ is reduced with more closely matched tissues, which are better tolerated by the immune system.

When bacteria evade protective barriers and chemicals to enter an animal's body, the animal's macrophages attack, engulfing and digesting the invaders. One bacterium may have thousands of different antigenic segments that can be recognized on its surface or inside the cell. Small parts of these digested cells, the individual antigens, are joined to the macrophage's newly formed MHC Class I and Class II before they are exposed on the cell surface. The foreign antigens fit into a space or pocket within the MHC molecule and are recognized by T cells that have the same MHC molecules and can respond specifically to the foreign antigen. Cytotoxic T cells (T) react to antigens held in the pocket of a Class I molecule, while helper T cells (T) respond to those presented by Class II molecules. T cells are the agents of cellular immunity, producing perforin molecules that puncture and kill cells bearing the foreign antigen against which the T cells are specific. T cells, when activated by encountering their specific antigens presented with Class II molecules, release cytokines that help to activate both T cells and B lymphocytes. Activated B cells divide to produce memory B cells and lymphocytes that mature into plasma cells, which secrete about two thousand antibody molecules per second over their active lifespan of four or five days.

Both T and B lymphocytes can react with their specific foreign antigens because the antigen-MHC complex binds to receptor molecules on the lymphocyte surfaces. Each clone of lymphocytes has the genetic ability to respond to a particular shape that fits its receptors. There may be millions of different receptors among the lymphocytes of a single animal, capable of binding millions of different antigens, even artificial chemicals not existing in nature. This enormous variability in response capability makes the immune system of each animal protective against many kinds of foreign invaders. Since each individual has its own set of immune responses, a population is less likely to have all its members die in an epidemic. Certain animals will be more resistant to the pathogens, so some will survive to reproduce and keep the population from extinction.

Primary and Secondary Immune Responses

When a foreign antigen is encountered by an animal for the first time, both T and B cells that can bind the antigen are activated, but not immediately. In a series of reactions, macrophages first break down the antigen-bearing cell, processing and presenting the antigen on its surface with MHC. The T cell specific to that antigen then encounters the antigen-MHC complex on the macrophage and divides to produce a clone of memory T cells and a clone of effector (activated) T cells. Activated T cells release cytokines that activate T and B cells so that they can attack the same foreign antigen. The first encounter with antigen produces a slow primary response, taking more than a week to reach peak effectiveness. During the time needed to generate this response, pathogenic bacteria or viruses can produce disease in the animal under attack. The memory T and memory B cells remain alive but inactive until the same foreign antigen is encountered again, even years later. The secondary response that results immediately when these memory cells are activated occurs so quickly that the disease process does not recur.

The importance of the immune system is seen in humans who lack its function, those with acquired immunodeficiency syndrome (AIDS). Human immunodeficiency virus (HIV) is the causative agent of AIDS, and is similar to viruses that attack other species in the same way. Most who die with AIDS really succumb to one of many opportunistic infections that cause diseases in HIV-positive individuals, but which are eradicated by the immune system in normal individuals.

—*Jean S. Helgeson*

See also: Anatomy; Beaks and bills; Bone and cartilage; Brain; Circulatory systems of invertebrates; Circulatory systems of vertebrates; Claws, nails, and hooves; Digestive tract; Diseases; Ears; Endoskeletons; Exoskeletons; Eyes; Fins and flippers; Kidneys and other excretory structures; Lungs, gills, and tracheas; Muscles in invertebrates; Muscles in vertebrates; Nervous systems of vertebrates; Noses; Physiology; Reproductive system of female mammals; Reproductive system of male mammals; Respiratory system; Sense organs; Skin; Tails; Teeth, fangs, and tusks; Tentacles; Wings.

Bibliography

Campbell, Neil A., Jane B. Reece, and Lawrence G. Mitchell. *Biology*. 5th ed. Menlo Park, Calif.: Benjamin/Cummings, 1999. Chapter 43 on "The Body's Defenses" is a clear presentation of how the immune system works, in a college textbook for science majors. Color diagrams explain more difficult concepts.

Paul, William E. *Fundamental Immunology*. 4th ed. Philadelphia: Lippincott-Raven, 1999. Despite the name, this medical text covers all aspects of human immune function in over 1,500 pages. Regulation of the immune system and its functions in fighting disease are discussed extensively.

Roitt, Ivan, Jonathan Brostoff, and David Male. *Immunology*. 5th ed. Philadelphia: C. V. Mosby, 1998. Both vertebrate and invertebrate immune functions are discussed in Chapter 15, "Evolution of Immunity." Good color photos and diagrams help explain cellular and molecular aspects of immunology.

Staines, Norman, Jonathan Brostoff, and Keith James. *Introducing Immunology*. 2d ed. St. Louis: C. V. Mosby, 1993. For readers with no previous knowledge of immunology, this primer covers the complex language without using jargon and explains the information clearly.

IMPRINTING

Type of animal science: Behavior
Fields of study: Ethology, neurobiology, zoology

Imprinting is the process of very rapid attachment between parents and offspring during a brief period that alters the offspring's behavior permanently. Imprinting involves behavior that is necessary to the survival of the animal under natural conditions.

Principal Terms

CONSPECIFICS: animals of the same species

CRITICAL PERIOD: a very brief period of time in the development of an animal during which certain experiences must be undergone; the effects of such experiences are permanent

INNATE: inborn, unchangeable

NIDIFUGOUS BIRDS: ground-nesting birds

PRECOCIAL: possessing, as infants, well-developed sense organs, the capability of locomotion, and the ability to contribute substantially to the establishment of bonds between them and their parents

RELEASER: a stimulus that releases a sequence of reflexes that always occur in the same order and manner

SCHEMA (pl. SCHEMATA): an innate releasing mechanism; a neural process that programs an animal for receiving a particular sign stimulus and causes a specific behavioral response

SENSITIVE PERIOD: a period during which a given event produces a stronger effect on development or in which a given effect can be produced more rapidly than it can earlier or later

Imprinting is the process of very rapid attachment between parents and offspring, during a brief sensitive period, that alters the offspring's behavior permanently. This period is found only in precocial birds and a very few mammals. The significance of imprinting is the development of behavioral patterns that will enable the animal, as it matures, to recognize its own species (for future mating), to socialize and cooperate with other conspecifics, and to locate food and territory.

Konrad Lorenz, the first scientist to investigate the phenomenon of imprinting thoroughly, drew attention to the process by which young duck-

Konrad Lorenz's studies of imprinting in young animals contributed to his sharing the Nobel Prize in Physiology or Medicine in 1973. (Nobel Foundation)

Konrad Lorenz

Born: November 7, 1903; Vienna, Austria
Died: February 27, 1989; Altenburg, Austria
Fields of study: Ethology, physiology, zoology
Contribution: Lorenz, the founder of ethology, developed a deep understanding of the behavior patterns in animals, notably imprinting in young birds.

Konrad Lorenz, the son of an orthopedic surgeon, received a medical degree in 1928 and a Ph.D. in zoology in 1933. Both were earned at the University of Vienna. Lorenz studied animal behavior by means of comparative zoological methods, a field now known as ethology. Many of his observations came from establishing and watching colonies of birds, particularly the jackdaw and greylag goose. He published numerous papers based on his observations of these birds.

In the mid 1930's, Lorenz reported his findings about learning behavior in young ducklings and goslings. The survival of a young bird requires an extremely rapid establishment of the behavioral bond between it and a parent. Lorenz discovered that a newly hatched bird learns within a few hours to follow real or foster parents. This process, known as imprinting, may occur with a human, or even an inanimate object, if the bird's parent is not present. Although the behavior itself is genetically determined, the bird learns the object. Through imprinting, many young animals follow the first moving object they encounter.

To demonstrate the process of imprinting, Lorenz imitated a mother duck's quacking sounds in the presence of newly hatched mallard ducklings. The ducklings responded to Lorenz as their mother and followed him accordingly. After numerous studies, Lorenz concluded that when a young bird is imprinted to a member of another species, the imprinted bird will adjust its own functional cycles to that of its adopted parent. Once accomplished, the behavior pattern is very stable and may be irreversible in some cases.

In 1937, Lorenz taught comparative anatomy and animal psychology at the University of Vienna and became the coeditor of the leading journal for ethology. After World War II, Lorenz headed the Institute of Comparative Ethology at Altenberg, Austria. From 1961 to 1973, he served as the director of the Max Planck Institute for Behavior Physiology in Seewiesen, Austria. Lorenz developed the concepts of animal behavior that led to the modern scientific understanding of how behavioral patterns evolve in a species. He advanced the concepts of how these patterns develop and mature during the lifetime of an individual organism.

In 1963, Lorenz published his book *Das sogenannte Böse* (1963; *On Aggression*, 1966), in which he argued that the aggressive behavior in humans may be modified or channeled, but in other animals, it is primarily survival motivated. Based on his insights and discoveries associated with animal behavioral patterns, Lorenz shared the 1973 Nobel Prize in Physiology or Medicine with Karl von Frisch and Nikolaas Tinbergen.

In 1973, Lorenz was appointed as the director of the department of animal sociology at the Institute for Comparative Ethology of the Austrian Academy of Sciences in Altenberg, where he continued to apply his ideas about the behavior of humans as members of a social species. In 1978, he published *Vergliechende Verhaltensforschung: Grundlagen der Ethologie* (*The Foundations of Ethology*, 1981), a comprehensive review and assessment of his major contributions to science.

—*Alvin K. Benson*

lings and goslings follow the parent soon after hatching. By doing this, the young become sensitive to the visual and auditory stimulus pattern presented by the parent. This process is remarkably confirmed when a young bird fostered to a parent of another species establishes a permanent bond with the foster species and excludes members of the bird's own kind. The bond is so lasting that Lorenz called the process which gives rise to it *Prägung*, or imprinting. This single experience imprints the pertinent object of the infantile instinctive behavior patterns in a young bird that does not recognize the object instinctively. The object can be imprinted only during a quite definite period in the bird's life. From this observation, Lorenz set forth two principles of imprinting: First, that there is a critical time period in which the young offspring must form the attachment and after which no imprinting stimulus is effective, and second, that once formed, the attachment to a specific object, whether it be to its parent, a person, or an inanimate object, is permanent and irreversible. Several ethologists who have since studied imprinting do not totally accept Lorenz's conclusions that imprinting is irreversible and unchangeable, believing instead that the "critical period" is more flexible and is not necessarily dictated by inheritance. Thus, the term, "sensitive period" has gained popularity and become more widely accepted.

The Imprinting Process

Imprinting processes occurring during this sensitive period are basically different from the processes that occur after the sensitive period is over. The differences can be observed both in the immediate behavior and in the long-range effects of these experiences. The order in which these imprinting events occur has little relevance to the sequence in which they are later manifested. In jackdaws, the sensitive period for sexual imprinting comes well before the sensitive period for imprinting of the approach and follow responses. In this bird, sexual imprinting occurs while the nestling sits in the nest still half-naked, showing no responses to conspecifics other than that of gaping at its parents when they arrive to feed it. The sensitive period for imprinting of following behavior occurs immediately before fledgling. The effect of sexual imprinting, however, does not become obvious until two years later, whereas the following response is observable only a few days after imprinting.

A variety of stimuli may elicit the imprinting responses; most responses are to either auditory or visual stimuli. Some authors have suggested that auditory imprinting is especially important in the song-learning of birds. In many species, the song is completely innate, while in others it is learned; in still others, a combination of learned and innate factors is necessary. In some species where there are learned components, there seems to be a sensitive or receptive period during which songs are normally learned permanently and irreversibly. Lorenz found that ducklings respond to both visual and auditory stimuli but that the strongest responses are to their mother's call. More often, however, when both auditory and visual cues are provided, the extent of successful imprinting exceeds that attained with either stimulus provided singly. Not to be ignored are those cases in which the olfactory sense plays a role in imprinting, particularly in those animals with high olfactory acuity. In these animals, the olfactory sense may regulate relationships between members of the same species (as in sexual behavior), promote predator or prey recognition, and help identify food or territory when conspicuous visual aspects are absent.

Experimenting with Imprinting

The law of effort says that the more energy a bird expends during the primary attachment formation, the stronger the imprinting will be. To observe this effect, researchers may manipulate the energy expenditure in at least three ways: by using a moving object rather than a nonmoving one, by providing continuous rather than spaced training sessions with the object being imprinted, and by having longer training sessions using a stimulus that moves around the arena more quickly. Ducklings placed in a runway with ten-centimeter hurdles were forced to exert more effort than animals that followed an object in an ordinary runway; it was found that birds compelled to go over the hurdles in order to keep up with the moving model later achieved higher imprinting scores than did control birds.

Other experiments seem to indicate that movement by the subject is not always necessary for imprinting to take place. In one such study, ducklings were exposed to a moving object but were not allowed to follow it, while control subjects were permitted to follow the object. Both experimental and control animals exhibited poor responses to the new object. The role of movement in imprinting was also tested by placing ducklings in wooden stocks that restricted movement, then exposing them individually either to a moving object or to a stationary one. Despite the restriction of their movement, the animals became imprinted. The way the object was moved made no difference to the extent of approach and follow-on testing, but ducklings that had been exposed to a stationary object did not follow as well. This experiment showed that, while movement of the subject was immaterial, movement of the object was important for imprinting. In most cases, birds that expend more energy are more successfully imprinted.

Recognition and discrimination are the primary means to determine whether imprinting has occurred: If a bird is given a choice between known and unknown figures, it will choose the known. This is considered the standard test for imprinting; birds will usually run to their mother. Observations other than recognition and discrimination may also be worthwhile in assessing the amount of imprinting. One of these is distress at separation. Young chicks and ducklings, when separated from the mother or companions, make specific distinctive calls. Another evaluation can be made upon recognition at reunion: Approach and follow responses are clearly shown by birds that, upon hatching, are visually exposed to that figure. When a chick or duckling is placed in a runway with a moving object and is left there for a number of hours or even days, it will eventually stop following the object and otherwise occupy itself, maybe eating or drinking or wandering about. Disruption of the familiar, however—by a noise, by placing a piece of paper on one of the runway walls, or by the appearance of a strange object—will cause the bird to run to the familiar object, where it will remain for a time.

Theories to explain the ending of the ability to imprint can be divided into four categories. The first category is end of sensitivity because of maturation; that is, the tendency to approach and follow naturally decreases as the animal ages, and this lessened tendency is internally determined, rather than being caused by any impact of experiences. The second category is inhibition through socialization. Evidence so far presented indicates that the sensitive period for imprinting is likely to continue until a firm imprinting experience has occurred. Once imprinting of the approach and follow responses to a particular stimulus has occurred, it tends to inhibit approach and imprinting to new figures. Third, growth of timidity or fear responses eventually represses approach and follow responses. This is perhaps the most popular theory. Last, the ending of imprinting may be shown by an end to varied, restless activity that is not performed to achieve a goal.

Innate Versus Learned Behavior

That imprinting behavior occurs is not debatable, but controversy has arisen pitting Lorenz and his followers, who believe that the process is wholly innate, against a host of later investigators who conclude that imprinting is actually a form of learning. Those who say that imprinting is innate base their opinion on observations that the drive to imprint is strikingly different from other instinctive behavior patterns, whose releasing schemata are not innately determined but are acquired like conditioned reflexes. Lorenz states that imprinting is different from learning in that imprinting can occur only during a very definite, and perhaps short, period in the animal's life and, therefore, is dependent on a specific physiological developmental condition in the young bird. He further asserts that imprinted recognition, even after the critical period, is the same as innate behavior because the recognition response cannot be forgotten (as opposed to learned processes, which can be forgotten). Others, however, believe that imprinting involves a behavior that is absolutely necessary for the survival of the animal under normal conditions. In the case of both social and food imprinting, this is clear. The desire to follow the parent or to want to eat a particular food (as well as to learn what objects are the targets) motivates these behaviors. The urge to learn these objects is so strong that the associated processes which fulfill this learning are contradictory to those associations usually involved in this rote-type learning.

If imprinting is a learned behavior, then can it be considered conditioning? Conditioning involves the building of associations between stimuli and responses. A wide variety of stimuli may initiate the imprinting process by innate unconditioned approach responses. The particular stimulus continues to elicit filial responses, but eventually any new stimuli begin to be ignored and later even feared. There is no selective pairing of stimuli and responses, as in conditioning. In imprinting, the primary bond between stimuli and responses continues to be strengthened and becomes exclusive.

The Study of Imprinting

The earliest documentation of the imprinting process was by the seventh century English monk Saint Cuthbert, who spoke of it as "object fixation." Sigmund Freud later discovered the process independently of ethologists, but several ethologists take credit for its discovery and subsequent investigations. Most means of study practiced by ethologists investigating this phenomenon have been patterned after techniques employed by Eckhard Hess. Experiments have usually been conducted using models that do not closely resemble any real animals, let alone the parent model. Each subject is individually "trained" with a particular figure. Then, all the subjects are individually presented with a situation in which they can choose between two figures. Imprinting has occurred when an animal is drawn to the figure that is familiar. (If only one figure were used to train all the animals, a later choice test would be invalid because, if the figure was favored on the test, the investigator would not know whether the attraction was caused by imprinting or by some other impression of the figure.) Earliest observations showed that moving animals, persons, or even inanimate objects could evoke approach and follow responses in newly hatched nidifugous birds. Goslings, for example, have been found to follow people if the mother goose is absent. Incubator-hatched ducklings will follow anybody who encourages them to do so by tapping the ground and/or calling while moving away from them.

Another much-copied method that Hess used employs ducklings hatched in the dark and kept isolated until trained to the test stimulus. They then are placed on a circular track and allowed to follow a moving, sound-producing object around the runway for a determined time or distance. Later, the ducklings are tested for adequacy of the imprinting by allowing them to choose between two models to see which is approached. It is important to know what external stimuli arouse the strongest following response. The stimulus may be size of the object: If the object is too large, the duckling will flee. The duckling will simply peck at an object that is too small. Another factor is color; blue objects are the first choice for chicks, followed by red, then yellow. A third aspect is shape; strangely, adding immovable wings, tail, and head to a sphere actually reduces its effectiveness in generating a response.

Early investigators attempted to model releasers after parent birds. They discovered that a noisy moving object tended to elicit stronger responses than a silent one. Approach and follow responses in ducklings succeed intermittent noises, but to be most effective the noises should be simple, rhythmical, and monosyllabic, and given in rapid succession. Such stimuli are, overall, quicker than visual stimuli in eliciting approach from ducklings. In fact, ducklings that fail to imprint to moving models occasionally do imprint to noise-making ones—enough so that they will continue to follow its movement even when the noise ceases. Generally speaking, it is necessary for sound to accompany visual stimuli, though ducklings and goslings do not appear to imprint on sound alone.

Some scientists wonder whether all this laboratory study has really been relevant to the study of natural imprinting. Laboratory procedures are undoubtedly useful in helping to determine the behaviors of young animals; however, because the animals are given such an overwhelming number of inappropriate stimuli, it may be difficult to assess how the animal would react under normal conditions. Observations of animals in their natural settings are difficult to make and analyze, but the known data reveal several differences from laboratory tests. For example, hatched ducklings normally stay in the nest with the female mallard long past the earlier laboratory-assigned critical period, and, as a result, they are both visually and auditorially attached to the parent by the time they leave the nest. This enhances their ability to survive on their own. Continued investigation into imprinting must occur, but it should take place in an animal's natural setting.

The Complexity of Imprinted Behaviors

The imprinting process has a much broader significance for the animal than merely giving it the tendency to follow moving objects. Behaviors usually associated with imprinting are following and approach responses, sexual identification, species recognition, and socialization among conspecifics. Following behavior is far from a simple behavior; there are many degrees, from close, nonfearful following, through hesitant following mixed with fear responses, to strong escape behavior in which only a minimum of following can be observed. Following can also be balanced by aggressive behavior, and the actual behavior of the young bird is largely determined by the interplay between these competing tendencies. Following behavior is important to those duck species that build their nests at a distance from the water, the eventual home of the birds; mother and young may move to the water within several weeks of hatching. The use of species-specific visual or auditory signals in imprinting may enable young birds to follow the mother successfully through dense vegetation from the nest to the water. When the young are ready to leave the nest, generally located several feet above the ground in a tree hole or other suitable site in a pond or swamp, the mother goes down to the water and calls her ducklings. In response to the call, the ducklings jump to the nest opening and then down to the water to follow the mother. The same imprinting patterns may also aid the ducklings in avoiding predators. Quickly following the mother to safety in a few seconds between the time the predator is sighted and an attack may mean the difference between life and death.

Imprinting may serve as a species-identifying mechanism and, even more narrowly, as a species-isolating mechanism. The sexual preferences of birds have been shown, by appropriate exposure to an imprinting stimulus and later testing, to be imprinted to stimuli to which they were exposed. For virtually all species in nature, encounters with birds of other species during the first few days or even weeks after hatching are limited; this alone ensures that young birds will imprint on members of their own species and that when, as mature adults, they begin to engage in reproductive activities they will court only conspecifics. Therefore, imprinting helps guarantee that reproductive energy investments and gametes are not wasted on nonproductive mating endeavors. The overall recognition of conspecifics that can occur as a result of imprinting may also be important for the socialization of young birds and for general cooperation of conspecifics in a social organization. Associating with conspecifics, in turn, may be a significant means of increasing survival, by aiding in processes such as locating food, finding shelter, and migrating.

—*Iona C. Baldridge*

See also: Communication; Ethology; Instincts; Learning; Mating; Rhythms and behavior.

Bibliography

Drickamer, Lee, and Stephen Vessey. *Animal Behavior: Concepts, Processes, and Methods.* Boston: Willard Grant Press, 1982. An intermediate-level college text with a segment on imprinting. The study of behavior, how it is controlled, the behavior of individuals and groups, behavioral ecology, and animal populations as well as the evolution of behavior are all presented. The authors are interested in answering the questions researchers ask about animal behavior and in the methods they use to answer these questions. Bibliography.

Heyes, Cecilia M., and Bennett G. Galef, Jr., eds. *Social Learning in Animals: The Roots of Culture*. San Diego, Calif.: Academic Press, 1996. Focuses on laboratory studies of the mechanisms of social learning and the functional significance of information transfer in natural settings.

Hinde, Robert A. *Ethology*. New York: Oxford University Press, 1982. The author's goal in this complex book is to deal with what he perceives as current problems facing ethologists and to look at how these questions are being solved. He treats imprinting as a learned behavior, primarily to tie ethology in with psychology. Glossary and bibliography.

Lehner, Philip N. *Handbook of Ethological Methods*. 2d ed. New York: Cambridge University Press, 1996. A comprehensive guide to the study of animal behavior, concentrating on the application of orthodox biological methods.

Lorenz, Konrad Z. *The Foundations of Ethology*. Translated by Konrad Z. Lorenz and Robert W. Kickert. New York: Springer-Verlag, 1981. The aim of this book, translated from German, is to discuss the framework of ethology—its history as perceived by one of the pioneers in the field. Only students with advanced ethological knowledge could truly appreciate the complex concepts that Lorenz sets forth. Bibliography.

_______. *Studies in Animal and Human Behavior*. Translated by Robert Martin. Vol. 1. Cambridge, Mass.: Harvard University Press, 1970. This is the first volume of the collected papers of Lorenz. Written during the years 1931 to 1942, these six papers provide a rich source of information on the development of his ethological approach. Lorenz defines a range of behavioral acts: innate, instinctive, conditioned, and insight-controlled. He then analyzes their manner of interrelation in the overall behavior of a species of animal.

Sluckin, W. *Imprinting and Early Learning*. 2d ed. Chicago: Aldine, 1973. A brief but good review of early historical literature and easily understood descriptions of the full range of experiments that have been done. Imprinting is considered side by side with related concepts and empirical studies. The second edition expands the first (1964) by adding more up-to-date developments in the field. Complete bibliographical references are cited at the conclusion of the book.

INFANTICIDE

Types of animal science: Behavior, evolution
Fields of study: Ecology, ethology, wildlife ecology, zoology

Infanticide, the killing of immature members of the same species, is one of the more disturbing phenomena seen in animals, including humans. However, it follows from the evolutionary logic of individuals striving to maximize reproductive success.

Principal Terms

CONSPECIFIC: a member of the same species

GENERA: plural of genus, a grouping of animals above the species level

GESTATION: the period when the young are nourished within the mother's body; pregnancy

LACTATION: the period when mammal mothers produce milk to nourish their infants

OVICIDE: killing of fertilized eggs

RAPTORS: predatory birds such as hawks and eagles

SIBLICIDE: infanticide committed by the siblings of the individual killed

Infanticide is the intentional killing of dependent immatures by a member of the same species, or a conspecific. Victims may be as young as fertilized eggs or nearing independence from their parents. The infanticidal attackers may be strangers or relatives. The motive is always selfish competition, but sometimes the young are direct competitors and sometimes indirect competitors with the killers. Infanticide occurs in every class of animal, but mammals and birds seem to have evolved the most pervasive and pernicious form, sexually selected infanticide. Wherever it is seen, infanticide reveals the darker side of evolutionary adaptation. There are three scientific explanations for infanticide, each of which applies to particular circumstances.

Infanticide by Genetic Relatives

The first circumstance occurs when related individuals kill dependent young. For example, extreme sibling rivalry occurs in spotted hyenas (*Crocuta crocuta*) and many raptors. The unpredictability of resources or extreme competition for parental care may lead siblings to fight to the death in the den or nest. Even parents may act infanticidally. When parents have larger numbers of offspring than they can feed, the parents themselves may neglect or inflict damage on the smallest or least vigorous young, so as to reduce the number of offspring and increase each survivor's chances of success. Sibling rivalry and parental neglect or abuse of young can lead to infanticide in extreme cases when resources, such as food or parental time and energy, are in short supply. The brutal logic of natural selection shows that selfishness and lethal competition can divide even close kin. However, in most animal species, unrelated individuals are more dangerous to the young.

Infanticide by Direct Competitors

The second circumstance in which infanticide arises is referred to as local resource competition. When immatures use resources that unrelated animals need, aggression may be severe and directed to killing the infants. All ages of infants are vulnerable, even independent juveniles. In the burying beetle (*Nicrophorus orbicollis*), male and female pairs compete for access to rotting meat in which to lay eggs. If defenders of such a resource are displaced by an intruding pair, the intruders proceed to kill and eat the eggs (ovicide) or larvae of the previous pair. Similarly, many birds, such as black-and-white casqued hornbills (*Bycanistes subcylindricus*), compete for rare tree hole nesting sites. If adults encounter another nest with eggs in such a tree hole, they will roll these eggs out or crush them and lay their own eggs. Finally, cannibalism of unrelated young has been seen in many amphibians, fish, reptiles and even in chimpanzees. Stepchildren suffer much higher rates of severe neglect and homicide than a parent's biological children. All of these cases represent extreme competition where resources are scarce and vulnerable young are eliminated or eaten by unrelated killers. The behavior patterns seen in this type of infanticide often resemble predation, are directed at young of any age, and can be performed by adults of either sex.

Sexually Selected Infanticide

The third circumstance in which infanticide occurs is motivated by sexual competition among adults. In a wide range of mammals, including over twenty genera of primates, adult males will kill unrelated infants if they can then mate with the mother. The mother of that infant does not usually reject the infanticidal male even though he has inflicted a tremendous cost on her. Typically, it takes female mammals many months—even years—to nurse young to independence. During this period of infant dependency, the mother is generally physiologically incapable of reproducing. An infanticidal male benefits if he can cut short this period so that he can fertilize the female's next egg. Therefore, a successful infanticidal male eliminates another male's offspring and advances his own reproductive career. Sexually selected infanticide is not restricted to mammals, although the best-documented cases come from lions (*Panthera leo*), langur monkeys (*Semnopithecus entellus*), and rodents (order Sciurognathi). Some birds may also behave infanticidally. For example, the wattled jacana (*Jacana jacana*) shows a fascinating reversal. In this species, the adult females may commit ovicide and infanticide against unrelated young. Adult male jacanas incubate eggs on the nest and protect the young from predators, so females strive to monopolize the parental care donated by males. To do this, a female must eliminate the young of another female and lay her own eggs in the care of the male. Typically, the behavioral patterns seen in sexually selected infanticide differ from those in local resource competition. The young are virtually never eaten and permanent separation of mother and young is the primary goal—whereupon attacks usually cease.

Evolutionary Consequences of Infanticide

Infanticide is widespread and may reach high frequencies in certain species, but it is not universal. Siblicide and parental neglect are found in a very restricted subset of birds and very few mammals. Local resource competition is more common, found in most classes of animals. Finally, sexually selected infanticide is most common in mammals, but even here there are many orders of mammals that never display infanticide. Species that have a very long period of lactation relative to their gestation period are most vulnerable to sexually selected infanticide. The reason for this is that infanticidal attacks are risky because of maternal defense and counterattacks by allies. Therefore, infanticidal behavior will only evolve when the males gain substantially by shortening lactational infertility of the mother.

Clearly, infants and their parents have a very strong motivation to avoid infanticide. Evolution has favored various mechanisms to reduce the risk. In general, infants avoid strangers, even in humans. In some primates, it seems that infants are born with unusual coat colorations that may impede males' efforts to determine which infants are theirs and which were fathered by other males. Parents are normally very protective of their young, both because of predators and because of the risk of attack by conspecifics. This parental protection is expressed through frequent proximity, carrying, and physical defense. One of the most fascinating consequences of a high risk of infanticide is the tendency for the father and mother to establish a long-lasting relationship whose primary benefit is protection of the young from conspecifics. This is seen in burying beetles and many primates. Alternatively, groups of mothers may cooperate in protection against infanticidal males, as in lions and langurs. Therefore, infanticide is one of the few evolutionary pressures that favors complex social relationships.

—*Adrian Treves*

See also: Altruism; Competition; Copulation; Courtship; Estrus; Fertilization; Lactation; Mating; Offspring care; Predation; Reproduction; Reproductive strategies; Sexual development.

Bibliography

Butynski, T. M. "Harem-Male Replacement and Infanticide in the Blue Monkey (*Cercopithecus mitis stuhlmanni*) in the Kibale Forest, Uganda." *American Journal of Primatology* 3 (1982): 1-22. A study of infanticide by male monkeys of infants not their own.

Hausfater, Glenn, and Sarah Blaffer Hrdy, eds. *Infanticide: Comparative and Evolutionary Perspectives*. New York: Aldine, 1984. This was the first book that compiled a mass of evidence that sexually selected infanticide was a real strategy with important evolutionary consequences. Although many of its chapters are scholarly and rich in data, there are also several review chapters that synthesize information and can be understood by an educated lay reader. This book also tackles and lays to rest the historical debate over infanticide as a social pathology.

Parmigiani, Stefano, and Frederik S. vom Saal. *Infanticide and Parental Care*. Langhorne, Pa.: Harwood Academic Press, 1994. This volume spans a variety of animal species that display infanticide. In particular, it shows the intricacy of the behavioral mechanisms underlying infanticide in rodents. It also presents the best information on infanticide in lions. As always, primates are well-represented in this book aimed at a scientific audience.

Van Schaik, Carel P., and Charles H. Janson, eds. *Infanticide by Males and Its Implications*. New York: Cambridge University Press, 2000. This most recent compilation devotes most chapters to primates. It is aimed at the scientific audience, but also has some review chapters that are accessible to any educated reader. This book presents the state of the science in its most up-to-date form, debunking misconceptions about infanticide and making a strong case that infanticide is an evolutionary pressure of equal importance to predation.

INGESTION

Type of animal science: Physiology
Fields of study: Biochemistry, ecology, neurobiology

Ingestion is the process of taking food into the body. The animal kingdom possesses myriad strategies for this process, reflecting the diversity of available food sources and how they are utilized.

Principal Terms

CARNIVORE: any organism that eats animals or animal tissues

DETRITUS: small bits of dead matter derived from decay of plants and animals

HERBIVORE: any animal that eats plants or plant material

INVERTEBRATE: any animal that lacks a backbone

OMNIVORE: any animal that eats both plants and animals or their tissues

PARASITE: any organism that lives on or in other living organisms and obtains its food from them

PLANKTON: microscopic plants and animals that float in water

PREDATOR: any organism that kills another living organism to eat it

PROTOZOAN: a single-celled animal-like organism

SAPROVORE: any organism that consumes dead or decaying plant or animal matter

TAXONOMY: a classification scheme for organisms based primarily on structural similarities; taxonomic groups consist of genetically related animals

Ingestion is the process of taking food into the body to satisfy nutritional and energy needs. Although basic nutritional requirements are remarkably similar for all animals, mechanisms of ingestion are exceedingly diverse. This diversity stems from the varied nature of available food sources and the resulting behavioral adaptations and specific body forms required to procure adequate nutrition. Because it is convenient to classify feeding strategies by the type of food consumed, organisms are often described as herbivores, carnivores, omnivores, and saprovores. Although very descriptive, these terms alone are not sufficient to describe fully the feeding adaptations used by animals, especially when considering invertebrates, which make up more than 97 percent of all animal species. Accordingly, this section will expand on these ideas by further categorizing ingestion by the type, size, and consistency of food, while also describing behavioral adaptations that lead to ingestion.

Small Particle Ingestion

Numerous animals live exclusively on a diet of very small particles that include bacteria, algae, plankton, and detritus. Although most small-particle consumers are relatively small animals, they range from the tiniest animal-like organisms (protozoans) to the largest (whales). Many mechanisms have evolved to permit ingestion of small food particles. One common method is by endocytosis. During this process, the outer membrane of a cell surrounds a food particle and engulfs it within the cytoplasm, forming a food vacuole (a cellular organelle used for digestion). This mode of ingestion is best illustrated by protozoans. The amoeba, for example, uses pseudopodia (extensions of its cell membrane) to surround and then ingest prey. Paramecia use cilia (tiny hairlike processes on cell membranes that beat in a coordinated manner) to guide food particles into an oral region prior to ingestion by endocytosis.

Multicellular animals may also ingest small food particles by endocytosis. Sponges use flagella (motile, whiplike structures resembling long cilia) to aid in the gathering of small particles of food. The body cavity of sponges is lined with flagellated cells called choanocytes, or collar cells. The beating action of their flagella creates currents that move water through the body cavity, where food particles are removed and incorporated by the choanocytes.

Filter feeding is another strategy that is commonly used to obtain small bits of food. Most often cilia, mucus sheets, flagella, tentacles, and nets are used as filtering devices. Rotifers (small aquatic invertebrates), for example, use a special double-banded ciliary system to transport water and filter out suspended food particles, which are conducted to the mouth by a third set of cilia. In general, sessile (immobile) organisms (such as sponges, rotifers, and some oysters) are called filter feeders, because they must wait for food to come to them, then extract it from the surrounding medium.

Numerous free-moving animals also use filtering devices to obtain food. Sea cucumbers (animals related to starfish) live on the bottom of the seabed. By extending and periodically retracting sticky tentacles, they capture and ingest small food items. Clams, mussels, and some snails produce a sticky mucus that covers ciliated cells. The mucus traps fine suspended food particles, which are transported to the mouth by ciliary action. Mussels, for example, continuously pass water between their mucus-covered gill filaments for respiration, simultaneously trapping food particles. Similarly, herring and mackerel (fast-swimming fish) possess special structures called gill rakers, which act as sieves to catch plankton from water that continually passes over their gills for the process of respiration. Basking sharks and whale sharks also feed on plankton that is strained from water that enters their mouth and flows over their gills. Baleen whales use a filter consisting of a curtain of parallel filaments (called baleen) attached to the upper jaw to feed. These whales en-

Elephants use their trunks to bring food and water to their mouths. (Corbis)

gulf a mouthful of water containing krill (small, shrimplike plankton) and use their tongue to force the water out, trapping krill in the hairlike edges of the baleen. The mouthful of krill is then swallowed. Flamingos feed in a similar manner. They use specially adapted beaks lined with filaments to strain plankton from the muddy bottoms of their aquatic habitat. Some fast-water caddis fly larvae spin tiny silk nets that are used to filter small food items. The nets are then periodically gleaned. Finally, spiders may be considered filter feeders because they use their webs to "filter" flying insects from their environment.

Large Item Ingestion

In contrast with small-particle consumers, numerous animals ingest large food items. Some of these organisms are saprovores, such as earthworms. These worms consume large masses of soil and dead leaves, from which they obtain usable organic matter. Some planarians (flatworms) extend a long extensible tube (called a pharynx) from their mouth to ingest decaying material. Other examples of saprovores include millipedes, wood-eating beetles, some sea cucumbers, many roundworms, and a few snails.

An extraordinary number of animals have adapted to eat plants. Eating plants requires special structures to free plant material for ingestion. Although invertebrates lack true teeth, they have other structures to obtain plants or plant parts. Snails have a unique structure in their mouth called a radula, which acts as a miniature rasping file that scrapes plant material from surfaces and rasps through vegetation. The freed plant material is then ingested. Sea urchins scrape algae by using a highly developed oral apparatus composed of five large pointed plates. Termites use strong jaws made of chitin, the hard structural component of their external skeleton, to cut tiny chunks of wood for ingestion.

Birds use horny beaks to obtain and ingest plant material. Cardinals and grosbeaks, for example, are well adapted to hull and then consume the nutritional portion of large seeds. Herbivorous mammals use specialized teeth to obtain and chew plant material. Rodents (such as beavers, porcupines, and mice), rabbits, and hares have chisel-like front teeth (called incisors) that are used to gnaw, slice, or pull off plant material. Other herbivorous mammals (including cows, sheep, deer, moose, elk, and giraffes) lack upper incisors and therefore use their lower incisors pressed against the roof of their mouth to pull off leaves. The ingested food is then chewed by grinding with premolars and molars.

In contrast to herbivores, many animals capture other animals and eat them. To be effective, these carnivorous predators must have appropriate behavioral adaptations to find and capture prey as well as specialized structures to seize and hold their victims. Jellyfish use tentacles that are equipped with stinging cells to grasp and subdue animals, whereas the tentacles of squid and octopuses have suction-cup-like structures to grasp and manipulate prey. The giant water bug is a carnivorous insect that hunts and captures small fish, relatively large prey for an insect. To do this, water bugs use their legs to seize and hold fish, and their piercing mouthparts to suck juices from their victims. Fish, amphibians, and reptiles have pointed teeth to seize and hold prey. It also is common for many of them to swallow their food whole. Snakes, for example, swallow whole items such as birds' eggs and small mammals. In addition, a snake's jaws are held together by elastic ligaments, permitting it to spread apart and ingest victims larger than its own head. Some large tropical snakes actually consume small pigs and deer. Chameleons and frogs swallow their prey whole, but in contrast to snakes, they use a long and sticky tongue that rapidly shoots out to capture insects. Some predatory carnivores (such as lions, tigers, bears, and dogs) have long, pointed, daggerlike teeth called canines to pierce and kill their prey. Carnivores also may have knifelike molars (called carnassials) that are used to slice flesh from bones. Carnivorous birds (such as hawks, eagles, and owls) use long, sharp talons to seize and kill small animals. Their beaks can be used to tear small pieces of food for ingestion. Further, many carnivorous animals possess specialized mechanisms for paralyzing victims. Jellyfishes, centipedes, spiders, scorpions, and some snakes possess structures that inject toxins that inhibit the nervous system of their prey. Finally, some electric eels may locate and stun their prey with electrical discharges.

Liquid Ingestion

Surprisingly, many animals live exclusively on a liquid diet. Herbivores such as bees, butterflies, moths, hummingbirds, and some bats derive nutrition by consuming plant nectar. As a consequence of their feeding, these animals help plants reproduce by dispersing pollen. Hummingbirds have specially adapted long, narrow beaks and long tongues to suck nectar from flowers. Nectar-consuming bats also have long narrow faces and tongues. Aphids, and many other insects that consume plant sap, have highly specialized mouthparts that pierce plants and act as miniature straws to suck sap.

A number of carnivores also are adapted to consume a liquid diet. As ghoulish as it might sound, ingesting blood is the most common mechanism of feeding for these animals. Mosquitoes, for example, are equipped with a syringelike mouthpart called a proboscis. Although the male sips nectar, the female mosquito uses her proboscis to pierce skin and suck blood. As with most bloodsucking animals, mosquitoes secrete an anticoagulant that prevents blood from clotting (and also makes people itch). Some flies use a similar mode of feeding, but the common housefly generally laps up food and sugary solutions. Leeches are also well adapted for bloodsucking: They have a suction-cup-like mouth that clings tenaciously to a host while their jaws make a Y-shaped cut in the skin. Further, leeches have a muscular pharynx (throat) that literally pumps blood from their host. Ticks, which are related to spiders and scorpions, have tiny heads that are designed to burrow into the skin of their host and suck blood. Unfortunately, they also are vectors for potentially serious diseases such as Lyme disease and Rocky Mountain spotted fever. Although vampire bats consume blood, they do not suck it. Instead, they lap blood with their tongue as it oozes from a shallow scrape in the skin made by their teeth.

Spiders also are adapted to live on a liquid diet; however, they prey on insects that have a tough external skeleton that is not easily ingested. To feed, they first pierce the insect's exoskeleton with hollow jaws and pump in strong digestive juices that liquefy the internal contents of their prey. Later, the spider sucks the insect empty. Some young birds, such as pigeons and emperor penguins, feed on a regurgitated milklike secretion (called crop milk) that is produced by their parents' crop. In addition, all mammals begin their life as fluid feeders, ingesting milk produced by their mother.

Finally, endoparasites comprise a group of animals that consume a liquid diet by living inside other organisms. Some eat host tissues, while others rely on their host to digest food for them and, as a result, lack a digestive system. Tapeworms, for example, have a specially adapted anterior end with hooks and suckers to maintain a fixed position in their hosts' gut while they consume predigested food. In contrast, hookworms (a type of roundworm) do have a digestive system, and they use their mouth opening and toothlike structures to draw and ingest blood from the inside of their host.

Studying Ingestion

Much of what has been learned about ingestion has come from simple observation of feeding behavior combined with careful note-taking. In fact, observation and data collection, along with analysis and interpretation, are the most fundamental of all scientific activities. Most people have seen a robin use its beak to pull an earthworm from the ground or a cat capture a mouse. By closely watching the lifestyle and daily activities of animals, biologists have discovered what food items are eaten and how they are ingested.

The naked eye is insufficient for observing very small animals, and for this activity microscopes aid biologists. These tools have permitted observation of endocytosis of paramecia and bacteria by the amoeba as well as other feeding mechanisms of protozoans. Microscopic, inert latex beads are used to study the direction and power of feeding currents generated by cilia and flagella and the formation of food vacuoles. For example, paramecia will direct beads into their oral region by ciliary action and then engulf them by endocytosis. Dyes are also commonly used to study the direction and action of feeding currents. This method has revealed that flagellated choanocytes of sponges move water in through small pores in the sides of the sponge and out a single larger opening at the top.

Videotape and photographic film are routinely used to supplement simple observation. These procedures not only provide a permanent record of the event but also permit additional analysis to be conducted at some future time. Further, feeding mechanisms that occur very quickly are difficult to analyze with the naked eye. These events can be recorded by high-speed cinematography and later played back at a slower speed for analysis. This method of study has been used to observe the lightning-fast movement of a chameleon's tongue and the way bats catch insects with their wing membranes while in flight. Alternatively, very slow feeding events, such as endocytosis or a snake swallowing a rat, can be recorded by time-lapse photography and later viewed at a faster speed for analysis.

Mechanisms of ingestion may also be inferred by carefully analyzing the body design of an animal. For example, birds possessing beaks that have an arrangement of tightly packed vertical filaments, as found in flamingos, feed by filtering. In contrast, birds with long, pointed beaks, such as woodpeckers, probe for food in narrow places. Finally, the contents of the stomach and fecal samples from animals may be analyzed to determine the type and size of the food items that were eaten.

The Necessity of Food

Animals have an absolute requirement for food. Animals must ingest food items because, unlike photosynthetic organisms, they cannot manufacture all the necessary nutrients they require from raw materials. Animals require food both as a fuel source to provide energy for locomotion and metabolism and as building blocks for growth, maintenance, and repair. Obtaining sufficient food is of paramount importance for survival. Therefore, the limited availability of food is selected by animals that have the most-successful feeding strategies and body designs for procurement and ingestion of nutrients. Because different sources of nutrition are utilized, the selection pressures for obtaining food may result in vastly different feeding mechanisms among closely related animals. For example, giant water bugs, termites, and aphids are all classified as insects. Yet, they rely on different diets and therefore possess divergent methods of feeding and structures for ingestion. In addition, and conversely, selection pressures for obtaining food may result in the development of similar body structures in distantly related species (convergent evolution). Baleen whales and flamingos, for example, are classified in different taxonomic groups (mammals and birds) but have similar feeding methods and therefore similar structures for ingestion. Thus, specific feeding behaviors and structures for ingestion are primarily shaped by the nature of the food items being utilized, and the vast diversity of feeding mechanisms reflects a similar diversity in food sources.

Ultimately, the source of energy to create food comes from the Sun. Photosynthetic organisms use light to synthesize energy-rich organic compounds, such as glucose, from energy-poor inorganic compounds, such as carbon dioxide and water. One exception to this scheme occurs in certain regions of the ocean floor, near the thermal vents. Far removed from sunlight and organic material derived by photosynthesis, the food chain of oceanic thermal vents is based upon certain bacteria that synthesize organic compounds from inorganic substances emitted by these undersea geysers.

Unfortunately, there is an unavoidable loss in usable material and energy between links in a food chain. This loss occurs because much of the energy stored in food is irreversibly lost when it is used by organisms for growth, repair, and maintenance. Therefore, it is more efficient to have a lower number of links in a food chain between its base (usually plants or plankton) and its end (typically large carnivores). With that in mind, it is interesting to note that some of the largest fish (whale sharks) and mammals (whales) in the world are filter feeders of tiny organisms. These huge animals, which require a large amount of energy, avoid extra links in their food chain by feeding on plankton instead of other large animals.

—*Douglas B. Light*

See also: Carnivores; Digestion; Digestive tract; Food chains and food webs; Lactation; Nutrient requirements; Omnivores; Predation.

Bibliography

Childress, James J., Horst Felbeck, and George N. Somero. "Symbiosis in the Deep Sea." *Scientific American* 256 (May, 1987): 115-120. This article describes how a unique society of animals survives deep below the ocean surface totally devoid of sunlight and organic material derived from the Sun. Authors present the methods these animals utilize to gain energy from deep-sea hydrothermal vents and to obtain nutrition from certain bacteria. Illustrated with diagrams.

Eckert, Roger, David Randall, and George Augustine. *Animal Physiology: Mechanisms and Adaptations*. 4th ed. New York: W. H. Freeman, 1997. An intermediate college textbook that presents a comparative approach to animal physiology. Covers ingestion as well as related topics such as digestion and absorption. The text is written in a clear and concise manner, delineating different mechanisms or strategies of feeding. Well-constructed diagrams display a variety of feeding strategies in both invertebrates and vertebrates. Evolutionary correlations are also discussed.

Hickman, Cleveland, Jr., Larry S. Roberts, and Frances M. Hickman. *Integrated Principles of Zoology*. 11th ed. Boston: McGraw-Hill, 2001. A textbook for introductory college zoology. Discusses feeding mechanisms of animals as well as related topics such as nutrition, digestion, and regulation of food intake. Takes a comparative approach to ingestion, classifying animals on the basis of food type and consistency. Selected references are provided.

Pearse, Vicki, John Pearse, Mildred Buchsbaum, and Ralph Buchsbaum. *Living Invertebrates*. Rev. 4th ed. Cambridge, Mass.: Blackwell Scientific, 1992. Designed for the beginning college student, this text summarizes the biology of invertebrates. Without going into excessive detail, it describes the essential features of invertebrate biology, including feeding and ingestion. Supplemented with many excellent figures and photographs, including a series of color plates, that are easily interpreted.

Schmidt-Nielsen, Knut. *Animal Physiology: Adaptation and Environment*. 5th ed. New York: Cambridge University Press, 1998. Takes a comparative approach to animal ingestion and covers related topics such as nutrition, digestion, and noxious compounds used for defense by plants and animals. Although lacking in illustrations, the text is easily understood by the beginning college student. Classifies animal feeding by the nature of the ingested food and provides numerous biological examples. A long list of references is given.

Wessels, Norman K., and Janet L. Hopson. *Biology*. New York: Random House, 1988. Specifically designed for the college freshman biology student, this textbook has numerous learning aids, including key terms, summary statements, questions, and suggested readings. Presents animal ingestion by taking a taxonomic approach for the lower animals and then dividing the vertebrates into herbivores and carnivores. Clear figures and photographs supplement the text.

INSECT SOCIETIES

Type of animal science: Behavior
Fields of study: Ethology, invertebrate biology

Ants, termites, and many kinds of bees and wasps live in complex groups known as insect societies. Studies of such societies have enriched scientific knowledge about some of the most successful species on earth and have provided insights into the biological basis of social behavior in other animals.

Principal Terms

BROOD: all the immature insects within a colony; these include eggs, larvae, and, in the Hymenoptera, the pupal stage

CASTE: one of the recognizable types of individuals within a colony, usually physically and behaviorally adapted to perform specific tasks

EUSOCIAL: referring to any of the truly social species characterized by division of labor, with a sterile caste, overlapping generations, and cooperative brood care

HAPLODIPLOIDY: sex determination found in the Hymenoptera, where males arise from unfertilized eggs and females from fertilized eggs

METAMORPHOSIS (COMPLETE): a transformation that occurs during the development of higher insects, in which a grublike immature form enters a resting (pupal) stage for major tissue reorganization; after pupation, the adult, which bears no resemblance to the larval form, emerges

PHEROMONE: a chemical produced by one member of a species that influences the behavior or physiology of another member of the same species

TROPHALLAXIS: the exchange of bodily fluids between nestmates, either by regurgitation or by feeding on secreted or excreted material

Many of the most robust, thriving species today owe their success in great part to benefits that they reap from living in organized groups or societies. Nowhere are the benefits of group living more clearly illustrated than among the social insects. Edward O. Wilson, one of the foremost authorities on insect societies, estimates that more than twelve thousand species of social insects exist in the world today. This number is equivalent to all the species of known birds and mammals combined. Although insect societies have reached their pinnacle in the bees, wasps, ants, and termites, many insects show intermediate degrees of social organization—providing insights regarding the probable paths of the evolution of sociality.

Ant, Wasp, and Bee Societies

Scientists estimate that eusociality has evolved at least twelve times: once in the Isoptera, or termites, and eleven separate times in the Hymenoptera, comprising ants, wasps, and bees. In addition, one group of aphids has been found which has a sterile soldier caste. Although the eusocial species represent diverse groups, they all show a high degree of social organization and possess numerous similarities, particularly with regard to division of labor, cooperative brood care, and communication among individuals. The organization of a typical ant colony is representative, with minor modifications, of all insect societies.

A newly mated queen, or reproductive female, will start a new ant colony. Alone, she digs the first nest chambers and lays the first batch of eggs. These give rise to grublike larvae, which are unable to care for themselves and must be nourished from the queen's own body reserves. When the larvae have reached full size, they undergo metamorphosis and emerge as the first generation of worker ants. These workers—all sterile females—take over all the colony maintenance duties, including foraging outside the nest for food, defending the nest, and cleaning and feeding both the new brood and the queen, which subsequently becomes essentially an egg-laying machine. For a number of generations, all eggs develop into workers and the colony grows. Often, several types of workers can be recognized. Besides the initial small workers, or minor workers, many ant species produce larger forms known as major workers, or soldiers. These are often highly modified, with large heads and jaws, well suited for defending the nest and foraging for large prey. Food may include small insects, sugary secretions of plants or sap-feeding insects, or other scavenged foods. After several years, when the colony is large enough, some of the eggs develop into larger larvae that will mature into new reproductive forms: queens and males. Males arise from unfertilized eggs, while new queens are produced in response to changes in larval nutrition and environmental factors. These sexual forms swarm out of the nest in a synchronized fashion to mate and found new colonies of their own.

With minor modifications, the same pattern occurs in bees and wasps. Workers of both bees and wasps are also always sterile females, but they differ from ants in that they normally possess functional wings and lack a fully differentiated soldier caste. Wasps, like ants, are primarily predators and scavengers; bees, however, have specialized on pollen and plant nectar as foods, transforming the latter into honey that is fed to both nestmates and brood. The bias toward females reflects a feature of the biology of the Hymenoptera that is believed to underlie their tendency to form complex societies. All ants, wasps, and bees have an unusual form of sex determination in which fertilized eggs give rise to females and unfertilized eggs develop into males. This type of sex determination, known as haplodiploidy, generates an asymmetry in the degree of relatedness among nestmates. As a consequence, sisters are more closely related to their sisters than they are to their own offspring or their brothers. Scientists believe that this provided an evolutionary predisposition for workers to give up their own personal reproduction in order to

Ant colonies can build huge mounds through their united efforts. (Corbis)

raise sisters—a form of natural selection known as kin selection.

Termite Society

The termites, or Isoptera, differ from the social Hymenoptera in a number of ways. They derive from a much more primitive group of insects and have been described as little more than "social cockroaches." Instead of the strong female bias characteristic of the ants, bees, and wasps, termites have regular sex determination; thus, workers have a fifty-fifty sex ratio. Additionally, termite development lacks complete metamorphosis. Rather, the young termites resemble adults in form from their earliest stages. As a consequence of these differences, immature forms can function as workers from an early age, and—at least among the lower termites—they regularly do so.

Termites also differ from Hymenoptera in their major mode of feeding. Instead of feeding on insects or flowers, all termites feed on plant material rich in cellulose. Cellulose is a structural carbohydrate held together by chemical bonds that most animals lack enzymes to digest. Termites have formed intimate evolutionary relationships with specialized microorganisms—predominantly flagellate protozoans and some spirochete bacteria—that have the enzymes necessary to degrade cellulose and release its food energy. The microorganisms live in the gut of the termite. Because these symbionts are lost with each molt, immature termites are dependent upon gaining new ones from their nestmates. They do this by feeding on fluids excreted or regurgitated by other individuals, a process known as trophallaxis. This essential exchange of materials also includes, along with food, certain nonfood substances known as pheromones.

Insect Communication

Pheromones, by definition, are chemicals produced by one individual of a species that affect the behavior or development of other individuals of the same species that come in contact with them. Pheromones are well documented throughout the insect world, and they play a key role in communication between members of nonsocial or subsocial species. Moth mating attractants provide a well-studied example. Pheromones are nowhere better developed than among the social insects. They not only appear to influence caste development in the Hymenoptera and termites but also permit immediate communication among individuals. Among workers of the fire ant (*Solenopsis saevissima*), chemical signals have been implicated in controlling recognition of nestmates, grooming, clustering, digging, feeding, attraction or formation of aggregations, trail following, and alarm behavior. Nearly a dozen different glands have been identified which produce some chemical in the Hymenoptera, although the exact function of many of these chemicals remains unknown.

In addition to chemical communication, social insects may share information in at least three other ways: by tactile contact, such as stroking or grasping; by producing sounds, including buzzing of wings; and by employing visual cues. Through combinations of these senses, individuals can communicate complex information to nestmates. Indeed, social insects epitomize the development of nonhuman language. One such language, the "dance" language of bees, which was unraveled by Karl von Frisch and his students, provides one of the best-studied examples of animal behavior. In the waggle dance, a returning forager communicates the location of a food resource by dancing on the comb in the midst of its nestmates. It can accurately indicate the direction of the flower patch by incorporating the relative angles between the sun, the hive, and the food. Information about distance, or more precisely the energy expended to reach the food source, is communicated in the length of the run. Workers following the dance are able to leave the nest and fly directly to the food source, for distances in excess of one thousand meters.

Benefits of Cooperation

Living in cooperative groups has provided social insects with opportunities not available to their solitary counterparts. Not only can more individuals cooperate in performing a given task, but also several quite different tasks may be carried out simultaneously. The benefits from such cooperation are considerable. For example, group foraging allows social insects to increase the range of foods they can exploit. By acting as a unit, species such as army ants can capture large insects and even fledgling birds.

A second benefit of group living is in nest building. Shelter is a primary need for all animals. Most solitary species use naturally occurring shelters or, at best, build simple nests. By cooperating and sharing the effort, social insects are able to build nests that are quite elaborate, containing several kinds of chambers. Wasps and bees build combs, or rows of special cells, for rearing brood and storing food. Subterranean termites can construct mounds more than six meters high, while others build intricate covered nests in trees. Mound-building ants may cover their nests with a thatch that resembles, in both form and function, the thatched roofs of old European dwellings. Colonial nesting provides two additional benefits. First, it enhances defense. By literally putting all of their eggs in one basket, social insects can centralize and share the guard duties. The effectiveness of this approach is attested by one's hesitation to stir up a hornet's nest. Nest construction also provides the potential to maintain homeostasis, the ability to regulate the environment within a desirable range. Virtually all living creatures maintain homeostasis within their bodies, but very few animals have evolved the ability to maintain a constant external living environment. In this respect, insect societies are similar to human societies. Workers adjust their activities to maintain the living environment within optimal limits. Bees, for example, can closely regulate the internal temperature of a hive. When temperatures fall below 18 degrees Celsius, they begin to cluster together, forming a warm cover of living bees to protect the vulnerable brood stages. To cool the hive in hot weather, workers initially circulate air by beating their wings. If further cooling is needed, they resort to evaporative cooling by regurgitating water throughout the nest. This water evaporates with wing fanning and serves to cool the entire hive. Other social insects rely on different but equally effective methods. Some ants, and especially termites, build their nests as mounds in the ground, with different temperatures existing at different depths. The mound nests of the African termite, *Macrotermes natalensis*, are an impressive engineering feat. They are designed to regulate both temperature and air flow through complex passages and chambers, with the mound itself serving as a sophisticated cooling tower.

Finally, group living allows the coordination of the efforts of individuals to accomplish complex tasks normally restricted to the higher vertebrates. The similarities between insect societies and human society are striking. An insect society is often referred to as a superorganism, reflecting the remarkable degree of coordination between individual insects. Individual workers have been likened to cells in a body, and castes to tissues or organs that perform specialized functions. Insect societies are not immortal; however, they often persist in a single location for periods similar to the life spans of much larger animals. The social insects have one of the most highly developed symbolic languages outside human cultures. Further, social insects have evolved complex and often mutually beneficial interactions with other species to a degree unknown except among human beings. Bees are inseparably linked with the flowers they feed upon and pollinate. Ants have actually developed agriculture of a sort with their fungus gardens and herds of tended aphids. On a more sobering note, ants are the only nonhuman animals that are known to wage war. These striking similarities with human societies have led researchers to study social insects to learn about the biological basis of social behavior and have led to the development of a new branch of science known as sociobiology.

Studying Insect Society

Because of the diversity of questions that investigators have addressed regarding insect societies, many methods of scientific inquiry have been employed. In Karl von Frisch's experiments, for example, basic behavioral observations were cou-

Edward O. Wilson

Born: June 10, 1929; Birmingham, Alabama
Fields of study: Ecology, entomology, ethology, evolutionary science, genetics, zoology
Contribution: Wilson is recognized as the world's chief authority on social insects, particularly ants. He is also an expert on the genetic basis of the social behavior of all animals, including humans.

After his graduation from high school in Decatur, Alabama, Edward O. Wilson earned his bachelor's and master's degrees in biology at the University of Alabama at Tuscaloosa, concentrating his efforts on the study of ants. Continuing his educational pursuits at Harvard University, he earned his doctor's degree in zoology and received an appointment as a professor of zoology and curator of entomology at the Museum of Comparative Zoology at Harvard. Based on their study of insect societies, Wilson and W. L. Brown proposed the idea of character displacement: that after the populations of two closely related species come into contact, they undergo rapid evolutionary differentiation so as to minimize the chances of competition and hybridization between the two species.

Wilson's definitive work on ants and other social insects, *The Insect Societies* (1971), treats the societal behavior patterns of many species. It covers numerous aspects of insect societies, ranging from paleontology to formal genetics and from ethology to biochemistry. Providing an account of the natural history of social insects, with their great proliferation of genera, species, and behavioral types, Wilson incorporated concepts from the fields of modern genetics, selection theory, and biomathematics to explain the evolution of insect societies and their diversity in size and longevity. In *The Ants* (1990), Wilson and Bert Hölldobler published a comprehensive summary of current knowledge on ants.

Wilson's thorough study of insect societies led to his proposal that the biological principles governing animal societies extend to humans, which he explored in *Sociobiology: The New Synthesis* (1975). Wilson theorized that the preservation of the gene, versus the preservation of the individual, is the fundamental concept in evolutionary development. His argument was that social animals, including humans, behave according to rules written in their genes. Since this theory contradicts the belief in free will and suggests that some human groups may be biologically superior to others, it sparked much controversy among scientists and nonscientists alike. In *The Diversity of Life* (1992), Wilson further investigated how the world's living species became diverse, as well as the massive species extinctions produced by human activities in the twentieth century.

Wilson became the first and only person to receive both the highest award for science in the United States, the National Medal of Science, and the premier literary award, the Pulitzer Prize in literature. He received the latter award for his book *On Human Nature* (1978) in 1979. In 1990, he was awarded the prestigious Crafoord Prize by the Swedish Academy of Science for his work in ecology. In 1995, *Time* recognized Wilson as one of the twenty-five most influential people in America. An international poll in 1996 ranked him as one of the hundred most influential scientists of all time. His present research is focused on the concept of consilience, a controversial attempt to unify all knowledge by means of science so that explanations of differing kinds of phenomena are connected and consistent with each other.

—*Alvin K. Benson*

pled with simple but elegant experimental design to unravel the dance language of bees. The bees were raised in an observation colony. This was essentially a large hive housed between plates of glass so that an observer could watch the behavior of individual bees. Researchers followed specific workers by marking them with small numbers placed on the abdomen or thorax. Sometimes the entire observation hive was placed within a small, darkened shed to simulate more closely the conditions within a natural hive.

Bees learned to find an artificial "flower"—a glass dish filled with a sugar solution. Brightly colored backgrounds and odors such as peppermint oil were added to the sugars to provide specific cues for the bees to associate with the reward. Feeding stations were set up at fixed distances; observers could follow the exact movements of known individuals both at the feeder and at the hive. In this way, von Frisch was able to describe several types of dances (the round dance for near food sources, the waggle dance for feeding stations that were farther from the hive) and show that a returning bee could share information regarding the location and quality of a source with her nestmates. Scientists subsequently have developed robot bees that can be operated by remote control to perform different combinations of dance behaviors. This allows them to determine which parts of the dance actually convey the coded information.

The investigation of forms of chemical communication requires application of a variety of techniques. Chromatography is useful for identifying the minute amounts of chemical pheromones with which insects communicate. Chromatography (which literally means "writing with color") is particularly suitable for separating mixtures of similar materials. A solution of the mixture is allowed to flow over the surface of a porous solid material. Since each component of the mixture will flow at a slightly different rate, eventually they will become separated or spaced out on the solid material. Once the components of the pheromone have been separated and identified, their activity is assessed separately and in combination using living insects. Such bioassays allow researchers to determine exactly which fractions of the chemical generate the highest response.

Other biochemical techniques, such as electrophoresis, have been used to determine subtle behavioral differences, such as kin discrimination among hive mates. Each individual carries a complement of enzymes or proteins that catalyze biological reactions in the body. The structure of such enzymes is determined by the genetic makeup of the individual, and it varies among individuals. Because enzyme structure is inheritable, however, much as eye color is, the degree of similarities between the enzymes can be used as a measure of how closely related two individuals are. The amino acids composing the enzyme differ in their electrical charges, so different forms can be separated using the technique of electrophoresis. When a liquid containing their enzymes is subjected to an electrical field, the proteins with the highest negative charge will move farthest toward the positive pole. This provides a tool to distinguish close genetic relatives for use in conjunction with behavioral observations to test, for example, whether workers can discriminate full sisters from half-sisters, or relatives from nonrelatives, as kin selection theory would predict.

The Success of Social Insects

Social insects are among the most successful groups of animals throughout the world, especially in the tropics. Although the number of species is low when compared to all insects (twelve thousand out of more than a million species), their relative contribution to the community may be unduly large. In Peru, for example, ants may make up more than 50 percent of the individual insects collected at any site.

The study of social insects has provided scientists with new ways of looking at social behavior in all animals. Charles Darwin described the evolution of sterile workers in the social insects as the greatest obstacle to his theory of evolution by natural selection. In attempts to explain this seeming paradox, William D. Hamilton closely examined the social Hymenoptera, where sociality had evolved eleven separate times. Realizing that the haplodiploid form of sex determination led to sisters being more closely related to one another than they would be to their own young, Hamilton developed a far-reaching new theory of social evolution: kin selection, or selection acting on groups of closely related individuals. This theory, which provides insights into the evolution of many kinds of seemingly altruistic behaviors, arose primarily from his perceptions regarding the asymmetrical relatedness of nestmates in the social Hymenoptera. These insects, then, should be credited with providing the model system that has led to a subdiscipline of behavioral ecology known as sociobiology, the study of the biological basis of social behavior. Moreover, given their central roles in critical ecological processes such as nutrient cycling and pollination, it would be hard to imagine life without them.

—*Catherine M. Bristow*

See also: Altruism; Ants; Arthropods; Bees; Communication; Communities; Endocrine systems in invertebrates; Ethology; Groups; Herds; Hierarchies; Home building; Insects; Instincts; Invertebrates; Mammalian social systems; Packs; Pheromones; Symbiosis; Wasps and hornets.

Bibliography

Crozier, Ross H., and Pekka Pamilo. *Evolution of Social Insect Colonies: Sex Allocation and Kin Selection*. New York: Oxford University Press, 1996. Investigates the genetic basis for insect sociality.

Frisch, Karl von. *The Dance Language and Orientation of Bees*. Translated by L. E. Chadwick. Cambridge, Mass.: The Belknap Press of Harvard University Press, 1967. A detailed description of the research conducted by von Frisch and his students that provided the basis for scientists' knowledge of honeybee communication and orientation. Well illustrated with pictures and photographs, this book provides a particularly good section on the classical experimental methods used to study insect behavior in the field.

Gordon, Deborah M. *Ants at Work: How an Insect Society Is Organized*. New York: W. W. Norton, 1999. Summarizes fifteen years of research on red harvester ants for a general audience.

Hoelldobler, Bert, and Edward O. Wilson. *Journey to the Ants: A Story of Scientific Exploration*. Cambridge, Mass.: Belknap Press, 1994. Written for the nonspecialist, with many illustrations and color photographs. Investigates the reasons why ants are such a successful species, and also uses ants to illustrate concepts and research methods of sociobiology.

Ito, Yoshiaki. *Behavior and Social Evolution of Wasps: The Communal Aggregation Hypothesis*. New York: Oxford University Press, 1993. Uses wasps to argue that mutual defense rather than kin selection is the basis of insect sociality. Many illustrations, charts, and graphs, as well as extensive bibliography. Written for college level through professional readers.

Moffett, Mark W. "Samurai Aphids: Survival Under Siege." *National Geographic* 176 (September, 1989): 406-422. The discovery in 1972, by Aoki in Japan, of aphid species that possess a sterile soldier caste has added another insect group (the Homoptera) to those commonly considered eusocial. This article provides a popular discourse on this unusual group, with details of their life cycles and interactions with other species, including a mutually beneficial association with another social insect, a tiny ant. Excellent photographs and diagrams of the life cycle.

Prestwich, Glenn D. "The Chemical Defenses of Termites." *Scientific American* 249 (August, 1983): 78-87. Discusses the effective defenses of a specialized caste of termites known as nasutes. These highly modified soldiers have heads resembling nozzles, with which they squirt sticky toxic chemicals onto enemies. Striking electron micrographs show close-up details of the defensive apparatus.

Seeley, Thomas D. "How Honeybees Find a Home." *Scientific American* 247 (October, 1982): 158-168. Describes the behavior of honeybees searching for and choosing a new nest site, including how scouts assess the suitability of a new nest cavity and how they share that information with nestmates through a dance similar to the waggle dance. Good illustrations of the exploratory movements and dances of scouts, as well as a good photographic treatment of swarming.

Wilson, Edward O. *The Insect Societies*. Cambridge, Mass.: The Belknap Press of Harvard University Press, 1971. This is the most comprehensive treatment of social insects. Wilson covers every aspect known up to the writing in a lucid and well-organized fashion. This is the book for the expert wanting a summary of ant mimics and guests or wanting to learn more about the genetics of social evolution. Nevertheless, the material remains largely accessible to the nonexpert as well, in large part because of the wealth of excellent drawings and summary tables. The bibliography is extensive and covers many of the classic works as well as more modern treatments.

_______. *Sociobiology: The New Synthesis*. Cambridge, Mass.: The Belknap Press of Harvard University Press, 1975. A discussion of the evolution and structure of societies—both those of insects and those of higher animals. The format is very similar to that of Wilson's *The Insect Societies*, and the drawings and illustrations are of equal quality. The book is almost seven hundred pages long but is well indexed and suitable for browsing. An excellent source for anyone interested in comparing different social animals or in tracing ideas on the evolution of sociality.

_______. *Success and Dominance in Ecosystems: The Case of the Social Insects*. Oldendorf/Luhe, Federal Republic of Germany: Ecology Institute, 1990. Wilson argues that evolutionary longevity is a measure of a species' success, and ability to control biomass and energy in its ecosystem is a measure of its dominance. Includes an introduction by Otto Kinne.

INSECTS

Type of animal science: Classification
Fields of study: Anatomy, ecology, entomology, invertebrate biology

More than one million species of insects have been identified, and countless others await discovery. Insects provide many valuable products and services within ecosystems, but some cause damage by eating crops or spreading disease to plants, animals, and humans.

Principal Terms

APTEROUS: insects without wings, such as fleas
ECDYSIS: molting; the shedding of the exoskeleton that allows for insect growth
HEXAPOD: six-footed, a general term for an insect
OVIPOSITOR: egg-laying apparatus on the female abdomen, modified into a stinger in bees
PUPA: intermediate stage between the larval and adult stages of the life cycle
SPIRACLES: openings on the outside of the insect abdomen that lead to breathing tubes
VECTOR: transmits pathogens from one host to another

The history of insects dates back to the Devonian period, about 400 million years ago. Today numbering more than one million different species, insects are the most diverse class of animals on earth. Insects populate almost every habitat except the deep oceans and permanently frozen land masses. Insects are specialized to live underground, in live or rotting trees, in fast-flowing rivers, or stagnant puddles. Parasitic insects live attached to the outside (hair, skin) or inside (stomach lining, respiratory tract) of other animals. Crafty insects modify their habitat by building their own houses. Spittlebugs mix air into slimy anal secretions to form shelters. Bagworms spin silk to hold leaves together around them. Some African termites glue soil particles together with saliva into colossal nests measuring six meters high and nearly four meters across.

Insect Anatomy

Along the relatively straight digestive tract of an insect there are regions specialized for food storage, grinding, chemical breakdown, and nutrient absorption. Before undigested remains leave the insect, water and salts are reclaimed. Insect waste, termed frass, is thus very dry. This last step of water conservation is very important to insects because they lose water rapidly due to their small size.

The insect respiratory system is also adapted to save water. The tracheal system is a collection of tubes that allows for gas exchange from the outside environment directly with individual tissues. The openings in the body wall, spiracles, can be closed to conserve water between breathing cycles.

The circulatory fluid of insects, hemolymph, functions in keeping tissues moist and transporting nutrients and hormones. There are no red blood cells or other oxygen-carrying molecules because the insect respiratory system is totally separate from the circulatory system. The circulatory system is open. Hemolymph enters a series of chambers, collectively termed the heart, through holes called ostia. Muscular pumping of the heart propels the fluid through the aorta, toward the insect's brain. In the vicinity of the brain, the hemolymph flows out into the general body cavity, or hemocoel. Hemolymph directly bathes internal organs on its way to and from the legs and wings, starting the cycle again by entering the heart ostia. There are distinct veins in the wings. Insects take advantage of fluid circulation in the wings to transfer heat, warming themselves or cooling off when needed.

Insects are noted for their jointed exoskeletons. Chitin makes this armor especially strong. The exoskeleton functions much like the internal bones of vertebrates, serving to protect internal organs and as sites of muscle attachment. The exoskeleton puts limits on the overall size that insects can attain and makes growth energetically costly. The insect must shed its old exoskeleton and form a new one in order to increase in size. The old exoskeleton may be eaten to reclaim some of the nutrients within it. While a freshly molted insect is waiting for its new outer covering to harden, it is particularly vulnerable to attack from predators. Molting insects tend to be white; the exoskeleton will also develop color as it hardens.

The Insect Diet

Phytophagous insects feed on living plants. Caterpillars are famous leaf eaters. Other insects feed on plant roots, shoots, flowers, stems, or fruits. When a large group of phytophagous insects occurs in the same area, the crop or forest damage that they cause can be extensive. Yet there is quite another side to the relationship between insects and plants. Insect-pollinated plants have evolved odors, shapes, and ultraviolet color patterns to attract these important visitors. When the butterfly, bee, or other pollinator stops to drink the plant's rich nectar, pollen from the male part of the flower sticks to the insect's body. When it visits another flower of the same species, it deposits the pollen on the female floral organ. Without this interaction, beans, tomatoes, tea, cocoa, and many other plants could not reproduce.

Scavengers also serve an important purpose by recycling nutrients found in dead plant and animal matter. Termites are a nuisance when they infest a house but are invaluable at breaking down dead wood in nature. Dung beetles have the curious habit of forming animal feces into balls and rolling them away to feed their young. Carrion feeders assist in the decomposition of animal corpses. A succession of flies and beetles reduce a corpse to bones.

Predators hunt and kill to eat. Some predators use powerful mouthparts to tear apart their prey. Other predators inject digestive enzymes into their prey, digest them externally and suck out the liquefied tissues through a specialized beak. The first pair of legs of the praying mantis are striking

Insect Facts

Classification:
Kingdom: Anamalia
Subkingdom: Bilateria
Phylum: Arthropoda
Subphylum: Uniramia
Class: Insecta
Orders: Common orders include Coleoptera (beetles); Diptera (flies, mosquitoes); Hemiptera (true bugs); Homoptera (cicadas), Hymenoptera (ants, bees, wasps); Isoptera (termites); Lepidoptera (butterflies, moths); Odonata (dragonflies, damselflies); Orthoptera (crickets, grasshoppers)
Geographical location: Every continent except Antarctica
Habitat: Mainly terrestrial, some aquatic, primarily freshwater
Gestational period: Highly variable; some insects produce one, two, or several generations per year; conversely, relatively large insects may take more than one year for larval development
Life span: Highly variable; adult mayflies live less than one week, while queen termites have been known to live for more than twenty years
Special anatomy: Major regions include the head, featuring one pair of antennae, the thorax, with three pairs of legs and up to two pairs of wings, and the abdomen, housing spiracles and genitalia; in some insects, one or both pairs of wings are modified for functions other than flight, such as protection or balance

Insects must repeatedly undergo molting in order to accommodate their increasing size. (Rob and Ann Simpson/ Photo Agora)

examples of raptorial modifications for grasping and holding struggling prey.

Dragonfly nymphs, the young aquatic stage, have a diet of insects, crustaceans, tadpoles, and even small fish. Adult dragonflies eat bees, butterflies, and mosquitoes. Harmless insects such as dragonflies, which consume pest insects such as mosquitoes, are good candidates for biological control programs. Biological control is a method of pest control that takes advantage of natural predators of the pest. Ladybugs are commonly used by gardeners to control plant-damaging aphid populations.

On the other hand, a good parasite does not kill its food source, the host. Parasitic insects typically ingest host blood, mucus, or tissues, with minimal irritation or harm to the host. Trouble begins if there is a heavy parasitic burden (a large number of parasites per host), or if the parasitic insect transmits disease-causing organisms to the host. Ectoparasites live on the outside of their hosts. In this category, fleas and lice are adapted to avoid detection by the host and deter removal during normal grooming processes. Endoparasitic fly larvae live in the digestive or respiratory tract, mainly of livestock. Endoparasites of invertebrates do not follow the rules. Because there is little difference in size between this parasite and host pair, the host routinely perishes in this relationship. Species of endoparasitic wasps inject their eggs into caterpillars. When the larvae hatch, they tunnel farther into the host and feed off of its tissues. Upon completion of larval development, the wasps emerge from the caterpillar, killing it in the process.

Some parasites spend their lifetimes closely associated with the host, but others only briefly visit the host. On the time continuum, lice not only form a constant, more-or-less permanent association with a single host, but many generations of lice may inhabit the same host. However, these parasites are transferred from host to host during mating, nesting, or other close contact between individuals. Unlike lice, fleas leave the host frequently between bloodmeals. On the other end of the spectrum, mosquitoes normally require multiple hosts to complete a single bloodmeal; they are here and gone before they can be slapped by hand or tail.

Insect Behavior

Much of what insects do, they do by instinct, a genetically preprogrammed response to environmental stimuli, none the less amazing in its elegance and effectiveness. Antennae and setae, hairlike projections through the exoskeleton, serve as two of the many receptors of external stimuli. The necessity of keeping these sensory organs clean is evidenced by preening, commonly using mouthparts and legs. Major categories of environmental stimuli signal friend or foe.

Many insects locate opposite sex conspecifics, individuals of the same species, by sight. Color patterns, especially in the ultraviolet (UV) spectrum attract potential mates. Fireflies (Coleoptera) use visual recognition of patterns of light flashes. In some species, females use signaling to attract males. In other species the male and female signal to each other with the proper code and response. Females of the genus *Photuris* mimic the flashing patterns of other firefly species. When the male approaches, she makes a meal of him.

Sound production and reception are also used for mate location. The sound generated by a female mosquito's beating wings in flight is picked up by the male mosquito's antennae. Male homopterans and orthopterans produce songs of courtship. Cicadas use abdominal muscles and a resonating chamber for sound production. Grasshoppers and crickets rub their wings and or legs together. Sound receptors, "eardrums" of females in these orders, are found on their forelegs or first abdominal segments. Sound production and reception vary with temperature.

Pheromones, externally broadcast chemical signals, are another means of attracting mates. In moths, glands near the tip of the female's abdo-

Vector-Borne Disease

Insects are carriers of major human diseases. Interaction between disease-causing microorganisms and the insect vector is an interesting and vital component of the transmission cycle. The bubonic plague, for example, is endemic (naturally occurring) in populations of rodents. The plague bacteria, *Yersinia pestis*, is transmitted to humans primarily by *Xenopsylla cheopis*, the Oriental rat flea. When a flea ingests infected blood, the bacteria multiply in its gut and stick together, rapidly forming a plug. When a blocked flea attempts to feed, it regurgitates infected blood into its host. Still hungry, a blocked flea will repeatedly attempt to feed, increasing the chances of it infecting many hosts.

While there are still small geographic pockets where plague is present today, malaria rages as a widespread, tropical epidemic. Malaria is caused by species of protozoa in the genus *Plasmodium*, and carried by mosquitoes in the genus *Anopheles*. These mosquitoes not only transport the disease from host to host, they provide the only environment in which the protozoa can complete its life cycle. *Plasmodium* exhibits both asexual and sexual reproduction, but it reproduces sexually within the mosquito gut and nowhere else. After development in the gut, the protozoa migrate, through the mosquito, to the salivary glands. While feeding, the mosquito then spits *Plasmodium* into a new host.

Chagas' disease is also caused by a protozoan, *Trypanosoma cruzi*, but the vector and mode of transmission are radically different from that of malaria. Like *Plasmodium*, *T. cruzi* requires residence in the insect gut in order to develop into an infective stage. However, the insects that *T. cruzi* are associated with are true bugs (Hemiptera, Reduviidae). The protozoan remains in the bug's digestive tract until excretion within fecal material. Transmission of Chagas' disease relies on the host scratching infected feces into the itchy bite wound.

Forensic Entomology

Scientists and law enforcement agents capitalize upon the predictable sequence of insect development on human corpses to provide clues in homicide cases. Species of *Calliphora* (blowflies) lay eggs on the body almost immediately postmortem. The eggs hatch in a given amount of time, then the larvae go through distinct growth stages and pupate. During these first weeks, other species of flies also colonize and develop in this specialized microenvironment. After the flies have done their work, beetles arrive. Forensic entomologist M. Lee Goff has done many experiments with the decomposition of pig carcasses. Using his work, data tables have been established to aid scientists in interpreting the story of the insects present, and in turn provide clues regarding the fate of the deceased victim. The size of the larvae of each species is adjusted for temperature and this data is used to estimate the length of time that the person has been dead. Larval development is calculated at 72 degrees Fahrenheit. Time is subtracted for exposure to warmer environments since insect development speeds up at higher temperatures. At 80 degrees Fahrenheit, a larval fly would appear to be older than it really is. Insect evidence can assist investigators in determining whether the body has been moved and yield clues about poisons or drugs present in the body at the time of death.

men release pheromones that are received by the antennae of the male moths. Certain male butterflies and moths have specialized scent-producing glands on their wings. The notable pair of black patches, or androconia, on the hindwings of the male monarch secrete aphrodisiac pheromones that increase the female's receptivity to mating. Pheromones can also make already-mated females unattractive to subsequent suitors.

Members of the class Insecta display numerous protective behaviors and defense mechanisms when threatened. Walkingsticks are capable of losing part of a leg in the grasp of a predator. The amputated appendage continues to wiggle when severed, thus distracting the predator, while the walkingstick gets away. Some beetles play dead until the predator loses interest. Blister beetles are capable of reflex bleeding. They squeeze drops of hemolymph through joints in their exoskeleton. Their hemolymph contains cantharadine, which irritates and repels the predator. Bombardier beetles spray a noxious repellant from their anal glands for distances up to one meter.

Another means of defense is coloration. Insects blend into the surroundings by resembling a leaf, twig, pebble, or flower. Some insects use bright colors to hide in plain sight. Large eyespots on wings can startle would-be predators, or at least trick them into taking a bite out of the wing rather than from the head or body. Color patterns of orange and black seem to warn vertebrate predators that the insects would not make a good meal. Some of these insects derive chemicals from their diet that make them distasteful, even harmful, to predators. Without possessing such chemicals, mimics derive protection from their resemblance to the group that does.

Social insects exhibit the most complex behavior, approaching learning. Honeybees can communicate direction of a food source, and distance from the hive, to other bees through a multipart waggle dance. Some ants are farmers, planting and nurturing fungus gardens. Mound-building harvester ants ensure that the colony's young are kept at the optimal temperature through vertical migrations. The young are transported to top levels within the mound to warm up in the morning and evening. During the heat of the day, the young are carried to lower levels to cool off. Kidnapping of ants from other colonies to serve as workers in the home colony may not be considered a socially advanced behavior, but it is a complex one. The other major group of social insects, the termites, work together to build huge mounds, called termitaria, some reaching heights of 6 meters and diameters of over 3.5 meters. They engineer series of chimneys into the structure that can be opened and closed to regulate airflow and maintain constant temperature.

—*Sarah Vordtriede*

Maria Sibylla Merian

Born: April 2, 1647; Frankfurt am Main, Germany
Died: January 13, 1717; Amsterdam, Netherlands
Fields of study: Developmental biology, ecology, entomology
Contribution: Merian's paintings are significant as natural history illustrations in seventeenth century Western art. Her detailed observations, particularly of insect metamorphosis, also contributed to the development of entomology as a science.

Maria Sibylla Merian, the German-born Dutch naturalist, was the youngest of three children born into a family of artists. Her father, Matthaus Merian the Elder, a publisher, engraver, and botanical artist, died when Maria was just three years old. Her older brother, Matthaus Merian the Younger, became a famous portrait painter. Maria's own artistic talent was noticed early on by her stepfather, Jacob Marrell, himself a trained painter. Her curiosity and observation of the natural world was also established early in life. She kept a detailed journal of her observations from her teenage years until just before her death at the age of sixty-nine.

Deviating from other seventeenth century artists, Maria Sibylla Merian sketched, painted, and engraved insects from real-life observations rather than as still-life symbols. In this way, she was able to put their behavior and development into ecological context. She bred and raised silkworms and, at age thirteen, illustrated the egg, larva, pupa, and adult stages of these creatures, as well as many other butterflies. Merian is credited with being the first person to record such detailed observations on insect metamorphosis. Although she did not realize it at the time, she also documented parasitoid infection of some of the caterpillars. The organisms that she described as small black flies emerging from butterfly pupa were likely small wasps that complete their larval development within the pupal case of the butterfly host.

In 1699, funded by the city of Amsterdam, Merian and her youngest, teenaged daughter traveled to South America to study and paint insects, birds, and plants. They spent two years exploring the natural wonders of the cities, plantations and rainforests of the country of Suriname (formerly Dutch Guiana). Observational records from this excursion were published in 1705 under the title "De generatione et metamorphosibus insectorum surinamensium" ("of the reproduction and the transformation of insects in Suriname"). Both before and after her trip, she published several books of illustrations and observations of plants and the insects associated with them. Two of her paintings are featured on thirty-two cent U.S. postage stamps. In addition to her artwork, she traded in preserved insects, and instructed others according to the proper preservation of specimens.

Maria Sibylla Merian left quite a legacy. Her daughters continued research on insects after their mother's death, and Merian's artwork and keen scientific observations, meticulously recorded, are preserved for all to learn from and enjoy. Six plants, nine butterflies, and two beetles bear her name, which is ultimately fitting, since Carolus Linnaeus is said to have consulted her work while establishing his binomial nomenclature system for naming the living organisms of the earth.

—Sarah Vordtriede

See also: Antennae; Ants; Bees; Beetles; Butterflies and moths; Cockroaches; Cold-blooded animals; Communication; Communities; Exoskeletons; Flies; Flight; Grasshoppers; Insect societies; Invertebrates; Metamorphosis; Mimicry; Molting and shedding; Mosquitoes; Praying mantis; Symbiosis; Termites; Wasps and hornets; Wings.

Bibliography

Arnett, Ross H., Jr., and Richard L. Jacques, Jr. *Guide to Insects*. New York: Simon & Schuster, 1981. Anyone interested in entomology should have multiple field guides. Descriptions of the insects pictured are on the same page, making identification easier.

Borror, Donald J., Charles A. Triplehorn, and Norman F. Johnson. *An Introduction to the Study of Insects*. 6th ed. Philadelphia: Saunders College Publishing, 1992. This is a classic for those interested in classification. An excellent compilation of family level identification keys for the major insect orders. Illustrations help students unfamiliar with necessary anatomical details.

Elzinga, Richard J. *Fundamentals of Entomology*. 5th ed. Upper Saddle River, N.J.: Prentice Hall, 2000. Very student-friendly college level introductory textbook. Early chapters introduce important themes across the class Insecta to lay the groundwork for later collection and classification.

Kettle, D. S., ed. *Medical and Veterinary Entomology*. 2d ed. Oxford, England: CAB International, 1995. An advanced treatment of insect and arthropod vectors and the diseases they carry. The comprehensive and well-formatted index makes this a great reference text for the serious student.

O'Toole, Christopher. *Alien Empire: An Exploration of the Lives of Insects*. New York: HarperCollins, 1995. This companion to the Nature public television miniseries lures readers in with breathtaking photographs and logical organization and does not disappoint; the text is equally exciting and educational. Appropriate for general audiences.

Turpin, F. Tom. *Insect Appreciation*. 2d ed. Lanham, Md.: Entomological Society of America, 2000. Written in a good-humored and understandable manner, this part workbook/part instructional supplement appeals to a wide audience.

INSTINCTS

Type of animal science: Behavior
Fields of study: Biochemistry, ethology, neurobiology

Instincts cover a range of inherited behaviors that are unlearned and are predictable for a species. Scientists are interested in locating the biochemical sources of instincts and analyzing the relationship between instincts and a species' biological success.

Principal Terms

ALTRUISM: a high degree of devotion to the interest of others that often includes self-sacrifice

ETHOLOGY: a branch of biology that studies behavior

FIXED ACTION PATTERN (FAP): a behavior whose timing and duration are invariable for all members of a species

INNATE: denotes an inherited and unalterable condition or ability in an organism

NEUROBIOLOGY: the study of the biology of the brain

NEUROETHOLOGY: the study of behavior as it relates to brain functions

PEPTIDE: a chemical combination of certain amino acids

SIGNAL: information transmitted through sound, such as bird calls, or through sight, such as body posture

STEREOTYPED BEHAVIOR: an unlearned and unchanging behavior pattern that is unique to a species

Instincts are patterns of behavior found in animals that are unlearned and that are inherited by successive generations of the species. Ethologists call these stereotyped behaviors fixed action patterns (FAPs). Over time, such patterns are shaped by evolution, but at any given moment, a species maintains a range of instincts that are unique to its members. A number of categories of stereotyped behavioral responses exist that apply to all organic forms. These include the basic drives such as reproduction, feeding, and protection from predators. The term "instinctive behavior," however, is generally reserved for animals and insects, while excluding plants, bacteria, and viruses. Simple forms of stereotyped responses include a variety of reflex actions in which sensory nerve cells are affected by conditions such as heat and light. The more complex forms of responses are studied in order to understand the sequence of responses and the process of evolution that selected these patterns.

The Study of Instincts

The study of instinctive behavior began in the nineteenth century with the development of clinical psychology. Scientists noticed a direct link between animal and human behavior, and since clinical experimentation of human subjects was not always possible or desirable, experiments were carried out on monkeys, mice, and guinea pigs. Experiments focused on the responses of test subjects to specific environmental conditions. An example of this type of experiment would be investigating the puzzle-solving ability of rats in a maze. The flexibility and nature of these responses were seen as part of a study of learning processes.

Meanwhile, the field of ethology began to formulate a different view of animal behavior. A European group including Konrad Lorenz, Nikolaas Tinbergen, and Karl von Frisch noticed that animals possess a specific innate capacity to perform complex activities in response to the environment. As a result, modern ethology shifted the study of animal behavior from learned responses to inherited patterns of behavior. From their beginning as "bird watchers," observing birds' courtship behavior, nest building, rearing of young, and territorial ownership, ethologists collected a body of scholarship that came to represent a respected field of study. A number of subfields have been created, ranging from neurology and genetics of behavior to ecology of species behavior. One element that links these diverse and distinctive fields of study is that of signals, language, and communications.

Charles Darwin gave the scientific world a theory of evolution, but he was also an avid observer of instinctive behavior. He observed domesticated pigeons and described how emotions are expressed by humans and animals. Other early pioneers in this field included Charles Whitman, who studied the family tree of a group of pigeons in Massachusetts, and Oskar Heinroth, who observed several species of waterfowls in Germany. Their observations led to the conclusion that members of a species not only share the same body functions and bone structures but also share behavior patterns.

The next step was to compare instinctive behavior across several closely related species. This work was accomplished between 1940 and 1960 by Lorenz and Tinbergen. Lorenz chose as his subject the courtship sequences of mallards, teals, and gadwalls. Recognition signals among ducks are a specific sequence of behaviors that include tail shaking, head flicking, and whistling. The exact sequence for different species is different, yet the components are the same. Through these studies, it is possible to conclude that courtship sequences originated in a single ancestral form, and crossbreeding between duck species produced variations in the pattern. Tinbergen arrived at a similar conclusion with his work on the calls, body postures, and movements of gulls. His study of the signaling behavior of more than fifteen species of gulls showed that their system of signals is similar. As a result of their work, Lorenz and Tinbergen provided a scientific basis both for the biological source of behavior and for an evolutionary source of instinctive reaction.

The Nature Versus Nurture Debate

Although few scientists of the time had objected to the notion that genetics contributed to instinctive behavior, there was an ongoing debate as to the extent of the genetic contribution. The scientific bias was on the side of learned behavior patterns in such areas as language, signals, and body postures. It seemed clear that higher animals learn how to transfer information among themselves. One example was that an isolated songbird, denied access to parental teaching, does not instinctively know the songs of its species. Consequently, the genetic component may play only a small part in the overall development of behavior. In this debate, the contribution of Karl von Frisch shifted the balance to the side of inheritance. Von Frisch showed that honeybees communicate both the direction and the

Nikolaas Tinbergen, one of the founders of the science of ethology and a Nobel laureate in 1973, conducted many experiments on the subject of instincts and behavior, especially in gulls. (Nobel Foundation)

distance of a source of nectar through a sequence of dance patterns. A scouting bee returning to the hive can provide exact information to other bees. While it was argued that the larger brain sizes of higher animals contributed substantially to learning a system of communication, it was difficult to make the same argument for insects.

The debate on nature versus nurture produced extensive efforts to determine the extent to which instinctive behavior is shaped by inheritance or by learning. In the study of sea gull chicks, Jack Hailman created a series of experiments that added evidence to the side of learned behavior. The parent of a sea gull chick can elicit pecking from the chick either by pointing its bill downward or by swinging its bill from side to side. The chick will respond not only to a parent's action, however, but also to a model that has the shape of the parent. The initial feeding behavior of a newly hatched chick is a "hit and miss" affair. With growth, the chick responds more precisely to the figure of the parent and acquires greater coordination. Thus, the initial instinct is for the chick to peck at a variety of motions, but with maturity a learning component takes place to create greater discrimination.

Sequences of Behavior

During the 1950's and 1960's, researchers in instinctive behavior studied a number of animals and insects. Scientists began to understand that instincts are not a simple response to the environment but are a complex sequence of behavior. Scientists also began to find evidence suggesting that natural selection acts on behavior patterns as well as on an organism's biological makeup. The result of this research came to be categorized into three broad groups of instinctive behavior. One was the response of a simple instinct on the part of a single organ to some stimulus. An example is a nerve cell responding to light that triggers a reflex response. Other reflex reactions include locomotion and movement. Another type of stereotyped behavior was called fixed action patterns (FAPs). For example, the courting behavior of ducks can be classified as FAPs, wherein the pattern and timing of the responses are invariable for all members of the species. Similar FAPs are found in spiders, crabs, and a number of other insects and lower animals. A final group of instinctive behavior was described as modifiable action patterns (MAPs) and included fixed patterns that could be modified by the environment or by learning. For example, in species of birds, the core of nest-building behavior is fixed, but actual nest building depends on the availability of preferred building materials, which can be altered by location and setting.

With the deciphering of the genetic code during the 1950's, research in instinctive behavior shifted toward the genetic basis of innate behavior. Research has shown that one group of chemicals called neuropeptides are produced by specific genes. Within the brain, these peptides have the ability to govern stereotyped behavior. For example, a specific peptide (angiotensin II), when injected into vertebrates, causes spontaneous drinking activity. Research on neuroactive chemicals that influence behavior patterns is still in its infancy; research on brain functions and neuron pathways has only begun. In the future, biochemists expect to isolate the link between the genetic code and brain function.

There remains one area of stereotyped behavior that has puzzled scientists since the time of Darwin: the question of altruistic behavior. Scientists have wondered how to explain behaviors such as ants drowning themselves in a stream so that others can cross over them or a parent animal risking its life so that offspring can survive. W. D. Hamilton, among others, began to use probability models, which show that cooperative behavior, which may risk the lives of individuals, results in greater survival of the rest of the group. Altruistic behavior not only is "for the good of the species" but also provides the greater probability that large numbers of that group will reproduce and therefore gain an advantage.

The Fieldwork of Ethology

The early students of ethology were often called "bird-watchers" because they began their work by observing the behavior of birds. Fieldwork involves repeated observations of a subject over long periods of time. Eventually, a sequence of behavior emerges, and then it is possible to read the language of the behavior. For example, Lorenz observed that the courtship sequence of the mallard duck involved some ten segmented parts, such as bill-shake, head-flick, tail-shake, and grunt-whistle.

In certain instances, fieldwork with insects and lower animals offers the possibility of direct experimentation. In his attempt to translate the dancing motion of bees, von Frisch made the food source available. Consequently, he was able to vary the distance of the food source, change its location, and alter the quantity of food. Each change in the variables produced some variation in the dance—perhaps a new "phrase" added to the language. In other areas of research, the test subject can be modified for direct experimentation. When William Keeton attempted to explain how pigeons found their way home, he used contact lenses to cover the eyes of the pigeon to block out the position of the sun. He also created secondary magnetic fields around the pigeon to test the subject's sensitivity to the earth's magnetic field.

While all studies of stereotyped behavior begin with observation, either in the field or in controlled settings, further exploration usually requires laboratory research. In Jack Hailman's study of the learning component of sea gull chicks, he constructed models of gull parents and correlated pecking accuracy with growth. He also modified features of the model sea gull to study possible changes in responses from the chicks. As the search for the causes of instinctive behavior moves further into the organism, the methodology follows—into areas of brain function (neuroethology), the chemistry of innate behavior, and the genetic component of behavior. Investigating the source of egg-laying behavior of a species of a large marine snail (*Aplysia*), Richard Scheller and Richard Axel found three genes that produce a number of peptides that govern this behavior.

Instincts and Genes

Instincts are a part of all living organisms, and observable instinctive responses are only a small part of an intricate pattern. The genetic makeup of an organism dictates specific unlearned patterns of responses and variations that, in turn, determine a favorable selection of individuals within a species. Clearly, instincts in sexual selection, reproduction, food gathering, and other basic needs are critical for the survival of members of a species. In higher animals, instinctive activities are often overshadowed (and sometimes disguised) by learned patterns of responses. For example, in dogs, the pulling back of facial muscles and the showing of teeth is a response to fear and attack. In humans, laughter is a similar response to surprise, embarrassment, and uneasiness. Because of social adaptation, however, laughter takes on additional behavioral conventions.

Until the early part of the twentieth century, instincts were thought to be learned responses to specific situations. Consequently, if aggression were learned, then it could be modified, changed, and unlearned. With the establishment of a genetic and biochemical foundation for instincts, research in stereotyped behavior has become part of a heated debate. In 1975, Edward O. Wilson published *Sociobiology: The New Synthesis*. This highly technical work found a surprisingly large audience; in it, Wilson attempted to place all social behavior on a biological basis. Although the work emphasized animal behavior, Wilson implied that all human history was also part of evolutionary biology and that his work would synthesize all the social sciences with biology. Since instincts such as aggression, selection of sexual partners, and care of the young play a prominent part in cultural activities, Wilson seems to suggest that in the future, the study of society will be grounded in neurobiology and sociobiology.

—Victor W. Chen

See also: Altruism; Brain; Communication; Competition; Courtship; Defense mechanisms; Ethology; Habituation and sensitization; Hormones and behavior; Imprinting; Learning; Mating; Migration; Reflexes; Rhythms and behavior; Territoriality and aggression.

Bibliography

Bateson, P. P. G., Peter H. Klopfer, and Nicholas S. Thompson, eds. *Behavior and Evolution*. New York: Plenum Press, 1993. A collection of essays focusing on description of behaviors in an attempt to explore the interaction of sameness and difference at the levels of individual, population, and species. Written for more advanced students of ethology.

Frisch, Karl von. "Dialects in the Language of Bees." *Scientific American* 130 (August, 1962): 2-8. The article describes a series of experiments undertaken to discover the "language" of honeybees. An excellent article for the reader with little technical background; highly recommended.

Goodall, Jane. *In the Shadow of Man*. Boston: Houghton Mifflin, 1971. An excellent view of fieldwork on primate behavior. Goodall gives a complete description of the difficulties and rewards of living and interacting with the subjects of her research. Highly recommended for the general reader.

Lorenz, Konrad Z. "The Evolution of Behavior." *Scientific American* 199 (December, 1959): 67-78. Lorenz suggests that instinctive behavior is inherited and that these traits are characteristic of a species. The article contains an excellent and brief summary of earlier scientific attitudes to instincts. Addressed to the general reader and highly recommended.

Scheller, Richard H., and Richard Axel. "How Genes Control an Innate Behavior." *Scientific American* 250 (March, 1984): 54-63. Describes how methods of recombinant DNA were used to find a set of genes that encode a neuropeptide and result in specific egg-laying behavior in a marine snail. Beyond the article's opening page, some background in college biochemistry and genetics is helpful.

Smith, John Maynard. "The Evolution of Behavior." *Scientific American* 239 (September, 1978): 176-192. Smith addresses the problem of how altruism can be favored by natural selection since the altruistic individual often does not survive. Some of the probability mathematics and game theory might prove difficult for the general reader, but the article is readable without the mathematics.

INTELLIGENCE

Types of animal science: Behavior, development
Fields of study: Developmental biology, ethology, evolutionary science

Animals are guided by more than instinct when interacting with their environment, yet the exact measurement of intelligence in various species remains problematic. While scientists devise numerous problem-solving tasks to assess intelligence, anthropomorphism leads to exaggerated claims of intelligence through anecdotal evidence.

Principal Terms

ANTHROPOMORPHISM: attributing human characteristics or states of mind to animals

COGNITION: transformation and elaboration of sensory input

COGNITIVE ETHOLOGY: scientific study of animal intelligence

LEXIGRAMS: symbols associated with objects or places in keyboard communication experiments with primates

PROTOGRAMMAR: word coined to signify the early foundation for grammar development found in primates

RECAPITULATION: stages of human development reappearing in different animal species

Both the general public and scientific community have long been intrigued with questions about how animals think and what are they thinking. Published reports of animal cognition increased dramatically in the last half of the twentieth century. Chimpanzees in the Ivory Coast have demonstrated extensive use of rocks as tools in cracking nuts. These primates have also been reported to hide undesirable expressions from their faces and act as if blind or deaf. Vervets have been found to use an elaborate system of alarm calls that seem to function as words. Parrots can demonstrate the ability to count, and birds exhibit the capacity to make and use tools to gather food. Dolphins apparently understand and follow simple commands. Primates have been trained to use signs in a symbolic fashion, communicating their needs, desires, and thoughts.

Theories of Cognitive Ethology

Cognitive ethology is a relatively new discipline that studies animal intelligence. Donald Griffin is considered to have founded this branch of study through the publication of *Animal Thinking* (1984) and *Animal Minds* (1992). Since the appearance of his books, numerous instances of animal intelligence have been gathered from observation and experimentation.

Traditionally, attitudes about animal intelligence can be sorted into those that place animals on a continuum with humans and those that see animals as distinct from humans. From the former perspective, animal behavior is readily interpreted as a definite sign of various cognitive skills and special abilities along a continuum of development. From a discontinuity perspective, only humans are considered to possess the higher cognitive skill of reasoning. The higher cognitive abilities are considered to be a uniquely human capacity that sets them apart from the lower animals, who are controlled by instinct.

Charles Darwin, in *The Descent of Man* (1871), defended the idea of the intelligence of animals existing on a continuum with humans. Since animals and humans have a common ancestry, animals would have the fundamental capacities for rational choice, reflection, and insight. Darwin concluded that the differences between the minds of humans and animals were of degree rather than of kind. Following Darwin's proclamation, a number of anecdotal studies concerning animal intelligence appeared that suggested extensive cognitive ability in animals. Unfortunately, many of the examples illustrated anthropomorphism. This is the process whereby humanlike characteristics are attributed to animal behavior.

Some interpretations of Darwin's statement created a distorted view about evolution that persisted long into the twentieth century. The idea that life on earth represents a chain of progress from inferior to superior forms began to influence the view of animal intelligence. The theory that ontogeny recapitulates phylogeny also became popular in the early years of the twentieth century. This theory, which does not have any scientific support, suggested that the advancement of life forms corresponded to the stages of development for humans. This stepladder approach to animal intelligence led to a ranking of animals compared to the developmental stages of human infants and children.

This approach to animal intelligence is flawed because it relies on the notion that some animals are more highly evolved than others are. Evolution does not have a single point of greatest evolution. The branches of the evolutionary tree have culminated with many different species occupying special niches. Thus, the "degree" of a species' evolution depends on the extent to which it successfully occupies its niche.

Animals Who Might Think

In addition to this tendency to attribute states of mind to animals that are found in humans, there were a number of cases of labeling trained behavior in animals as signs of reasoning skills. One of the most famous examples was the case of the horse, Clever Hans, in the early 1900's. Wilhelm von Osten owned a horse that demonstrated extensive arithmetic skills. When von Osten presented a written arithmetic problem to Hans, the horse would tap out the answer with his forefoot. Clever Hans also appeared adept at telling time, and answered questions about sociopolitical events by nodding or shaking his head yes or no. The horse's abilities suggested to many individuals the similarity between animal and human minds. Eventually, the Prussian Academy of Sciences discovered that Hans was not answering the

Measuring Arithmetic Skills in Primates

The chimpanzee Sheba was involved in a number of widely reported experiments that purported to demonstrate arithmetic ability in primates. Sheba was taught to associate a tray containing one, two, or three pieces of candy with cards containing the corresponding number of marks. If two candies were present on the tray, Sheba learned to pick the card with two marks. If the correct card was selected, Sheba would be rewarded by being allowed to eat the candy. In the next stage of the experiment, the marks on the cards were replaced by the numerals 1, 2, and 3. Then Sheba had to match the number of candy pieces with the correct number. After Sheba showed success in this phase of the experiment, the candy pieces were replaced with other inedible objects. If Sheba chose the number that matched the number of objects, she was rewarded with the corresponding number of candies. Eventually the numbers were expanded to include 0 and 4. Next Sheba was given the challenge of counting the total number of candies presented on a series of trays. Sheba demonstrated a rudimentary ability for addition showing correct responses 75 percent of the time. Initially the sums were restricted to one, two, or three. Sheba was eventually able to exhibit the skill to add numbers when the candies were replaced with the numerals 1, 2, and 3. Although Sheba was able to show mathematical ability, even the experimenter acknowledged some important limitations. Sheba required extensive training and "heroic" effort on the part of the trainer to accomplish the counting performances. A true mathematical ability should generalize to other situations, yet animals such as Sheba do not demonstrate easy or automatic transfer of numerical performance from one realm to another.

questions by means of any reasoning skills, but was an astute observer of the behavior of his owner and those around him. When questions were posed to Hans, cues were provided unconsciously to the horse about the correct answer. Since horses have evolved to ascertain subtle visual cues from others in their herd, Hans was able to form a number of cued associations which led to a reward. The owner of Clever Hans was not attempting to perpetrate fraud. He believed in the possibility that a horse could have reasoning ability, but von Osten was not sophisticated in how he tested for the skills. The inadvertent cueing of an animal to respond in a certain fashion is one of the major confounding factors found in the investigation of animal intelligence.

The case of Clever Hans illustrates two other problems that confound reports concerning the level of intelligence in animals. First is the problem of anthropomorphism. People develop an emotional bond with animals and interpret behavior in order to enhance the closeness they feel to them. The second problem concerns the methods used to measure intelligence. The classic case of Kohler's chimpanzees illustrates this problem.

In the early part of the twentieth century, Wolfgang Köhler assessed the reasoning ability of chimpanzees to obtain food outside of an enclosure. After a rake was left in the enclosure, food was placed out of reach of the caged chimpanzees. The chimpanzees were able to use the rake to bring food to the cage. Köhler concluded that the animals had insight into the nature of the problem and used reasoning to achieve a solution. A further study, requiring the fitting together of two sticks in order to reach the food, also supported Köhler's conclusions. However, later experimentation has revealed that chimpanzees without a history of playing with sticks could not solve the problem. Apparently, in order to solve the problem, the chimpanzees needed an extensive history of playing with sticks, which enabled them to learn how sticks could be used at a later time. In solving the problem, they were using an instinctual tendency to play with sticks and scraping them over the ground.

Primates and Sign Language

A contemporary example of the problem of measurement can be provided with the case of Washoe, the first chimpanzee to be taught sign language. Because of physical inability to vocalize human speech, chimpanzees were taught sign language as a mode of communication with humans. Soon Washoe and another signing chimpanzee, Nim Chimsky, were reported to have spontaneously created novel sentences through their signing. For example, Washoe was reported to have signed the combination water and bird after seeing a swan. Being a novel combination of signs, the trainers of Washoe explained the behavior as creative insight. Unfortunately, Washoe had also shown repeated signing of meaningless combinations, leading to the conclusion that a significant pairing of signs would eventually appear not because of the primate's cognitive reasoning but as a result of chance. Inevitably, these early attempts to demonstrate animal intelligence were widely discredited as exaggeration or self-delusion on the part of the animal's trainers, and this animal language research from the late 1970's fell into disrepute.

In order avoid the ambiguities of sign language, later researchers used keyboards that related symbols to a variety of objects, people, and places. Much of this research has taken place at the Language Research Center at Georgia State University in Atlanta under the guidance of Dr. Sue Savage-Rumbaugh. In the first experiments, two chimpanzees, Austin and Sherman, were familiarized with a system of symbols or lexigrams. Each was abstract and arbitrarily associated with an object, person, place, or situation. Eventually Austin and Sherman learned to communicate with symbols illustrated on a keyboard. For example, an experiment was devised where one chimpanzee was shown where food was being deposited in a certain container while the other had control of a tool to open the container. With the keyboard present, the chimpanzees were able to communicate with one another to use the tool on the correct container.

Soon a bonobo chimpanzee, Kanzi, became the star pupil of this technique and learned a vocabulary of two hundred symbols. Kanzi eventually showed the capacity to construct rudimentary sentences that were generated spontaneously. The chimpanzees trained using the keyboards appear to be exhibiting a protogrammar. This is a term to indicate the beginnings of grammar, roughly equivalent to the verbal skills seen in a human child about two to three years old.

In the late 1990's, another bonobo chimpanzee, Panbanisha, surpassed the capacities evidenced by Kanzi. Panbanisha has been reported to understand complex sentences and use the keyboard to communicate spontaneously with the outside world. Although the results have been impressive, critics of the Center's activities remain. The question remains whether the chimpanzees are demonstrating extremely effective training or some level of abstract reasoning.

—Frank J. Prerost

See also: Brain; Emotions; Imprinting; Instincts; Language; Learning; Reflexes; Rhythms and behavior; Tool use.

Bibliography

Budiansky, Stephen. *If a Lion Could Talk*. New York: Free Press, 1998. A good collection of contemporary and historical cases of animal intelligence. The stories cover a wide range of examples seen in various animal species.

Moss, Cynthia. *Elephant Memories*. New York: William Morrow, 1988. An interesting account of thirteen years of field observations concerning the behavior of elephants in the Amboseli National Park in Kenya.

Page, George. *Inside the Animal Mind*. New York: Doubleday, 1999. The author begins with a historical account of the popular and scientific views about animal intelligence. He provides good details about the various attempts to communicate with primates by teaching them sign language or through the use of keyboards.

Savage-Rumbaugh, Sue. *Kanzi: The Ape at the Brink of the Human Mind*. New York: John Wiley & Sons, 1994. This book presents an apparent breakthrough in the communication with chimpanzees using symbols and a keyboard. It includes a number of incidences of the spontaneous construction of sentences by the primates.

Savage-Rumbaugh, Sue, Stuart G. Shanker, and Talbot J. Taylor. *Apes, Language, and the Human Mind*. New York: Oxford University Press, 1998. A book written from an academic and scientific perspective about the ability of chimpanzees to communicate by means of symbols and a keyboard display.

INVERTEBRATES

Type of animal science: Classification
Fields of study: Anatomy, developmental biology, ecology, entomology, histology, invertebrate biology, marine biology, physiology

Over 98 percent of all known animal species are invertebrates, animals without backbones. At the base of many food webs, invertebrates provide nutrients to a wide range of other animals on land, underground, and in the seas. The diversity within the invertebrates is amazing, ranging from single-celled protozoans to cephalochordates possessing nerve cords, and claiming close relationship to vertebrates.

Principal Terms

COELOM: a true body cavity, lined by mesoderm

COLONY: a cluster of genetically identical individuals formed asexually from a single individual

CYST: a secreted covering that protects small invertebrates from environmental stress

GONOCHORISTIC: having separate sexes; an individual is either male or female

HYDROSTATIC SKELETON: a system in which fluid serves as the support by which muscles interact

MESODERM: a middle layer of embryonic tissue between the ectoderm and endoderm

With the exception of insects, which have extensively colonized terrestrial environments, most invertebrates are aquatic, many of those being marine. Even some of the land dwellers start life as aquatic larvae. Invertebrates follow one of three types of general body plan. Some marine sponges are asymmetrical, lacking an ordered pattern to their structure. Cylindrical organisms, such as sea anemones, are radially symmetric; any cut through the center of the organism divides it in equal halves. Asymmetrical and radially symmetric animals tend to stay in one place; thus their body plan helps them to collect environmental stimuli from every direction. Animals that exhibit bilateral symmetry have right and left halves that are mirror images of each other. They are mobile and usually have a distinct head end. This area of cephalization, concentrated nerve and sensory tissues, is directed forward in their travels, giving them new information about where they are going.

Invertebrate Feeding

Among the protozoans there are a variety of diets and feeding mechanisms. Organisms in the phylum Ciliophora have a cytostome, a cell mouth that can be found anteriorly, laterally, or ventrally, depending on species, on these single-celled creatures. Ciliates feed primarily on bacteria, algae, and other protozoans. Members of the phylum Amoebozoa have a similar diet to the ciliates but, in the absence of mouths, use pseudopodia to wrap around a food item, engulfing it. *Euglena* (phylum Euglenozoa), a flagellate protozoan commonly used in biology laboratories, is a self-feeder (autotroph). It contains chloroplasts and uses light energy to produce sugars, photosynthesized as in plants. It is interesting to note that some euglenids can and do ingest solid food if they are exposed to darkness for too long. Another group of interesting flagellates is the hypermastigotes. Species such as *Trichonympha campanula* live in the guts of termites. *T. campanula* breaks down the high cellulose content present in the termite diet of wood products, something that the termite cannot do for itself. In return these protozoans keep some of the nutrients for themselves.

Sponges, phylum Porifera, do not have any or-

Nematodes Are Everywhere

Members of the phylum Nematoda are typically one to two millimeters in length and tapered at both anterior and posterior ends. It is difficult to tell the head from the tail, externally. Except for the mouth and the genital pore, there are no obvious body openings to the outside. Nematodes do not have eyes, ears, or noses. However, they respond to temperature changes, light, mechanical stimulation, and chemical cues because they do possess special organs located in tiny pits that are sensitive to chemicals, and small papillae that are sensitive to touch. They breathe through their cuticle, skin, which is permeable to water and gases. With such a covering, desiccation (drying out) is a constant threat to their well-being. Thus, nematodes live in moist environments. There are some that are fully aquatic; those that are terrestrial surround themselves in a film of water; and those that are parasitic take advantage of the high moisture levels within the body of the host.

As far as numbers of individuals are concerned, nematodes are regarded as the most abundant multicellular animals alive today. There can be literally millions of these small worms per square meter of soil or shallow water sediment. Free-living nematodes are important for decomposition and nutrient cycling. Of the 40 percent of species that are parasitic, much of the toll of the resulting disease in the host is due to the concentration of individual worms competing with the host for nutrients and blocking vessels or entirely filling the heart. One ounce of undercooked pork can contain one hundred thousand worms, and each female can produce thousands of young in just a few days. Nematodes are prevalent in research labs as well. *Caenorhabidits elegans*, a soil nematode, is a model organism that has been studied to unlock many of the secrets of developmental biology and genetics of humans.

gans, but they do possess specialized cell types. Choanocytes, also termed collar cells, line the inside of the sponge and capture small food particles present in the circulating water. The phylum Cnidaria is well known for quite another specialty. Organisms in this phylum, such as jellyfish, have specialized stinging cells called nematoblasts. One of the functions of cnidae within the nematoblasts is secretion of toxins used to paralyze and kill prey. Cnidarians are patient hunters, lying in wait until their next meal contacts a tentacle or two, triggering the toxic sting.

Tapeworms (phylum Platyhelminthes, class Cestoda) have no mouth or digestive tract. They are highly adapted to a parasitic way of life. Swimming in nutrients that their host, usually a vertebrate, is in the process of digesting, tapeworms absorb nutrients through their outer surface, and return nothing besides waste products to their host.

Many rotifers (phylum Rotifera) are omnivores, meaning they eat anything that will fit into their mouth. After being ingested, prey passing into the muscular pharynx encounters grinding, crushing jaws called trophi. Rotifers have a one-way digestive tract. Wastes pass out the anus rather than being expelled through the mouth, unlike many of the animals discussed thus far.

There is a wide range of feeding styles among the mollusks (phylum Mollusca). Some feed on plants; others feed on animals, and others feed on particles suspended in the water or mud that they themselves inhabit. There are also mollusks that are ectoparasites, living on the exterior of their host rather than inside it. An interesting molluscan feeding structure is the radula. The teeth on this tonguelike organ are replaced as they wear down or break.

The phylum Arthropoda is divided into two subphyla based on mouthpart structure. The Chelicerates, including horseshoe crabs, and arachnids (spiders, mites, ticks, and scorpions) have fanglike oral appendages used to grab and shred food items. Conversely, insects and crustaceans generally have mandibles, a pair of jaws for crushing food. This basic plan can be highly modified. The specialized coiled straw proboscis of a butterfly is such an example.

Starfish (phylum Echinodermata) prey on large invertebrates and small fish. These echinoderms have two stomachs and can protrude the cardiac stomach through the mouth and begin to digest prey externally. This is an especially useful maneuver when feeding on particularly large prey or when eating a clam through the small opening between its shells.

Respiration and Circulation

As single cells, protozoans have high body surface to volume ratio. They inhabit wet or at least moist environments and are able to take care of gas exchange through simple diffusion. Even though flatworms, phylum Platyhelminthes, are multicellular, they also posses a large external surface area relative to their internal volume, so they, too, rely on simple diffusion across their body surface area for gas exchange. A highly branched gastrovascular cavity allows most cells to be in contact with the digestive system, which means that nutrients do not have to circulate to remote parts of the body. Metabolic wastes generally diffuse out across the body surface.

Sponges rely on their body cavity (the spongocoel) for circulation and gas exchange. Seawater with dissolved oxygen is pulled into the spongocoel through pores in the body wall called ostia. Flagella in specialized collar cells lining the spongocoel set up an internal current to provide circulation. Water and waste products then pass out of a larger opening, called an osculum.

Many invertebrates are radially symmetrical marine organisms, such as these sea anemones. (PhotoDisc)

Cnidarians also lack specialized respiratory structures. They rely on epidermal and gastrodermal surfaces for gas exchange. The gastrodermis lines the gastrovascular, main body cavity. As its name suggests, the gastrovascular cavity functions in both gastric capacity of digestion and the vascular role of circulation.

Most aquatic mollusks have comblike gills that function in gas exchange and also in filter feeding in some. Both aquatic and terrestrial snails in the subclass Pulmonata have highly vascularized structures that function as modified lungs. Mollusks typically posses a heart or similar pump that circulates fluid through an open circulatory system. In an open circulatory system the conduits, or vessels, are limited, and most organs are bathed directly in the circulatory fluid. Cephalopods, such as squid and octopuses, are mollusks with closed circulatory systems. In a closed circulatory system, blood is contained in vessels. The squid actually has three hearts. The systemic heart receives oxygenated blood from the gills and sends it to the tissues. The two branchial hearts pump deoxygenated blood back to the gills.

Arthropods have open circulatory systems. The circulatory fluid, called hemo-

The Coral Reef Community

Coral reefs are colonies of small invertebrate animals (phylum Cnidaria, class Anthozoa). Some species of coral are solitary; the colonial reef builders are called hermatypic. The reef grows as new members of the colony take up residence on top of the calcium carbonate skeletons of previous generations. In this way, layer by layer, reefs can attain huge dimensions. Australia's Great Barrier Reef is over 2,000 kilometers long and 145 kilometers wide. Coral reefs are very diverse and fragile ecosystems. Reefs consisting of hundreds of species of corals, not to mention the countless species of other organisms, occur only in clear, shallow, tropical marine environments. Corals eat plankton but also rely on photosynthetic symbiotic algae for nutrient provision. Corals provide shelter and feeding grounds for many organisms: sponges, mollusks, echinoderms, tube dwelling worms, and fish, just to name a few reef inhabitants and visitors. Reefs also affect life on land by absorbing wave force, thus mitigating beach erosion.

There are four major types of reef formations. Fringing reefs border the shoreline. A lagoon separates a barrier reef from land. Atolls form on top of submerged volcanoes and platform reefs may form near atolls or be part of a larger barrier reef system. Stresses that threaten reef communities include wide fluctuations in water temperature, sedimentation, coral mining, blast fishing, and souvenir hunting. Natural disturbances, such as storms, can also disrupt life in the reef but may help propagate the reef. The breaking of coral branches during a storm is one way that coral asexually reproduces.

lymph, enters the heart through holes, called ostia. It is then pumped through short arteries and into the body cavity. There are a number of adaptations for gas exchange among the arthropods. Spiders have book lungs consisting of multiple stacked plates, like pages in a book. Insects have trachea, tubes that connect to the outside through holes (spiracles) in the exoskeleton. The trachea transport air directly between body tissues and the environment. Rounding out the arthropods, aquatic crustaceans respire using filamentous gills.

The water vascular system (WVS) is unique to the phylum Echinodermata. This system of canals services thousands of tube feet, in the starfish, for example. These feet extend through the body wall and have many other functions in addition to locomotion. Due to circulating fluid within the WVS, tube feet also are involved in gas exchange, waste excretion, chemoreception, and food collection. Another echinoderm, the sea cucumber, possesses internal rather than external respiratory structures. These respiratory trees attach to the cloaca, the common collection area for the exit of digestive and reproductive systems from the body. Gases are exchanged via the pumping of seawater in and out of the cloaca.

Reproduction and Development

Reproduction among the invertebrates is almost as varied as the animals themselves. Cnidarians and sponges are just two of the many phyla that reproduce asexually, offspring arising by breaking off from the parent organism. Groups such as rotifers and some arthropods alternate between asexual and sexual reproduction, depending upon environmental conditions. The term parthenogenesis is used to describe development of an egg in the absence of fertilization. Parthenogenesis tends to occur during stable, favorable conditions. Sexual reproduction, the mixing of genes from two parents through uniting of egg and sperm, produces individuals with new combinations of genes, which may adapt them for survival under stressful conditions.

There are variations in sexual reproduction as well. Some invertebrates are hermaphrodites. These animals, such as earthworms, possess both male and female reproductive systems. Some can fertilize themselves; others cannot. When hermaphrodites mate with other members of their species, each individual donates and receives sperm, resulting in twice as many offspring per mating. Gonochoristic species have separate sexes, an individual being either male or female. This is the case for most insects.

In marine invertebrates, fertilization may be external. This involves broadcasting sperm and eggs into the surrounding seawater and relying on the ocean currents to bring the two together. Internal fertilization is the rule for freshwater and terrestrial invertebrates. Sperm may be transferred directly into the female's reproductive tract through a copulatory organ such as a penis. An indirect method of sperm transfer involves a package called a spermatophore. This chemical packet commonly provides nutrients for the female and her resultant offspring.

Aquatic larvae are commonly the dispersal stage in the life history of the particular invertebrate. Many times the adults are sedentary or even sessile, anchoring themselves to the ocean floor or remaining in self-constructed burrows. The larvae, on the other hand, are planktonic, relying on the ocean currents for transportation. A similar scenario exists with spiders. The young spin silken parachutes and use the wind currents to disperse away from the web of the mother spider. However, for many terrestrial invertebrates, the young represent the feeding stage. Lepidopteran caterpillars eat voraciously, and some adults, such as the luna moth, do not eat at all, living strictly off food stores acquired as a larva. Especially with the winged insects, it is the adults who colonize new areas.

—Sarah Vordtriede

See also: Ants; Arthropods; Bees; Beetles; Butterflies and moths; Circulatory systems of invertebrates; Cockroaches; Cold-blooded animals; Coral; Crustaceans; Echinoderms; Endocrine systems of invertebrates; Exoskeletons; Flatworms; Flies; Grasshoppers; Insects; Jellyfish; Mollusks; Mosquitoes; Muscles in invertebrates; Octopuses and squid; Praying mantis; Protozoa; Roundworms; Snails; Starfish; Tentacles; Termites; Vertebrates; Wasps and hornets; Worms, segmented; Zooplankton.

Bibliography

Barnes, Robert D. *Invertebrate Zoology*. 5th ed. Philadelphia: Saunders College Publishing, 1987. This detailed, morphologically oriented textbook is a classic. There are excellent drawings and black and white photographs.

Campbell, Neil A., Jane B. Reece, and Lawrence G. Mitchell. *Biology*. 5th ed. Menlo Park, Calif.: Benjamin/Cummings, 1999. This is a general biology text used in many universities and some high schools. The chapter on invertebrates is an excellent survey of the group. Clear text and an abundance of color figures and photographs help to communicate important concepts.

Knutson, Roger M. *Fearsome Fauna: A Field Guide to the Creatures That Live in You*. New York: W. H. Freeman, 1999. This little volume and its companion, *Furtive Fauna: A Field Guide to the Creatures Who Live on You* (1996), are humorous but factual depictions of the invertebrate parasites of humans.

Pechenik, Jan A. *Biology of the Invertebrates*. 4th ed. Boston: McGraw-Hill, 2000. This is a very accessible book for studying the diversity of the invertebrates. Highlights include research focus boxes within the chapters, and questions for discussion at the end of each chapter, complete with pertinent, introductory-level references from major journals.

Robertson, Matthew, ed. *The Big Book of Bugs*. New York: Welcome Enterprises, 1999. Complete with 3D glasses, this book is fun for everyone. Contrary to what the title suggests, the book deals with a wide range of invertebrates. There are fact boxes, activities, quizzes, and many pictures. It lacks a table of contents, which makes it difficult to look up specific items, but it is fun to flip through the pages.

ISOLATING MECHANISMS IN EVOLUTION

Type of animal science: Evolution
Fields of study: Ecology, evolutionary science, genetics, population biology, reproduction science, zoology

Isolating mechanisms act to prevent interbreeding and the exchange of genes between species. The establishment of isolating mechanisms between populations is a critical step in the formation of new species.

Principal Terms

ALLOZYME: one of two or more forms of an enzyme determined by different alleles of the same gene; usually analyzed by gel electrophoresis

CHLOROPLAST DEOXYRIBONUCLEIC ACID: a circular DNA molecule found in chloroplasts; chloroplasts are cytoplasmic organelles of green plants and some protists that carry out photosynthesis

FISSION: the division of an organism into two or more essentially identical organisms; an asexual process

HERMAPHRODITE: an individual with both male and female organs; functions as both male and female

HYBRID: a term that in the broad sense can apply to any offspring produced by parents that differ in one or more inheritable characteristics; used to denote offspring produced by a cross between different species

MITOCHONDRIAL DEOXYRIBONUCLEIC ACID (DNA): a circular molecule of DNA found in the mitochondria; mitochondria are cytoplasmic organelles that function in oxidative respiration

NUCLEAR RIBOSOMAL DEOXYRIBONUCLEIC ACID (DNA): nuclear DNA that codes for the ribosomal DNAs; ribosomes are small cytoplasmic particles that function in protein synthesis

Isolating mechanisms (reproductive isolating mechanisms) prevent interbreeding between species. The term, which was first used by Theodosius Dobzhansky in 1937 in his landmark book *Genetics and the Origin of Species*, refers to mechanisms that are genetically influenced and intrinsic. Geographic isolation can prevent interbreeding between populations, but it is an extrinsic factor and therefore does not qualify as an isolating mechanism. Isolating mechanisms function only between sexually reproducing species. They have no applicability to forms that reproduce only by asexual means, as by mitotic fission, stoloniferous or vegetative reproduction, or egg development without fertilization (parthenogenesis in animals). Obligatory self-fertilization in hermaphrodites (rare in animals) is a distortion of the sexual process that produces essentially the same results as asexual reproduction. Many lower animals and protists regularly employ both asexual and sexual means of reproduction, and the significance of isolating mechanisms in such forms is essentially the same as in normal sexual species.

Premating Mechanisms

Reproductive isolating mechanisms are usually classified into two main groups. Premating (prezygotic) mechanisms operate prior to mating, or the release of gametes, and, therefore, do not result in a wastage of the reproductive potential of the individual. Postmating (postzygotic) mechanisms come into play after mating, or the release of gametes, and could result in a loss of the genetic contribution of the individual to the next generation.

This distinction is also important in the theoretical sense in that natural selection should favor genes that promote premating isolation; those that do not presumably would be lost more often through mismatings (assuming that hybrids are not produced, or are sterile or inferior), and this could lead to a reinforcement of premating isolation.

Ethological (behavioral) isolation is the most important category of premating isolation in animals. The selection of a mate and the mating process depends upon the response of both partners to various sensory cues, any one of which may be species-specific. Although one kind of sensory stimulus may be emphasized, different cues may come into play at different stages of the pairing process. Visual signals provided by color, pattern, or method of display are often of particular importance in diurnal animals such as birds, many lizards, certain spiders, and fish. Sounds, as in male mating calls, are often important in nocturnal breeders such as crickets or frogs but are also important in birds. Mate discrimination based on chemical signals or odors (pheromones) is of fundamental importance in many different kinds of animals, especially those where visual cues or sound are not emphasized; chemical cues also are often important in aquatic animals with external fertilization. Tactile stimuli (touch) often play an important role in courtship once contact is established between the sexes. Even electrical signals appear to be utilized in some electrogenic fish.

Ecological (habitat) isolation often plays an important role. Different forms may be adapted to different habitats in the same general area and may meet only infrequently at the time of reproduction. One species of deer mouse, for example, may frequent woods, while another is found in old fields; one fish species spawns in riffles, while another spawns in still pools. This type of isolation, although frequent and widespread, is often incomplete as the different forms may come together in transitional habitats. The importance of ecological isolation, however, is attested by the fact that instances in which hybrid swarms are produced between forms that normally remain distinct have often been found to be the result of disruption of the environment, usually by humans. Mechanical isolation is a less-important type of premating isolation, but it can function in some combinations. Two related animal species, for example, may be mismatched because of differences in size, proportions, or structure of genitalia.

Finally, temporal differences often contribute to premating isolation. The commonest type of temporal isolation is seasonal isolation: Species may reproduce at different times of the year. A species of toad in the eastern United States, for example, breeds in the early spring, while a related species breeds in the late spring, with only a short period of overlap. Differences can also involve the time of day, whereby one species may mate at night and another during the day. Such differences, as in the case of ecological isolation, are often incomplete but may be an important component of premating isolation.

Postmating Mechanisms

If premating mechanisms fail, postmating mechanisms can come into play. If gametes are released, there still may be a failure of fertilization (intersterility). Spermatozoa may fail to penetrate the egg, or even with penetration there may be no fusion of the egg and sperm nucleus. Fertilization failure is almost universal between remotely related species (as from different families or above) and occasionally occurs even between closely related forms.

If fertilization does take place, other postmating mechanisms may operate. The hybrid may be inviable (F1 or zygotic inviability). Embryonic development may be abnormal, and the embryo may die at some stage, or the offspring may be defective. In other cases, development may be essentially normal, but the hybrid may be ill-adapted to survive in any available habitat or cannot compete for a mate (hybrid adaptive inferiority). Even if hybrids are produced, they may be partially to totally sterile (hybrid sterility). Hybrids between closely related forms are more likely to be fertile than those between more distantly related species, but the correlation is an inexact one. The causes for hybrid sterility are complex and can involve genetic factors, differences in gene arrangements on the chromosomes that disrupt normal chromosomal pairing and segregation at meiosis, and incompatibilities between cytoplasmic factors and the chromosomes. If the hybrids are fertile and interbreed or backcross to one of the parental forms, a more subtle phenomenon known as hybrid breakdown sometimes occurs. It takes the form of reduced fertility or reduced viability in the offspring. The basis for hybrid breakdown is poorly understood but may result from an imbalance of gene complexes contributed by the two species.

It should be emphasized that in most cases of reproductive isolation that have been carefully studied, more than one kind of isolating mechanism has been found to be present. Even though one type is clearly of paramount importance, it is usually supplemented by others, and should it fail, others may come into play. In this sense, reproductive isolation can be viewed as a fail-safe system. A striking difference in the overall pattern of reproductive isolation between animals and plants, however, is the much greater importance of premating isolation in animals and the emphasis on postmating mechanisms in plants. Ethological isolation, taken together with other premating mechanisms, is highly effective in animals, and postmating factors usually function only as a last resort.

Field Studies and Experimental Studies

Field studies have often been employed in the investigation of some types of premating isolating mechanisms. Differences in such things as breeding times, factors associated with onset of breeding activity, and differences in habitat distribution or selection of a breeding site are all subject to direct field observation. Comparative studies of courtship behavior in the field or laboratory often provide clues as to the types of sensory signals that may be important in the separation of related species.

Mating discrimination experiments carried on in the laboratory have often been employed to provide more precise information on the role played by different odors, colors, or patterns, courtship rituals, or sounds in mate selection. Certain pheromones, for example, which act as sexual attractants, have been shown to be highly species-specific in some insects. The presence or absence of certain colors or their presentation has been shown experimentally to be important in mate discrimination in vertebrates as diverse as fish, lizards, and birds. Call discrimination experiments, in which a receptive female is given a choice between recorded calls of males of her own and another species, have demonstrated the critical importance of mating call differences in reproductive isolation in frogs and toads. Synthetically generated calls have sometimes been used to pinpoint the precise call component responsible for the difference in response.

Studies on postmating isolating mechanisms have most often involved laboratory crosses in which the degree of intersterility, hybrid sterility, or hybrid inviability can be analyzed under controlled conditions. In instances in which artificial crosses are not feasible, natural hybrids sometimes occur and can be tested. The identification of natural backcross products can attest incomplete postmating, as well as premating isolation. Instances of extensive natural hybridization are of special interest and have often been subjected to particularly close scrutiny. Such cases often throw light on factors that can lead to a breakdown of reproductive isolation. Also, as natural hybridization more often occurs between marginally differentiated forms in earlier stages of speciation, new insights into the process of species formation can sometimes be obtained. Finally, such studies may yield information on the evolutionary role of hybridization, including introgressive hybridization, the leakage of genes from one species into another. Morphological analysis has long been used in such cases, and chromosomal studies are sometimes appropriate. In recent years, allozyme analysis by gel electrophoresis has become a routine tool in estimates of gene exchange, and molecular analysis of nuclear deoxyribonucleic acid (DNA), or mitochondrial DNA, have been useful. As mitochondria are normally passed on only maternally, their DNA can also be used to identify cases in which females of only one of the two species has been involved in the breakdown of reproductive isolation.

Investigations of the role of natural selection in the development and reinforcement of reproductive isolation have employed two different approaches. One has involved the measurement of geographic variation in the degree of difference in some signal character (call, color, or pattern, for example) thought to function in premating isolation between two species that have overlapping ranges. If the difference is consistently greater within the zone of overlap (reproductive character displacement), an argument can be made for the operation of reinforcement. Another approach has involved laboratory simulations, usually with the fruit fly *Drosophila*, in which some type of selective pressure is exerted against offspring produced by crosses between different stocks, and measurement is made of the frequency of mismatings through successive generations. The results of such studies to this time are contradictory, and the role of selection with regard to development of reproductive isolation requires further study.

Enhancing Reproductive Efficiency

The efficiency of reproduction in most animals is enhanced immeasurably by premating isolating mechanisms. Clearly, in animals a random testing of potential mates without regard to type is totally unacceptable for most species in terms of reproductive capacity and time and energy resources. Premating isolation in this sense is a major factor in promoting species diversity in animal communities.

Both premating and postmating isolating mechanisms are also critical to the maintenance of species diversity in that they act to protect the genetic integrity of each form: A species cannot maintain its identity without barriers that prevent the free exchange of genes with other species. Furthermore, a species functions as the primary unit of adaptation. Every species in a community has its own unique combination of adaptive features that enable it to exploit the resources of its environment and to coexist with other species with a minimum of competition. The diversity of different species that can coexist in the same area depends upon the unique "niche" that each occupies; adaptive features that determine that niche are based on the unique genetic constitution of each species, and this genetic constitution is protected through reproductive isolation.

The development of reproductive isolating mechanisms is also critical to the formation of new species (speciation), and ultimately to the development of new organic diversity. The most widely accepted, objective, and theoretically operational concept for a sexual species is the biological species concept. Such a species can be defined as population or group of populations, members of which are potentially capable of interbreeding but which are reproductively isolated from other species. The origin of new species, therefore, depends upon the development of reproductive isolating mechanisms between populations. A major focus of research in evolutionary biology and systematics has been, and continues to be, on the various factors that influence the development of reproductive isolating mechanisms.

—John S. Mecham

See also: Adaptive radiation; Biodiversity; Clines, hybrid zones, and introgression; Convergent and divergent evolution; Ecological niches; Ecosystems; Evolution: Historical perspective; Extinctions and evolutionary explosions; Fauna: Australia; Fauna: Galápagos Islands; Gene flow; Genetics; Hardy-Weinberg law of genetic equilibrium; Mutations; Natural selection; Population analysis; Population genetics; Punctuated equilibrium and continuous evolution; Systematics.

Bibliography

Baker, Jeffrey J. W., and Garland E. Allen. *The Study of Biology*. 4th ed. Reading, Mass.: Addison-Wesley, 1982. Some of the more complete college-level biology texts provide a good overview of evolution and related topics for the beginner. Coverage in this text is particularly complete, including treatment of isolating mechanisms and speciation. A list of suggested readings should be helpful.

Dobzhansky, Theodosius. *Genetics of the Evolutionary Process*. New York: Columbia University Press, 1970. Dobzhansky's contributions over the years have had a major impact on genetics and evolutionary biology. The present book represents a revision and expansion of his *Genetics and the Origin of Species*, which first appeared in 1937. The book is directed at the advanced reader and presupposes a foundation in genetics. Much of chapter 10, however, which is devoted to an excellent review of reproductive isolation, can be understood by the more general reader. An extensive technical bibliography is provided.

Dobzhansky, Theodosius, Francisco J. Ayala, G. Ledyard Stebbins, and James W. Valentine. *Evolution*. San Francisco: W. H. Freeman, 1977. An excellent, if somewhat uneven textbook by four distinguished scientists. Although intended for advanced undergraduates and graduate students, the material is presented in such a way as to benefit those with a less complete background. Coverage of molecular techniques is particularly good but somewhat dated. Twelve pages are devoted to isolating mechanisms, and many references are given under literature cited.

Futuyma, Douglas J. *Evolutionary Biology*. 3d ed. Sunderland, Mass.: Sinauer Associates, 1998. This process-oriented textbook assumes some undergraduate training in biology, especially genetics, but most topics are prefaced with background information that should help the less prepared. Treatment of isolating mechanisms is brief, but the book provides one of the best introductions to current evolution theory presently available. Major references are given at the end of each chapter, and there is an extensive list of references under literature cited.

Mayr, Ernst. *Populations, Species, and Evolution*. Cambridge, Mass.: The Belknap Press of Harvard University Press, 1970. This text is essentially an abridgment of Mayr's *Animal Species and Evolution* (1963). Ernst Mayr has been a prime mover and shaker in the development of evolution theory over the past fifty years, and anyone with a serious interest in the subject should be exposed to some of his writings. This book, although directed toward the more advanced student, is lucidly written and can benefit those with a more limited background. Chapter 5 deals with isolating mechanisms, particularly in animals. A technical bibliography is provided.

JAGUARS

Type of animal science: Classification
Fields of study: Anatomy, physiology, wildlife ecology

Jaguars are the largest cats found in North or South America, originally occupying a wide diversity of habitats from the southwestern United States to Patagonia.

> **Principal Term**
>
> VIBRISSAE: stiff hairs, projecting as feelers from nose and head

Among big cats, jaguars, much larger than leopards, are exceeded in size only by lions and tigers. Males weigh from 125 to 250 pounds, are 6 to 9 feet long (including a tail up to 2.5 feet long), and stand twenty-four to thirty inches tall at the shoulder; females tend to be 20 percent smaller. Jaguar heads are massive and rounded; their bodies compact and heavily muscled. Individuals living in densely forested areas of the Amazon basin are significantly smaller than those inhabiting open terrain.

Tawny or yellow, with black rings and spots, jaguar coats resemble those of the leopard; however, jaguar coat rosettes are larger and usually contain black spots in their centers. Examples of melanism occur in Amazon regions, where jaguars are often called black panthers.

Behavior

Jaguar litters usually contain one to four cubs, which remain with their mother for eighteen months to two years while learning how to hunt. Other than during mating periods, adults live solitary lives, patrolling their own distinctly marked territories. Jaguar hunting ranges vary in size from five square miles, where prey is abundant, to two hundred square miles, where it is scarce. Male territories usually overlap the smaller ranges of several females.

Jaguars are crepuscular hunters, preferring dim light in which to stalk and surprise victims by leaping on their backs. The name jaguar comes from the Guarani word *yaguara*, meaning "wild beast that can kill its prey in a single bound." Large eyes and sensitive vibrissae permit jaguars to maneuver in the dark. They are opportunistic hunters, taking armadillos, peccaries, deer, capybaras, anteaters, caimans, turtles, and fish. Jaguars possess the most powerful bite among big cats; large canine teeth easily crush skulls and

> **Jaguar Facts**
>
> **Classification:**
> *Kingdom:* Animalia
> *Subkingdom:* Bilateria
> *Phylum:* Chordata
> *Subphylum:* Vertebrata
> *Class:* Mammalia
> *Subclass:* Eutheria
> *Order:* Carnivora
> *Family:* Felidae (cats)
> *Genus and species: Panthera onca*
> **Geographical location:** Originally ranged from the southwestern United States to southern Argentina
> **Habitat:** Forests, jungles, and grassy plains
> **Gestational period:** About fourteen weeks
> **Life span:** Ten to twelve years in the wild, over twenty years in captivity
> **Special anatomy:** Large eyes with excellent night vision; jaws adapted to seizing and gripping prey, teeth designed for tearing and slicing flesh

The jaguar's spots, or rosettes, help it blend into the mottled shadows of the forest. (Kenneth Layman/Photo Agora)

penetrate armadillo armor or turtle shells. Sharp carnassial teeth and rasplike papillae soon clean their victims' bones.

Relations with Humans

Pre-Columbian Indian societies called the jaguar "Master of Animals," associating it with success in hunting and warfare, invoking it in religious rituals, and assigning it high social status. Jaguar thrones and jaguar skins were power symbols for rulers; hunters and warriors wore necklaces and bracelets of jaguar teeth or claws. Preferred foods of the elite were the meats that jaguars ate: venison, peccary, capybara, and armadillo.

European settlers viewed jaguars as dangerous competitors to be hunted and killed. When Europeans arrived, sixteen subspecies of jaguars inhabited a continuous area stretching from the southwestern United States to Patagonia in Argentina. Before the end of the twentieth century, all subspecies of jaguar were endangered, their territories reduced to a series of disconnected areas lying between southern Mexico and northeast Argentina. In North and Central America, the jaguar lost 67 percent of its range, in South America about 38 percent. Most of the estimated fifteen thousand jaguars remaining exist in a few relatively undisturbed jungle regions of Central America and the Amazon basin.

Hunting, destruction of habitat, and competition with ranchers and farmers all threaten the survival of the jaguar. In 1968, the United States imported 13,516 jaguar skins. The number of cats slain declined after the 1975 Convention on International Trade in Endangered Species banned traffic in jaguar pelts. However, illegal trade continues; it is profitable because the beautiful skins are as greatly prized by today's high-status women as they were by Inca monarchs.

Jaguars tend to avoid open areas such as pastures and villages, and rarely cross into fenced fields. However, when ranchers and farmers permit their animals to wander into jaguar hunting territory, they provide an easy meal. Such depredations, along with rare attacks on humans, create demands for the extirpation of the offenders.

Continued destruction of habitat, as forests and jungles are leveled for timber or for farm and ranch land, is the greatest threat to jaguar survival. As remaining territory becomes ever more discontinuous, populations becomes less dense, and reproductive success becomes problematic. Whether jaguars can survive in wild, free-ranging populations, or will be found only in zoos and carefully protected national parks, remains an unanswered question.

—*Milton Berman*

See also: Carnivores; Cats; Cheetahs; Fauna: Central America; Fauna: South America; Leopards; Lions; Mountain lions; Predation; Tigers.

Bibliography

Brakefield, Tom, ed. *Kingdom of Might: The World's Big Cats*. Stillwater, Minn.: Voyageur Press, 1993. Documents habits, natural history, biology, and challenges that recent changes in population and range pose for the jaguar. Extensive bibliography.

Kitchener, Andrew. *The Natural History of the Wild Cats*. Ithaca, N.Y.: Comstock, 1991. Summarizes recent biological and ecological research concerning jaguar habits and social life. Twenty-three page bibliography of scientific articles.

Lumpkin, Susan. *Big Cats*. New York: Facts on File, 1993. Written for younger readers, the work describes jaguar's physical characteristics and behavior.

Rabinowitz, Alan. *Jaguar: Struggle and Triumph in the Jungles of Belize*. New York: Arbor House, 1986. First-person account of two years spent studying jaguars in their native habitat. Rabinowitz established the world's first jaguar preserve.

Saunders, Nicholas J., ed. *Icons of Power: Feline Symbolism in the Americas*. New York: Routledge, 1998. Explores ritual and symbolic use of jaguars by Aztec, Maya, and Inca Indians.

JELLYFISH

Type of animal science: Classification
Fields of study: Anatomy, invertebrate biology, marine biology, physiology

Over two hundred species of jellyfish are known; they are abundant, especially in shallow, warm, or subtropical waters. The jellyfish body plan is simple, but they possess unique stinging structures that are used for foraging and defense.

Principal Terms

CILIATED: bearing short, hairlike organelles on the surface of cells, used for motility

CNIDOBLASTS: specialized cells on the body or tentacles of jellyfish that contain nematocysts

MEDUSA: adult umbrella- or bell-shaped forms of jellyfish, with mouth facing downward

MESOGLEA: gelatinous material lying between the inner and outer layers of a jellyfish

NEMATOCYST: stinging structures containing barbs and/or poison

PLANULA: free-swimming, ciliated jellyfish larva

POLYP: immature cylindrical forms of jellyfish with mouth facing upward

The gelatinous jellyfish are widespread in marine environments, although they are most common in tropical and subtropical regions. These ancient animals first appeared on earth over 650 million years ago. The smallest jellyfish are difficult to see without a microscope, while the largest known jellyfish is 2.5 meters in diameter; some jellyfish may have tentacles over 100 feet in length. The body plan of jellyfish is relatively primitive and contains less than 5 percent solid organic matter, the remaining bulk coming from water. They completely lack internal organs. The bell-like jellyfish bodies are composed of an outer layer of epidermis and an inner layer of gastrodermis that lines the gut. The gut has a single oral opening. Between the two layers is the mesoglea, which contains few cells and has a low metabolic rate. Four to eight oral arms are located near the oral opening and are used to transport food that has been captured by the tentacles.

Jellyfish are able to exert minimal control over their movement, being largely at the mercy of ocean currents. Jellyfish do, however, have some regulation over vertical movement. They possess a ring of muscles embedded on the underside of the bell that pulses rhythmically, pushing water out of the hollow bell. Using this jet propulsion, jellyfish can change their position in the water column, moving in response to light and prey.

Feeding Strategy of Jellyfish

Jellyfish are simple but specialized carnivores. Although jellyfish have a low metabolic rate, they have the ability to capture large prey. These two characteristics allow jellyfish to survive in environments where prey are scarce. Jellyfish are equipped with a specialized apparatus, the cnidoblast, for defense and feeding. Cnidoblasts are found by the hundreds or thousands on the tentacles and sometimes on the body surface. Within each cnidoblast is a coiled harpoonlike nematocyst that is discharged by the presence of potential prey. The nematocyst injects poison into the prey as spines on the nematocyst anchor it to the prey. The trapped, paralyzed prey is pulled back by the tentacles and stuffed into the gastrovascular cavity to be digested. Jellyfish do not attack humans, but humans may receive stings if they encounter jellyfish. The effects of jellyfish poison on humans can range from a mild, itchy rash to death.

Jellyfish Facts

Classification:
Kingdom: Animalia
Subkingdom: Eumetazoa
Phylum: Cnidaria
Classes: Scyphozoa (jellyfish), Cubozoa (box jellyfish)
Orders: Stauromedusae (sessile cup-shaped forms), Coronatae (deep-dwelling jellyfish), Semaeostomae (disc jellyfish), Rhizostomae (tentacle-less jellyfish), Chirodropidae and Carybdeidae (box jellyfish)
Geographical location: All the Earth's oceans
Habitat: Almost entirely marine, although a few freshwater species are known
Gestational period: Not well studied, but many species appear to reproduce once a year
Life span: Polyps usually develop over a period lasting a few months but may live for several years producing clones; adult medusa forms live two to six months
Special anatomy: Umbrella-like body; no head or skeleton; composed of outer epithelial layer and inner gastrodermis layer with thick elastic jellylike substance between them; gastrovascular cavity; specialized ring of epitheliomuscular cells that pulses rhythmically to propel the animal through the water; four to eight oral arms; tentacles bearing cnidoblasts containing nematocysts

Jellyfish Reproduction

Most jellyfish proceed through several distinct stages in their life cycles. Male medusae produce sperm that are released from the oral opening into the oral opening of the female. The female then releases fertilized eggs which develop into slipper-shaped, solid masses of ciliated cells called planula which move through the water and even-

The tentacles of jellyfish contain cnidoblasts, organs that shoot harpoonlike, venomous nematocysts into prey; the paralyzed victim is then reeled back to the jellyfish and consumed. (PhotoDisc)

tually settle onto a solid surface. From these settled planula develop polyps that have cylindrical stalks attached to the substrate, with tentacles surrounding their mouths. At this stage, the polyps resemble sea anemones. The polyps divide and bud into tiny young jellyfish (ephyra) which are often carried far from the parent polyp by ocean currents. The ephyra develop into mature medusae over several weeks. The medusae normally live three to six months.

—*Lisa M. Sardinia*

See also: Carnivores; Deep-sea animals; Digestion; Ingestion; Invertebrates; Marine animals; Marine biology; Reproduction.

Bibliography

Brusca, Richard C. C., and Gary J. Brusca. *Invertebrates*. Sunderland, Mass.: Sinauer Associates, 1997. Introductory textbook that includes functional body architecture, developmental patterns and life history strategies, and evolution and phylogenetic relationships

Buchsbaum, Ralph Morris, Vicki Pearse, John Pearse, and Mildred Buchsbaum. *Animals Without Backbones*. Chicago: University of Chicago Press, 1987. An introductory textbook for advanced high school students, college students, and the motivated general reader. Extensive section on jellyfish.

Landau, Elaine. *Jellyfish*. Dublin, Ireland: Children's Press, 1999. Written for young adolescents, this book contains beautiful photographs and information on physiology, morphology, and life history of jellyfish.

Stefoff, Rebecca. *Jellyfish*. Boston: Benchmark Books, 1997. This book is for very young readers and integrates color photographs into text on habitats, life cycles, and unusual features and behaviors of jellyfish.

Twig, George C. *Jellies: The Life of Jellyfish*. Brookfield, Conn.: Millbrook Press, 2000. Aimed at a young audience, this book describes the basic processes of life in jellyfish, accompanied by numerous color photographs.

KANGAROOS

Type of animal science: Classification
Fields of study: Anatomy, conservation biology, ethology, physiology, reproduction science, wildlife ecology

Kangaroos, comprising six species, are the largest marsupials. They evolved as the major grazing animals in Australia. Alone among large vertebrates, they hop as their preferred form of locomotion.

Principal Terms

DIAPAUSE: an interruption in embryonic development
HERBIVORE: animal solely dependent on plant material for its nutrition
QUADRUPED: animal with four feet

The ancestors of kangaroos differentiated from small, tree-dwelling, possumlike marsupials fifty million years before the present. About thirty million years ago, they came down from the trees of the rain forest that covered most of Australia at the time. Around five million years ago, the continent had started to dry out, and species closely related to modern kangaroos appeared. When the Aborigines first came to Australia forty thousand years ago, the continent included some larger marsupials. The latter apparently were not as fast and mobile as kangaroos, were easier prey, were overhunted, and as a result became extinct. The kangaroos were smaller and swifter, permitting them to survive. The six living species of true, large kangaroos differ in their habitats (temperate woodlands, tropical grasslands, arid outback) and size; the red kangaroo is the largest, with some males weighing two hundred pounds and standing six feet tall; the smallest, the black wallaroo, can weigh sixty pounds. Smaller relatives of kangaroos include wallabies (up to sixty pounds), pademelons (up to fifteen pounds), and rat kangaroos (less than one pound). While kangaroos are only found in Australia, some smaller relatives are also found in New Guinea, which was contiguous with Australia in the distant past.

Physical Characteristics of Kangaroos

Kangaroos stand on large rear legs, using their long tail for added support. They have small front legs, with handlike paws that lack an opposable thumb. Hopping is their most unusual characteristic. Besides kangaroos and their relatives, no vertebrate bigger than ten pounds hops. At slow speeds, kangaroos walk awkwardly and inefficiently, using their front legs and tail. However, at speeds over fifteen miles per hour, they hop upright in a graceful motion that can be more energetically efficient than running by quadrupeds, whose energy use is proportional to their speed. Kangaroos increase their speed by lengthening their stride, while keeping their hop frequency constant, at little increased energy expenditure. They propel themselves by virtue of highly elastic legs which move in unison and use their long tails to provide balance.

As marsupials, kangaroos nurse their young (called joeys) in a pouch. Female kangaroos, half the size of males, have one-month gestations, which can be interrupted if a young is still suckling in the pouch or under adverse nutritional conditions. In these cases, the embryo goes into diapause, a form of "suspended animation," until hormonal signals permit development to resume. The newborn is highly immature, pink and naked, resembles a slug, and weighs less than 0.03 ounces. Using its front legs and a good sense of smell, it crawls from the birth canal into the pouch

The kangaroo's strong tail assists it in maintaining an upright, bipedal posture. It is the only mammal with a weight over ten pounds that hops as its mode of locomotion. (Adobe)

and attaches itself to one of four teats of the mammary gland. Over the next three months, it remains permanently attached to that teat and becomes fully developed. Depending on the species, joeys leave the pouch for the first time at six to ten months, permanently leave the pouch at eight to eleven months, and are weaned at eleven to eighteen months. Females are sexually mature at eighteen months to two years, although some males do not become so until they are four years old.

Kangaroos are herbivores, and all six species are grazing animals. Their teeth are suited to grasses rather than shrubs and trees. They are also very efficient in their use of water, making them suitable for the arid regions of Australia. Some species consume less than 10 percent the water sheep do under the same conditions. When temperatures are moderate, they can get all of their water from the plants that they eat. They are inactive in the heat of the day and cool themselves by panting, sweating, and licking; the latter refers to the fact that they cover their front legs with saliva, which by evaporation cools not only their extremities but also their bodies via a dense network of blood vessels close to the surface. Kangaroos are among the most heat-tolerant of mammals. In addition, they have large, padded feet that compact the soil less than domesticated livestock.

Future of Kangaroos

Totaling over twenty million, the six species of large kangaroos are not presently endangered. This is in contrast with the risk to survival faced by some wallabies and smaller marsupials that are preyed upon by introduced wild animals, such as foxes, or by feral cats. The habitats of most kangaroos have been reduced by human activities, housing, industry, and agriculture, although the range of some of the less arid-tolerant species (eastern gray and western gray) has been increased when water is provided for livestock in remote regions. In addition, areas where sheep are protected from dingoes, as with the patrolled fence that stretches across Australia, have increased numbers of kangaroos. While they do compete with sheep and cattle for food and water, the extent of competition is limited, except during drought. Farming kangaroos for meat and hides remains a possibility but has not been developed. Some extensive preserves would be desirable not only to display these large animals in their natural environment but also to conserve their smaller, endangered relatives.

—*James L. Robinson*

See also: Embryology; Fauna: Australia; Fertilization; Koalas; Marsupials; Opossums; Pregnancy and prenatal development; Reproduction; Reproductive strategies; Reproductive system of female mammals.

Kangaroo Facts

Classification:
Kingdom: Animalia
Subkingdom: Bilateria
Phylum: Chordata
Subphylum: Vertebrata
Class: Mammalia
Order: Marsupialia
Family: Macropodidae (kangaroos and wallabies)
Genus and species: Macropus, with fourteen species including *M. giganteus* (eastern gray kangaroo), *M. fuliginosus* (western gray kangaroo), *M. rufus* (red kangaroo), *M. robustus* (wallaroo, four subspecies), *M. bernardus* (black wallaroo), *M. antilopinus* (antilopine kangaroo)
Geographical location: Australia
Habitat: Grasslands
Gestational period: One month, followed by six to eleven months in marsupium
Life span: Twelve to eighteen years in the wild; twenty-eight years in captivity
Special anatomy: Hops instead of running, by virtue of elastic legs and large tail for balance; gives birth to highly immature young that nurse and develop in a pouch

Bibliography

Burt, Denise. *Kangaroos*. Minneapolis, Minn.: Carolrhoda Books, 2000. Written for adolescents, this accurate and well-photographed book discusses the life cycle and future of kangaroos, wallabies, and other related marsupials.

Dawson, Terence J. *Kangaroos: Biology of the Largest Marsupials*. Ithaca, N.Y.: Comstock Publishing, 1995. Written by an internationally recognized expert on kangaroos, this book details the natural history of all six species, including their habitat, nutrition, and reproduction. It also deals with conservation efforts and the potential of managing them as ranched animals.

McCullough, Dale, and Yvette McCullough. *Kangaroos of Outback Australia*. New York: Columbia University Press, 2000. This book documents a fifteen-month study of the ecology and behavior of the three largest kangaroos, the red, eastern gray, and western gray on a reserve where their ranges overlap.

Watts, Dave. *Kangaroos and Wallabies of Australia*. Sydney: New Holland, 1999. Written and photographed by one of Australia's leading wildlife photographers, this book describes and pictures kangaroos and their relatives, including wallabies, pandemelons, and rat kangaroos.

KIDNEYS AND OTHER EXCRETORY STRUCTURES

Type of animal science: Physiology
Fields of study: Anatomy, biochemistry, cell biology

All cells must live in a relatively constant environment. The kidney and other excretory organs play important roles in the regulation of the processes that maintain this necessary homeostasis. Although excretion is often assumed to be the only role of the kidney, it also regulates blood pressure, produces hormones, and affects the body's pH.

Principal Terms

ANTIDIURETIC HORMONE (ADH): a hormone produced in the hypothalamus that controls reabsorption of water in the loop of Henle

CONTRACTILE VACUOLE: the excretory organ of several one-celled organisms

FILTRATION: the process of diffusion of plasma from the blood to the glomerulus and nephron

GLOMERULUS (pl. GLOMERULI): a capsule fitting around capillary blood vessels that receives the filtrate from the blood and passes it into the tubule

LOOP OF HENLE: a slender hairpin turn in the tubule where most adjustment of the water balance of the body occurs

MALPHIGIAN TUBULE: the primitive excretory organ of insects

NEPHRIDIA: the primitive forms of kidneys found in worms and lower organisms

NEPHRON: the basic excretory unit of the kidney

TUBULE: the long, slender part of the nephron that is the location of almost all kidney function

UREA: a substance formed from by-products of protein metabolism and excreted by the kidney

All cells, from the single-celled animals to the highly diversified cells of higher mammals, must maintain a constant internal environment with regard to the kinds and amounts of specific ions, the pH of the protoplasm, the osmotic pressure, the water content of the cells, and the excretion of wastes. The kidney in vertebrates and the varied types of excretory organs found in lower animals perform this function. Although the excretory function is very important, an animal will die more quickly from a disturbance of the composition of the internal environment of the cell than it will from an accumulation of wastes. Fortunately, the kidney performs functions that keep both of these from happening.

Maintaining Fluid Homeostasis

All cells live in a watery environment, which is maintained constant through the processes of homeostasis. Maintaining homeostasis presents different problems for different organisms. For a primate, living in the air, water loss is a constant problem, and water must be conserved. A freshwater fish, on the other hand, takes in large quantities of water that has a lower concentration of salts than its body fluids, and it must excrete the excess. An ocean fish, living in water with a much greater concentration of salts than its body fluids, must obtain water from its environment and still avoid increasing the concentration of salts in the body fluids.

A kidney is able to perform all these functions, although in each of these cases it acts in a different way. The freshwater fish must excrete large quantities of very dilute fluid in order to maintain salt conservation, whereas the saltwater fish must excrete a high concentration of salt in a very concentrated solution. The primate must be able to regulate the output of fluid as a function of water intake. Evolution has adapted the kidney of each animal to its environment.

In addition, animals survive by metabolizing foodstuffs to provide the energy for movement. One of the major metabolic processes is the breakdown of protein to produce energy for the synthesis of other proteins and the rebuilding of body structures. In the breakdown of protein, nitrogen is freed from the organic molecule and must be excreted. One of the resultant nitrogen products is ammonia, which is toxic. The ammonia is converted into a less toxic material, urea (or uric acid, in some animals), before it is excreted by the kidney. In addition, the metabolism of the body usually results in the production of acids, particularly if fat is metabolized. The body functions well only within a narrow range of acidity, so unless the excess acid is removed, serious problems quickly arise. The kidney serves the function of maintaining the pH at a constant value.

The Excretory Organs

The organs of excretion have taken many forms. Simple single-celled organisms such as amoebas are able to form contractile vacuoles, or walled-off spaces within the cell, in which water can be stored and waste products deposited. These vacuoles are periodically transported through the cytoplasm and excreted through the external cell membrane. The size and number of vacuoles are determined by water intake and the organism's need to eliminate water as well as by the accumulation of waste materials.

As animals developed more cell types and the number of cells increased dramatically, the need to provide a constant internal environment around the cells arose. The excretory organs became, of necessity, more complex in nature. In addition to excreting waste products, they developed abilities to retain some ions and excrete others, to retain or excrete water, and to retain or excrete acids or bases to maintain a constant environment.

Many organisms have no obvious means of regulating water and salt balance, and they apparently accomplish this feat through the skin or the gut; others have rudimentary organs of excretion. Many lower animals have nephridia, primitive versions of the kidney that excrete water and wastes and regulate ion concentration. These are simple tubes into which body fluids pass; the fluids are excreted after chemical alteration. In animals such as worms, that have segments, a pair of nephridia may be located in each segment. Some of them open into the body cavity, while others are closed. In some animals, these are well differentiated and are called flame cells. These tubular structures serve to regulate the internal environment of the body. For example, if the sodium concentration in the coelom, or internal cavity, is high, the nephridia excrete sodium; if it is low, they reabsorb it. In the insect, the organ of excretion is the Malphigian tubule, which is able to regulate ion and water exchange. The accumulated fluid is flushed into the gut, where absorption of ions and water takes place.

The Kidney

In vertebrates, the kidney is the organ responsible for eliminating water and waste products. The kidney is a bean-shaped organ that receives a large blood supply from the heart. It has several auxiliary structures: the ureter, which collects fluid or urine from all the tubules; the bladder, which acts as a storage organ; and the urethra, which opens from the bladder to the outside of the body. The kidney consists of more than a million nephrons arranged symmetrically, with the lower part of each nephron pointing toward the hilus (the pole of the kidney where the ureter arises).

The kidney maintains homeostasis of the body through four basic mechanisms: filtration, reabsorption, secretion, and concentration. In filtration, a liquid portion of the blood is transferred to the tubule. There the cells proceed to reabsorb necessary materials, to secrete additional materials into the tubular liquid from the blood, and to concentrate the fluid in the tubule.

The nephron is the fundamental structure of the kidney. The nephron is a long, slender tube with different parts that are capable of secreting or reabsorbing ions, water, and other substances either to remove materials from the blood or to return materials to the blood, depending upon the needs of the body. One process it performs is the elimination of ammonia products and other waste materials.

The nephron consists of two major parts: a glomerulus and a tubule. The process of urine formation begins with filtration. The blood enters the kidney and then the glomeruli, under high pressure from the heart. The pressure forces fluid from the blood into the tubule. The amount is tremendous; in humans, every day some 180 liters (about 40 gallons of fluid) pass from the blood into the nephron. All the blood in the body passes through the nephron about thirty times per day. During this same period, seven hundred liters of blood pass through the kidneys, so only a small portion is actually filtered or transferred to the tubule. Most remarkable of all, only about 1.5 liters of urine are produced each day. The rest of this large volume is reabsorbed by the tubule.

As the fluid passes into the nephron, it enters a section of tubule at the beginning that reabsorbs much of the material needed by the body. Such things as the glucose needed for energy, amino acids for protein building, vitamins, and ions needed to maintain the correct concentration of the blood are removed and transported by the cells of the nephron back into the blood. At the same time, about 85 percent of the water of the filtrate is also transported back into the blood.

The cells of the nephron are able to secrete materials from the blood to the tubular fluid. This process is exactly the opposite of the reabsorption of substances. One of the most readily secreted substances is sodium. The cells of the body are high in potassium and low in sodium; since sodium is a constituent of every diet, the removal of excess sodium is necessary. Sodium is picked up from the blood that circulates around the nephron and is secreted into the tubule. The process of secretion also extends to other materials. The potent antibiotic penicillin was ineffective when it was first used to treat systemic infections, because it was rapidly secreted by the tubules. A high enough concentration could not be accumulated in the blood to destroy bacteria. It became useful only when a derivative that was not secreted could be found.

The Loop of Henle

The tubular fluid that has been adjusted in concentration, volume of fluid, and concentration of ions and other materials now passes into a hairpin-shaped portion of the tubule called the loop of Henle. The loop of Henle adjusts the volume of filtrate. A hormone is produced by the brain (in the hypothalamus) that is capable of altering the permeability of the cells of the loop of Henle to water. The substance, a protein hormone called antidiuretic hormone (ADH), causes the reabsorption of water from the loop. Cells of the hypothalamus respond to the concentration of particles in the blood (its osmotic pressure) and adjust the amount of water that is reabsorbed from the tubule by secreting more or less ADH as necessary to maintain a constant concentration in the blood. The range of adjustment is remarkable. The volume of urine produced can range from about 0.5 liter to more than 30 liters per day, depending upon the need. If water is administered or restricted, the water concentration of the body changes. This causes a change in the production of ADH, which in turn increases or decreases the excretion of water to return the level to normal.

Sweating also causes water loss and thus decreases urine flow. Intake of large amounts of fluid will dilute the body fluids and cause an increased urine output. There is a constant adjustment, because water is lost by breathing, through the skin, and through excretions, and the kidney must make the proper corrections. Losses have been reduced to a minimum in animals such as the kangaroo rat, which lives in the desert and must conserve water. All of its water intake is from seeds and other foods containing some water, and excretion is almost zero. The desert rat is able to concentrate urine to a level about five times that of the human.

As the urine passes from the loop of Henle, it enters upon a final adjustment in the latter portion of the tubule. Volume, concentration of material, and the like are adjusted to maintain homeostasis. Other alterations of the fluid are also made in the passage down the tubule. If the body becomes acidic, the cells of the tubule exchange sodium ions, which are neutral, for acid ions (H+), thus causing the body to lose acid. Conversely, if the body becomes basic (lacks acid or hydrogen ions), the reverse is true.

Studying Kidney Function

The kidney can be studied on many different levels. The output of the kidney, the urine, and the input, the blood, can be analyzed in order to determine function of both kidneys. In some animals, such as the frog, the individual nephrons are visible under a microscope. Through the use of the stop-flow technique, the behavior of an individual portion of the nephron can be studied. In this technique, a very small needle is introduced into one portion of the nephron, and a small drop of oil is injected. Another drop is injected in an adjacent section of the nephron, thus isolating a section between the two drops. The exchanges that occur in that portion of the nephron can then be measured by taking samples of the fluid at intervals.

The whole kidney can be studied by examination of the kidney in relation to a reference substance. For example, the chemical inulin goes through the nephron without any alteration, so its excretion can be compared with other substances: If more of another substance appears in the urine in proportion to the inulin, the substance must have been excreted from the blood into the urine and secreted. If less is present, some material must have been reabsorbed from the tubule into the blood. This is called the clearance technique, and it is widely used to predict kidney function in health and disease. For less sophisticated testing, substances—such as certain dyes—that can be taken orally and are excreted by the kidney can be used to measure the rate of excretion. Radioopaque substances, substances that will appear on X rays, can be used to detect overt kidney malfunction.

—J. H. U. Brown

See also: Circulatory systems of vertebrates; Endocrine systems of vertebrates; Metabolic rates; Osmoregulation; pH maintenance; Water balance in vertebrates.

Bibliography

Brenner, Barry M., ed. *Brenner and Rector's the Kidney*. 6th ed. Philadelphia: W. B. Saunders, 1999. Authoritative coverage of the kidney, focusing on human anatomy. Covers normal renal function, fluid and electrolyte abnormalities, pathogenesis, pathophysiology, and management of renal failure. Written for medical students and professionals.

Hainsworth, F. Reed. *Animal Physiology*. Reading, Mass.: Addison-Wesley, 1981. This is an excellent book, with a wealth of examples of kidney function in various animals, focusing on the processes by which it adjusts pH, ion concentration, and the like. Excellent illustrations, bibliographies, and notes.

Smith, Homer. *From Fish to Philosopher*. Garden City, N.Y.: Doubleday, 1961. This book exists in many paperback editions. It is a classic work for the layman on the function of the kidney. Smith is a final authority on the kidney, and he is also an excellent writer, an essayist of note, and a raconteur. There are few references, as this is a series of essays.

Starr, Cecie, and Ralph Taggart. *Biology: The Unity and Diversity of Life*. 9th ed. Pacific Grove, Calif.: Brooks/Cole, 2001. An elementary biology text. It has good material on the kidney, dealing with all of its functions and using good examples of kidney performance. References are cited, and questions and examples are given.

Valtin, Heinz. *Renal Function*. 3d ed. Boston: Little, Brown, 1995. A more advanced book on the function of the kidney that is relatively easy to read. It is detailed and deals at length with the artificial kidney and with diseases of the kidney.

KOALAS

Type of animal science: Classification
Fields of study: Anatomy, physiology, zoology

Koalas are solitary animals existing on a low-energy diet of eucalyptus leaves. Since koalas are marsupials, the females have a pouch where their young develop. They are the sole member of the family Phascolarctidae.

Principal Terms

ARBOREAL: living in trees
BELLOWING: guttural sound used in communication
CAECUM: structure in the digestive tract that aids in digestion and water retention
MARSUPIAL: animals whose young develop in a pouch
NOCTURNAL: active at night

Although they are commonly referred to as "koala bears" because of the resemblance to teddy bears, koalas are not bears. The koalas are marsupials. This is one of the oldest classes of animals, existing since over fifty thousand years ago. Koalas average about twenty-six pounds in weight and thirty-one inches in length. The coat of the koala is the thickest among the marsupials and has a gray to tawny color. White coloration appears on the chin, chest, and forelimbs. The fur is short, soft, densely packed, and springy to the touch. It is the most effective fur insulation among animals. The koalas do not rely on fat beneath the skin for insulation; rather, blood flow to the extremities can be reduced as a means to conserve heat. In the rain, water runs off the koala's fur. Only sick koalas will appear to be wet when it rains.

Koala Life

Koalas are nocturnal and highly arboreal, living solitary lives high up in eucalyptus trees. Koalas are known as phalangers, because they use their hands and hind feet to effectively grip tree trunks and branches when tree climbing and jumping from tree to tree. They walk with a clumsy gait in the following sequence: front right foot then back left foot, front left foot, back right foot. They have a very specialized diet, feeding almost exclusively on the leaves of a few species of eucalyptus. The leaves provide most of their water intake; in fact, the word "koala" means "no drink" in Aboriginal languages. For an average day, a koala will consume about a pound of leaves. They are very fussy

Koalas are extremely specialized feeders, eating primarily the leaves of the eucalyptus trees in which they live. (Digital Stock)

eaters, typically being very careful in selecting which leaves from a bough to ingest. The koala uses a set of thirty teeth, comprising incisors, canines, premolars, and molars, to chew the eucalyptus leaves. Each day the koala spends approximately eighteen to twenty-one hours sleeping or resting.

In order to communicate, the koala uses a range of vocalizations. The male koala uses a deep, grunting bellow to communicate its physical and social position. The sound can resemble a far-off rumbling, like a motorcycle or pig snorting. During the mating season, the koala will use the bellowing as a means to locate and accurately pinpoint potential mates. The mating call is a deep, loud, guttural sound that can be heard for long distances. Female koalas do not show the same level of bellowing. Their calls communicate aggression and are part of the mating ritual. Both the male and female koalas share a similar call, sounding like a baby screaming. This is often accompanied by shaking and signals fear. Mother and cubs make soft squeaking noises to one another, as well as humming or murmuring.

Koala Facts

Classification:
Kingdom: Animal
Phylum: Vertibrata
Class: Mammalia
Subclass: Marsupialia
Order: Diprotodontia
Suborder: Vombatiformes
Infraorder: Phascolarctomorphia
Family: Phascolarctidae
Genus and species: Phascolarctos cinereus
Subspecies: P. c. victor (Victoria), *P. c. cinereus* (New South Wales), *P. c. adustus* (Queensland)

Geographical location: Two main groups living out in the wild are in eastern Australia in an area extending from Cooktown in northern Queensland to southwestern Victoria; have been introduced into western and southern Australia

Habitat: Wild eucalyptus forests and woodlands; they are found only in pockets with suitable vegetation of a relatively small number of eucalyptus species that they prefer to ingest

Gestational period: Thirty-five days

Life span: Thirteen to eighteen years

Special anatomy: Females have a pouch that faces the rear and has a drawstring type muscle which can be tightened to close the opening; mammary glands are located along the abdomen within the pouch; females have a duplicate reproductive system with two vaginas; males have dual-pronged, forked penises; each hand has two opposable thumbs, which are crucial for the ability to climb and cling to trees; digestive system includes a caecum, a structure used to digest eucalyptus leaves and assist in water extraction

Koala Reproduction

Females of the species have a pouch in which the young develop. The young are born in nearly embryonic form about the size of a human's little finger. After birth, the infant travels to the mother's pouch, where it attaches to teat. The teat then becomes engorged and forms a seal with the newborn. A single offspring is usually born. It is not until twenty-four weeks after birth that the young is covered with fur and develops teeth. The first six months of life are spent in the mother's pouch. The cub remains with the mother until about twelve months after birth. Mating is a brief event that takes place about once a year. Male koalas are nomadic and play virtually no part in the raising of the young. Extensive chlamydia infection has caused widespread infertility in female koalas and is a major contributing factor in their declining numbers.

—Frank J. Prerost

See also: Embryology; Fauna: Australia; Fertilization; Kangaroos; Marsupials; Opossums; Pregnancy and prenatal development; Reproduction; Reproductive strategies; Reproductive system of female mammals.

Bibliography

Hunter, S. *Koala Handbook*. London: Chatto & Windus, 1987. The book has a variety of good color photographs and basic information about habitat and behavior.

Lee, A., and R. Martin. *The Koala: A Natural History*. Kensington, Australia: New South Wales University Press, 1988. A source of basic information with some black and white pictures.

Phillips, K. *Koalas: Australia's Ancient Ones*. New York: Macmillan, 1994. This is an excellent source of color pictures of koalas in various settings, with good information about their life and habits.

Sharp, A. *The Koala Book*. Gretna, Calif.: Pelican, 1995. An excellent resource book with interesting color photographs. Covers a wide variety of topics about koalas.

Wexo, J. B. *Koalas*. Mankato, Minn.: Creative Education, 1989. Part of a series of books on various animals, with easy-to-read text and lifelike illustrations. Activities for teachers are also included.

LACTATION

Type of animal science: Reproduction
Fields of study: Biochemistry, developmental biology, embryology, physiology

Lactation is the process by which female mammals produce milk for the nourishment of their offspring. For all species except humans, successful lactation is essential for the survival of the young.

Principal Terms

ALVEOLI: the milk-producing areas within the mammary glands

COLOSTRUM: the precursor to milk that is formed in the mammary gland during pregnancy and immediately after birth of the young

DUCTS: the tubular structures that carry milk from the alveoli to the outside through the nipple or teat

LACTATION: the process of producing and delivering milk to the young; also, the time period during which milk is produced

MAMMARY GLANDS: the milk-producing glands found in all mammals; for example, the cow's udder contains the mammary glands

MILK EJECTION: also known as milk letdown, this is the reflex response of the mammary gland to suckling of the nipple; the hormone oxytocin mediates this reflex

MYOEPITHELIAL CELLS: the specialized cells within the mammary gland that surround the alveoli and contract to force milk into the ducts during milk ejection

NIPPLE: the raised area on the surface of the skin over the mammary gland that contains the duct openings

TEAT: an elongated form of nipple that contains one duct opening

Lactation is the process by which female mammals produce milk to feed their young. The ability to produce milk is one of the defining characteristics of the class Mammalia: All mammals, but no other animals, possess the highly specialized glands necessary for lactation. In evolutionary terms, the appearance of lactation coincides with the tendency of mammals to produce only a few offspring at a time; the provision of milk for these offspring helps to ensure their survival while removing competition between the adults and the young for food.

The Mammary Glands

The mammary glands are the milk-producing organs. The number varies among species from two to about twenty, with a rough correlation between the number of young born and the number of glands present. The glands are located on the ventral surface of the body, either in the thoracic (in humans, for example) or abdominal region (in horses and cows) or in two lines extending almost the length of the body (in dogs and rodents). Both male and female mammals have mammary glands, because in early mammalian development the basic body plan of male and female embryos is identical. The mammary glands of males are nonfunctional, however, since they lack the hormonal stimulation necessary for lactation.

Internally, the mammary glands of all mammals follow the same basic plan, consisting of alveoli that produce milk and ducts that carry the milk to openings on the surface of the skin. The alveoli are surrounded by myoepithelial cells that contract to squeeze the milk into the ducts during suckling by the young.

Externally, considerable variation exists among the mammals in the appearance of the mammary glands and their associated openings. In the spiny anteater and platypus, the many lobes of the mammary glands each open directly to the surface of the abdominal skin through individual ducts, and the young suck the hair-covered skin to obtain the milk. In other mammals, the mammary glands are more obvious as swellings beneath the skin, with a raised area, the nipple or teat, that contains the duct openings. In some four-legged animals (cows, horses, and goats) the mammary glands are located in a baglike structure called the udder, from which are suspended the elongated teats. In humans, the nipple, which contains the openings of fifteen to twenty-five ducts, is surrounded by pigmented skin, the areola. The areola contains glands (tubercles of Montgomery) that secrete a lubricating fluid.

Milk Production

Lactogenesis (milk production) does not begin until a female has produced young. During pregnancy, a complex of hormones prepares the mammary glands for milk production by promoting their growth and internal development. These hormones include prolactin from the mother's anterior pituitary gland, placental lactogen from the placenta within the uterus, and estrogen and progesterone, which are produced in the corpus luteum of the mother's ovary and in the placenta. Other hormones, including cortisol from the adrenal gland, thyroxine from the thyroid gland, and insulin from the pancreatic islets, may also be involved. Progesterone appears to participate in the induction of mammary development, but, paradoxically, it also prevents milk secretion during pregnancy.

Although true milk is not produced during pregnancy, a precursor to milk, colostrum, can be produced in small amounts by the mammary glands of most species. Colostrum is a sticky, yellowish, transparent liquid. Colostrum secretion continues in the first few days after birth of the young; there is then a gradual transition to production of true milk.

Milk contains water, proteins, fats, vitamins, minerals, and a unique sugar, lactose. The exact concentration of the various components varies greatly between species according to the nutritional demands of the young. The milk of seals is high in fat and other solids that contribute to rapid weight gain in the pups, a strategy that appears to be essential for their survival.

Noteworthy among the constituents of milk are antibodies produced by the mother. These antibodies help protect the newborn from disease in the period when the newborn's own immune system is immature and incapable of providing significant defense. The antibody concentration of colostrum is higher than that of true milk, and for this reason the first few days of nursing are considered the most important for immunological protection of the newborn.

The transition in production from colostrum to true milk is brought about by a change in the hormonal status of the mother. At the time of birth, the placenta is expelled from the mother's body, thus removing the source of progesterone, estrogen, placental lactogen, and other hormones. The decrease in progesterone levels is thought to be essential for the onset of lactogenesis. In addition, at the time of birth, there are changes in prolactin secretion that may play a role in initiating milk secretion.

Suckling

Once lactogenesis is established, a set of hormonal reflexes act to match milk production and delivery to the needs of the newborn. Suckling of the nipple involves motions similar to chewing as the infant takes the nipple between the tongue and the palate. This suckling motion stimulates nerve endings in the mother's nipple that relay signals about the stimulation back to the mother's brain. Within thirty to sixty seconds, these signals result in the release of prolactin from the mother's anterior pituitary gland and oxytocin from her posterior pituitary gland. Prolactin causes continued production of milk by the alveolar cells of the mammary glands. Oxytocin acts immediately on the myoepithelial cells of the mammary gland, causing them to contract and push milk from the alveoli into the ducts and thence through the nipple into the infant's mouth. Thus, the infant does not actually remove milk from the mammary gland by suction, but instead is responsible for promoting a hormonal reflex that results in active milk ejection, or letdown, from the mammary gland.

Because of the operation of the prolactin and oxytocin reflexes during suckling, lactation is a biological example of the principle of supply meeting demand. All that is necessary to increase milk production is to increase the suckling stimulus by nursing the young more often. Once established, lactation in some species can be sustained in this manner for years, assuming the nutritional needs of the mother are met. On the other hand, if the mother fails to nurse her offspring, the absence of the suckling stimulus will cause the mammary glands gradually to cease milk production.

The exact composition of the milk is altered as lactation continues to meet the changing nutritional needs of the growing offspring. The most extreme example of the ability of the mammary gland to change the composition of milk is seen in the kangaroo. In this animal, the newborn attaches to a teat in the mother's pouch shortly after birth and remains there for a month or more. A mother kangaroo may nurse offspring of different ages from separate teats, and each teat supplies a milk with the appropriate nutritional composition for that young.

Species vary in time spent suckling the young. The rabbit nurses her litter for only about five minutes once a day, while the rat nurses for about half an hour at a time, at intervals throughout the day. Lactation lasts about ten days in rodents, but it may persist for months in large species such as horses and cows. Continued lactation has a suppressive effect on ovulation that is thought to be attributable to interference by prolactin with the normal hormonal mechanisms that cause ovulation.

The Study of Lactation

Although it has always been clear that milk is expelled from the mammary glands, the realization

All mammals produce milk to feed their offspring. (Robert Maust/Photo Agora)

that the glands themselves actually produce the milk is a relatively recent one. Early anatomists erroneously assumed that milk must be a product of the uterus, since the uterus is involved in support and nourishment of the fetus. Thus, much of the early anatomical work attempted to show some sort of connection between the uterus and the mammary glands. It was not until the late 1800's that the light microscope clearly demonstrated that milk is formed within the mammary glands. In the twentieth century, electron microscopy showed that during pregnancy the intracellular organization of the alveoli becomes increasingly more complex as the cells become capable of milk secretion.

Various techniques for labeling compounds with radioactive or fluorescent markers have been used in conjunction with electron microscopy to examine how milk is synthesized in the alveoli. The alveoli cells extract necessary precursors from blood flowing through the mammary gland, assemble the precursors into milk components, and then secrete the constituents of milk into the mammary gland ducts. Specific routes of secretion have been identified for the major components of milk.

More recently, researchers have used cell-free systems to study the biochemical pathways involved in milk synthesis. These systems use isolated fragments of deoxyribonucleic acid (DNA), ribonucleic acid (RNA), and perhaps some cell organelles to examine the intermediate chemical steps in the synthesis of milk components. Using these techniques, researchers have been able to "watch" as complex milk proteins and constituents are assembled step by step. The knowledge of how the components of milk are assembled is leading to a fuller understanding of how the amounts of these substances in milk are hormonally regulated.

Knowledge of the hormones involved in inducing and maintaining lactation has come about through systematic assembly of information from several lines of research. Test animals can be treated with a specific hormone to determine if that hormone causes or suppresses lactation. The test animals may be males or immature females, with the goal being the duplication of the specific mix of hormones that cause lactation in the adult female. The opposite approach may also be taken: An endocrine gland can be removed from a lactating female to determine if the hormonal products of that gland are necessary for lactation. Another approach is to make careful measurements of the levels of hormones circulating in the blood as the lactational state changes; any hormone that shows a correlated change may be a good candidate for further investigation by treatment or removal from test animals. These methods have led to an understanding of the importance of prolactin, placental lactogen, oxytocin, estrogen, and progesterone in promoting lactation, but researchers still do not understand how the system is fine-tuned. For example, considerable variation exists in the volume and quality of milk produced by different individuals—or by the same individual at different times—but these differences cannot currently be explained by any known change in hormone levels. Research is focusing not only on describing changes in circulating levels of hormones but also on elucidating the exact effects of these hormones on the biosynthetic pathways within the mammary gland.

—Marcia Watson-Whitmyre

See also: Birth; Digestion; Endocrine systems of vertebrates; Hormones in mammals; Ingestion; Mammals; Nutrient requirements; Offspring care; Placental mammals; Pregnancy and prenatal development; Reproductive system of female mammals; Sexual development.

Bibliography

Larson, Bruce L., and Vearl R. Smith, eds. *Lactation: A Comprehensive Treatise*. 4 vols. New York: Academic Press, 1974. Every aspect of lactation is covered: development, structure, and diseases of the mammary gland; hormonal control; lactogenesis; milk components; and maintenance of lactation. Some of the chapters are specific to humans, and others cover domestic and laboratory animals.

Mepham, T. B., ed. *The Biochemistry of Lactation*. Amsterdam: Elsevier, 1985. Concentrates on ruminant lactation.

Peaker, M., R. G. Vernon, and C. H. Knight, eds. *Physiological Strategies in Lactation*. London: Academic Press, 1984. The purpose of this book is not only to explore mechanisms of lactation but also to analyze the impact of evolution on lactational patterns in different mammal species.

LAKES AND RIVERS

Types of animal science: Classification, ecology, reproduction
Fields of study: Anatomy, ecology, zoology

Lakes and rivers are large bodies of water produced by the hydrologic cycle. Lakes are stable bodies of water surrounded by land, while rivers are moving bodies of water flowing downhill toward oceans. They comprise important, animal-rich ecosystems.

Principal Terms

CARNIVORE: any animal that eats only the flesh of other animals

COURSE: the pathway of a river from its source to its entry into an ocean

ECOSYSTEM: an ecological community, which together with its environment is perceived as a unit

EROSION: the processes, including weathering, dissolution, and abrasion, by which earth or rock is removed from a part of earth's surface

ESTUARY: part of river's lower course, near entry to an ocean, where slow river flow and tidal action forms mud flats and sand banks

HYDROLOGIC CYCLE: earth's cycle of evaporation and condensation of water, which produces rain and maintains oceans, rivers, and lakes

LAKE: a large area of water surrounded by land

A lake is a large body of water surrounded by land. Lake waters move, partly due to rivers that enter and leave them. In addition, lake water circulates vertically, especially near its surface, where sunlight heats the water and lowers its density. Seasonal warming and cooling of water, evaporation, and other factors contribute to annual cycles of vertical lake water movement.

Most lakes form in hollows dug out of the ground by glaciers. Some huge lakes are really inland seas (such as the North American Great Lakes). Lake water is supplied by rivers and accumulated from rain draining off nearby land. River water contains minerals obtained by running over rocks, and rain holds airborne chemicals. Continued water flow through a lake keeps it fresh, enabling animal and plant survival. When water does not flow out of a lake, minerals are trapped and concentrated, as solar heat evaporates water. In time, salt accumulation turns such lakes into bodies of salt water (such as the Dead Sea, ten times saltier than the oceans) in which few organisms can live.

Animals Found in Lakes

Lakes provide rich habitats for many kinds of wildlife. They are also excellent examples of ecosystems, self-contained ecological units wherein all plants and animals present depend on each other for survival. Removing or damaging one plant or animal species can cause a chain of events that affect, even ruin the ecosystem.

Land beside a lake is usually marshy. Rushes, reeds, willow, and alder trees, as well as water lilies, grow in the shallows around a lake edge. The plant roots prevent lakeshore erosion. Farther from shore are submerged and floating plants which produce oxygen and provide food for small underwater animals such as insects, insect larvae, snails, shellfish, and worms. These small animals are eaten by fish, which are then eaten

Lakes offer rich habitats for complex ecosystems comprising organisms that live on both land and water. (Digital Stock)

by birds. Land mammals, such as martens and bears, also eat the fish. Overall, lake plants and animals make up a food chain that supports the lake ecosystem.

A great number of animals inhabits lakes and live around them. Animals present vary greatly depending on the lake's geographic location. For example, birds, which usually eat fish and aquatic insects, inhabit lakes throughout the world. Lake-dwelling bird species range from ducks, geese, and loons in Canada's cold lakes, to ducks, kingfishers, and herons throughout the United States, to the flamingos of the southern United States, Central and South America, and Africa. Marabou storks and other carrion eaters also inhabit African lakes.

Flamingos may be among the most dramatic of lake-dwelling birds. There are six species of these gorgeous water birds, which have long legs for wading, webbed feet, swanlike necks, and red, flame-colored, pink, or white feathers. Their bills bend abruptly in the middle and the upper mandible fits tightly into the lower one. Flamingos dip their heads upside down under water and scoop backward to feed. Their bill edges have ridges, and when the tongue pushes against the inner bill, the water runs out, leaving behind shellfish or other food. Flamingos live along lakes in nests made of mud, hollow on top, and able to hold one egg. The largest species, the greater flamingo, has two subspecies, one bright red, the other pale red. Males reach five feet tall. The smallest, most abundant species is the lesser flamingo.

Reptiles such as snakes and salamanders, amphibians such as frogs, and many turtle species are also present in lakes. Depending on size, they eat insects or each other, and are themselves eaten by birds. Many different fish species are also present, from minnow to sunfish to trout. They are parts of food chains, eating insects, plants, small reptiles, and each other, depending on size.

River Animals

Only animals adapted to cold, fast-flowing water survive in the upper courses of rivers, which are fine habitats, clean and full of oxygen. Some plants survive, but limpets, snails, and hardy water insects are preponderant. They become foods for the carnivorous fish, such as trout and salmon, found in rivers' upper courses.

The current slows along a river's middle course, and sediment settles to the river bed as earth and rock particles holding minerals on which plants thrive. These plants, from microscopic algae up, are used as food by wildlife inhabiting river banks, such as otters, martens, and raccoons, as well as fish and other animals which live in river waters. Small fish and insects, such as water striders, dragonflies, and aquatic beetles, are plentiful in the water. Reeds and trees are also homes and nests for birds of many kinds, as well as small to medium-sized mammals, such as rodents and skunks, snakes such as water moccasins, and other reptiles.

Depending on location, the size and danger associated with middle course river-animal life varies. However, food chains in rivers also assure continuation of the balance of nature, as small plants and animals are eaten by medium-sized animals and those creatures are, in turn, eaten by bigger animals. The largest, most dangerous river animals are found in the hot southwestern United States, Central and South America, Africa, and Asia.

Crocodiles, alligators, and caimans are among the most dangerous river animals. They are all crocodilians, differing in size, position of the fourth mandibular tooth, and snout shape. Crocodiles have lizardlike bodies up to twenty-five feet

Rivers

Roughly 70 percent of Earth's surface is covered with water. About 97 percent of this water is salt water in oceans, and 2 percent is freshwater in glaciers and groundwater. Only 1 percent is freshwater in rivers, lakes, and water vapor. Freshwater is recycled again and again by evaporation and condensation in a hydrologic cycle which creates and maintains rivers and lakes.

When water vapor condenses and falls as rain, it seeps into the ground, filling pores in soil and rocks. After all pores fill, excess water pools on the surface, flowing downhill due to gravity. Then it becomes streams that coalesce into rivers. Each river is a large body of water, regenerated over and over by the hydrologic cycle, flowing through low ground areas that become channels (river courses) and valleys. Because Earth's oceans are lower than the land, rivers flow down into them.

A river and the composition of its banks change as movement toward the ocean both creates and destroys land. A river's course has upper, middle, and lower parts. The upper course, near the river's source, flows downhill and carves deep valleys, as well as waterfalls wherever tough rock resists erosion. Next, the river flows onto plains in its middle course, and the gently sloping land slows it, leading to sideways erosion and snakelike bends. The last part of a river, its lower course, near entry to an ocean, is often an estuary. Here, the slow flow of the river across virtually flat land causes settling of sediment, forming mud flats and sandbanks.

long, with short limbs, long, thick tails that facilitate swimming, huge teeth, eyes atop their heads, and short, broad snouts in which the fourth mandibular teeth are visible. They are cold-blooded and inhabit salty, brackish, and fresh water in Africa, Australia, Asia, the Indies, and the Americas. They are carnivores, eating birds, amphibians, fish, mollusks, crustaceans, small mammals, and large mammals ambushed while drinking. Crocodiles mate in the water. After two-month gestation periods, females lay eggs on land. The eggs hatch in three months. Crocodiles can live for up to one hundred years.

Alligators, physically and reproductively similar to crocodiles, are only up to sixteen feet long. Their fourth mandibular teeth are hidden by their upper lips, they are more comfortable on land than crocodiles, and they have short, broad snouts. During the day, alligators sun themselves on river banks in the southeastern United States, China, and Central and South America. Caimans of Central and South America are small alligators, only up to eight feet long (although the black caiman can be twice as large). They inhabit rivers and swamps. Alligators and caimans live for forty-five to sixty-five years.

The largest river mammal is the herbivorous African hippopotamus, which can reach lengths of fifteen feet and body weights of four tons. All sorts of fish live in rivers, including minnows, catfish, sunfish, bass, trout, and river dolphins. Most such fish are not very dangerous to organisms other than their specific prey. However, the carnivorous piranha of South American rivers eat any animals that stray into their territory.

Many wild animals found in a river's middle course are even more plentiful downriver in its lower course, where the current is quite slow. However, in estuaries, which are salty due to the admixture of ocean water with the river's freshwater, few plants and animals can survive. There, worms, crustaceans, and gastropods such as snails inhabit mud banks and provide food for birds such as herons, gulls, and kingfishers.

River Ecosystems

Rivers contain series of ecosystems whose compositions vary with conditions in a given portion of the upper, middle, or lower course. In river ecosystems, like those in and around lakes, all plants and animals depend on each other for survival, and removing or harming one species can ruin the ecosystem. Beyond this, existence of multiple ecosystems on land and their connection to aquatic ecosystems points out that damage to any living organism or its local ecosystem will have ripple effects, which can spread out to damage much wider areas than at first perceived. Hence, reasonable actions of humans toward all other living organisms should be attempted and achieved.

—Sanford S. Singer

See also: Ecosystems; Food chains and food webs; Forests, coniferous; Forests, deciduous; Habitats and biomes; Marine biology; Rain forests; Tidepools and beaches.

Bibliography

Bramwell, Martyn. *Rivers and Lakes.* London: Franklin Watts, 1986. Describes rivers, lakes, and their inhabitants.

Love, Bill, William W. Lamar, and Tim Ohr, eds. *The World's Most Spectacular Reptiles and Amphibians.* Tampa, Fla.: Carmichael, 1997. Covers many reptiles and amphibians, including snakes, lizards, turtles, frogs, and crocodilians.

Ogilvie, Malcolm, and Carol Ogilvie. *Flamingos.* Gloucester, England: A. Sutton, 1986. This book on flamingos describes their natural history, reproduction, and distribution.

Pringle, Laurence. *Rivers and Lakes*. Alexandria, Va.: Time-Life Books, 1985. This well-illustrated book discusses rivers and lakes, as well as their inhabitants.
Wade, Nicholas, ed. *The Science Times Book of Birds*. New York: Lyons Press, 1997. This illustrated book discusses all sorts of birds, their lives, and their habitats.

LAMPREYS AND HAGFISH

Type of animal science: Classification
Fields of study: Anatomy, evolutionary science, physiology, systematics (taxonomy), zoology

Jawless vertebrates, or Agnatha, are represented today by the lamprey and hagfish. These organisms are of great interest as the survivors of a once-flourishing group of Paleozoic vertebrates that includes the earliest vertebrates in the fossil record.

Principal Terms

AGNATHA: a class of vertebrates that includes all forms in which jaws are not developed; the group to which the earliest vertebrates belong

AMMOCOETE: the larval form of lamprey, which lives in river silts

CLADISTICS: a method of determining relationships in which shared derived (advanced) characters exhibited by the organism are used

CYCLOSTOMES: the modern agnathans, comprising lampreys and hagfish

GNATHOSTOMATA: all vertebrates in which jaws are developed

NASOHYPOPHYSIAL OPENING: an opening in the head of modern agnathans leading to a sac that aids in olfaction

OSTRACODERMS: armored fossil agnathans that flourished during the Paleozoic

PALEOZOIC ERA: time period from 570 to 245 million years ago, which comprises the Cambrian, Ordovician, Silurian, Devonian, Carboniferous, and Permian periods

The vertebrates are normally divided into two groups based on the presence or absence of jaws. The vast majority possess jaws and are termed Gnathostomata; jawless forms are termed Agnatha. The Agnatha are represented today by the lampreys and hagfish, organisms of little economic importance but of great interest to biologists as the sole survivors of an extensive group of Paleozoic (245 to 570 million-year-old) agnathans, which includes the earliest known vertebrates. The extant forms are often referred to as the cyclostomes, the fossil forms as ostracoderms; however, these do not constitute monophyletic groups (groups including the common ancestor and all descendants), and some fossil Agnatha are closely related to cyclostomes.

The Physiology of Jawless Fishes

Lampreys and hagfish are similar in general appearance despite far-reaching differences in internal organization. Both are elongated fish, without paired fins, that swim with an eel-like motion and reach lengths of one meter and weights of up to one kilogram. They possess an entirely cartilaginous skeleton; however, the organization of the body conforms to the basic vertebrate pattern. The lamprey has more active life habits and, therefore, has a more developed median fin. The propulsive section of the tail is longer than that of the hagfish. The lamprey's mouth is surrounded by a sucker-like oral disc covered by small teeth on its internal surface. This rasping organ allows the lamprey to attach itself to prey, on which it feeds by rasping away surface tissues and ingesting blood. Above and behind the oral disc is the nasohypophysial opening, which leads into an olfactory pouch and ends blindly in a dilated sac above the anterior gill pouches. Lampreys have seven gill pouches opening through seven gill ports.

In hagfish, the mouth is ventral and overhung

The lamprey's mouth is shaped like a suction cup and filled with pointed teeth to hold on to its prey. (AP/Wide World Photos)

by a rostrum, and it contains a dental plate that can be rapidly retracted, enabling the animal to bite off fragments of tissue from the dead and decaying fish on which it feeds. The nasohypophysial duct passes backward below the brain to join the pharynx, and the respiratory current passes along it. Hagfish are quite variable in their number of gill pouches and openings. Pacific hagfish have twelve pouches with separate openings (this number may vary up to fifteen), while Atlantic hagfish have five pouches with one common opening. The eye is vestigial and covered by surface tissue, so it is not normally visible.

Lampreys

There are about thirty-five species of lampreys, and all breed in freshwater and spend the major part of their life cycle in a freshwater larval stage before transformation to an adult stage. The eggs develop into a blind larval form, the ammocoete, which burrows into silt banks and remains for several years filtering microscopic particles from the water. The ammocoete is very unlike the adult, as no suctorial disc is present; a ciliated groove on the floor of the pharynx, together with a tubular endostyle, forms the feeding apparatus. After metamorphosis, three different types of adult lamprey are known: nonparasitic brook lampreys; freshwater parasitic forms; and anadromous forms (living in marine conditions but ascending rivers to spawn). The nonparasitic forms, which constitute about half the known genera, have the mechanisms for parasitic feeding but do not use them. The gonads mature immediately after metamorphosis, the intestine atrophies, and the foregut remains as a solid rod. These nonparasitic forms remain dwarf as they cannot feed, and six to nine months after metamorphosis they reach sexual maturity, spawn, and die.

The parasitic freshwater forms live entirely in river systems, where they feed on fish. The anadromous forms move downstream to the estuary or sea after metamorphosis. There, they feed voraciously and grow rapidly, eventually returning to the river on an upstream spawning migration after one to three years. As in all lampreys, death follows shortly after spawning. In some cases, it appears that populations of originally anadromous lampreys may have become landlocked and are now entirely freshwater. That is what appears to have happened with the lampreys (*Petromyzon marinus*) that now inhabit the Great Lakes. They are capable of causing significant damage to fish populations and almost eradicated trout and whitefish from the Great Lakes between 1950 and 1960, when they peaked in that system, resulting in the collapse of a flourishing fishery.

Hagfish

Hagfish are purely marine fish, in contrast, and as they are benthonic (live on the bottom) and often inhabit deep water, little is known of their reproductive biology beyond the fact that eggs may be laid at any time of year and there appears to be no larval stage. They remain buried in mud during the day, emerging at night to feed. It appears that they normally attack only dead or dying fish, lacking the suction apparatus that makes the lamprey so successful in its attacks on living organisms; however, they will consume small invertebrates if they are available. When feeding, they may show "knotting behavior," during which the flexible body ties into a knot in order to gain a better purchase for the tearing off of food fragments. One characteristic of all twenty species of hagfish is the development of mucus pores along the body. If roughly handled or irritated, these pores produce copious quantities of mucus, resulting in the name "slime eels" for these animals. Neither hagfish nor lampreys are of more than local economic importance.

The Fossil Record

The fossil record of hagfish and lampreys is sparse, presumably because of the lack of an ossified endoskeleton. Both are known only from the Pennsylvanian period (280 million years ago) of Illinois and show close similarity to modern forms. These sediments also include two other agnathans of unknown affinities, indicating a wide diversity of forms of this type. There is, however, an extensive record of armored agnathans (ostracoderms) from rocks of Ordovician to Devonian age (360 to 470 million years old). These have been divided into two main groups, the Osteostraci and the Heterostraci, together with some smaller groups.

The Osteostraci were a group of fish that lived in the Silurian and Devonian periods (370 to 425 million years ago) and were characterized by the presence of an armored head-shield, the rest of the body being covered by large bony scales. The large eyes were dorsally placed, and between them were the pineal foramen (an opening for a light-sensitive organ) and the nasohypophysial opening. Located laterally on the head-shield were sensory fields that probably were sensitive to pressure waves in the surrounding water. The ventrally placed mouth and dorsoventrally flattened head indicate that the Osteostraci were benthonic fish, possibly feeding by sucking organic debris or small organisms into the mouth. Paired pectoral fins may have acted to move the animals by rhythmic undulations. Because the head-shield surrounded the brain, a considerable amount of detail of brain structure and cranial nerve pattern has been preserved. It has shown that the general pattern was very similar to that of the lamprey, implying a close relationship.

The Heterostraci were a long-ranging group whose earliest representatives occurred in the Middle Ordovician period (470 million years ago) and represent the first known record of vertebrates. Typical Heterostraci were armored over the anterior part of the body by variable numbers of bony plates and were further characterized by having only one pair of external gill openings. They were common in shallow marine and freshwater environments during the Upper Silurian and Devonian periods (360 to 425 million years ago) but became extinct at the end of the Devonian. They seem to have been adapted to a variety of modes of life, from benthonic detritus feeding to cropping algae and filter feeding. The Ordovician forms are known from rocks in North and South America and Australia and are united with later Heterostraci by the presence of the same type of acellular bone, aspidin, in the armor. Their exact position is uncertain, however, as they do not appear to have a series of branchial (gill) openings on each side.

Although it is clear that the Osteostraci were closely related to modern lampreys, no relationship has been determined yet between hagfish and any ostracoderm group. The relationship of the fossil and modern agnathans to gnathostomes is also still the subject of considerable debate, though in broad terms it is accepted that lampreys are the sister group of gnathostomes and hagfish the sister group of lampreys and gnathostomes. Much further information is needed before the details of their phylogeny can be elucidated, and continuing work on these organisms will aid understanding of both the origin of vertebrates and the early development of major vertebrate groups.

The Ichthyology of Jawless Fishes

Fish are generally excellent subjects for study and for the demonstration of anatomy, physiology, ecology, evolution, and other aspects of science. The detailed anatomy of the cyclostomes has been known for many years, and knowledge in this area is dependent on careful dissection of specimens. General knowledge of the life cycle, ecology, and feeding habits of the modern forms is based on observation either in the natural habitat or in aquariums. As lampreys spend most of their life cycle in freshwater, understanding of their development is fairly complete. Hagfish, however, are marine benthonic fish that move and feed mostly at night, and hence the ability to observe them is somewhat limited and understanding of their life cycle is incomplete. Sophisticated laboratory techniques now make it possible to analyze the biochemistry and physiology of these organisms and compare them to other chordates. Studies have also been made of their swimming methods, using high-speed cameras and electromyography, a technique that allows the tracing of the electrical changes that take place in muscles when they are active.

Fossil Agnatha cannot be studied as completely, because only the hard parts are preserved in the sediments. Techniques for studying these remains have changed very little in the past one hundred years. The bones are removed from the rock by chipping or dissolving the surrounding sediment away. Bone is composed of calcium phosphate and thus will resist some acids that can break down carbonate rocks. The acids most commonly used are acetic acid and formic acid, and the specimen to be dissolved out is often backed with plastic so that it does not disintegrate when the supporting matrix (surrounding sediment) is removed. The bone from these ancient agnathans is often so well preserved that thin ground sections can be made and viewed through a microscope using transmitted light.

The characters determined by these means are used to determine relationships. Studies on the relationships of the fossil and recent Agnatha have relied in recent years on the methodology termed "phylogenetic systematics," or "cladistics." Cladistics is distinguished from other taxonomic methods (taxonomy is the study of interrelationships) by the fact that it is a rigorous system in which only shared advanced characters are used to show relationships. These relationships are expressed as branching diagrams termed cladograms (from the Greek *klados*, "branch"), hence the name cladistics. Although studies using this methodology have improved understanding of the relationship between modern Agnatha and the gnathostomes, the relationships of the fossil forms are still poorly understood.

Modern Jawless Fishes

Modern Agnatha, or cyclostomes, are relatively rare and unimportant organisms, although they are representatives of a group that was important in the early history of vertebrate evolution. For example, hagfish can be a nuisance to fishermen, attacking and destroying bait and even the catch itself. Lampreys, because of their active parasitic mode of life, can be a serious menace to fisheries, as evidenced by the sea lamprey depredations in the Great Lakes. The construction of a canal bypassing Niagara Falls unfortunately allowed lampreys to enter the upper Great Lakes from Lake Ontario, where they had been established. Once in the upper lakes, they underwent a population explosion, probably as a result of the abundance of prey fishes, lack of predators, and suitability of the system for spawning and maintenance of larvae. The establishment of the lamprey resulted in a serious decline of a number of fish species and the collapse of a flourishing commercial fishery that has only been reversed by the establishment of control measures and the use of larvicides.

The fossil Agnatha, or ostracoderms, are generally poorly known as a result of their incomplete preservation. Yet they do throw some light on the earliest stages of vertebrate evolution, indicating that, as far back as 470 million years ago, humankind's earliest ancestors were small, rather tadpolelike fish with an external armor of bony plates. Although they appear in the fossil record before the earliest jawed vertebrates, or gnathostomes, it appears that the separation into jawed and jawless forms had already occurred. Many gaps remain, but it is to be hoped that further discoveries will enable humankind to develop a clearer picture of its earliest vertebrate ancestry.

—*David K. Elliott*

See also: Amphibians; Chordates, lower; Evolution: Historical perspective; Extinction; Fish; Gill, trachea, and lung anatomy; Invertebrates; Marine animals; Marine biology; Reptiles; Symbiosis; Systematics.

Bibliography

Alexander, R. McNeill. *The Chordates*. 2d ed. Cambridge, England: Cambridge University Press, 1981. This text provides information on Agnatha in the third chapter and describes them in relation to other chordates. It provides an excellent overview of the whole phylum. Suitable for high school and college students.

Bond, C. E. *Biology of Fishes*. Philadelphia: W. B. Saunders, 1979. This text provides a good overview of the modern and fossil Agnatha and their relationship to other fishes. Suitable for high school and college-level students.

Carroll, Robert L. *Vertebrate Paleontology and Evolution*. New York: W. H. Freeman, 1987. This book on vertebrate paleontology is aimed at the college student. Deals with the origin of vertebrates and the diversity of jawless forms, but also provides much useful information on evolution and the uses of taxonomy.

Colbert, E. H. *Evolution of the Vertebrates*. 4th ed. New York: John Wiley & Sons, 1990. This volume is not as detailed as Carroll's and is suitable for high school as well as college-level students. The first three chapters on early vertebrates were extensively rewritten and updated for the fourth edition.

Elliott, D. K. "A Reassessment of *Astraspis disiderata*, the Oldest North American Vertebrate." *Science* 237 (July, 1987): 190. This article provides an overview of knowledge of early vertebrates together with a redescription of the earliest North American forms. It contains a rare illustration of a North American Ordovician vertebrate.

Gagnier, P. Y. "The Oldest Vertebrate: A 470 Million Year Old Jawless Fish, *Sacabambaspis janvieri*, from the Ordovician of Bolivia." *National Geographic Research* 5 (Spring, 1989): 250. An article on one discovery of early vertebrates, providing good illustrations of the appearance of these animals. This journal provides articles in a variety of scientific disciplines for the general reader as well as the scientist. Suitable for high school and college-level students.

Hardisty, M. W. *Biology of the Cyclostomes*. London: Chapman and Hall, 1979. Almost the only publication that deals only with Agnatha, this text covers all aspects of the biology of the modern forms together with their relationship to the fossil forms. Suitable for high school and college students.

Parker, Steve. *The Beginner's Guide to Animal Autopsy: A "Hands-In" Approach to Zoology, the World of Creatures and What's Inside Them*. Brookfield, Conn.: Copper Beach Books, 1997. Written for children, a guide to animal physiology. Illustrated with drawings and cartoons.

LANGUAGE

Type of animal science: Behavior
Fields of study: Evolutionary science, human origins

The origins of human language and animals' communicative abilities help to achieve an understanding of humans' position in the natural order of evolution.

Principal Terms

CLOSED-CLASS VOCABULARY: typically including the structural and functional words, such as prepositions, determiners, quantifiers, and morphological markers, closed in the sense of resisting the introduction of new members

DISPLACEMENT: language's power to refer to or describe things and events beyond the constraints of the here and now

GRAMMAR: the structure of a language, consisting of systematic rules to specify word formation, such as inflection, derivation, and compound words (morphology), and systematic rules to specify how words should be ordered in combination to form phrases and sentences (syntax)

OPEN-CLASS VOCABULARY: content words such as nouns, verbs, and adjectives, open in the sense of its readiness to admit new members

PRODUCTIVITY OR GENERATIVITY: language's power to produce or generate an infinite number of understandable words and sentences from a finite number of symbols and rules

SEMANTICITY: meaning in language

SYMBOL: something that stands for something else, the connection between symbol and object being arbitrary in nature

One way to understand humans' position in the natural order of evolution is to locate the origins of language. Some scholars think that language is probably the product of the "mental mutation" of the large brain of *Homo sapiens*, because human language is the only animal communicative system that possesses all of the fundamental characteristics of arbitrary symbols, semanticity, grammar, productivity, duality of patterning, and displacement. Furthermore, human children are able to become effective users of such complex symbolic systems without formal teaching and within a fairly short period of time, in striking contrast to the limited expressions of animals, even after lengthy and extensive training. Linguist Noam Chomsky, for example, has proposed that human beings have a unique language forming capacity and human babies are innately equipped with a "Language Acquisition Device," which resides somewhere in the brain.

Many others, however, disagree with this discontinuity theory of the human language origins. Instead, they argue for an evolutionary ground (neuroanatomical, behavioral, cognitive, social, and cultural) to cultivate language formation. They believe that language is no exception from the governing of the law of evolution. For example, Philip Lieberman rejects the existence of linguistic genes or a language organ in the human brain. Rather, language had its first sprout at the intersection of the evolutionary products of neural mechanisms, communication, and cognition. These are the two main theories of language origins. The first line builds on the protolanguage theory, the second on the notion of behavior determinism in evolution.

Protolanguage Theory

"Protolanguage" means that utterances are not yet full language, although they serve symbolic referential and other communicative functions. The telegraphic speech of toddlers (such as "Mommy cookie!") has been used as an example. A "baby talk" was formed first, only to be shaped and refined later into a full language. It was hypothesized that modern speech could have been possible about 100,000 years ago, when the earliest *Homo sapiens* started to migrate from Africa to other places. John McCrone reasons that toolmaking and tool use, moving in troops to other regions, and collective hunting called for more group actions, which in turn promoted social interaction. These activities required joint attention and intentional communication. Using eye contact and gesture to direct attention as a means beyond reflexive behaviors to achieve joint attention might thus have been a major step toward human speech. The prolonged period for taking care of dependent human infants, a consequence of brain growth, afforded opportunities to cultivate the intimacy between mother and child. Meanwhile, social ties in a colony began to form. All these changes might have encouraged what could be called personalized noises. To meet their communicative needs, early *Homo* species were pressured to refine and stabilize their coarse communicative noises into protowords. These protowords were then passed down to the next generations. Practice of such vocalizations in turn further promoted vocal structure refinement.

The evolutionary principle of economy and efficiency was at work, so that concept categories (words) and combinatorial rules (grammar) were naturally selected, because words and grammar are far more cognitively economic and efficiently generative than mechanical one-to-one referential associations. Martin Nowak and David Krakauer, based on their mathematical and computational modeling using computer simulations, have contended that protolanguages can evolve in a nonlinguistic society. At first, signal-object associations were established, and later, combinations of sounds to form words and combinations of words to form sentences evolved into semantic and syntactic systems through natural selection to reduce mistakes in communication.

In addition to the supportive results of computer simulations, other empirical evidence came from studies involving primates and human children. In Leavens and Hopkins's study, 115 chimpanzees (*Pan troglodytes*) in captivity (aged three to fifty-six years), without any explicit training whatsoever, commonly employed gaze alterations and a pointing gesture in face-to-face communicative interactions with humans and among themselves. This demonstrated the presence of communicative intent and gestural precursors of language among humans' closest relatives, chimpanzees. The bonobo (*Pan paniscus*) Kanzi understands spoken English sentences and knows how to use human-designated lexigrams to announce his intention, all through laissez-faire learning without explicit teaching. The chimpanzee Washoe spontaneously taught her acquired American Sign Language (ASL) to her adopted son Loulis. Human babies are like chimpanzees in many ways. As McCrone describes it, human infants are born with the standard ape vocal plan—it is after six months that the voice box descends down into the throat. Human infants first play with vowels (coo) and then babble (combining vowels with consonants)—so vocalization comes before producing true words. They also employ gaze alterations and gestures to communicate before they say words. In the second year, they produce telegraphic speech, typically composed of two to three content words, which happens to be the mode of expression in the utterances of language-trained chimps (such as "Shirt hide" by Kanzi).

Behavior Determinism

William Noble and Iain Davidson do not think that a protolanguage existed in evolution. They have also cautioned people against accepting the performance of animals in captivity and human interaction experiences as evidence to back up evolutionary arguments, because these environments are drastically different from the ecologies of the *Homo* ancestors millions of years ago. These environments are not the same as those of the free-living primates, either. Language-trained animals' performance is like language emergence in human infants, who learn through interaction with other humans who already have language, a learning process quite different from the prehistoric origins of symbols and language from scratch.

Noble and Davidson agree that natural selection favored bipedalism, leading to neuroanatomical changes including larger brains. They do not believe, however, in biological determination of behaviors. Instead, they believe the opposite. The behavior of standing upright led to larger brains that needed to consume more energy. Meat as a good energy source had already been increased in the diet of *Homo erectus*. As meat-eaters, hominids had to run fast (either to catch prey or to escape predators). Running brought about better control of the breathing system (necessary for speaking) and adjustments to the thermoregulatory system (leading to the selection of the feature "hairlessness," which could have fostered face-to-face adult-infant interaction). In addition, hunting for meat facilitated coordinated group actions as well as tool creation. To expand food sources hominids began migrating, which further promoted groups and interaction. Thus the social context was present for the emergence of language.

In their discussion, Noble and Davidson have emphasized one important behavior responsible for the emergence of language: stone throwing. To be effective, manual control and timing control had to be achieved. As the timing control behaviors were bettered, the neuroanatomical structures improved too. These positive adjustments, together with other contextual changes, led hominids on an increasingly divergent behavioral path from their chimpanzee relatives. Better control of the forearm could develop into a pointing form. Hairlessness made it easy to carry an infant hominid in front, increasing the likelihood of adult-baby mutual observation and imitation. One such likely behavior could be arm extensions for referring. Later, better-controlled movements of forearms and fingers developed into a pointing form. Any vocalizations in company with manual gestures were first associated with the referred objects, and later became symbols, once it was realized that the sounds alone could stand for the targets themselves, even when they were out of sight.

There is consensus that the human vocal structure is a necessary condition for human speech. It has been noticed that animal vocalizations are graded in nature. A graded system contains only variations of vowel sounds but no consonants. Variations of vowels, although functional in communication, lack distinctive boundaries to mark different categories. Sue Savage-Rumbaugh and Roger Lewin have concluded that language is unlikely to emerge unless an organism can produce consonants, no matter how large its brain. How the human vocal tract acquired the ability to pronounce consonants unfortunately remains a mystery.

The evolution of language must have benefited the *Homo* species tremendously. Noble and Davidson have speculated how language could have contributed to human mentality. For example, colonization in different places would cause isolated groups to have trouble in understanding each other. Such failures and misunderstandings could contribute to the awareness of "us" vs. "them." This appreciation would lead to the realization of the possibility for a group to use their own symbolic system as a means for social control. Thus, human mentality, with language in use, is itself an evolving feature of the natural world. It is only logical that, with language available, mental representation of the world became possible, which eventually made abstract, imaginary, retrospective, hypothetical, and metacognitive thinking a reality. No wonder these modes of thinking reflect themselves in the characteristics of human language.

Animal Communication

Animals do communicate, at least in a broad sense. Animals use vocalizations, facial expressions, gestures, body postures and movements, and even odors, to warn peers, to attract attention, to find food, to care for the young, to mark territories, and to maintain social structures. However, animal communicative systems are typically not recognized as language because of they lack the key features of a true language. Many have argued that animal communication, even among chimpanzees, is in essence instinctive and reflexive. McCrone says that these behaviors are not under conscious control, and are triggered only by an event in the immediate environment, with both parties present. Hence, chimpanzees have no true arbitrary symbols or displacement. In addition, animal communicative systems are "closed," with no combinatorial rules to create new meanings; hence they lack duality of patterning, syntax, and productivity. Edward Kako has pointed out that no animals so far, including the language-trained ones, have demonstrated the ability to understand closed-class lexical items. Despite the criticisms, animals' language-learning achievements have been acknowledged.

Talking Animals

The most famous talking parrot is Alex, an African gray parrot (*Psittacus erithacus*) trained by Irene Pepperberg. Alex is able to speak many words and phrases referring to objects, materials, actions, colors, shapes, numbers, and locations. He can answer questions that require labeling objects, classifying objects (color and substance), comparing objects ("bigger than" or "same as/

American Sign Language Among Primates

Primates are members of the mammalian order Primates, which includes prosimians, or lower primates, and anthropoids (monkeys, apes, and humans). Apes are any primates of the subfamily Hominoidea, including the small apes and the great apes. The gorilla (*Gorilla gorilla*) is the largest of the great apes. Another great ape is the chimpanzee. The larger member of the chimpanzee family is the common chimpanzee (*Pan troglodytes*), and the smaller member is the bonobo (*Pan paniscus*) or the pygmy chimpanzee.

To obtain an understanding of their evolutionary roots, human beings have turned to the great apes, especially chimpanzees, because they are humans' closest living relatives, sharing as much as 98 percent of identical genetic material. Many attempts have been made to teach apes to learn language. Early attempts to raise apes along with human babies in the hope that these human-reared apes would learn to speak human language ended in failure, but researchers came to realize that apes do not have the vocal structure necessary for producing human speech sounds. Enlightened by the Braille system of writing for the blind and sign language for the deaf, researchers began to employ nonverbal symbolic systems, including plastic chips (David Premack's chimp Sarah), computer lexigrams (Duane Rumbaugh's chimp Lana), and American Sign Language (ASL). This line of research has yielded interesting results. Allen and Beatrice Gardner's chimp Washoe acquired ASL, and, moreover, she began to combine words and taught her adopted son, Loulis, to use ASL. Twelve percent of the ASL utterances produced by Herbert Terrace's chimp Nim were spontaneous. Francine Patterson's gorilla Koko mastered six hundred ASL words. Through natural observation and social interaction, Sue Savage-Rumbaugh's bonobo Kanzi understands spoken English and uses sign language and computer buttons in communication.

Language research with primates creates a path for humans to search for their links to animals. There are many potential research areas, such as body language, cultural transmission of learned language over generations, and the impact of language-trained chimpanzees living with naïve peers in their natural environment, just to mention a few. On the other hand, scientists are concerned about the well-being and the rights of the primates, especially in the areas of early separation from the mother, or lab controls. Many questions have no answers yet, but one thing is certain—scientific inquiry will go on, and it will be done with good ethical concerns.

—Ling-Yi Zhou

different from"), and counting (from one to six).

Two Hawaii bottle-nosed dolphins (*Tursiops truncatus*), trained by Louis Herman and his colleagues, are Phoenix, with an acoustic language, and Akekamai (Ake), with a gesture-based language. They can correctly carry out commands in varied word orders (syntax) with different meanings. They have further demonstrated their semantic and grammatical knowledge by either not executing grammatically incorrect and semantically nonexecutable orders or by extracting an executable segment from an anomalous string and then completing the task according to the meaning provided in that specific syntactic structure.

Sue Savage-Rumbaugh's bonobo Kanzi is an exciting language star. Kanzi understands spoken English and uses hand signals as well as geometric symbols, and moreover, he has learned all that without explicit instruction but by mere social observation and interaction, just as a human baby learns a language. This natural learning process was successfully replicated with Kanzi's sister, Mulika.

Comparative Language Research

Research methodology in the comparative language field has been improved greatly over the years. Designation of "language" takes into consideration the biological constraints of the species involved. Social interaction in a natural way is underscored. Possible experimenters' and trainers' biases are controlled through blind techniques such as blindfolding the eyes of the person who gives commands, using one-way mirrors and remote cameras, or separating the person who does recording and interpretation from the one who gives the command. Recent data are thus more scientifically sound than before. Yet some people are not happy with the fact that humans have imposed their dialogue on the animals. These people are now using the playback technique to decode the meaning of the signals in animals' own communicative systems. Many species have been studied in their natural ecological niches, including vervet monkeys, tigers, humpback whales, orcas or killer whales, and elephants. These animals' wild calls in nature are recorded, and then are played back to the animals to see their differentiated reactions, thus making the message decoding possible. The playback studies are very encouraging in confirming the symbolic nature of animals' natural "languages" in the wild.

It is very important to study the animal's own "language" for its own sake, for without such knowledge, commenting on nonhuman species' linguistic abilities in the frame of human language is at least prejudiced. As primatologist and psychologist Roger Foutes put it: "The best approach to science is a humble one. We are humble enough to take the animals we are studying on their own terms and allow them to tell us about themselves. Too often science takes an arrogant approach." According to him, "Someday we'll realize that the human voice is not a lone violin but part of an orchestra. We're not playing a solo; instead it's a symphony."

—Ling-Yi Zhou

See also: Apes to hominids; Chimpanzees; Communication; Communities; Displays; Emotions; Ethology; Gorillas; Groups; Hominids; *Homo sapiens* and human diversification; Human evolution analysis; Insect societies; Intelligence; Learning; Mammalian social systems; Neanderthals; Pheromones; Tool use; Vocalizations.

Bibliography

Chomsky, Noam. *Aspects of the Theory of Syntax*. Cambridge, Mass.: MIT Press, 1965. Discusses categories and relations in syntactic theory (the deep structure and grammatical transformations, syntactic features, and types of base rules) as well as the inflectional and derivational processes in the lexicon structure.

Gardner, R. Allen, Beatrix T. Gardner, and Thomas E. Van Cantfort, eds. *Teaching Sign Language to Chimpanzees*. New York: State University of New York Press, 1989. A report of Project Washoe and ASL data of Washoe and other chimpanzees.

Kako, Edward. "Elements of Syntax in the Systems of Three Language-Trained Animals." *Animal Learning and Behavior* 27, no. 1 (1999): 1-14. Systematically analyzes the language performance of a parrot, two dolphins, and a bonobo according to the four criteria of syntax.

Leavens, David A., and William D. Hopkins. "Intentional Communication by Chimpanzees: A Cross-Sectional Study of the Use of Referential Gestures." *Developmental Psychology* 34, no. 3 (1998): 813-822. Reports that chimpanzees spontaneously used a pointing gesture and gaze alterations for intentional communication with humans and peers.

Lieberman, Philip. *The Biology and Evolution of Language*. Cambridge, Mass.: Harvard University Press, 1984. The book presents evolutionary models and data to propose that only comparative studies can ensure an understanding of the nature and evolution of language and cognition.

McCrone, John. *The Ape That Spoke: Language and the Evolution of the Human Mind*. New York: William Morrow, 1991. The author uses plain language to discuss the origins of language and the interwoven relationships between language and the evolution of human mind.

Noble, William, and Iain Davidson. *Human Evolution, Language, and Mind: A Psychological and Archaeological Inquiry*. New York: Cambridge University Press, 1996. A rigorous evaluation of various theories of the origins of language against archaeological evidence, with the authors' conclusion that behaviors are the bases of evolution, accounting for the emergence of human language and mentality through natural selection.

Nowak, Martin A., and David C. Krakauer. "The Evolution of Language." *Proceedings of the National Academy of Sciences of the United States* 96, no. 14 (July, 1999): 8028-8033. Describes how it is possible for language to evolve from a nonlinguistic society based on mathematical and computational modeling.

Patterson, Francine, and Eugene Liden. *The Education of Koko*. New York: Holt, Rinehart and Winston, 1981. A detailed report of Gorilla Koko's ASL training and her performance.

Premack, David. "Language in Chimpanzee?" *Science* 172 (1971): 808-822. Discusses what language is and whether Chimp Sarah can be taught a "plastic chip language."

Rumbaugh, Duane M., and E. Sue Savage-Rumbaugh. "Language in Comparative Perspective." In *Animal Learning and Cognition*, edited by N. J. Mackintosh. San Diego, Calif.: Academic Press, 1994. A chronological review of the language studies with the great apes.

Savage-Rumbaugh, Sue, and Roger Lewin. *Kanzi: The Ape at the Brink of the Human Mind*. New York: John Wiley & Sons, 1994. A brief review of perspectives on human-animal connections, a detailed description of Kanzi's and other apes' cognitive and linguistic abilities, and the authors' thoughts on the origin of language.

Starr, Doug. "Calls of the Wild." *Omni* 9, no. 3 (December, 1986): 52-59, 120-123. A very interesting article that reviews animal and interspecies communications.

LEARNING

Type of animal science: Behavior
Fields of study: Ethology, neurobiology

Learning is any change or modification in behavior that involves the nervous system and cannot be attributed to the effects of development, maturation, or fatigue. Learning takes a number of forms, including habituation, sensitization, associative learning, perceptual learning, and insight.

Principal Terms

ADAPTATION: any heritable characteristic that increases the probability that an animal will survive and reproduce in its natural environment

CONDITIONING: the behavioral association that results from the reinforcement of a response with a stimulus

INNATE: any inborn characteristic or behavior that is determined and controlled largely by the genes

INSTINCT: any behavior that is completely functional the first time it is performed

NATURAL SELECTION: the process of differential survival and reproduction that leads to heritable characteristics that are best suited for a particular environment

STIMULUS: any environmental cue that is detected by a sensory receptor and can potentially modify an animal's behavior

Learning, as defined by ethologists, is simply any change or modification in behavior that is directed by previous experience and involves the nervous system but cannot be attributed to the effects of development, maturation, fatigue, or injury. These latter phenomena contribute to changes in behavior that generally do not constitute learning. Learning takes a number of forms, including habituation, sensitization, associative learning, perceptual or programmed learning, and insight. Each type has its own basic characteristics and adaptive significance. Habituation and sensitization are the simplest and most widespread forms of learning, and insight is the most complex and least understood form. Insight involves the ability to put two previous experiences together to solve an unrelated problem.

Habituation and Sensitization

Habituation and sensitization are considered the simplest forms of learning. Habituation involves a decrease in a behavioral response that results from repeated presentation of a stimulus. A young, naïve duck, for example, will exhibit an innate startle response when any hawk-shaped object is passed overhead. With repeated presentation of the hawk model, however, the intensity of the bird's reflex declines as the animal becomes habituated, or learns that the stimulus has no immediate significance. Habituation learning is common throughout the animal kingdom and has tremendous adaptive significance in that it prevents repeated response to irrelevant stimuli that could otherwise overwhelm the animal's senses and prevent it from accomplishing other critical tasks. One of the common characteristics of habituation is that after a short period, usually defined by the particular species and the stimulus in question, the animal will completely recover from the habituation experience and will again exhibit a full response to the stimulus. This, too, has important survival implications, especially for species that rely on stereotypic alarm responses for avoiding predation.

In contrast to habituation, sensitization is the

increase in intensity of a response that results from the repeated presentation of a stimulus. A good example is the heightened sensitivity to even relatively soft sounds that results from the initial presentation of a loud, startling noise, such as a gunshot. Sensitization differs from habituation in important ways. First, the specific stimulus that elicits a sensitization response is different from the stimulus to which the animal becomes sensitized. Second, the underlying physiological mechanisms that control these two processes are fundamentally different.

The third and broadest category of learning is associative learning. In this type of learning, an animal makes a connection between some primary environmental stimulus (that involves either a reward or punishment) and a novel or neutral stimulus that is paired with the first stimulus.

Classical and Operant Conditioning

The simplest form of associative learning is classical conditioning, first studied by Ivan Pavlov. Pavlov observed that when a dog is presented with food, the dog will begin to salivate. He referred to the food in this case as an unconditioned stimulus (US), and to the salivation reflex as the unconditioned response (UCR). When the unconditioned response is effectively paired with a second, novel stimulus, such as a light or bell (called the conditioned stimulus, or CS), the dog will, after several trials, associate this second stimulus (CS) with the US and begin to salivate whenever the CS is presented. The salivation reflex that occurs following presentation of the CS is termed the conditioned response (CR).

Although classically conditioned learning is often associated with the controlled experiments of psychologists, it undoubtedly occurs throughout the animal kingdom, and it may be one of the most common ways by which animals learn about their immediate environment. A good example of this is the phenomenon of taste-aversion learning. Taste-aversion learning occurs when an animal associates a specific odor or visual stimulus with an unpleasant experience resulting from the consumption of an unpalatable or poisonous food item. After even a single experience with the distasteful food, the animal will subsequently avoid ingestion, even if it means starvation. Taste-aversion learning is especially important for nonspecialist feeders that forage on a variety of foods and must periodically sample unfamiliar food items. This learning phenomenon also serves as the basis for the evolution of many warning signals in animals.

Through natural selection, many distasteful prey have evolved distinctive marks, colorations, odors, or behavioral characteristics that serve as a reminder (a CS) to predators that it is distasteful or harmful. After one negative experience with this prey, the predator learns to associate these characteristics with the sight or smell of the animal. Such characteristics have obvious survival benefits for the prey. Taste-aversion learning differs from classical conditioning in that the critical time between the CS and the US is usually much longer and in that only one trial is necessary for the learning to occur. This latter effect has important implications for animals that rely on aversion learning to avoid poisoning. Many animals, such as blue jays and rats, wait a specific length of time after ingesting a novel food item to determine whether they will become ill.

A second major form of associative learning is operant, or instrumental, conditioning. Unlike classic conditioning, in which the animal is passively involved in the learning experience, in operant conditioning the animal learns by manipulating some part of its environment. In traditional operant learning experiments the animal, for example, presses a lever or rings a bell in order to receive some reward. Because this kind of learning usually improves with practice, it is often referred to as trial-and-error learning. This kind of learning has obvious adaptive significance under natural conditions. Perhaps the best example of this is the reinforced trial-and-error learning that is necessary for many young vertebrates to perfect their feeding techniques. The naïve young of many mammals and birds greatly enhance their feeding efficiency when repeatedly allowed to manipulate their food. Similarly, many animals, including insects, use this reinforced practice to learn their way around in their habitat, home range, or territory, much the way a rat learns its way around in a maze.

Programmed Learning

In addition to associative learning, there are a number of types of learning that seem to involve mechanisms more complex than simple association. The most common examples include song learning (in birds) and imprinting. Ethologists often refer to these types of learning as programmed learning, since they only take place at certain times and under very restricted circumstances.

Imprinting is the process whereby a young animal develops a behavioral attachment to some other animal or object. Animals have been observed to imprint naturally on their parents, individuals of the opposite sex, food items, preferred habitats, and home streams (in the case of salmon). All such types of imprinting have two general features in common. First, the imprinting must occur during some critical period. The most familiar type of critical period is that which occurs in parental imprinting, a specific imprinting routine whereby a newborn becomes behaviorally fixed on a parent. First described by Konrad Lorenz, this type of learning requires a critical period shortly after birth, in which the young learns to recognize and follow the parent. Outside this period the learning simply cannot occur. The second characteristic common to all types of imprinting is that the young animal must be actively involved in the learning process. In fact, the strength of the imprinting seems to depend largely on the degree of this involvement.

Song learning in birds is fundamentally quite similar to imprinting in that it too requires a specific learning period. White-crowned sparrows, for example, learn their song from their fathers, usually from one to six weeks after birth. During this critical period, these young birds learn to imitate the song that is specific to their species as well as the variations and dialect characteristic of their population. When young birds are raised in isolation and prevented from hearing their own species' song, they develop an abnormal vocalization. If given the opportunity to hear a recording of a normal adult song of its species during the critical learning period, the young bird will learn to sing normally. If, on the other hand, the animal is exposed to the song of some other closely related species, the animal will not develop a normal song. This suggests that birds are somehow innately programmed to learn their species-specific songs. Thus, it seems that both imprinting and song learning are in many ways quite similar and may be controlled by the same underlying mechanisms. The tendency to classify these as complex behaviors, however, may be attributable in part to ethologists' lack of understanding of these mechanisms.

Insight

Perhaps the most advanced and least understood form of learning is insight. Insight is said to differ from other forms of learning in that it is characterized by a modification in behavior that is not contingent on some particular recent experience. Instead, insight behavior involves the ability to put two independent ideas together to solve a third, unrelated, problem. Wolfgang Köhler's classic observations on learning in chimpanzees illustrate the phenomenon of insight. He observed that when a preferred food item (such as a banana) was placed out of reach of a caged chimpanzee, the animal quickly learned to use a pole as an extension of its arm to pull in the food; when the food was hung overhead, the animal would learn to stack boxes to reach the food. Examples of tool use by chimpanzees observed under natural field conditions include the use of sticks as probes for gathering insects and the use of small branches for warding off potential predators.

Although this type of problem-solving behavior seems fundamentally more complex than any other type of learning, it has been suggested that many of the specific behaviors cited as examples of insight may be nothing more than extensions of associative learning. Pigeons, for example, can be conditioned to perform certain activities that they use later in solving more complex problems. A pigeon conditioned at one time to push a box across its cage floor and at another time to climb on a box and peck at a food lever will later push and position a box under a lever so that it can peck at the lever and receive a reward. While this seems to reflect some type of problem-solving ability, it is interesting that birds that are not previously conditioned cannot solve the problem. Thus, insight may build on some form of associative learning.

Insight learning has also been invoked to explain the origin of many types of cultural learning. Cultural learning occurs when one animal in a group discovers a unique or novel behavior and the other members learn to copy the behavior through the process of observational learning. One of the classic examples of this kind of learning was observed in the blue tit, a small European bird that was observed to strip the caps off milk bottles in order to drink the cream that surfaced at the top. In relatively little time, the behavior spread and was exhibited by this species all across Western Europe. Although there is little doubt that such cultural transmission involves nothing more than the simple imitation of another animal's behavior, it is not clear whether the origin of such behaviors reflects some form of innovation.

Ethological and Psychological Approaches

The study of behavior and learning has long been characterized by two very different methodological and philosophical approaches: those of ethology and psychology.

Ethology, the study of animal behavior, is built on several very specific assumptions and principles that clearly distinguish it from the field of psychology. First, the study of ethology involves objective, nonanthropomorphic (that is, not biased by human expectations or interpretations) descriptions and experiments of the learning process within a natural context. Konrad Lorenz, one of the founders of the field, insisted that the only way to study behavior and learning was to make objective observations under completely natural field conditions. Building on Lorenz's purely descriptive approach, Nikolaas Tinbergen conducted rigorous field experiments, similar to those that now characterize modern ethology. The classic work of early ethologists helped demonstrate how an animal's sensory limitations and capabilities can shape its ability to learn. For example, in a series of classic learning experiments, Karl von Frisch convincingly documented the unusual visual capabilities of the honeybee. He first trained honeybees to forage at small glass dishes of sugar water and then, by attaching different visual cues to each dish, provided the animals with an opportunity to learn where to forage through the simple process of association. From these elegant (but simplistic) experiments, von Frisch found that bees locate and remember foraging sites by the use of specific colors, ultraviolet cues, and polarized light, a discovery that revolutionized how scientists view the sensory capabilities of animals.

A second important feature of ethology is that it is built on the assumption that learning depends not only on environmental experience but also on a variety of underlying physiological, developmental, and genetic factors. The work of countless neurobiologists, for example, clearly demonstrates how behavioral changes are linked to modifications in the function of nerves and neuronal pathways. By observing the response of individual nerves, neurobiologists can observe changes that occur in the nerves when an animal modifies its behavior in response to some stimulus. In a similar way, they can show how learning and behavior are affected when specific nerve fibers are experimentally cut or removed. Unfortunately, however, neurobiologists' understanding of the physiological control of learning is limited to simpler kinds of learning such as habituation and sensitization.

Like the neurobiologists, behavioral geneticists have shown that much of learning, and behavior in general, is intimately tied to internal mechanisms. The results of hybridization experiments and artificial breeding programs clearly demonstrate a strong genetic influence in learned behaviors. In fact, it has been well documented that many animals (including both invertebrates and vertebrates) are genetically programmed (or at least have a genetic predisposition) to learn only specific kinds of behaviors. Finally, the most important characteristic of ethology is that it places tremendous importance on the evolutionary history of an organism. It assumes that an animal's ability to learn is shaped largely by its evolutionary background, and it emphasizes the adaptive significance of the various types of learning.

In comparison with ethology, the field of psychology emphasizes the importance of rigorously controlled laboratory experiments in the study of learning. The most widely used methods in this field are those of classic and operant conditioning. The primary objective in these approaches is to eliminate and control as many variables as possible and thereby to remove any doubt as to the factors responsible for the behavioral changes. These approaches have met with considerable success at identifying specific external mechanisms responsible for learning. These techniques, however, tend to focus only on the input (stimulus) and output (response) of an experiment and, as a result, de-emphasize the importance of proximate mechanisms, such as physiology and genetics. In addition, these approaches generally ignore the evolutionary considerations that ethologists consider so fundamental to the study of behavior.

Understanding the Learning Process

Although the approaches used to study learning vary tremendously, nearly all such studies are directed at two goals: to understand the adaptive value of learning in the animal kingdom and to understand the physiological, genetic, and psychological mechanisms that control learning. For any animal, the adaptive advantages of learning result primarily from the increase in behavioral plasticity that learning provides. This plasticity (ability to be flexible) provides the animal with a greater repertoire of responses to a given stimulus and thereby increases the chances that the animal will survive, reproduce, and pass the genes that control the learning process on to the next generation. In comparison, the value of an innate behavior lies primarily in its ability to provide a nearly stereotypic response to a stimulus on the very first occasion on which it is encountered. Innate reflexes are especially important in situations in which there may not be a second chance for the animal to learn an appropriate response. The best examples are basic feeding responses (for example, the sucking reflex in newborn mammals) and predator-escape behaviors

Monkeys, especially chimpanzees, can be taught to do many humanlike tasks, such as riding on bicycles or using sign language, but the complete extent of their learning capacity and how it compares with that of humans is still uncertain. (Ken Schwab/Photo Agora)

(alarm calls in young birds). It is a common misconception, however, that a learned behavior is attributable entirely to the animal's environment, whereas instinct is completely controlled by the genes. Many studies have demonstrated that numerous animals are genetically programmed to learn only certain behaviors. In contrast, it has been shown that instinct need not be completely fixed, but can be modified with experience. Thus, learning and instinct should not be considered two mutually exclusive events.

In addition to its evolutionary implications, the study of learning has provided considerable insight into the internal mechanisms that control and regulate behavior. These mechanisms are the cellular and physiological factors that provide the hardware necessary for learning to occur. As neurobiologists and geneticists learn more about these types of control, it is becoming increasingly evident that learning, at nearly all levels, may involve the same basic mechanisms and processes. In other words, the only difference between simple and complex behaviors may be the extent to which the learning is physically constrained by the biology of the animal. Thus, many invertebrates, by virtue of their simple body plan and specific sensory capabilities, are limited to simple learning experiences. Vertebrates, on the other hand, live longer and are not as rigorously programmed for specific kinds of behavior.

—*Michael Steele*

See also: Brain; Ethology; Habituation and sensitization; Imprinting; Instincts; Language; Primates; Reflexes.

Bibliography

Alcock, John. *Animal Behavior: An Evolutionary Approach*. 6th ed. Sunderland, Mass.: Sinauer Associates, 1998. This is one of the most clearly written texts on the subject of animal behavior currently available. The chapter on the diversity of behavior describes different kinds of learning within an ethological and ecological context. One small section is devoted entirely to insight learning. It is well illustrated and is especially helpful at explaining the experimental evidence for various learning phenomena.

Bonner, John T. *The Evolution of Culture in Animals*. Princeton, N.J.: Princeton University Press, 1980. This small handbook provides a comprehensive but lucid survey of the process of cultural learning in the animal kingdom. Chapter 6 offers a solid introduction to the evolution of learning.

Donahoe, John W. *Learning and Complex Behavior*. Edited by Vivian Packard Dorsel. Boston: Allyn & Bacon, 1994. A biobehavioral approach to complex behavior in humans and animals, arguing that complex behavior is the product of selection. Requires some background in psychology.

Gould, James, L. *Ethology: The Mechanisms and Evolution of Behavior*. New York: W. W. Norton, 1982. This well-illustrated text provides an excellent introduction to the basic concepts of ethology. Three chapters are devoted specifically to the process of learning, and three additional chapters concentrate entirely on human ethology.

Grier, James W. *Biology of Animal Behavior*. 2d ed. St. Louis: Times Mirror/Mosby, 1992. This college-level text provides an excellent treatment of the study of animal behavior. It is clearly written and well illustrated; it should provide a good introduction for the layperson. The text integrates information from a variety of disciplines, including ethology, behavioral ecology, psychology, and neurobiology.

Hickman, Cleveland P., Jr., Larry S. Roberts, and Frances M. Hickman. *Integrated Principles of Zoology*. St. Louis: Times Mirror/Mosby, 1988. A general text on animal biology with an excellent introduction to the general concepts of ethology and animal behavior, with a specific emphasis on learning and instinct. A concise summary, additional references, and review questions appear at the end of each chapter.

McFarland, David, ed. *The Oxford Companion to Animal Behavior.* New York: Oxford University Press, 1987. Intended as a reference guide for both nonspecialists and people in the field, this comprehensive survey of the study of animal behavior was written by a team of internationally known biologists, psychologists, and neurobiologists, and it contains more than two hundred entries covering a variety of behavior topics. More than a dozen of the entries are specifically concerned with the process of learning, and a detailed index provides cross-references organized by both subject and species lists.

MAGILL'S ENCYCLOPEDIA OF SCIENCE

ANIMAL LIFE

ALPHABETICAL LIST

Volume 1

Volume 2

Volume 3

Volume 4

CATEGORY LIST

Amphibians
Amphibians
Cold-blooded animals
Frogs and toads
Metamorphosis
Salamanders and newts
Vertebrates

Anatomy
Anatomy
Antennae
Beaks and bills
Bone and cartilage
Brain
Cell types
Claws, nails, and hooves
Digestive tract
Ears
Endoskeletons
Exoskeletons
Eyes
Feathers
Fins and flippers
Fur and hair
Heart
Horns and antlers
Hydrostatic skeletons
Kidneys and other excretory structures
Lungs, gills, and tracheas
Muscles in invertebrates
Muscles in vertebrates
Noses
Scales
Sense organs
Shells
Skin
Tails
Teeth, fangs, and tusks
Tentacles
Wings

Arthropods
Arachnids
Arthropods
Centipedes and millipedes
Cold-blooded animals
Crabs and lobsters
Crustaceans
Exoskeletons
Horseshoe crabs
Invertebrates
Scorpions
Spiders
Vertebrates

Behavior
Adaptations and their mechanisms
Altruism
Camouflage
Cannibalism
Carnivores
Communication
Communities
Competition
Copulation
Courtship
Death and dying
Defense mechanisms
Displays
Domestication

Neutral mutations and evolutionary clocks
Nonrandom mating, genetic drift, and mutation
Nutrient requirements
Osmoregulation
pH maintenance
Regeneration
Reproduction
Sense organs
Water balance in vertebrates

Classification
Aardvarks
Allosaurus
American pronghorns
Amphibians
Animal kingdom
Antelope
Ants
Apatosaurus
Arachnids
Archaeopteryx
Armadillos, anteaters, and sloths
Arthropods
Baboons
Bats
Bears
Beavers
Bees
Beetles
Birds
Brachiosaurus
Butterflies and moths
Camels
Cats
Cattle, buffalo, and bison
Centipedes and millipedes
Chameleons
Cheetahs
Chickens, turkeys, pheasant, and quail
Chimpanzees
Chordates, lower
Clams and oysters
Cockroaches
Coral
Crabs and lobsters
Cranes
Crocodiles
Crustaceans
Deer
Dinosaurs
Dogs, wolves, and coyotes
Dolphins, porpoises, and toothed whales
Donkeys and mules
Ducks
Eagles
Echinoderms
Eels
Elephant seals
Elephants
Elk
Fish
Flamingos
Flatworms
Flies
Foxes
Frogs and toads
Geese
Giraffes
Goats
Gophers
Gorillas
Grasshoppers
Grizzly bears
Hadrosaurs
Hawks
Hippopotamuses
Hominids
Horses and zebras
Horseshoe crabs
Hummingbirds
Hyenas
Hyraxes
Ichthyosaurs
Insects
Invertebrates
Jaguars
Jellyfish
Kangaroos
Koalas
Lampreys and hagfish
Lemurs
Leopards
Lions
Lizards
Lungfish
Mammals
Mammoths
Manatees
Meerkats
Mice and rats
Moles
Mollusks
Monkeys
Moose
Mosquitoes
Mountain lions
Neanderthals
Octopuses and squid
Opossums
Orangutans
Ostriches and related birds
Otters
Owls
Pandas
Parrots
Pelicans
Penguins
Pigs and hogs
Platypuses
Polar bears
Porcupines
Praying mantis
Primates
Protozoa
Pterosaurs
Rabbits, hares, and pikas
Raccoons and related mammals
Reindeer
Reptiles
Rhinoceroses
Rodents

Ecology

Evolution

Fields of Study

Category List

Monkeys
Monotremes
Moose
Mountain lions
Neanderthals
Orangutans
Otters
Pandas
Pigs and hogs
Placental mammals
Platypuses
Polar bears
Porcupines
Primates
Rabbits, hares, and pikas
Raccoons and related mammals
Reindeer
Reproductive system of female mammals
Reproductive system of male mammals
Rhinoceroses
Rodents
Ruminants
Seals and walruses
Sheep
Shrews
Skunks
Squirrels
Tasmanian devils
Tigers
Ungulates
Vertebrates
Warm-blooded animals
Weasels and related mammals
Whales, baleen

Marine Biology

Clams and oysters
Coral
Crabs and lobsters
Crustaceans
Deep-sea animals
Dolphins, porpoises, and toothed whales
Echinoderms
Eels
Elephant seals
Fins and flippers
Fish
Horseshoe crabs
Ichthyosaurs
Jellyfish
Lakes and rivers
Lampreys and hagfish
Lungfish
Manatees
Marine animals
Marine biology
Mollusks
Octopuses and squid
Otters
Penguins
Reefs
Salmon and trout
Scales
Seahorses
Seals and walruses
Sharks and rays
Sponges
Starfish
Tentacles
Tidepools and beaches
Turtles and tortoises
Whale sharks
Whales, baleen
White sharks
Zooplankton

Marsupials

Fauna: Australia
Kangaroos
Koalas
Marsupials
Opossums
Reproduction
Reproductive strategies
Reproductive system of female mammals
Tasmanian devils

Omnivores

Baboons
Bears
Chimpanzees
Food chains and food webs
Gorillas
Hominids
Lemurs
Monkeys
Omnivores
Orangutans
Raccoons and related mammals
Shrews
Skunks
Weasels and related mammals

Physiology

Adaptations and their mechanisms
Aging
Asexual reproduction
Bioluminescence
Birth
Camouflage
Circulatory systems of invertebrates
Circulatory systems of vertebrates
Cleavage, gastrulation, and neurulation
Cold-blooded animals
Determination and differentiation
Digestion
Digestive tract
Diseases
Endocrine systems of invertebrates
Endocrine systems of vertebrates
Endoskeletons
Estrus
Exoskeletons
Fertilization
Gametogenesis
Gas exchange

Population Biology

Prehistoric Animals

Primates

Reproduction and Development